Sustainable Water Purification Technologies for Multiple Applications

Sustainable Water Purification Technologies for Multiple Applications

Topic Editors

Marco Pellegrini
Cesare Saccani
Alessandro Guzzini

MDPI

Basel • Beijing • Wuhan • Barcelona • Belgrade • Novi Sad • Cluj • Manchester

Topic Editors
Marco Pellegrini
Department of Industrial Engineering (DIN)
University of Bologna
Forli
Italy

Cesare Saccani
Department of Industrial Engineering (DIN)
University of Bologna
Bologna
Italy

Alessandro Guzzini
Department of Industrial Engineering (DIN)
University of Bologna
Bologna
Italy

Editorial Office
MDPI AG
Grosspeteranlage 5
4052 Basel, Switzerland

This is a reprint of the Topic, published open access by the journals *BioTech* (ISSN 2673-6284), *Molecules* (ISSN 1420-3049), *Separations* (ISSN 2297-8739), *Sustainability* (ISSN 2071-1050) and *Water* (ISSN 2073-4441), freely accessible at: https://www.mdpi.com/topics/2O4FGN81Q1.

For citation purposes, cite each article independently as indicated on the article page online and as indicated below:

Lastname, A.A.; Lastname, B.B. Article Title. *Journal Name* **Year**, *Volume Number*, Page Range.

ISBN 978-3-7258-8419-3 (Hbk)
ISBN 978-3-7258-8420-9 (PDF)
https://doi.org/10.3390/books978-3-7258-8420-9

Contents

About the Editors

Marco Pellegrini

Marco Pellegrini is an Associate Professor at the Department of Industrial Engineering, University of Bologna. His research focuses on industrial equipment design and optimization, renewable energy, sustainable wastewater treatment, waste recycling, multiphase flow systems and industrial sustainability. He has collaborated with numerous renowned international research institutions, served as a Guest Editor for several of MDPI's Special Issues, including *Sustainable Water Treatment Technologies for Multiple Applications*, and has devoted himself to the research and engineering applications of low-carbon technologies. His works have been published in many high-quality academic journals and included in relevant monographs.

Cesare Saccani

Cesare Saccani is a Full Professor at the Department of Industrial Engineering, University of Bologna, Italy. Since 1983, he has been engaged in research and teaching at the University of Bologna, and became a Full Professor of Industrial Mechanical Plants in 1999. His main research interests focus on multiphase flow, industrial plant design, sustainable waste and wastewater treatment, and environmental equipment. He has coordinated numerous national and international projects, authored over 110 peer-reviewed papers, and holds 10 patents.

Alessandro Guzzini

Alessandro Guzzini is a fixed-term Junior Assistant Professor at the Department of Industrial Engineering, University of Bologna, Italy. He earned his Bachelor's and Master's degrees in Energy Engineering from the same university, followed by a PhD in Mechanics and Advanced Engineering Sciences. His research focuses on industrial mechanical systems, sustainable wastewater treatment, energy systems, and natural gas network safety. Working closely with Cesare Saccani and Marco Pellegrini, he has contributed to over 50 peer-reviewed publications and participated in multiple national and international projects related to the water–energy–food nexus.

Preface

The Reprint "Sustainable Water Treatment Technologies for Multiple Applications" is intended to provide a comprehensive overview of innovative, sustainable, and cost-effective approaches to treating and purifying natural, wastewater, and saline waters. Its scope encompasses a broad range of technologies, including sorption-based processes, membrane systems, biological treatments, phytoremediation, and water conditioning solutions for domestic, agricultural, industrial, and energy-related applications.

The primary aim of the Reprint is to advance knowledge of sustainable water treatment technologies and to highlight their role in addressing global challenges related to water scarcity, environmental protection, public health, and the water–energy–food nexus. The motivation for compiling this collection stems from the growing need to ensure reliable access to clean water amid climate change, increasing anthropogenic pressures, and the rising demand for water resources across multiple sectors.

The contributions gathered in the Reprint present recent scientific developments, practical applications, and emerging perspectives that support the transition towards more resilient and resource-efficient water management systems. The Reprint is addressed to researchers, engineers, policy makers, industry professionals, and stakeholders involved in water treatment, environmental sustainability, energy production, and agricultural development. By bringing together multidisciplinary expertise, it aims to foster scientific exchange and encourage the implementation of sustainable solutions to meet present and future water challenges worldwide.

Marco Pellegrini, Cesare Saccani, and Alessandro Guzzini
Topic Editors

sustainability

MDPI

Systematic Review

Zeolites and Activated Carbons in Hydroponics: A Systematic Review of Mechanisms, Performance Metrics, Techno-Economic Analysis and Life-Cycle Assessment

Dana Akhmetzhanova [1,2], Aitugan Sabitov [2,3], Yerlan Doszhanov [1,2], Meiram Atamanov [2,3], Karina Saurykova [2], Arman Zhumazhanov [1,2], Tolganay Atamanova [2], Almagul Kerimkulova [2,4], Leticia F. Velasco [5], Assem Zhumagalieva [6], Jakpar Jandosov [2] and Ospan Doszhanov [7,*]

1 Faculty of Geography and Environmental Sciences, Al-Farabi Kazakh National University, Al-Farabi Ave. 71, Almaty 050040, Kazakhstan; adana128128@gmail.com (D.A.); doszhanov_yerlan@mail.ru (Y.D.); jumajanarman@gmail.com (A.Z.)
2 Institute of Combustion Problems, Bogenbay Batyr 172, Almaty 050012, Kazakhstan; aitugans@mail.ru (A.S.); amk1310@mail.ru (M.A.); saurykova.karina@mail.ru (K.S.); tolganay.temirgaliyeva@gmail.com (T.A.); almusha_84@mail.ru (A.K.); jandosov.j@kaznmu.kz (J.J.)
3 Faculty of Natural Sciences, Kazakh National Women's Teacher Training University, Gogol Str., 114, k. 1, Almaty 050000, Kazakhstan
4 Department of Materials Science, Nanotechnology and Engineering Physics, Satbayev University, Satbayev Str., 22a, Almaty 050013, Kazakhstan
5 Department of Chemistry, Royal Military Academy, Avenue de la Renaissance 30, 1000 Brussels, Belgium; leti_fv@hotmail.com
6 School of Engineering and Digital Science, Nazarbayev University, Kabanbai Batyr 53, Astana 010000, Kazakhstan; assemzhumagali@gmail.com
7 Department of Automation and Robotics, Almaty Technological University, Tole bi Str. 100, Almaty 050012, Kazakhstan
* Correspondence: ospan.doszhanov@mail.ru

Abstract: The sustainable operation of hydroponic systems depends on maintaining the chemical stability of circulating nutrient solutions and preventing the accumulation of toxic compounds. The accumulation of phytotoxic ammonium, heavy metals, and organic metabolites in recirculating nutrient solutions remains one of the key challenges limiting the efficiency, sustainability, and scalability of hydroponic cultivation. This review provides a comprehensive comparative analysis of zeolites, activated carbons (ACs), and their functionalized and composite forms as key sorbents for nutrient management, contaminant removal, and environmental safety in hydroponic cultivation. Natural zeolites, with their well-defined crystalline structure and high ion-exchange selectivity toward ammonium and heavy metal cations, enable effective NH_4^+/K^+ balance regulation and phytotoxicity mitigation. ACs, characterized by high specific surface area and tunable surface chemistry, complement zeolites by offering extensive adsorption capacity for organic compounds, root exudates, and pesticide residues, thereby extending the operational lifespan of nutrient solutions and improving overall system performance. Further advancements include the integration of zeolites and ACs with two-dimensional (graphene, g-C_3N_4) and three-dimensional (MOF, COF) frameworks, yielding multifunctional materials that combine adsorption, ion exchange, photocatalysis, and nutrient regulation. Transition-metal modification, particularly with Fe, Mn, Cu, Ni, and Co, introduces redox-active centers that enhance sorption, catalysis, and phosphate stabilization. The comparative synthesis reveals that the combined application of zeolite- and carbon-based composites offers a synergistic strategy for developing adaptive and low-waste hydroponic systems. From a techno-economic and environmental standpoint, the judicious application of these materials paves the way for more resilient, efficient, and circular hydroponic systems, reducing fertilizer and water consumption, lowering contaminant discharge, and enhancing food security.

This systematic review was conducted according to the PRISMA 2020 guidelines. Relevant studies were identified through Scopus, Web of Science, and Google Scholar databases using specific inclusion and exclusion criteria.

Keywords: zeolites; activated carbon; hydroponic systems; 2D/3D composites; transition-metal functionalization; sorption and catalysis

1. Introduction

Hydroponics is a technology for growing plants without the use of soil, in which the root system is immersed in a nutrient solution of strictly controlled composition. This approach ensures higher yields and accelerated plant growth by optimizing the supply of macro- and microelements. Unlike traditional agriculture, hydroponic systems make it possible to reduce water consumption by 70–90%, minimize fertilizer losses, and decrease the carbon footprint of production [1].

The advantages of hydroponics are manifested in the possibility of year-round cultivation, efficient use of enclosed spaces, and integration into urban ecosystems, including vertical farms and green building systems (Figure 1). Recent studies demonstrate that such systems can simultaneously serve as wastewater treatment and biofiltration units, making them promising components of the circular economy [2].

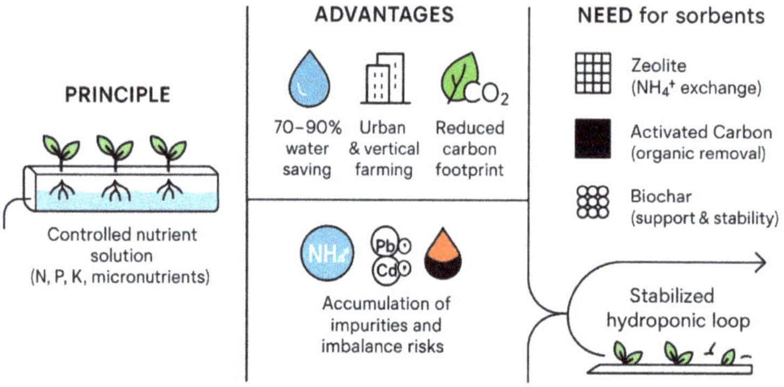

Figure 1. Principles, Advantages, and Challenges of Hydroponics.

Despite its numerous advantages, hydroponic technology faces several challenges. Among them are the need for strict control of nutrient solution quality, prevention of the accumulation of undesirable impurities (ammonium, heavy metals, organic metabolites), and ensuring substrate stability for long-term use. In this regard, increasing attention is being paid to the application of natural and modified sorbents, including zeolites, ACs, and biochar, which can enhance moisture retention, bind toxic compounds, and stabilize the chemical composition of nutrient solutions [3,4].

1.1. The Relevance of the Use of Sorbents in Hydroponic Systems

One of the key factors determining the stability of hydroponic systems is maintaining the chemical composition of the nutrient solution within the optimal range for plant growth. During intensive solution circulation, undesirable compounds—such as ammonium ions, heavy metals, organic metabolites, and secondary microbial products—tend to accumulate, which may cause phytotoxicity and reduce crop productivity. Consequently, increasing attention has been directed toward the use of sorbent materials capable of regulating the solution composition without disrupting the balance of macro- and microelements [5].

Natural zeolites are considered among the most effective materials due to their high ion-exchange capacity and selectivity toward ammonium ions. A number of studies have shown that incorporating zeolites into substrates or using them in column systems facilitates the removal of ammonium nitrogen and heavy metals, stabilizes pH, and enhances the yield of vegetable crops [6–8]. In particular, clinoptilolite has been effectively applied in aquaponics for simultaneous maintenance of water quality and plant growth [9].

ACs and biochar exhibit complementary properties, ensuring efficient adsorption of organic compounds, root exudates, and dissolved organic carbon (Figure 2). Their use helps prevent the accumulation of toxic metabolites, reduce the risk of pathogenic microbial development, and extend the lifespan of nutrient solutions [10,11]. Moreover, the combined use of zeolites and carbon-based materials opens new opportunities for comprehensive solution purification, as demonstrated by studies on integrating such sorbents into vertical hydroponic systems and when using desalinated water [12,13].

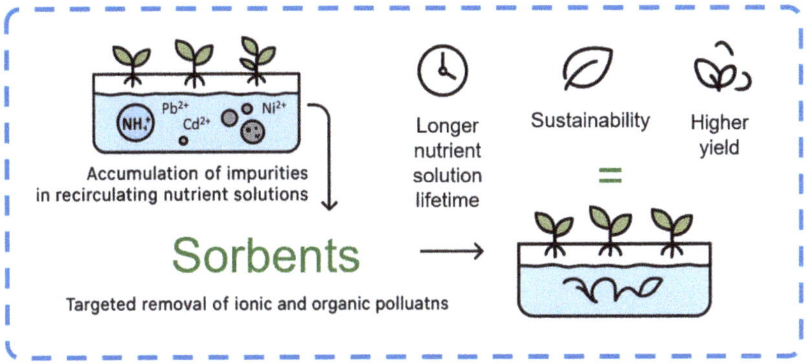

Figure 2. Relevance of Sorbents in Hydroponic Systems.

Thus, the integration of sorbents into hydroponic technologies not only improves the chemical stability of the growing environment but also enhances resource efficiency, ensuring the resilience of production systems to both biotic and abiotic stresses. However, a comprehensive comparative synthesis of zeolites and activated carbons in hydroponic systems, integrating performance, techno-economic, and environmental aspects, remains insufficiently explored.

1.2. Scientific and Practical Challenges in Nutrient Solution Quality Management

Despite the apparent advantages of hydroponics, its large-scale implementation faces several fundamental and practical constraints. The primary challenge is associated with the accumulation of undesirable compounds in the circulating nutrient solution. Excess ammonium nitrogen, produced during the mineralization of organic residues or as a result of fertilizer overuse, can exert phytotoxic effects and reduce crop yields [14]. In parallel, heavy metals (Pb^{2+}, Cd^{2+}, Ni^{2+}) originating from source water or substrates may be present in the solution; their accumulation in plant tissues poses a threat to food safety [15].

Another important limitation is the accumulation of dissolved organic carbon (DOC), which includes root exudates, microbial metabolites, and pesticide residues. These compounds alter the chemical composition of the solution, stimulate the growth of pathogenic microorganisms, and shorten the lifespan of nutrient mixtures. Several studies have shown that organic metabolites themselves can become the key destabilizing factor and a major cause of secondary infections in hydroponic systems [16,17].

Scientific challenges include the need to study competitive ionic interactions. For instance, zeolites exhibit high selectivity toward NH_4^+; however, they can also sorb K^+ and Ca^{2+} ions, thereby reducing the availability of these essential elements for plants. Carbon-based materials are effective in binding organic compounds, but their sorption capacity is limited by surface saturation and biofouling, which necessitates optimization of operating conditions [18].

Practical challenges are associated with ensuring the long-term performance of sorbents. Key issues include the high cost of production, the need for regeneration, disposal, or replacement of materials after saturation [19]. Moreover, the integration of filtration modules into a closed-loop system requires assessment of hydraulic resistance and its impact on the overall energy efficiency of the system [20]. A particular difficulty lies in developing universal sorbents capable of simultaneously retaining ions, organic compounds, and microbial metabolites, while maintaining structural and chemical stability over prolonged periods of operation [21].

Zeolites have long been recognized as versatile adsorbents, catalysts, and ion exchangers with well-defined structures that enable precise control of molecular partitioning and surface reactivity. Thus, managing the quality of nutrient solutions in hydroponics remains an interdisciplinary challenge, integrating agronomic, chemical, and engineering approaches. Recent reviews have highlighted growing interest in sustainable soilless cultivation technologies and precision control systems [22], as well as the potential role of bioelectrochemical systems in improving nutrient recycling and monitoring [23]. However, these studies have primarily focused on system innovations and automation, with insufficient discussion of sorbent materials, which play a crucial role in nutrient retention and contaminant removal. In contrast, research on adsorption materials has primarily focused on their general properties and application in separation processes [24] or specific experimental applications for nutrient recovery from unconventional sources, such as human urine [25]. However, a comprehensive synthesis linking the physicochemical characteristics of these materials to hydroponic performance and sustainability indicators remains absent.

Activated carbons, on the other hand, are widely used for the adsorption of organic compounds and nutrient recovery in wastewater and agricultural systems. For example [25], demonstrated the use of coconut shell-derived activated carbon for adsorption–desorption processes that convert organic components of human urine into nutrient solutions suitable for hydroponic cultivation. Although such studies reveal the potential of carbon adsorbents for nutrient recycling, a comprehensive analysis comparing their mechanisms of action, performance indicators, and environmental impacts with zeolites is lacking.

Thus, managing the quality of nutrient solutions in hydroponics remains an interdisciplinary challenge that integrates agronomic, chemical, and engineering approaches. Addressing these challenges requires a systematic synthesis of current data on sorbent materials and their interactions in closed hydroponic systems.

The aim of this review is to conduct a comparative analysis of zeolites and ACs used in hydroponic systems, with emphasis on their structural characteristics, mechanisms of ion exchange and adsorption, and roles in the removal of ammonium, heavy metals, and organic contaminants. In addition, this review examines the potential for material modification, including their integration with 2D and 3D structures and transition metal nanoparticles functionalization, which expand their sorption and catalytic properties. Furthermore, the study analyzes the techno-economic and environmental aspects of their application, as well as the prospects for developing combined systems that enhance the stability and efficiency of hydroponic technologies. Compared with previous reviews [22–25], this work provides a broader synthesis that integrates mechanistic, material, and sustainability perspectives, highlighting opportunities for the development of adaptive and low-waste hydroponic systems).

2. Methods

This review was conducted in accordance with the PRISMA 2020 (Preferred Reporting Items for Systematic Reviews and Meta-Analyses) guidelines. The protocol for this systematic review was preregistered on the Open Science Framework (OSF), an open-access research registry that serves as an accepted platform for systematic review registration. The registered protocol includes details of the research question, eligibility criteria, search strategy, screening workflow, and planned synthesis procedures. The registration can be accessed through its DOI: [https://doi.org/10.17605/OSF.IO/FE5JZ]. The registration was finalized prior to conducting study selection and data extraction, ensuring transparency and methodological rigor. No amendments were made to the registered protocol, and the review followed the preregistered procedures exactly as planned.

Studies were included if they addressed the use of zeolites, activated carbons, biochar, or similar sorbents in hydroponic systems or comparable aquatic environments, with a focus on adsorption, ion exchange, or techno-economic and environmental assessment. Only peer-reviewed journal articles in English published between 2001 and 2025 were considered.

A literature search was conducted in Scopus, Web of Science, Science Direct, and Google Scholar. The last search was performed on 30 September 2025. Relevant references from key articles were also manually screened to identify additional sources.

Data was extracted manually, and any missing or unclear summary statistics were marked as "not presented". The data were compiled into structured tables and visualized using diagrams to summarize the sorbent's characteristics and applications.

The formal risk of bias assessment was not performed because this study is a descriptive systematic review summarizing experimental and applied studies rather than a quantitative meta-analysis. Sensitivity analyses were not performed, as this review focuses on summarizing trends and key findings rather than statistical effect estimates.

A PRISMA flowchart summarizing the literature search, selection, and inclusion process is provided in the Supplementary Materials.

2.1. Study Selection

The study selection process is presented in Figure 3 PRISMA 2020 flow diagram. A total of 600 records were identified through database searches, and an additional 25 records were found through citation searching. After removing duplicates, the remaining records were screened based on titles and abstracts, followed by a full-text assessment for eligibility.

In total, 102 studies met the inclusion criteria and were incorporated into the final review. Full list of included studies along with their key characteristics is provided in Appendix A Table A1. This supplementary table contains detailed information such as publication year, study design, location, sample size, methods, and main findings, allowing readers to explore the characteristics of each study in depth.

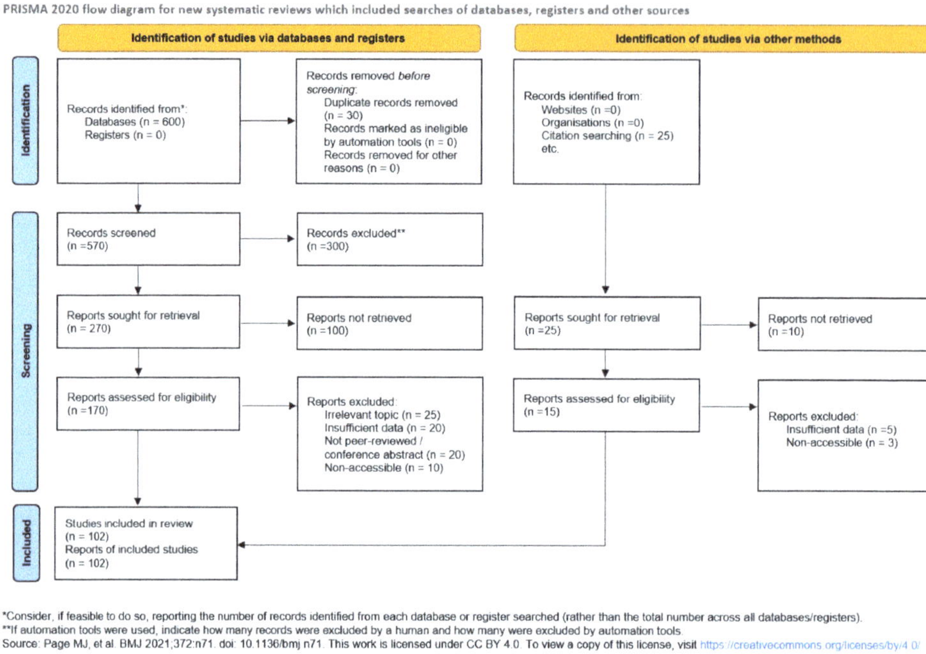

Figure 3. PRISMA 2020 flow diagram illustrating the study selection process [26].

3. Zeolites: Properties and Application in Hydroponics

3.1. Physicochemical Characterization of Zeolite

Natural zeolites are microporous, natural aluminosilicate materials with a specific structure, which is represented by a tetrahedral crystal lattice consisting of SiO^{4-} and AlO^{4-} ions. The SEM image of the natural zeolite (Figure 4a) shows an irregular and heterogeneous surface morphology composed of agglomerated particles of various sizes. The grains exhibit a porous structure with uneven edges, typical of clinoptilolite minerals formed under natural conditions [27].

In the FTIR spectrum of the unmodified natural zeolite (Figure 4c), the stretching vibrations of O–H bonds are not clearly observed around 3436 cm^{-1}. However, a weak absorption band appears near 1632 cm^{-1}, which can be attributed to the bending vibrations of adsorbed water molecules or hydroxyl groups present on the zeolite surface. This indicates that only a small amount of physically adsorbed moisture remains in the structure of the natural clinoptilolite. The degree of resolution of the characteristic absorption bands in the FTIR spectra of mineral forms of zeolites depends on the complexity of their structure. Absorption band in the zeolite spectrum at 1002 cm^{-1}, a silicon–oxygen tetrahedron of Si–O–Si bonds, which is associated with bending vibrations of SiO_4 [27].

The XRD pattern of the zeolite sample confirms that clinoptilolite is the dominant crystalline phase (Figure 4b). Minor diffraction peaks corresponding to albite and mordenite are also detected, indicating their low content in the natural mineral [28].

Table 1 presents the comparative chemical composition of natural zeolites from different deposits. As shown, the SiO_2 and Al_2O_3 contents vary slightly depending on the origin of the samples, with silica generally dominating the composition. The zeolites from Australia and Ukraine exhibit the highest SiO_2 values, indicating a high degree of framework stability typical for clinoptilolite-rich materials (Table 1) [29]. The contents of Fe_2O_3, Na_2O, K_2O, CaO, and MgO differ between deposits, reflecting the influence of local geological conditions and mineral impurities. The composition of the Kazakhstan sample falls within the general range reported for natural zeolites, confirming its comparable chemical nature.

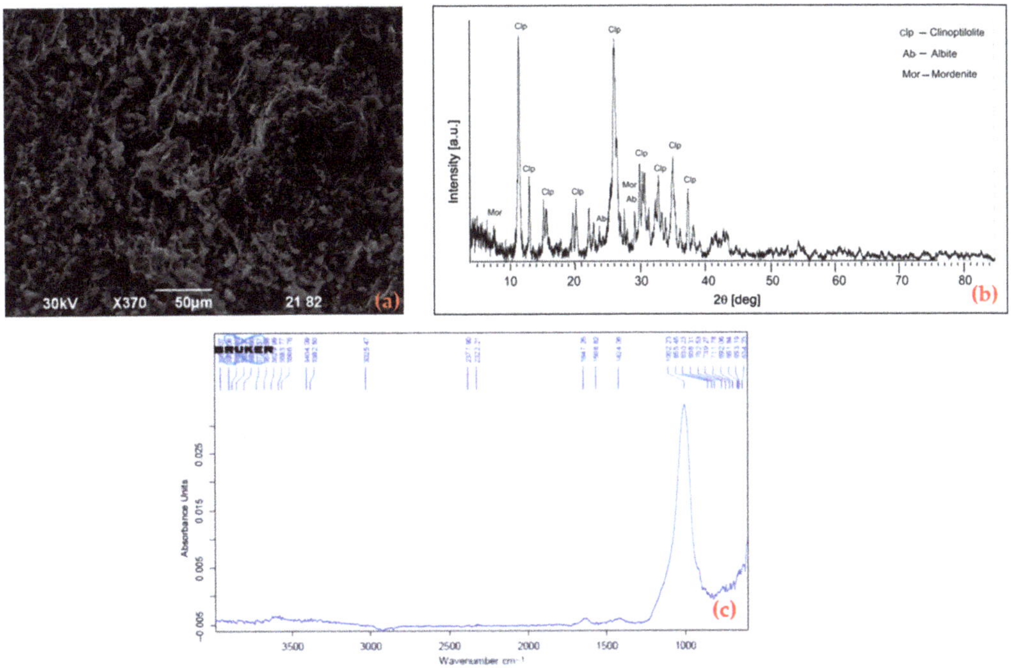

Figure 4. Structural and Physicochemical Properties of Zeolite: (**a**) SEM image of natural zeolite [27], (**b**) XRD pattern of natural zeolite [28], (**c**) FTIR spectrum of natural zeolite [27].

Table 1. Comparative chemical composition of zeolites from different deposits.

Countries	SiO_2 (%)	Al_2O_3 (%)	Fe_2O_3 (%)	Na_2O (%)	K_2O (%)	CaO (%)	MgO (%)
Australia	68.26	12.99	1.37	0.64	4.11	2.09	0.83
China	65.72	13.50	1.30	1.16	3.14	3.10	0.63
Ukraine	68.64	11.50	1.57	0.29	3.12	2.38	0.89
Ecuador	65.80	11.32	3.42	4.10	0.45	1.23	0.96
USA	66.61	12.91	1.70	0.39	2.44	3.18	1.54
Iran [28]	68.17	11.05	0.53	0.64	1.11	3.93	0.62
Kazakhstan [30]	62.2	13.4	5.9	1.6	6.5	5.3	2.2

3.2. Structure, Ion Exchange Mechanisms and Sorption Selectivity

Zeolites are crystalline aluminosilicates with a regular system of micropores formed by SiO_4 and AlO_4 tetrahedra. The substitution of silicon with aluminum in the crystal lattice leads to the formation of negative charges balanced by mobile cations such as Na^+, K^+ and Ca^{2+} [31]. These cations can be exchanged with ions from solution, which determines the high ion exchange capacity and selectivity of zeolites [32].

The selectivity of zeolites with respect to ammonium ion is of particular importance in hydroponic systems. Due to the size of the channels and their high affinity for NH_4^+, clinoptilolite and related minerals effectively remove ammonium from nutrient solutions, reducing the risk of its toxic accumulation and stabilizing the nitrogen balance [33,34]. In addition, zeolites are able to bind heavy metal cations (Pb^{2+}, Cd^{2+}, Ni^{2+}), which prevents their entry into plants and increases the safety of final products [35].

Ion exchange processes in zeolites are reversible, which makes it possible to regenerate them using saline solutions without destroying the structure. Studies show that even after multiple cycles of use, zeolites retain a significant part of their sorption capacity. This combination of structural stability, high selectivity, and the possibility of reuse makes zeolites promising materials for long-term use in hydroponic systems [36].

Figure 5 shows the mechanism of ion exchange and the factors determining the selectivity of zeolites. Panel (a) shows the crystalline structure of zeolite, in which the negative charge of the frame, resulting from the inclusion of aluminum, determines the ability to bind cations (the conceptual scheme shown in this figure has been adapted and scientifically redesigned by the Avogadro application based on the graphic style and structural principles illustrated in the reference information) [37]; panel (b) shows a SEM snapshot of the morphology of the zeolite particle. Panel (c) illustrates the process of substitution of Na^+ by K^+ and Sr^{2+}: at exchange rates above 80%, the lattice expands, accompanied by a decrease in selectivity. Panel (d) reflects the influence of pH: the competition between H_3O^+ and $Na^+/K^+/Sr^{2+}$ cations lead to the fact that, with increasing pH, hydronium is displaced and the selectivity of zeolite to metals increases. Finally, panel (e) shows the distribution of cations at different crystallographic positions–Site I (D6R ring) and Site II (supercage), which explains the differences in selectivity.

3.3. The Role of Zeolites in the Removal of Ammonium, Cations and Toxic Metals

One of the key functions of zeolites in hydroponic systems is the efficient removal of ammonium ions from nutrient solutions. Excess concentrations of $NH_4{}^+$ disrupt the nitrogen balance and can lead to reduced plant productivity, particularly under conditions of high salinity stress. Clinoptilolite, owing to its high selectivity toward ammonium, is considered one of the most promising materials for regulating nitrogen content. Several studies have shown that the addition of zeolites to the substrate or their use in filtration columns results in a significant decrease in $NH_4{}^+$ concentration and stabilization of the nitrogen cycle [38].

In addition to ammonium, zeolites demonstrate high efficiency in binding potassium, calcium, and magnesium cations. These processes have a dual effect: on one hand, they prevent the accumulation of excessive concentrations of individual ions, while on the other, they may alter the availability of nutrients for plants. Therefore, the practical value of zeolite application depends on precise control of dosage and operating conditions to avoid deficiencies of essential macronutrients These metals not only exert toxic effects on plants but also pose a threat to human health through the consumption of contaminated produce. The use of zeolites effectively reduces their concentrations, thereby ensuring the safety and environmental sustainability of hydroponic technologies [39].

Figure 6 shows the mechanism and efficiency of ion exchange and heavy metal sorption by zeolite. Zeolites are aluminosilicates with a 3D structure containing exchangeable cations (Na^+, K^+, Ca^{2+}, Mg^{2+}), which can be replaced by heavy metal ions Pb^{2+}, Cd^{2+}, Cu^{2+}, Ni^{2+}, Cr^{3+}, etc. during adsorption (Figure 6a). This selectivity is based on the electrostatic interaction between the charged surfaces of the zeolite and the cations in solution.

As can be seen from the diagram (Figure 6b), the heavy metal removal efficiency using natural zeolite without additives reaches 50–65%, with the highest values recorded for Cu and Cr (64.7% and 56.3%, respectively). The pH dependence graph (Figure 6c) shows that at low pH values (<3), the removal efficiency of all metals is minimal due to competition between H^+ ions for the active sites of the zeolite. With an increase in pH to 5–6, sorption increases sharply, reaching 90–100%, due to deprotonation of the zeolite surface and increased ion exchange. In a neutral and slightly alkaline environment (pH ≈ 7–8), maximum sorption efficiency is achieved, after which the process stabilizes due to saturation of available exchange sites.

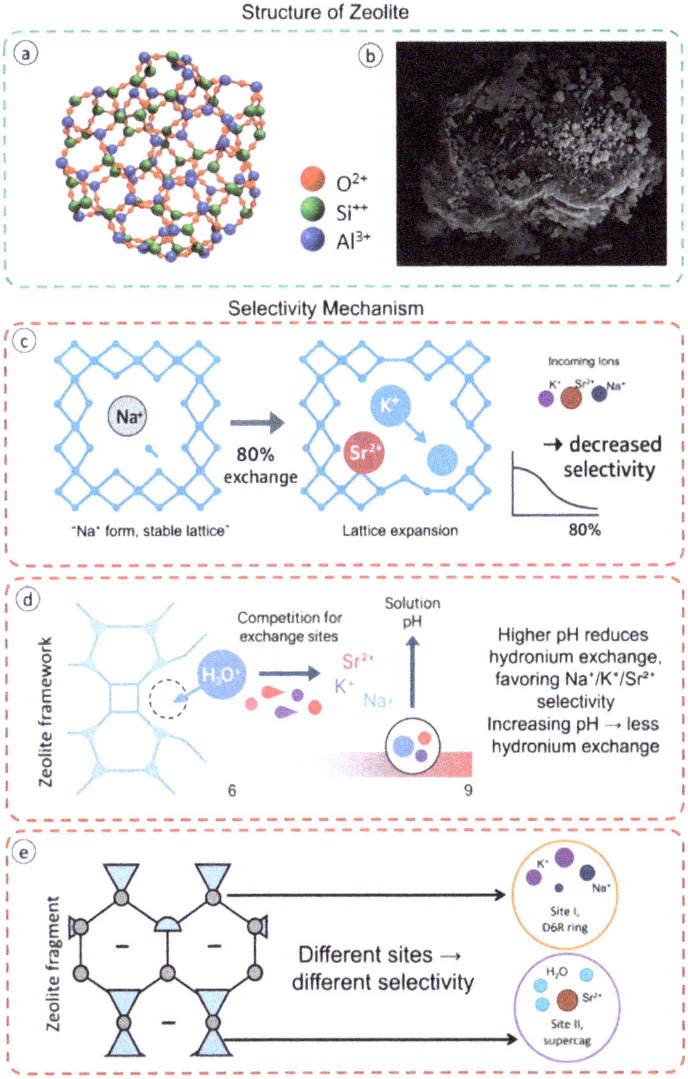

Figure 5. Structure and Ion-Exchange Selectivity of Zeolites materials. (**a**) Crystal structure of zeolite, (**b**) Morphology of zeolite (SEM image), (**c**) Mechanism of selective ion exchange of zeolite, (**d**) Effect of solution pH on ion exchange selectivity of zeolite framework, (**e**) Site-specific cation selectivity in zeolite fragments.

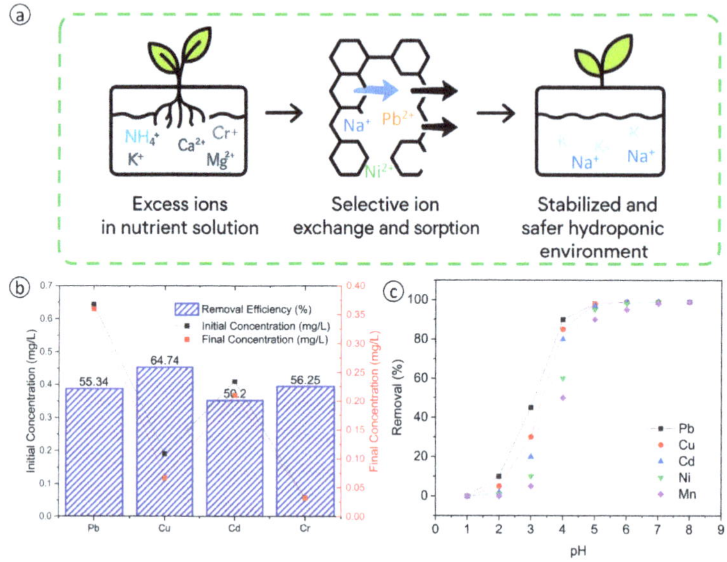

Figure 6. Ion exchange and heavy metal adsorption by zeolite: (**a**) Schematic diagram of the selective ion exchange and nutrient solution stabilization process; (**b**) metal removal efficiency (Pb, Cu, Cd, Cr) using zeolite without additives [40]; (**c**) effect of pH on the removal efficiency of heavy metals (Pb, Cu, Cd, Ni, Mn) [41].

It has been established that zeolites provide selective binding of heavy metal ions under optimal pH conditions of 5–7, simultaneously reducing the concentration of excess cations (NH_4^+, K^+, Ca^{2+}, Mg^{2+}) in hydroponic solutions and stabilizing the chemical composition of the nutrient medium.

3.4. Limitations and Prospects of Zeolite Application

Despite the wide range of favorable properties, the use of zeolites in hydroponic systems is associated with several limitations. One of the key drawbacks is the selectivity of ion exchange, which, under complex solution conditions, may lead to competition for active sites. For example, the high affinity of clinoptilolite for NH_4^+ is accompanied by the sorption of K^+ and Ca^{2+}, which can reduce the availability of macronutrients and cause deficiencies in plants [42].

Another important limitation is the restricted sorption capacity and saturation of the material. Under conditions of intensive operation, the pores of zeolites become rapidly filled, necessitating regular regeneration or replacement. At the same time, conventional methods for restoring sorption capacity–such as washing with saline solutions–are often accompanied by the generation of salt-containing effluents, which impose an additional environmental burden [43].

The physico-mechanical properties of zeolites also impose certain limitations on their use. Their low mechanical strength and tendency to degrade after prolonged contact with water may reduce their effectiveness as an independent substrate. Consequently, zeolites are more commonly used in mixtures with inert materials (such as perlite or vermiculite) or as additives in filtration cartridges [44].

Economic factors likewise play an important role. Although natural zeolites are widely available, their efficiency strongly depends on the deposit characteristics, degree of purification, and post-processing. In some cases, the costs of transportation and modification significantly increase the overall application cost [45].

Future prospects for the use of zeolites in hydroponics are primarily associated with their modification and functionalization. The development of nanostructured zeolites and composites with metal oxides enhances their selectivity and resistance to saturation. A promising direction involves integrating zeolites with 2D and 3D materials such as graphene, g-C_3N_4, and metal–organic frameworks (MOFs), which opens new possibilities for simultaneous sorption and catalytic functions [46].

In addition, the emerging concept of "zeoponics"–the use of zeolites as a long-term substrate for plant cultivation in closed systems, including space programs–has been actively developing. This approach highlights the potential of zeolites as a strategic material for sustainable agriculture under extreme conditions [47].

4. ACs: Properties and Application in Hydroponics

4.1. Morphology, Specific Surface Area and Adsorption Characteristics

ACs are highly porous carbonaceous materials characterized by a developed morphology and a broad pore-size distribution that includes micropores (<2 nm), mesopores (2–50 nm), and macropores (>50 nm). Such a structure provides an exceptionally high specific surface area (often exceeding 1000 m^2/g), which enables the efficient adsorption of a wide range of organic and inorganic compounds [48].

The morphology of ACs directly depends on the raw material (e.g., wood, nutshells, agro-industrial residues) and the activation method used–either physical activation with CO_2 or steam, or chemical activation using phosphates and alkalis. These factors determine both the pore size distribution and the presence of surface functional groups, which in turn influence hydrophilicity, acid–base properties, and electrostatic interactions with ions in solution [49].

A key characteristic of ACs in the context of hydroponics is their high capacity for binding dissolved organic compounds, including root exudates, phenolic metabolites, and pesticide residues. Through a combination of hydrophobic interactions and π–π associations, the carbon matrix effectively retains organic molecules, thereby reducing the risk of pathogenic microbial growth and extending the lifespan of nutrient solutions [50].

Figure 7 illustrates the structural and functional properties of ACs that ensure their high sorption capacity.

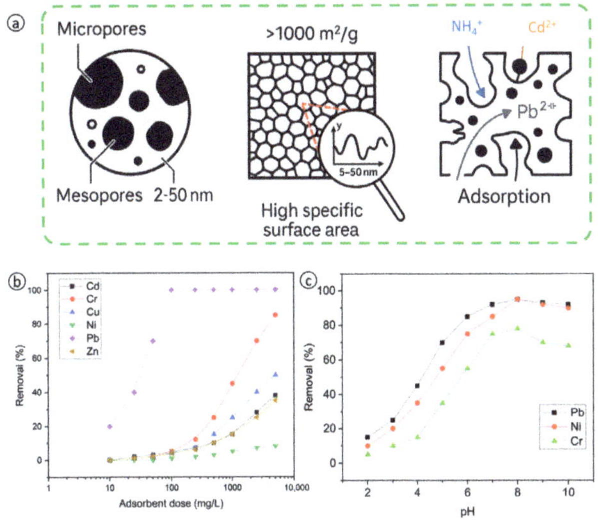

Figure 7. Morphology, specific surface area, and adsorption characteristics of ACs: (**a**) Schematic representation of the pore structure (micro- and mesopores 2–50 nm) and the sorption mechanism of

NH_4^+, Cd^{2+}, and Pb^{2+} ions on a surface with a specific area > 1000 m^2/g; (**b**) Removal efficiency of heavy metals (Cd, Cr, Cu, Ni, Pb, Zn) depending on the adsorbent dose [51]; (**c**) Effect of pH on the degree of metal removal (Pb, Ni, Cr) [52].

In addition, the highly developed surface of ACs facilitates the adsorption of heavy metal cations and ammonium through ion exchange and complexation with oxygen-containing functional groups. Thus, ACs exhibit a high degree of versatility: on one hand, they act as filters for organic contaminants, while on the other, they complement the function of zeolites in stabilizing the inorganic composition of nutrient solutions. The micro- and mesoporous structure creates a large specific surface area (>1000 m^2/g), which facilitates the adsorption of both organic compounds and heavy metal cations. Experimental curves show that with increasing adsorbent dose, the removal efficiency of Pb, Cr, and Cu increases to ~100%, and the optimal pH range (6–8) ensures maximum metal sorption due to reduced competition with H^+ ions and stabilization of complexation on oxygen-containing functional groups on the carbon surface [53].

4.2. The Use of ACs for the Removal of Organic Pollutants and Metabolites

The use of ACs in hydroponic systems is primarily associated with their ability to effectively remove dissolved organic compounds and secondary plant metabolites. Due to the combination of hydrophobic interactions, π–π associations, and hydrogen bonding, AC provides selective binding of a wide range of organic contaminants. This functionality is particularly important in closed-cycle systems, where root exudates and microbial metabolites tend to accumulate and become phytotoxic factors [54].

Compounds that negatively affect plant growth include low-molecular-weight organic acids, phenolic derivatives, and allelochemicals. Their presence in nutrient solutions can inhibit root growth, reduce nutrient uptake, and promote the development of pathogenic microflora. The use of AC helps to lower the concentrations of such substances, thereby stabilizing the biochemical balance of the nutrient medium and extending its service life [55].

The effectiveness of AC is also evident with respect to exogenous organic pollutants such as pesticide residues, pharmaceutical compounds, and endocrine disruptors that may enter the system through water or fertilizers. The removal of these compounds is critical for environmental safety and product quality [56].

In recent years, attention has been focused on modified forms of activated carbon. Surface functionalization with oxygen- or nitrogen-containing groups enhances the hydrophilicity of the material and strengthens its affinity toward polar organic compounds. Nanostructured forms, possessing an increased proportion of meso- and macropores, demonstrate accelerated adsorption kinetics due to reduced diffusion barriers [57].

4.3. Limitations and Prospects of Using ACs

Despite a wide range of positive properties, the use of ACs in hydroponic systems is associated with several limitations. The main issue lies in their low adsorption selectivity: the carbon surface effectively retains a broad spectrum of organic molecules but is not always capable of selectively removing only phytotoxic metabolites. This may lead to competition between target pollutants and essential organomineral components in the solution [58].

Another important factor is the limited durability and the need for regeneration. During prolonged operation, AC becomes saturated with organic matter, undergoes biofouling, and loses its adsorption capacity. Conventional regeneration methods (thermal or chemical) are energy-intensive and may lead to the formation of secondary pollutants. Therefore, under conditions of continuous nutrient solution circulation, the development of low-cost and environmentally friendly regeneration methods is of great importance [59].

The production cost also remains a significant constraint. Although various agro-industrial wastes are widely used as precursors for AC synthesis, the high activation temperatures and the use of chemical agents increase the overall cost compared with zeolites and other mineral sorbents. Furthermore, carbon materials are prone to mechanical degradation and leaching of fine particles, necessitating the development of granulated or composite forms with improved structural stability [60].

Future prospects for AC application in hydroponics are primarily related to surface functionalization and composite design. Modification with oxygen-, nitrogen-, and sulfur-containing groups enhances affinity toward polar organic compounds and ammonium ions. The integration of AC with 2D materials (graphene, g-C_3N_4) and 3D structures (MOFs, COFs) provides not only adsorption but also catalytic degradation of organics, thus expanding material functionality [61].

Special attention is being given to the development of biochar-based composites that combine the low cost and stability of biochar with the high adsorption capacity of AC. Such materials exhibit improved performance in nutrient solution purification and can serve as long-lasting substrates. Another promising direction involves the creation of multifunctional sorption systems combining AC with zeolites or hydrogels, thereby integrating the advantages of different contaminant removal mechanisms [62].

5. Composites of Zeolites and ACs with 2D/3D Materials

5.1. Zeolites and ACs Combined with 2D Materials (Graphene, g-C_3N_4, etc.)

The integration of zeolites and ACs with two-dimensional (2D) materials opens new opportunities for enhancing the efficiency of sorption and catalytic processes in hydroponic systems. Two-dimensional structures such as graphene, graphene oxide (GO), partially reduced graphene (prGO), and graphitic carbon nitride (g-C_3N_4) possess a high specific surface area, rich electronic structure, and tunable surface functionality, providing a synergistic effect when combined with traditional adsorbents [63,64].

Zeolites modified with graphene-based materials exhibit enhanced sorption capacity toward ammonium ions and heavy metals due to the combined effects of ion exchange and π–π interactions. In such composites, graphene layers improve electrical conductivity and create additional diffusion pathways, whereas the microporous framework of zeolites ensures selectivity and structural stability. These hybrids are considered promising materials for managing nutrient balance and removing xenobiotics in recirculating hydroponic solutions [65].

ACs, when modified with GO or g-C_3N_4, acquire enhanced catalytic and photocatalytic properties. The incorporation of g-C_3N_4 into the carbon matrix promotes the photodegradation of organic metabolites and phenolic compounds in hydroponic solutions through the formation of heterojunctions that suppress electron–hole recombination. Graphene layers in AC-based composites increase sorption capacity toward hydrophobic organic molecules while simultaneously improving the mechanical strength and structural stability of granular forms [66].

A promising direction involves the development of multifunctional composites combining zeolites, ACs, and 2D materials. Such hybrid systems integrate ion-exchange, hydrophobic, and π–π interaction mechanisms with photocatalytic degradation of organics, thereby addressing the simultaneous challenges of purification, regeneration, and stabilization of nutrient solutions (Figure 8). Furthermore, the incorporation of 2D structures enhances the electrical conductivity of the composites, paving the way for their application in sensing and intelligent monitoring of hydroponic systems [67].

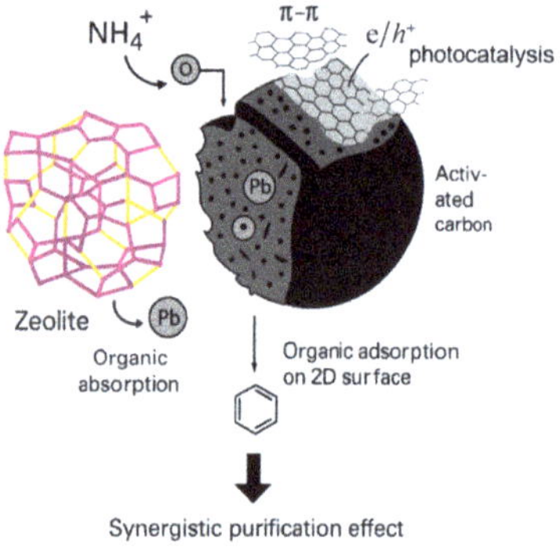

Figure 8. Zeolites and ACs in Combination with 2D Materials (Graphene, g-C_3N_4).

5.2. Composites with 3D Structures (MOF, COF) and Their Functional Features

MOFs and covalent organic frameworks (COFs) are considered promising 3D materials due to their exceptionally high specific surface area, well-defined pore geometry, and tunable functional groups. The integration of zeolites and ACs with these frameworks enables the combination of macro- and microporous carriers with the unique selectivity and adjustable pore architecture of MOF/COF structures [6,68].

Zeolites are characterized by high cation-exchange selectivity and excellent mechanical stability, making them an effective matrix for the growth of MOF structures. Such hybrids exhibit improved performance in removing heavy metals and oxyanions from nutrient solutions, as well as in the controlled release of micronutrients–an especially important feature for closed-loop hydroponic systems. For instance, zeolite–MOF composites have demonstrated higher sorption efficiency for Pb^{2+} and Cd^{2+} compared to their individual components, which is attributed to the synergistic effect between the ion-exchange centers of zeolites and the high surface energy of MOFs [69].

In the case of ACs, integration with MOF and COF is aimed at increasing the adsorption of organic metabolites and resistance to biofouling. Carbon surfaces provide electrochemical conductivity and hydrophobic interactions, while MOF/COF structure the adsorption centers and enhance selectivity. This provides advantages in the removal of phenolic compounds and root exudates, which can negatively affect plant growth [70,71].

COF composites based on zeolites and ACs are characterized by increased chemical stability and the possibility of introducing functional groups (–NH_2, –SO_3H), which enhances their sorption capacity to phosphates and nitrates. In addition, such materials are promising as polishing filters in integrated hydroponic systems, where fine-tuning of the ionic composition of the solution is required [72].

The combination of zeolites and ACs with MOF and COF forms hybrid sorbents of a new generation, which combine the mechanical stability and ion-exchange properties of zeolites, high specific surface area and hydrophobicity of carbon materials, as well as the selectivity and functional variability of 3D frameworks. This opens up opportunities for the creation of intelligent sorption units that adapt to the changing composition of hydroponic solutions [73].

5.3. Comparative Efficiency in Hydroponic Systems

A comparison of zeolites, ACs and their composites with 2D/3D structures under hydroponic conditions shows that the effectiveness of sorbents is determined not only by their internal characteristics, but also by the chemical composition of the nutrient solution. For zeolites, the key factor is the competition of cations: at high concentrations of K^+, characteristic of nutrient media, the exchange capacity for NH_4^+ is noticeably reduced, which accelerates the breakthrough and reduces buffering capacity (Table 2). The use of the Na-form of clinoptilolite and the selection of optimal granulometry partially compensate for these limitations, increasing the efficiency of ammonium retention and stabilizing plant nutrition [74–76].

ACs have advantages in removing dissolved organic matter, including phenolic acids and root exudates, which can inhibit growth. Their use in hydroponic systems is associated with improved yields by reducing the toxic load. At the same time, the service life of coal is limited, and after achieving a breakthrough in organic carbon, replacement or regeneration is necessary, which increases operating costs [77].

Integration with 3D structures introduces both new features and limitations. ZIF-8-type frameworks, despite their high specific surface area, are unstable in phosphate solutions and degrade rapidly. In contrast, zirconium MOFs (UiO-66) exhibit stability in water, but their high affinity for phosphates can lead to depletion of macronutrients and plant nutrition disorders. This limits the scope of MOF application in hydroponics to bypass "polishing" modes with strict control of the solution composition [78].

Photocatalytic composites based on g-C_3N_4 and other 2D materials have a pronounced ability to destroy a wide range of organics under visible light, which is promising for the removal of stable metabolites. However, their non-specificity is fraught with the destruction of useful chelates (for example, EDTA–bound iron), which requires careful implementation, mainly in bypass circuits or during episodic sanitation [79–81].

It can be noted that maximum efficiency is achieved with a functional separation of tasks: zeolites are advisable to use for ion balance management (NH_4^+/K^+), while carbon materials and their composites are more effective for controlling organic metabolites. Combined use makes it possible to stabilize the quality of the solution and increases plant productivity in closed hydroponic cycles.

The synergistic effect of zeolite and activated carbon is due to their complementary adsorption mechanisms and surface characteristics. Zeolites with their negatively charged aluminosilicate structure effectively trap cationic compounds such as NH_4^+, heavy metal ions, and other inorganic pollutants. This selective ion exchange process can also reduce biofouling and microbial growth on the surface of activated carbon while maintaining its adsorption capacity for a long time. Meanwhile, activated carbon with its high specific surface area and hydrophobic micro- and mesopores effectively adsorbs organic compounds that might otherwise clog the pores of the zeolite. As a result, the combined use of zeolite and activated carbon increases the overall cleaning efficiency, providing long-term stability and regeneration potential compared to their individual use.

Table 2. Materials and their applicability in hydroponic systems.

The Environment Condition	The Main Risk	The Most Suitable Material	Justification (+Links)
Accumulation of root exudates and organic matter	Growth inhibition, toxicity	Activated carbon (GAC, mesoporous carbon)	Effective reduction in DOC and increase in yield [82]
High concentration of K^+ in the presence of NH_4^+	Competition and reduction in ammonium capacity	Zeolite (Na-form of clinoptilolite)	Higher selectivity for NH_4^+; stabilization of nutrition [83]
Solutions with high phosphate content	Degradation of ZIF-8, depletion of PO_4^{3-} at UiO-66	COF or carbon sorbents; exclude	ZIF-8 ZIF-8 is unstable in PBS; UiO-66 binds PO_4^{3-} [42]
Destruction of stable metabolites is required	Potential degradation of beneficial chelates	Photocatalytic composites (g-C_3N_4) in bypass	High photocatalytic activity, but low selectivity [84]
Long-term operation and low OPEX	Increased operating costs	Zeolite with in situ regeneration with NaCl solutions	Multiple cycles without loss of mechanics [85]
Maintaining the NH_4^+/K^+ balance and reducing DOC	Deficiency/toxicity risks	Combination of zeolite and activated carbon	Separation of functions stabilizes the system and improves productivity [86]

Table 3 synthesizes the key performance characteristics of zeolites, activated carbons, and composite sorbents, highlighting differences in adsorption mechanisms and suitability for various contaminants.

Table 3. Comparative Evaluation of Zeolites, Activated Carbons, and Their Composites Across Key Performance Metrics.

Indicator	Zeolites	Activated Carbon	Zeolite-Carbon Composites
primary mechanism	ion exchange; molecular sieve	physical adsorption; π–π interactions	ion exchange + adsorption; catalytic effects
key target pollutants	NH_4^+, K^+, heavy metals	DOC, pesticides, organic compounds	NH4+, heavy metals, organic matter
advantages	high selectivity; structural stability; low cost	large surface area (~1000 m^2/g); high organic capacity	synergy, increased capacity, sustainability
disadvantages/limitations	ion competition; weak adsorption of organic substances	surface contamination; possible high cost	complexity of synthesis; variability of properties
regeneration potential	light (chemical regeneration)	complex (thermal regeneration degrades properties)	average; depends on the composition

6. Sorbents as Carriers for Transition Metal Nanoparticles

6.1. Zeolites Modified with Transition Metals: Catalytic Functions

Zeolites are among the most extensively studied carriers for ions and nanoparticles of transition metals due to their high cation-exchange capacity, well-defined porous structure, and stability under aggressive conditions. The incorporation of cations such as Cu^{2+}, Fe^{3+}, Mn^{2+}, Co^{2+}, or Ni^{2+} into the channels and cavities of zeolites imparts pronounced catalytic properties, enabling these materials to function not only as sorbents but also as active centers for redox reactions [87].

According to literature data, Cu- and Fe-containing zeolites exhibit high activity in the catalytic oxidation of ammonium and organic contaminants in aqueous media. These materials operate via the generation of hydroxyl radicals (OH), promoting deep oxidation to harmless end products (N_2, CO_2, H_2O) [88]. In hydroponic systems, this opens up prospects for the use of metal-modified zeolites in controlling ammonium nitrogen and organic metabolites that are poorly removed by purely sorptive methods.

Particular attention has been given to Mn-modified zeolites, which exhibit catalytic activity due to various forms of MnOx embedded in the structure or in contact with the surface. The various oxidation states of manganese ensure efficient oxidation of organic compounds, and the combination of Mn with zeolites improves adsorption and catalytic purification of solutions, reducing the load on carbon sorbents and increasing the stability of hydroponic systems [89].

Figure 9 presents a schematic overview of possible mechanisms in transition-metal-modified zeolites. The incorporation of Fe^{3+}, Cu^{2+}, and Mn^{2+} ions via ion exchange introduces redox-active sites that may enhance adsorption, ion exchange, and limited oxidation reactions under mild conditions. Although complete catalytic oxidation of ammonium and organic molecules requires higher temperatures, these modified zeolites can still promote redox buffering, nutrient stabilization, and suppression of harmful compound accumulation in hydroponic nutrient solutions.

An additional advantage lies in the ability of zeolites to stabilize metal nanoparticles within their micropores, preventing aggregation and leaching of active centers. For example, the incorporation of Pd nanoparticles into zeolite-alumina systems have been shown to improve catalytic performance in hydrogenation and oxidation reactions of organic compounds. These approaches can be adapted to aqueous environments, including hydroponic systems, where selective degradation of organic toxicants is required [90].

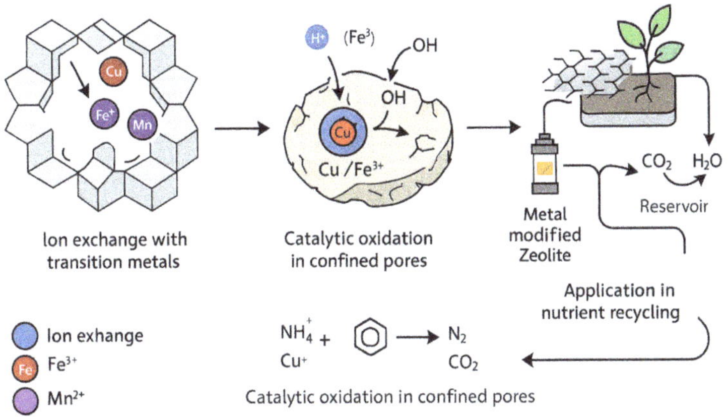

Figure 9. Transition-Metal-Modified Zeolites: Sorption and Catalytic Functions in Hydroponic Systems.

The modification of zeolites with transition metals leads to the development of multifunctional materials that combine sorptive, ion-exchange, and catalytic properties. Their integration into hydroponic systems can significantly expand the functionality of filtration units-from simple ion composition correction to active management of organic and nitrogen-containing contaminants.

6.2. Transition-Metal-Modified ACs for Environmental and Hydroponic Applications

ACs functionalized with transition metals represent a promising class of hybrid materials that combine adsorption, ion exchange, and redox catalytic properties. The integration of metal species such as Fe, Mn, Ni, Cu, and Co into the porous carbon matrix creates redox-active sites capable of simultaneously binding, transforming, and stabilizing inorganic and organic species in aqueous systems [91]. Metal incorporation is typically achieved by impregnation or co-precipitation, followed by thermal activation, resulting in uniformly distributed nanoparticles or ionic clusters anchored to oxygen-containing surface groups (–OH, –COOH, –C=O) [92].

Among these composites, Fe-modified ACs (Fe–ACs) have received particular attention due to their multifunctionality. The presence of Fe^{2+}/Fe^{3+} redox pairs enhances the material's capacity to remove oxyanions such as PO_4^{3-} and AsO_4^{3-} through complexation and surface precipitation, while also promoting oxidation-reduction reactions that stabilize nutrient composition in hydroponic media. Studies have shown that Fe_3O_4-coated ACs exhibit strong affinity toward phosphate and ammonium ions, preventing their excessive accumulation and thus mitigating nutrient imbalance and microbial overgrowth in closed hydroponic systems. Furthermore, the magnetic nature of Fe_3O_4 facilitates rapid separation and reusability of the composite, improving operational safety and sustainability [93].

Other transition metals impart complementary functions. Mn- and Cu-modified carbons catalyze Fenton-like reactions and enhance the degradation of organic metabolites and pesticide residues, which commonly accumulate in nutrient solutions during plant growth. Ni- and Co-doped carbons, while primarily studied for electrochemical applications, also demonstrate redox behavior that may contribute to the regulation of reactive oxygen species and ionic balance in hydroponic environments [94].

The transition-metal-modified ACs function as multifunctional materials, integrating adsorption, catalysis, and magnetic recovery. Their implementation in hydroponic systems offers an effective means of maintaining nutrient stability, minimizing toxic ion accumulation, and enabling the controlled removal of organic and inorganic contaminants, thereby supporting sustainable and high-yield plant production [95].

The results illustrated in Figure 10 demonstrate the multifunctional performance of Fe-modified AC (Fe–AC) as both an adsorbent and a reactive barrier for phosphate control in aqueous and soil systems. In Figure 10a, the adsorption isotherms reveal that Fe incorporation significantly increases the phosphate affinity compared to pristine activated carbon, confirming the formation of Fe^{2+}/Fe^{3+} redox sites that enhance surface complexation and electrostatic binding with PO_4^{3-} species. This mechanism aligns with the dual adsorption–redox functionality described earlier, where transition-metal centers stabilize anionic nutrients through coordination and partial redox conversion.

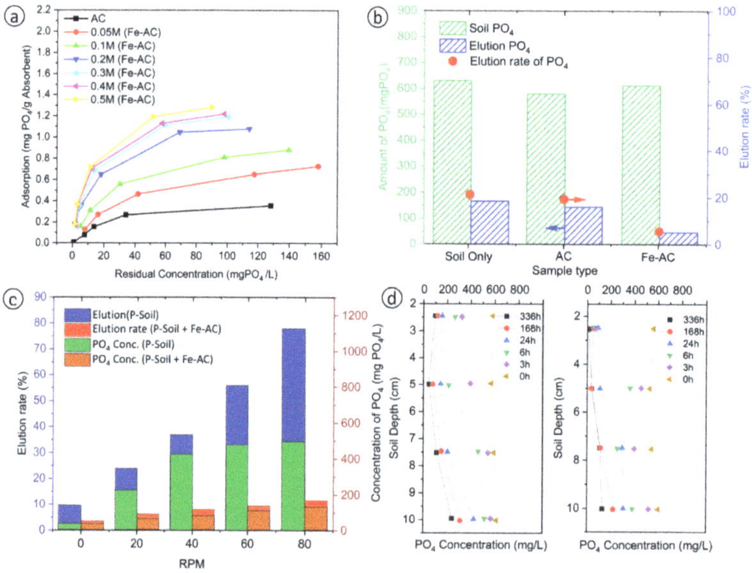

Figure 10. Phosphate adsorption and release characteristics of Fe-modified activated carbon [96]: (**a**) adsorption isotherms; (**b**) static soil system; (**c**) hydraulic flow conditions; (**d**) phosphate profile with soil depth.

Under static soil conditions (Figure 10b), the Fe–AC layer drastically suppressed phosphate leaching relative to unmodified soil or bare activated carbon, indicating its capacity to immobilize soluble phosphorus and prevent eutrophication. When hydraulic flow was introduced (Figure 10c), the Fe–AC maintained low elution rates (<10%) even at high turbulence, highlighting its structural stability and persistent adsorption efficiency under dynamic water movement–conditions typical for hydroponic circulation or irrigation systems. The vertical concentration profiles in Figure 10d further confirm that Fe–AC effectively restricted phosphate migration through the sediment column, maintaining a steep concentration gradient and minimizing nutrient release to the overlying water.

Collectively, these results emphasize that Fe-modified carbons function not only as high-capacity phosphate sorbents but also as reactive capping and stabilization agents, combining adsorption, ion exchange, and Fe-centered redox processes. In hydroponic applications, such properties can mitigate nutrient losses, suppress microbial overgrowth by controlling phosphate bioavailability, and sustain ionic equilibrium in recirculating nutrient solutions—illustrating the broader potential of transition-metal-modified ACs as multifunctional materials for water purification and sustainable plant cultivation.

7. Techno-Economic Analysis and Life-Cycle Assessment

7.1. Techno-Economic Considerations

Natural zeolite is widely regarded as a low-cost, readily available adsorbent due to abundant mineral deposits and minimal processing (Table 4) [95]. By contrast, chemically modified zeolites (acid-treated, ion-exchanged or metal-impregnated) incur higher capital cost because of extra reagents and thermal energy in their manufacture [97]. AC production is energy-intensive, so high-grade carbons tend to have moderate-to-high costs. In practice AC from waste biomass can be relatively inexpensive per kg, but its true cost should account for higher adsorption capacity of engineered carbons [98].

Table 4. Economic and environmental trade-offs of zeolites and activated carbon in hydroponics.

Feature	Natural Zeolite	Modified Zeolite	Activated Carbon (AC)
Cost	Low: abundant clinoptilolite is inexpensive [99]	Higher: chemical or thermal treatment raises cost [97]	Moderate–High: carbonization and activation energy cost [100]
Production footprint	Lower: mainly mining and milling	Moderate: plus reagents/heat for modification	Highest: high-temp pyrolysis and chemical activation [101]
Adsorption (nutrients)	Selectively exchanges cations (NH_4^+, K^+, etc.), retaining nutrients [100]	Enhanced capacity/selectivity (e.g., metal-loaded for P removal) [99]	Non-selective: strong organics removal, but also binds NO_3^-/PO_4^{3-} [23]
Regeneration	Easy: salt or base wash, low-energy; structure stable [99]	Similar approach; may need stronger reagents for heavy modifications	Difficult: thermal (>600 °C) or chemicals (high energy/cost) [101]
Operational benefits	Extends solution life (slow-release of nutrients); reduces fertilizer needs [100]	As natural, often higher removal rates; tailored pollutant uptake	Extends reuse by removing organics; but depletes nutrients [23]
End-of-life	Non-hazardous; can be used as nutrient-rich soil amendment [100]	Similar reuse as soil amendment; caution if loaded with added metals	Spent AC (if organics only) can be reactivated; otherwise disposed (generally inert if no toxic load)

From a techno-economic perspective, a critical consideration lies in evaluating the relative feasibility and cost-effectiveness of sorbent regeneration as opposed to complete material replacement. Regeneration (e.g., saline or basic wash for zeolites, thermal or chemical treatment for AC) avoids frequent purchase of fresh media but demands energy and labor. In full-scale water treatment, regenerated GAC typically retains 60–80% of virgin capacity after one cycle, and its cost can be roughly half that of new carbon [102]. Zeolites are often regenerated with NaCl or NaOH brines, which is relatively low-cost and can recycle the media many times [97]. However, each regeneration cycle degrades performance modestly, and eventual replacement is needed.

7.2. Agricultural and Water-Saving Benefits

Natural zeolites have high cation-exchange capacities for NH_4^+, K^+, etc., acting as slow-release reservoirs in solution. In practice this means nutrient solutions can be recycled longer before major fertilizer top-ups. For example, studies in soil/soilless culture show zeolites reduce nutrient leaching and volatilization, and gradually release adsorbed N or P back to plants [100]. By contrast, AC is non-selective: it effectively removes phytotoxic organics, but it can also adsorb significant amounts of nitrate and phosphate [23]. Thus, AC polishing can improve hydroponic solution quality, but often at the cost of depleting nutrients (necessitating more fertilizer). Overall, system models suggest that even modest extension of solution life due to sorbent use can yield substantial fertilizer cost savings, since hydroponic fertigation can represent a large part of operating expenses.

7.3. Environmental Footprint and Life-Cycle Assessment

Activated carbon manufacture (especially with KOH or steam) is energy- and emission-intensive: recent LCA data report ~27–28 MJ and ~1.2 kg CO_2 equivalent per kg of AC produced [101]. High-temperature pyrolysis is the main contributor to energy use and emissions. Zeolite production varies: simply mining and crushing clinoptilolite has a modest footprint, but synthesizing zeolites from chemical precursors is more impactful. Indeed, LCA studies show natural-mineral derived zeolites have substantially lower impacts across categories than purely synthetic routes [102]. Any chemical modification adds upstream footprint from reagents and processing. On balance, natural zeolite media tend to have a lower cradle-to-gate environmental cost than activated carbon.

In use, both sorbents deliver environmental benefits: by recycling nutrient solution, they avoid manufacturing and transporting extra fertilizer and reduce nutrient runoff. For instance, zeolites can adsorb tens of mg N/kg (even >30 mg/g NH_4^+ [97]) and similar P levels, preventing their release. The water savings in closed hydroponics can be further improved by sorbent-based solution recirculation. Studies also note that sorbent-aided systems emit less N_2O and NO_x and leach less nitrate to the environment [100].

At end-of-life, spent sorbents have distinct fates. Loaded zeolite containing plant nutrients is generally non-hazardous and can be recycled as a slow-release soil amendment or incorporated into fertilizer blends [100]. For example, nutrient-charged clinoptilolite has been used as a soil conditioner to release NH_4^+ and K^+ over time. In contrast, spent AC) can often be thermally reactivated for one or two more cycles or, if contamination (e.g., phenolics) is low, simply discarded without special hazard. If AC has sequestered heavy metals or organic toxins, disposal must follow regulations (often incineration or landfilling). Importantly, studies highlight that zeolite media can be regenerated and reused multiple times [97], while AC's regeneration efficiency falls after several cycles. Ultimately, nutrient-saturated zeolite tends to be beneficially reused, whereas exhausted AC is usually recycled (reactivated) or safely disposed.

8. Conclusions

A comparative analysis of zeolites, activated carbons, and their modified transition metals and composite derivatives shows their important role in maintaining the chemical stability and environmental safety of hydroponic systems. Natural zeolites, with their crystalline structure and high ion exchange capacity, demonstrate excellent efficiency in regulating the content of ammonium and cations, as well as in removing heavy metals. On the contrary, activated carbon has greater efficiency in the adsorption of organic compounds, root secretions, and secondary metabolites due to its large surface area and regulated pore structure. Thus, while zeolites dominate in the fight against inorganic ions, activated carbons outperform them in the fight against organic pollutants, which makes their combined use particularly effective.

The integration of zeolites and ACs with 2D (graphene, g-C_3N_4) and 3D (MOF, COF) frameworks leads to the creation of multifunctional sorbents that combine adsorption, ion exchange, photocatalysis, and nutrient regulation. Materials modified with transition metals (Fe-, Mn-, Cu-, Ni-, and cofunctionalized carbons and zeolites) additionally increase productivity by introducing redox centers that stabilize nutrients, inhibit phosphate leaching, and prevent eutrophication. These hybrid systems represent a new generation of adaptive "intelligent" materials capable of responding to dynamic hydroponic conditions.

From a technical and economic point of view, zeolite and carbon-based composites contribute to the introduction of cyclic and sustainable hydroponics methods, reducing fertilizer and water consumption, extending the service life of nutrient solutions, and minimizing waste emissions. However, existing studies still lack consistent long-term data on field productivity, regeneration efficiency, and environmental impacts. Therefore, future research should focus on in situ regeneration mechanisms, long-term nutrient stability, and life cycle assessment of these multifunctional sorbents. Overall, this review shows that a reasonable combination of zeolites and activated carbons, especially when combined with 2D/3D frameworks, provides a synergistic effect to create more sustainable, efficient, and circular hydroponic systems. Further development of adaptive, self-healing sorbents will not only increase agricultural productivity but also reduce the negative impact on the environment, paving the way for sustainable food production in conditions of limited resources.

Supplementary Materials: The following supporting information can be downloaded at https://www.mdpi.com/article/10.3390/su172410977/s1. PRISMA 2020 Checklist [26].

Author Contributions: Conceptualization D.A. and M.A.; methodology O.D., L.F.V. and Y.D.; validation A.S., K.S., A.Z. (Assem Zhumagalieva) and D.A.; formal analysis A.Z. (Arman Zhumazhanov); investigation T.A. and A.K.; writing—original draft preparation M.A.; writing—review and editing A.Z. (Arman Zhumazhanov) and A.Z. (Assem Zhumagalieva); visualization A.K., A.S. and J.J.; supervision O.D.; project administration Y.D., L.F.V. and J.J.; All authors have read and agreed to the published version of the manuscript.

Funding: This work was supported by the Ministry of Science and Higher Education of the Republic of Kazakhstan grant (AP23489070).

Institutional Review Board Statement: Not applicable.

Informed Consent Statement: Not applicable.

Data Availability Statement: The data used to support the findings of this study are available from the corresponding author upon reasonable request.

Conflicts of Interest: The authors declare that they have no conflicts of interest.

Abbreviations

The following abbreviations are used in this manuscript:

ACs	Activated carbons
GO	Graphene oxide
prGO	Partially reduced graphene
MOFs	Metal–organic frameworks
COFs	Covalent organic frameworks

Appendix A

Following the rigorous study selection process detailed in Section 2.1, a total of 102 individual studies met all inclusion criteria and form the basis of this review. Full characteristics of each included study are provided in detail below in Appendix A Table A1. This comprehensive table serves as a supplementary resource, detailing the citation, study type/design, system/model used, intervention/exposure, key outcomes measured, and a summary of the relevant findings for every source.

Table A1. Study characteristics.

Ref. No	Study ID (Citation, Year)	Study Type/Design	System/Plant Model	Intervention/Exposure Details	Key Outcomes Measured	Relevant Findings (Summary)
[1]	Chatzigeorgiou et al., 2022	Experimental Study	Vine leaves in a Mediterranean greenhouse (Hydroponic NFT/Perlite)	Comparison of substrates (Perlite, Perlite-Attapulgite, Perlite-Zeolite)	Yield, Carbon Footprint (CFP), Phenols, Nutrients	Perlite treatment yielded highest leaf number and lowest CFP.
[2]	Li et al., 2022	Experimental Study	Hydroponic Vertical Greening System (VGS)	Optimization of operating conditions for blackwater treatment	Removal efficiencies for C, N, P; plant growth	System effectively treated blackwater; removal rates optimized at specific conditions.
[3]	Awad et al., 2021	Experimental Study	Cabbage, dill, red lettuce (NFT Hydroponics)	Substrate comparison: Perlite (PL) vs. Rice Husk Biochar (RB) vs. PL+RB mix	Shoot length, fresh/dry mass, leaf nutrient content	PL+RB mixture doubled the yield and enhanced micronutrient content compared to PL alone.
[4]	Dyer, 2001	Book Chapter/Technical Note	N/A (Theoretical/Lab context)	N/A	Definition of Ion Exchange Capacity (IEC)	Explains the fundamental concept and how to estimate IEC in zeolites.
[5]	Sambo et al., 2019	Review Article	N/A (Conceptual/Review)	Review of hydroponic solutions and smart agriculture tech	Opportunities and limitations of soilless systems	Highlights issues and opportunities in nutrient management for soilless production.
[6]	Bindra et al., 2023	Experimental Study (Lab/Pot)	Tomato plant	MOF/Zeolite composite for targeted nutrient release	Nutrient release profile, plant uptake	Developed a smart fertilizer mechanism using zeolite composites.
[7]	Mussina et al., 2025	Experimental Study (Soil)	*Aesculus hippocastanum* (Horse chestnut)	Evaluation in artificial *soil* contamination	Phytoremediation potential	Study focused on soil remediation, not directly hydroponics.
[8]	Mosa et al., 2019	Experimental Study	Tomato (*Lycopersicon esculentum* L.) in NFT	Biochar filters added to nutrient solution with Nickel (Ni) contamination	Ni toxicity reduction, plant growth	Biochar filters significantly reduced toxic effects of Ni on tomato plants.
[9]	Olasehinde, 2025	Review/Perspective Article	N/A	Review of biodegradable growth media	Suitability of alternatives (e.g., coconut coir, peat, biochar)	Discusses sustainable alternatives to traditional substrates.

Table A1. *Cont.*

Ref. No	Study ID (Citation, Year)	Study Type/Design	System/Plant Model	Intervention/Exposure Details	Key Outcomes Measured	Relevant Findings (Summary)
[10]	Calabria et al., 2019	Experimental Study/Pilot	Lettuce (*Lactuca sativa*) in vertical hydroponics	Zeolite ion exchange column for treating anaerobic wastewater	N recovery efficiency, lettuce fertigation success	Demonstrated successful recovery of N and subsequent safe reuse for plant growth.
[11]	El-Shal. et al., 2022	Experimental Study	Alfalfa plants (hydroponics)	Application of NPK nano-fertilizers under different water qualities	Plant growth metrics, nutrient uptake	Nano-fertilizers improved plant performance and nutrient utilization.
[12]	Pan et al., 2016	Experimental Study	Rice and Wheat seedlings (Hydroponic)	Varying ammonia/ammonium concentrations	Root morphology changes, N concentration in leaves/roots	Established threshold levels for toxicity symptoms induced by high ammonium concentration.
[13]	Nurbaity et al., 2019	Conference Paper/Study	N/A (Methodology focus)	Optimization of hydroponic technology	Production yield of mycorrhiza biofertilizer	Study focused on the production of biofertilizer using hydroponics methodology.
[14]	Vehar et al., 2025	Experimental Study	Tomatoes (hydroponic)	Exposure to contaminants of emerging concern (CECs)	Primary and secondary metabolites in fruit	Assessed the uptake and influence of pharmaceuticals/pollutants on crop safety/quality.
[15]	Li et al., 2020	Experimental Study	N/A (Zeolite material study)	Ammonium-modified clinoptilolite ion exchange experiments	Cation exchange capacity (CEC), selectivity for methane/nitrogen	Lab study characterizing zeolite material properties for gas/liquid separation.
[16]	Korotta-Gamage, S.M., Sathasivan, A., 2017	Experimental Study	N/A (Water treatment study)	Biologically activated carbon (BAC) treatment of surface water	Organic carbon removal efficiency (DOC, UV-254)	BAC effectively removed organic carbon; process optimization was studied.
[17]	Yang et al., 2022	Experimental Study	Hydroponic Rice and Wheat seedlings	Irrigation with "activated water" (magnetic treatment)	Root morphology metrics (length, volume, surface area)	Activated water significantly improved root development in both crops.
[18]	Suryaningtyas et al., 2023	Conference Paper/Review	N/A (Medium development)	Zeoponic media review (Zeolite/compost mix)	Material characteristics, application potential	Discusses potential of zeoponics as a sustainable growing medium.

Table A1. *Cont.*

Ref. No	Study ID (Citation, Year)	Study Type/Design	System/Plant Model	Intervention/Exposure Details	Key Outcomes Measured	Relevant Findings (Summary)
[19]	Zahid et al., 2021	Experimental Study	N/A (Water purification study)	Modification of natural clinoptilolite and mordenite	Heavy metal removal efficiency	Demonstrated effective application of local Kazakhstan zeolites for water treatment.
[20]	Rashed & Pal.anisamy, 2018	Book Chapter/Review	N/A (Review of Zeolites)	Review of adsorption/ion exchange properties of zeolites	General properties and application overview	Provides foundational information on zeolite use in water treatment.
[21]	Adam et al., 2025	Conference Paper/Study	N/A (Wastewater treatment focus)	Application of natural zeolite clinoptilolite for ammonia removal	Ammonia removal efficiency	Zeolite was effective for ammonia removal in lab-scale wastewater treatment.
[22]	Fuentes-Peñailillo et al., 2024	Review Article	N/A (Review of tech)	Review of sustainable technologies for soilless production	Technology assessment	Discusses new tech including potential of substrates and nutrient solutions.
[23]	Wang et al., 2024	Review Article	N/A (Review of systems)	Review of current hydroponic practices and bioelectrochemical systems	System assessment and future roles	Evaluates current systems and innovative solutions for sustainable production.
[24]	Pérez-Botella et al., 2022	Review Article	N/A (Review of Zeolites)	Review of zeolites in adsorption processes	Adsorption mechanisms	Comprehensive review of zeolite use in various adsorption applications.
[25]	Van et al., 2021	Experimental Study/Pilot	Lettuce	Production of hydroponic solution from human urine with activated carbon	N, P, K recovery efficiency; plant growth	Developed a viable method to recycle nutrients from urine for fertigation.
[27]	Seitzhanova et al., 2023	Experimental Study	N/A (Material science)	Influence of heat treatment on zeolite sorption characteristics	Sorption capacity, physical properties	Focused on how heat treatment affects zeolite properties for water purification.
[28]	Mansouri et al., 2013	Experimental Study	N/A (Material characterization)	Characterization of natural zeolite-clinoptilolite	Porosity, structural properties	Studied the fundamental properties of a specific zeolite material.

Table A1. *Cont.*

Ref. No	Study ID (Citation, Year)	Study Type/Design	System/Plant Model	Intervention/Exposure Details	Key Outcomes Measured	Relevant Findings (Summary)
[29]	Elshorbagy & Chowdhury, 2013	Book	N/A (General topic)	N/A	N/A	An edited book providing a general overview of water treatment.
[30]	Kenzhal.iyev et al., 2020	Experimental Study	N/A (Material modification)	Modification of natural Kazakhstan sorbents	Material characteristics, sorption properties	Focused on improving the properties of local sorbent materials.
[31]	Wasielewski et al., 2020	Experimental Study	N/A (Wastewater sludge)	Application of natural clinoptilolite for ammonium removal	Ammonium removal from sludge water	Zeolite proved effective in lab-scale ammonium removal from specific wastewater.
[32]	Cheng & Ding, 2015	Experimental Study	N/A (Water treatment study)	Natural and modified zeolite for ammonium removal	Kinetic, equilibrium, thermodynamic parameters	Characterized the mechanisms and efficiency of ammonium removal by zeolite.
[33]	Nogueira et al., 2025	Experimental Study	N/A (Wastewater treatment study)	Zeolite-SPION nanocomposite use	Ammonium and heavy metal removal efficiency	Developed a novel composite material effective at removing multiple pollutants.
[34]	Rahmani et al., 2009	Experimental Study	N/A (Water treatment study)	Clinoptilolite regeneration by air stripping	Regeneration efficiency	Focused on the practical aspect of reusing zeolite material after saturation.
[35]	Ofiera & Kazner, 2025	Experimental Study	N/A (Water treatment study)	Influence of ammonium on heavy metal adsorption	Heavy metal removal efficiency	Investigated competitive adsorption dynamics between ammonium and heavy metals on zeolites.
[36]	Murkani et al., 2015	Experimental Study	N/A (Water treatment study)	Natural clinoptilolite for removal of ammonium and nitrate	Removal efficiency (%)	Zeolite was highly effective for ammonium removal, but not nitrate removal.
[37]	Nakano et al., 2013	Experimental Study	N/A (Material science)	Potassium metals loaded into zeolite LSX cages	Magnetic and electronic properties	Study focused purely on material physics and magnetism.
[38]	Velarde et al., 2023	Review Article	N/A (Review of zeolites)	Adsorption of heavy metals on natural zeolites	Review of mechanisms and applications	Comprehensive review of using zeolites for heavy metal remediation.

Table A1. *Cont.*

Ref. No	Study ID (Citation, Year)	Study Type/Design	System/Plant Model	Intervention/Exposure Details	Key Outcomes Measured	Relevant Findings (Summary)
[39]	Montégut et al., 2016	Experimental Study	N/A (Wastewater study)	Use of clinoptilolite, chabazite, and faujasite zeolites	Ammonium and Potassium removal efficiency	Compared different zeolite types for nutrient recovery from swine manure.
[40]	Hal.imoon & Yin, 2010	Experimental Study	N/A (Wastewater study)	Zeolite use for heavy metal removal from textile wastewater	Metal removal efficiency	Demonstrated effectiveness in specific industrial wastewater context.
[41]	He et al., 2016	Experimental Study	N/A (Material synthesis)	Zeolite synthesized from fly ash for heavy metal removal	Synthesis process, metal removal efficiency	Created effective zeolite material from waste products.
[42]	Castro et al., 2021	Experimental Study (Field Scale)	Non-sewered sanitation system	Zeolite application and regeneration in a field unit	Ammonium removal from real-world wastewater	Field study showing practicality and regeneration methods for large-scale use.
[43]	Guida et al., 2020	Experimental Study (Lab/Pilot)	N/A (Wastewater treatment study)	Evaluation of different zeolites for ammonium removal	Ammonium removal capacity (mg/g)	Compared several commercial zeolites for efficiency in municipal wastewater treatment.
[44]	Subramanian et al., 2023	Review Article	N/A (Material synthesis review)	Zeolite/graphene oxide composites synthesis review	Future perspectives/applications	Discusses advanced material development (not a direct hydroponic study).
[45]	Kim et al., 2023	Review Article	N/A (Material science review)	2D MOFs and zeolites for composite membranes	Review of material properties/applications	Focuses on gas separation and material science (not direct hydroponics).
[46]	Zhao et al., 2024	Review Article	N/A (Material science review)	Synthesis and photoelectrocatalysis of zeolite-based composites	Review of advanced material properties	Focuses on catalytic properties of novel zeolite composites.
[47]	Doszhanov et al., 2024	Experimental Study (Soil/Lab)	Parsley (*Petroselinum crispum*)	Organic I and Se fertilizer based on biochar	Biofortification levels, plant growth	Developed a new slow-release fertilizer for soil application.
[48]	Lesbayev et al., 2024	Experimental Study	N/A (Material synthesis)	Preparation of nanoporous carbon from rice husk	Textural characteristics, hydrogen sorption capacity	Focused on material science; creating high-quality sorbent material.

Table A1. *Cont.*

Ref. No	Study ID (Citation, Year)	Study Type/Design	System/Plant Model	Intervention/Exposure Details	Key Outcomes Measured	Relevant Findings (Summary)
[49]	Mansurov et al., 2022	Experimental Study	N/A (Material science)	Modified carbon sorbents based on walnut shell	Sorption of toxic gases	Focused on gas phase environmental remediation (not water/hydroponics).
[50]	Sanchez et al., 2025	Experimental Study	Lettuce (Hydroponic and Soil systems)	Walnut shell biochar and fertilizer application	Yield, nutrient uptake, plant growth parameters	Direct comparison showing influence of biochar on yield in both systems.
[51]	Zhang et al., 2021	Experimental Study	N/A (Water treatment study)	Magnetic activated carbon for heavy metal adsorption	Adsorption efficiency Pb(II) and Cd(II), mechanism	Lab study characterizing novel magnetic sorbent performance.
[52]	Rahman et al., 2014	Experimental Study	N/A (Water treatment study)	Acid activated carbons from oil palm and coconut shells	Heavy metal removal efficiency	Lab study comparing activation methods and removal capacities.
[53]	Asaduzzaman & Asao, 2020	Review/Experimental Study	Strawberry in recycled hydroponics	Autotoxicity mitigation methods (e.g., activated carbon)	Plant growth, accumulation of allelochemicals	Identified causes and solutions for autotoxicity in closed-loop systems.
[54]	Qiu et al., 2022	Experimental Study	N/A (Water treatment study)	Activated carbon adsorption of heavy metals in presence of NOM	Adsorption efficiency, regeneration methods	Investigated the influence of natural organic matter on heavy metal removal.
[55]	Foo & Hameed, 2010	Review Article (Theoretical)	N/A	Review of adsorption isotherm modeling systems	Modeling approaches	Theoretical paper focusing on mathematical models of adsorption.
[56]	Jones et al., 2009	Review Article (Conceptual)	N/A (Soil/Plant biology context)	Review of carbon flow in the rhizosphere	Root exudate mechanisms	Foundational review on plant–soil interactions (soil focus).
[57]	Bertin et al., 2003	Review Article (Conceptual)	N/A (Plant biology context)	Role of root exudates and allelochemicals	Mechanisms of allelopathy	Foundational review on chemical signaling and allelopathy.
[58]	Badri & Vivanco, 2009	Review Article (Conceptual)	N/A (Plant biology context)	Regulation and function of root exudates	Genetic/biological regulation	Foundational review on biological control of exudation.

Table A1. *Cont.*

Ref. No	Study ID (Citation, Year)	Study Type/Design	System/Plant Model	Intervention/Exposure Details	Key Outcomes Measured	Relevant Findings (Summary)
[59]	Rivera-Utrilla et al., 2013	Review Article	N/A (Environmental context)	Pharmaceuticals as emerging contaminants and removal methods	Review of contamination and treatment technologies	General review on pharmaceutical pollution (not specific to hydroponics).
[60]	Nam et al., 2014	Experimental Study	N/A (Water treatment study)	Activated carbon adsorption of micropollutants	Adsorption characteristics	Study on removing trace organic contaminants using activated carbon.
[61]	Sevilla & Fuertes, 2009	Experimental Study	N/A (Material science)	Hydrothermal carbonization of saccharides	Material properties, porosity	Study focused on synthesis of carbon materials.
[62]	Yang & Xing, 2010	Theoretical/Review	N/A (Theoretical context)	Adsorption of organic compounds by carbon nanomaterials	Polanyi theory application, modeling	Theoretical paper applying Polanyi theory to carbon nanomaterial adsorption modeling.
[63]	Rivera-Utrilla et al., 2011	Review Article	N/A	Activated carbon modifications for water treatment	Review of modification techniques	Overview of chemical and physical modifications to enhance AC performance.
[64]	Foo & Hameed, 2010	Review Article	N/A	Review of dye removal via activated carbon adsorption	Adsorption efficiency, mechanisms	Comprehensive review focusing on using AC for textile dye removal.
[65]	El Gamal. et al., 2018	Review Article	N/A	Bio-regeneration of activated carbon	Review of regeneration methods	Overview of biological methods for restoring AC capacity.
[66]	León et al., 2020	Design/Cost Analysis	N/A (Industrial process design)	Production of AC from waste nutshells (physical activation)	Design parameters, cost estimation	Engineering study on the industrial production of AC.
[67]	Ahmad et al., 2014	Review Article	N/A (Environmental context)	Biochar as a sorbent for contaminant management	Review of application in soil/water	Overview of biochar use for remediation of various pollutants.
[6]	Bindra et al., 2023	Experimental Study (Lab/Pot)	Tomato plant	Zeolite/MOF composite for targeted nutrient release	Nutrient release profile, plant uptake	Developed a smart fertilizer using composite materials.

Table A1. *Cont.*

Ref. No	Study ID (Citation, Year)	Study Type/Design	System/Plant Model	Intervention/Exposure Details	Key Outcomes Measured	Relevant Findings (Summary)
[68]	Safarpour & Khataee, 2019	Book Chapter/Review	N/A (Material science context)	Graphene-based materials for water purification	Material properties, purification efficiency	Overview of graphene applications in water treatment.
[69]	Liu et al., 2017	Experimental Study	N/A (Material science)	Preparation of graphene/zeolite composites	Adsorption of pollutants (dyes) in water	Developed a composite material and tested its pollutant adsorption capacity.
[70]	Seitzhanova et al., 2024	Experimental Study	N/A (Material synthesis/desalination)	Graphene membranes from rice husk waste	Desalination efficiency	Study focused on membrane technology for water desalination.
[71]	Ong et al., 2016	Review Article	N/A (Material science/catalysis)	Graphitic carbon nitride (g-C3N4) photocatalysts	Catalytic properties, sustainability applications	Review focused on advanced catalytic materials.
[72]	Hua et al., 2023	Experimental Study (Lab/Sensor Dev)	N/A (Biosensor development)	3D graphene oxide–CNT composite electrochemical sensor	Ammonium detection limit/sensitivity	Developed a new sensor for detecting ammonium in human sweat.
[73]	Horcajada et al., 2006	Review Article	N/A (Biomedical context)	Metal–Organic Frameworks (MOFs) for drug delivery	Drug delivery efficacy	Review focused on biomedical applications of MOFs.
[74]	Li et al., 2009	Review Article	N/A (Material science context)	MOFs for selective gas adsorption and separation	Gas separation efficiency	Review focused on industrial gas separation using MOFs.
[75]	Khan et al., 2013	Review Article	N/A (Environmental context)	MOFs for removal of hazardous materials from water	Review of adsorption mechanisms	Overview of using MOFs for water remediation.
[76]	Cui et al., 2016	Review Article	N/A (Material science context)	MOFs as platforms for functional materials	Material synthesis and applications	General review of MOF material science.
[77]	Mohammad, H.S et al., 2022	Review Article	N/A (Environmental context)	Carbon-Based Materials for Adsorption of Organic Pollutants	Review of recent advances	Overview of various carbon materials for environmental remediation.
[78]	Hasan & Jhung, 2015	Review Article	N/A (Environmental context)	MOFs for removal of hazardous organics from water	Adsorption mechanisms	Review detailing potential mechanisms for organic pollutant removal using MOFs.

Table A1. *Cont.*

Ref. No	Study ID (Citation, Year)	Study Type/Design	System/Plant Model	Intervention/Exposure Details	Key Outcomes Measured	Relevant Findings (Summary)
[79]	Ding & Wang, 2013	Review Article	N/A (Material science context)	Covalent Organic Frameworks (COFs)	Design principles, applications	General review of COF material science.
[80]	Ahmadijokani et al., 2024	Review Article	N/A (Material science/environmental context)	COF and MOF hybrids for wastewater treatment	Review of performance	Overview of advanced hybrid materials for water treatment applications.
[81]	Hedström, 2001	Review Article	N/A (Water treatment context)	Ion exchange of ammonium in zeolites	Literature review of mechanisms	Foundational review summarizing zeolite performance for ammonium removal.
[82]	Teutli-Sequeira et al., 2009	Experimental Study	N/A (Water treatment study)	Influence of Na^+, Ca^{2+}, Mg^{2+}, NH_4^+ on Cd^{2+} sorption by zeolite	Sorption dynamics of heavy metals	Investigated competitive adsorption dynamics on zeolite materials.
[83]	Rahavi et al., 2024	Experimental Study	Tomato in soilless cultivation (Hydroponic/Perlite)	Activated carbon and K-enriched NanoZeolite	Yield, nutrient content, water use efficiency	Demonstrated improved tomato growth using combined amendments under low-quality water conditions.
[84]	Velásquez-Hernández et al., 2019	Experimental Study	N/A (Material science)	Degradation of ZIF-8 (MOF) in buffered media	Stability in physiological solutions	Studied the stability of a specific MOF (ZIF-8) in biological/aqueous buffers.
[85]	Luzuriaga et al., 2019	Experimental Study	N/A (Material science)	ZIF-8 degradation in cell media, serum, and buffers	Material stability analysis	Confirmed the degradation of ZIF-8 in certain application contexts.
[86]	Wang et al., 2015	Experimental Study	N/A (Water treatment study)	Zirconium MOF UiO-66 for arsenic removal	Arsenic removal efficiency	Demonstrated high efficiency of a specific MOF for removing heavy metals.
[87]	Lin et al., 2024	Experimental Study	N/A (Water treatment study)	Alginate/UiO-66-NH2 nanocomposite	Phosphate removal efficiency	Developed a novel composite for nutrient/pollutant sequestration.

Table A1. *Cont.*

Ref. No	Study ID (Citation, Year)	Study Type/Design	System/Plant Model	Intervention/Exposure Details	Key Outcomes Measured	Relevant Findings (Summary)
[88]	Ghaffari et al., 2019	Experimental Study	N/A (Catalysis study)	Fe-loaded zeolites and silica for phenol degradation	Catalytic degradation efficiency	Study focused on catalytic breakdown of organic pollutants.
[89]	Wu et al., 2024	Experimental Study	N/A (Catalysis study)	Mn3O4 catalysts for oxidation of phenolic contaminants	Catalytic oxidation efficiency	Study focused on advanced oxidation processes for water treatment.
[90]	Hosseinzadeh et al., 2017	Experimental Study	Closed Hydroponic System (Tomato)	Comparison of GAC, ion exchange, and ozonation processes	Water quality parameters, nutrient reuse feasibility	GAC and ion exchange were effective treatments for water reuse in closed systems.
[91]	Hosseinzadeh et al., 2019	Experimental Study	Closed Hydroponic System	UV/H_2O_2 advanced oxidation process	Degradation kinetics of root exudates, cost analysis	UV/H_2O_2 effectively degraded organic pollutants, ensuring water quality for reuse.
[92]	Shah et al., 2015	Experimental Study	N/A (Water treatment study)	Iron-impregnated activated carbon adsorbent	Methylene blue removal, regeneration efficiency	Developed an effective modified AC material for dye removal.
[93]	Xu et al., 2022	Experimental Study	N/A (Environmental remediation)	Electroactive Fe-biochar for remediation	Removal efficiency of arsenic and chromium	Focused on using biochar in redox-related soil/water remediation strategies.
[94]	Okpara et al., 2025	Review Article	N/A (Material science review)	Copper-carbon nanostructured heterojunctions	Pollutant removal mechanisms	Review of advanced material science in water purification.
[95]	Alam et al., 2021	Experimental Study	N/A (Water treatment study)	Pd–Ni nanoparticles supported on activated carbon	Basic blue 3 dye removal efficiency	Developed modified AC for efficient dye removal.
[96]	Seo et al., 2025	Experimental Study	N/A (Water treatment study)	Iron-modified activated carbon application	Phosphate removal efficiency	Study focused on targeted removal of phosphate using modified AC.
[97]	Biliani et al., 2025	Experimental Study	N/A (Water treatment study)	Zeolites of different origin	Eutrophication control (N, P removal) from freshwater	Compared natural zeolites for effectiveness in nutrient management.

Table A1. *Cont.*

Ref. No	Study ID (Citation, Year)	Study Type/Design	System/Plant Model	Intervention/Exposure Details	Key Outcomes Measured	Relevant Findings (Summary)
[99]	Król, 2020	Review Article	N/A (Material science review)	Natural vs. Synthetic Zeolites	Material properties, applications	Comparative review of natural and synthesized zeolite materials.
[101]	Saleem et al., 2025	Life Cycle Assessment	N/A (Industrial assessment)	High-value activated carbon production	Environmental impact, LCA metrics	Engineering/sustainability study on AC production impacts.
[98]	Wang et al., 2024	Review Article	N/A (Review of AC technology)	Deactivation mechanisms and regeneration methods of AC	Review of industrial practices	Overview of practical challenges and solutions in AC use.
[100]	Mondal. et al., 2021	Review Article	N/A (Soil context review)	Zeolites enhance soil health, crop productivity	Review of agricultural application (soil-based)	Discusses benefits of zeolites primarily in traditional agriculture.
[102]	Chen et al., 2024	Environmental Assessment	N/A (Material science/synthesis)	Zeolites synthesized from chemicals vs. natural minerals	Environmental impact, Green Chemistry metrics	Comparative life cycle assessment of different zeolite production methods.

References

1. Chatzigeorgiou, I.; Liantas, G.; Spanos, P.; Gkriniari, V.; Maloupa, E.; Ntinas, G.K. Hydroponic Cultivation of Vine Leaves with Reduced Carbon Footprint in a Mediterranean Greenhouse. *Sustainability* **2022**, *14*, 8011. [CrossRef]
2. Li, X.; Zhou, J.; Tang, Y.; Li, Y.; Jin, Z.; Kong, H.; Zhao, M.; Zheng, X.; Bei, K. A hydroponic vertical greening system for disposal and utilization of pre-treated Blackwater: Optimization of the operating conditions. *Ecol. Eng.* **2022**, *183*, 106739. [CrossRef]
3. Awad, Y.; Lee, S.-E.; Vu, N.T.; Farooq, M.; Kim, S.; Kim, H.S.; Vithanage, M.; Usman, A.; Al-Wabel, M.; Meers, E.; et al. Biochar, a potential hydroponic growth substrate, enhances the nutritional status and growth of leafy vegetables. *J. Clean. Prod.* **2017**, *156*, 581–588. [CrossRef]
4. Dyer, A. Ion exchange capacity. In *Verified Syntheses of Zeolitic Materials*; Elsevier Science: Amsterdam, The Netherlands, 2001; pp. 67–68. [CrossRef]
5. Sambo, P.; Nicoletto, C.; Giro, A.; Pii, Y.; Valentinuzzi, F.; Mimmo, T.; Lugli, P.; Orzes, G.; Mazzetto, F.; Astolfi, S.; et al. Hydroponic solutions for soilless production systems: Issues and opportunities in a smart agriculture perspective. *Front. Plant Sci.* **2019**, *10*, 923. [CrossRef]
6. Bindra, P.; Sharma, S.; Sahu, B.K.; Bagdwal, H.; Shanmugam, V.; Singh, M. Targeted nutrient application to tomato plant with MOF/Zeolite composite wrapped with stimuli-responsive biopolymer. *Mater. Today Commun.* **2023**, *34*, 105264. [CrossRef]
7. Mussina, A.; Baitasheva, G.; Rakhimova, Z.; Raimbekova, B.; Gubasheva, B.E.; Medeuova, G.; Baymurzina, Z.; Tynyshbek, D. Evaluation of tree species Aésculus hippocastanum as a phytoremediator in conditions of artificial soil contamination. *Casp. J. Environ. Sci.* **2025**, *23*, 707–720. [CrossRef]
8. Mosa, A.; El-Banna, M.F.; Gao, B. Biochar filters reduced the toxic effects of nickel on tomato (*Lycopersicon esculentum* L.) grown in nutrient film technique hydroponic system. *Chemosphere* **2019**, *149*, 254–262. [CrossRef]
9. Olasehinde, A.A. Biodegradable Growth Media Alternatives for Sustainable Hydroponic Farming. *Curr. J. Appl. Sci. Technol.* **2025**, *44*, 147–152. [CrossRef]
10. Calabria, J.L.; Lens, P.N.; Yeh, D.H. Zeolite ion exchange to facilitate anaerobic membrane bioreactor wastewater nitrogen recovery and reuse for lettuce fertigation in vertical hydroponic systems. *Environ. Eng. Sci.* **2019**, *36*, 690–698. [CrossRef]
11. El-Shal, R.M.; El-Naggar, A.H.; El-Beshbeshy, T.R.; Mahmoud, E.K.; El-Kader, N.I.A.; Missaui, A.M.; Du, D.; Ghoneim, A.M.; El-Sharkawy, M.S. Effect of Nano-Fertilizers on Alfalfa Plants Grown under Different Salt Stresses in Hydroponic System. *Agriculture* **2022**, *12*, 1113. [CrossRef]
12. Pan, W.L.; Madsen, I.J.; Bolton, R.P.; Graves, L.; Sistrunk, T. Ammonia/ammonium toxicity root symptoms induced by inorganic and organic fertilizers and placement. *Agron. J.* **2016**, *108*, 2485–2492. [CrossRef]
13. Nurbaity, A.; Istifadah, N.; Haryantini, B.A.; Ilhami, M.F.; Habibullah, M.I.; Arifin, M. Optimization of hydroponic technology for production of mycorrhiza biofertilizer. *IOP Conf. Ser. Earth Environ. Sci.* **2019**, *347*, 012017. [CrossRef]
14. Vehar, A.; Marsic, N.K.; Železnikar, Š.; Potočnik, D.; Strojnik, L.; Petkovšek, M.M.; Šircelj, H.; Kastelec, D.; Kovačič, A.; Pintar, M.; et al. Influence of contaminants of emerging concern on the primary and secondary metabolites of hydroponically grown tomatoes. *Food Chem.* **2025**, *491*, 145226. [CrossRef] [PubMed]
15. Li, J.; Ullah, R.; Jiao, J.; Sun, J.; Bai, S. Ion exchange of cations from different groups with ammonium-modified clinoptilolite and selectivity for methane and nitrogen. *Mater. Chem. Phys.* **2020**, *256*, 123760. [CrossRef]
16. Korotta-Gamage, S.M.; Sathasivan, A. Potential of a biologically activated carbon treatment to remove organic carbon from surface waters. *Int. Biodeterior. Biodegrad.* **2017**, *124*, 82–90. [CrossRef]
17. Yang, X.; Fan, J.; Ge, J.; Luo, Z. Effect of irrigation with activated water on root morphology of hydroponic rice and wheat seedlings. *Agronomy* **2022**, *12*, 1068. [CrossRef]
18. Suryaningtyas, D.T.; Fatiha, Y.G.N.; Oktariani, P. Zeoponic, a plant growing medium from zeolite mineral. *IOP Conf. Ser. Earth Environ. Sci.* **2023**, *1133*, 012020. [CrossRef]
19. Zahid, M.; Doszhanov, Y.; Saurykova, K.; Ahmadi, N.; Bolatova, D.; Kurmanbayeva, M.; Aydarbek, A.; Ihsas, R.; Seitzhanova, M.; Akhmetzhanova, D.; et al. Modification and Application of Natural Clinoptilolite and Mordenite from Almaty Region for Drinking Water Purification. *Molecules* **2021**, *30*, 2021. [CrossRef]
20. Rashed, M.N.; Palanisamy, P.N. Introductory chapter: Adsorption and ion exchange properties of zeolites for treatment of polluted water. In *Zeolites and Their Applications*; IntechOpen: London, UK, 2018. [CrossRef]
21. Adam, M.R.; Othman, M.H.D.; Hubadillah, S.K.; Abd Aziz, M.H.; Jamalludin, M.R. Application of natural zeolite clinoptilolite for the removal of ammonia in wastewater. *Mater. Today Proc.* **2025**, *in press*. [CrossRef]
22. Fuentes-Peñailillo, F.; Gutter, K.; Vega, R.; Silva, G.C. New Generation Sustainable Technologies for Soilless Vegetable Production. *Horticulturae* **2024**, *10*, 49. [CrossRef]
23. Wang, S.; Kleiner, Y.; Clark, S.M.; Raghavan, V.; Tartakovsky, B. Review of current hydroponic food production practices and the potential role of bioelectrochemical systems. *Rev. Environ. Sci. Biotechnol.* **2024**, *23*, 897–921. [CrossRef]

24. Pérez-Botella, E.; Valencia, S.; Rey, F. Zeolites in Adsorption Processes: State of the Art and Future Prospects. *Chem. Rev.* **2022**, *122*, 17647–17695. [CrossRef]
25. Van, Q.N.; Van, H.T.; Le, S.H.; Nguyen, T.H.; Nguyen, H.T.; Pham, Q.T.; Nguyen, T.T.; Tran, T.N.H.; Nguyen, T.B.H.; Hoang, T.K. Production of hydroponic solution from human urine using adsorption–desorption method with coconut shell-derived activated carbon. *Environ. Technol. Innov.* **2021**, *23*, 101708. [CrossRef]
26. Page, M.J.; McKenzie, J.E.; Bossuyt, P.M.; Boutron, I.; Hoffmann, T.C.; Mulrow, C.D.; Shamseer, M.; Tetzlaff, J.M.; Akl, E.A.; Brennan, S.E.; et al. The PRISMA 2020 statement: An updated guideline for reporting systematic reviews. *BMJ* **2021**, *372*, n71. [CrossRef]
27. Seitzhanova, M.A.; Doszhanov Ye, O.; Kuldeev, E.I.; Mansurov, Z.A.; Taju, K.; Tanirbergenova, S.K.; Tazhkenova, G.K. Influence of heat treatment on the sorption characteristics of zeolite used in water purification. *Combust. Plasma Chem.* **2023**, *21*, 173–179. (In Russian) [CrossRef]
28. Mansouri, N.; Rikhtegar, N.; Panahi, H.A.; Atabi, F.; Shahraki, B.K. Porosity, characterization and structural properties of natural zeolite-clinoptilolite-as a sorbent. *Environ. Prot. Eng.* **2013**, *39*, 139–152. [CrossRef]
29. Elshorbagy, W.; Chowdhury, R. (Eds.) *Water Treatment*; IntechOpen: London, UK, 2013. [CrossRef]
30. Kenzhaliyev, B.K.; Surkova, T.Y.; Berkinbayeva, A.N.; Dosymbayeva, Z.D.; Abdikerim, B.E. Revisiting the Kazakhstan natural sorbents modification. *Metalurgija* **2020**, *59*, 117–120.
31. Wasielewski, S.; Rott, E.; Minke, R.; Steinmetz, H. Application of natural clinoptilolite for ammonium removal from sludge water. *Molecules* **2020**, *26*, 114. [CrossRef]
32. Cheng, Z.; Ding, W. Ammonium removal from water by natural and modified zeolite: Kinetic, equilibrium, and thermodynamic studies. *Desalination Water Treat.* **2015**, *55*, 978–985. [CrossRef]
33. Nogueira, H.P.; Toma, S.H.; Silveira, A.T., Jr.; Araki, K. Zeolite-SPION nanocomposite for ammonium and heavy metals removal from wastewater. *J. Braz. Chem. Soc.* **2025**, *31*, 2342–2350. [CrossRef]
34. Rahmani, A.R.; Samadi, M.T.; Ehsani, H.R. Investigation of clinoptilolite natural zeolite regeneration by air stripping followed by ion exchange for removal of ammonium from aqueous solutions. *J. Environ. Health Sci. Eng.* **2009**, *6*, 167–172.
35. Ofiera, L.M.; Kazner, C. Influence of Ammonium on the Adsorption and Desorption of Heavy Metals in Natural Zeolites. *Processes* **2025**, *13*, 2647. [CrossRef]
36. Murkani, M.; Nasrollahi, M.; Ravanbakhsh, M.; Bahrami, P.; Jaafarzadeh Haghighi Fard, N. Evaluation of natural zeolite clinoptilolite efficiency for the removal of ammonium and nitrate from aquatic solutions. *Environ. Health Eng. Manag. J.* **2015**, *2*, 17–22.
37. Nakano, T.; Thi Hanh, D.; Owaki, A.; Nozue, Y.; Nam, N.H.; Araki, S. Insulator-to-metal transition and magnetism of potassium metals loaded into regular cages of zeolite LSX. *J. Korean Phys. Soc.* **2013**, *63*, 512–516. [CrossRef]
38. Velarde, L.; Nabavi, M.S.; Escalera, E.; Antti, M.L.; Akhtar, F. Adsorption of heavy metals on natural zeolites: A review. *Chemosphere* **2023**, *328*, 138508. [CrossRef] [PubMed]
39. Montégut, G.; Michelin, L.; Brendlé, J.; Lebeau, B.; Patarin, J. Ammonium and potassium removal from swine liquid manure using clinoptilolite, chabazite and faujasite zeolites. *J. Environ. Manag.* **2016**, *167*, 147–155. [CrossRef] [PubMed]
40. Halimoon, N.; Yin, R.G.S. Removal of heavy metals from textile wastewater using zeolite. *Environ. Asia* **2010**, *3*, 124–130.
41. He, K.; Chen, Y.; Tang, Z.; Hu, Y. Removal of heavy metal ions from aqueous solution by zeolite synthesized from fly ash. *Environ. Sci. Pollut. Res.* **2016**, *23*, 2778–2788. [CrossRef]
42. Castro, C.J.; Shyu, H.Y.; Xaba, L.; Bair, R.; Yeh, D.H. Performance and onsite regeneration of natural zeolite for ammonium removal in a field-scale non-sewered sanitation system. *Sci. Total Environ.* **2021**, *776*, 145938. [CrossRef]
43. Guida, S.; Potter, C.; Jefferson, B.; Soares, A. Preparation and evaluation of zeolites for ammonium removal from municipal wastewater through ion exchange process. *Sci. Rep.* **2020**, *10*, 12426. [CrossRef]
44. Subramanian, N.; Perumal, T.; Mangesh, V.L.; Chinnadurai, R.; Sakthinathan, S.; Chiu, T.; Selvaraj, M.; Madhavan, J. Future perspectives on zeolite/graphene oxide composite synthesis and applications. *Energy Fuels* **2023**, *37*, 17013–17051. [CrossRef]
45. Kim, M.; Choi, W.; Lee, C.H.; Kim, D. 2D MOFs and zeolites for composite membrane and gas separation applications: A brief review. *ACS Mater.* **2023**, *4*, 148–161. [CrossRef] [PubMed]
46. Zhao, Y.; Liu, M.; Guo, Y.; Wu, Z. Recent Advances in the Synthesis and Photoelectrocatalysis of Zeolite-Based Composites. *Catalysts* **2024**, *14*, 938. [CrossRef]
47. Doszhanov, Y.; Atamanov, M.; Jandosov, J.; Saurykova, K.; Bassygarayev, Z.; Orazbayev, A.; Turganbay, S.; Sabitov, A. Preparation of Granular Organic Iodine and Selenium Complex Fertilizer Based on Biochar for Biofortification of Parsley. *Scientifica* **2024**, *2024*, 6601899. [CrossRef]

48. Lesbayev, B.; Rakhymzhan, N.; Ustayeva, G.; Maral, Y.; Atamanov, M.; Auyelkhankyzy, M.; Zhamash, A. Preparation of-Nanoporous Carbon from Rice Husk with Improved Textural Characteristics for Hydrogen Sorption. *J. Compos. Sci.* **2024**, *8*, 74. [CrossRef]
49. Mansurov, Z.A.; Velasco, L.F.; Lodewyckx, P.; Doszhanov, E.O.; Azat, S. Modified carbon sorbents based on walnut shell for sorption of toxic gases. *J. Eng. Phys. Thermophys.* **2022**, *95*, 1383–1392. [CrossRef]
50. Sanchez, E.; Zabaleta, R.; Navas, A.L.; Maldonado, V.N.F.; Fabani, M.P.; Mazza, G.; Rodriguez, R. Influence of Walnut Shell Biochar and Fertilizer on Lettuce Production in Hydroponic and Conventional Systems. *Agronomy* **2025**, *15*, 658. [CrossRef]
51. Zhang, Z.; Wang, T.; Zhang, H.; Liu, Y.; Xing, B. Adsorption of Pb(II) and Cd(II) by magnetic activated carbon and its mechanism. *Sci. Total Environ.* **2021**, *757*, 143910. [CrossRef]
52. Rahman, M.M.; Adil, M.; Yusof, A.M.; Kamaruzzaman, Y.B.; Ansary, R.H. Removal of heavy metal ions with acid activated carbons derived from oil palm and coconut shells. *Materials* **2014**, *7*, 3634–3650. [CrossRef]
53. Asaduzzaman, M.; Asao, T. Autotoxicity in strawberry under recycled hydroponics and its mitigation methods. *Hortic. J.* **2020**, *89*, 124–137. [CrossRef]
54. Qiu, L.; Suo Ch Zhang, N.; Yuan, R.; Chen, H.; Zhou, B. Adsorption of heavy metals by activated carbon: Effect of natural organic matter and regeneration methods of the adsorbent. *Desalination Water Treat.* **2022**, *252*, 148–166. [CrossRef]
55. Foo, K.Y.; Hameed, B.H. Insights into the modeling of adsorption isotherm systems. *Chem. Eng. J.* **2010**, *156*, 2–10. [CrossRef]
56. Jones, D.L.; Nguyen, C.; Finlay, R.D. Carbon flow in the rhizosphere: Carbon trading at the soil–root interface. *Plant Soil* **2009**, *321*, 5–33. [CrossRef]
57. Bertin, C.; Yang, X.; Weston, L.A. The role of root exudates and allelochemicals in the rhizosphere. *Plant Soil* **2003**, *256*, 67–83. [CrossRef]
58. Badri, D.V.; Vivanco, J.M. Regulation and function of root exudates. *Plant Cell Environ.* **2009**, *32*, 666–681. [CrossRef] [PubMed]
59. Rivera-Utrilla, J.; Sánchez-Polo, M.; Ferro-García, M.A.; Prados-Joya, G.; Ocampo-Pérez, R. Pharmaceuticals as emerging contaminants and their removal from water. *Chemosphere* **2013**, *93*, 1268–1287. [CrossRef]
60. Nam, S.W.; Cho, H.H.; Han, J.; Her, N.; Yoon, Y. Adsorption characteristics of selected hydrophilic and hydrophobic micropollutants in water using activated carbon. *J. Hazard. Mater.* **2014**, *270*, 144–152. [CrossRef]
61. Sevilla, M.; Fuertes, A.B. Chemical and structural properties of carbonaceous products obtained by hydrothermal carbonization of saccharides. *Chem. Eur. J.* **2009**, *15*, 4195–4203. [CrossRef]
62. Yang, K.; Xing, B. Adsorption of organic compounds by carbon nanomaterials in aqueous phase: Polanyi theory and its application. *Chemosphere* **2010**, *79*, 681–688. [CrossRef]
63. Rivera-Utrilla, J.; Sánchez-Polo, M.; Gómez-Serrano, V.; Álvarez, P.M.; Alvim-Ferraz, M.C.M.; Dias, J.M. Activated carbon modifications to enhance its water treatment applications. An overview. *J. Hazard. Mater.* **2011**, *187*, 1–23. [CrossRef]
64. Foo, K.Y.; Hameed, B.H. An overview of dye removal via activated carbon adsorption process. *Desalination Water Treat.* **2010**, *19*, 255–274. [CrossRef]
65. El Gamal, M.; Mousa, H.A.; El-Naas, M.H.; Zacharia, R.; Judd, S. Bio-regeneration of activated carbon: A comprehensive review. *Sep. Purif. Technol.* **2018**, *197*, 345–359. [CrossRef]
66. León, M.; Silva, J.; Carrasco, S.; Barrientos, N. Design, cost estimation and sensitivity analysis for a production process of activated carbon from waste nutshells by physical activation. *Processes* **2020**, *8*, 945. [CrossRef]
67. Ahmad, M.; Rajapaksha, A.U.; Lim, J.; Zhang, M.; Bolan, N.; Mohan, D.; Vithanage, M.; Lee, S.; Ok, Y. Biochar as a sorbent for contaminant management in soil and water: A review. *Chemosphere* **2014**, *99*, 19–33. [CrossRef]
68. Safarpour, M.; Khataee, A. Graphene-based materials for water purification. In *Nanoscale Materials in Water Purification*; Elsevier: Amsterdam, The Netherlands, 2019; pp. 383–430. [CrossRef]
69. Liu, H.; Yu, A.; Liu, H.; Chu, S.; Tan, S. Preparation of graphene/zeolite composites and the adsorption of pollutants in water. *Russ. J. Appl. Chem.* **2017**, *90*, 1171–1180. [CrossRef]
70. Seitzhanova, M.; Azat, S.; Yeleuov, M.; Taurbekov, A.; Mansurov, Z.; Doszhanov, E.; Berndtsson, R. Production of graphene membranes from rice husk biomass waste for improved desalination. *Nanomaterials* **2024**, *14*, 224. [CrossRef]
71. Ong, W.J.; Tan, L.; Ng, Y.; Yong, S.; Chai, S. Graphitic carbon nitride (g-C_3N_4)-based photocatalysts for artificial photosynthesis and environmental remediation: Are we a step closer to achieving sustainability. *Chem. Rev.* **2016**, *116*, 7159–7329. [CrossRef]
72. Hua, Y.; Guan, M.; Xia, L.; Chen, Y.; Mai, J.; Zhao, C.; Liao, C. Highly stretchable and robust electrochemical sensor based on 3D graphene oxide–CNT composite for detecting ammonium in sweat. *Biosensors* **2023**, *13*, 409. [CrossRef]
73. Horcajada, P.; Serre, C.; Vallet-Regí, M.; Sebban, M.; Taulelle, F.; Férey, G. Metal–Organic Frameworks as Advanced Materials for Drug Delivery. *Angew. Chem. Int. Ed.* **2006**, *45*, 5974–5978. [CrossRef]

74. Li, J.-R.; Kuppler, R.J.; Zhou, H.-C. Selective Gas Adsorption and Separation in Metal–Organic Frameworks. *Chem. Soc. Rev.* **2009**, *38*, 1477–1504. [CrossRef]
75. Khan, N.A.; Hasan, Z.; Jhung, S.H. Adsorptive Removal of Hazardous Materials Using Metal–Organic Frameworks (MOFs): A Review. *J. Hazard. Mater.* **2013**, *244–245*, 444–456. [CrossRef] [PubMed]
76. Cui, Y.; Li, B.; He, H.; Zhou, W.; Chen, B.; Qian, G. Metal–organic frameworks as platforms for functional materials. *Acc. Chem. Res.* **2016**, *49*, 483–493. [CrossRef] [PubMed]
77. Mohammad, H.S.; Maryam, A.T.; Toraj, M. One-dimensional graphene for efficient aqueous heavy metal adsorption: Rapid removal of arsenic and mercury ions by graphene oxide nanoribbons (GONRs). *Chemosphere* **2020**, *253*, 126647. [CrossRef]
78. Hasan, Z.; Jhung, S.H. Removal of Hazardous Organics from Water Using Metal–Organic Frameworks (MOFs): Plausible Mechanisms for Selective Adsorptions. *J. Hazard. Mater.* **2015**, *283*, 329–339. [CrossRef] [PubMed]
79. Ding, S.-Y.; Wang, W. Covalent Organic Frameworks (COFs): From Design to Applications. *Chem. Soc. Rev.* **2013**, *42*, 548–568. [CrossRef]
80. Ahmadijokani, F.; Ghaffarkhah, A.; Molavi, H.; Dutta, S.; Lu, Y.; Wuttke, S.; Kamkar, M.; Rojas, O.J.; Arjmand, M. COF and MOF hybrids: Advanced materials for wastewater treatment. *Adv. Funct. Mater.* **2024**, *34*, 2305527. [CrossRef]
81. Hedström, A. Ion exchange of ammonium in zeolites: A literature review. *Water Res.* **2001**, *35*, 1599–1610. [CrossRef]
82. Teutli-Sequeira, A.; Solache-Ríos, M.; Olguín, M.T. Influence of Na^+, Ca^{2+}, Mg^{2+} and $NH_4{}^+$ on sorption of Cd^{2+} by zeolitic material. *Hydrometallurgy* **2009**, *97*, 46–52. [CrossRef]
83. Rahavi, A.; Shahriari, M.H.; Hedayat, M.; Dindarlou, A. Application of activated carbon and Potassium Enriched NanoZeolite on tomato in soilless cultivation under the conditions of using low quality water. *Iran. Water Res. J.* **2024**, *18*.
84. Velásquez-Hernández, M.J.; Ricco, R.; Carraro, F.; Limpoco, F.T.; Linares-Moreau, M.; Leitner, E.; Wiltsche, H.; Rattenberger, J.; Schröttner, H.; Frühwirt, P.; et al. Degradation of ZIF-8 in phosphate buffered saline media. *CrystEngComm* **2019**, *21*, 4538–4544. [CrossRef]
85. Luzuriaga, M.A.; Benjamin, C.E.; Gaertner, M.W.; Lee, H.; Herbert, F.C.; Mallick, S.; Gassensmith, J.J. ZIF-8 degrades in cell media, serum, and some—But not all—Buffers. *ACS Appl. Mater. Interfaces* **2019**, *11*, 10119–10126. [CrossRef]
86. Wang, C.; Liu, X.; Chen, J.P.; Li, K. Superior removal of arsenic from water with zirconium MOF UiO-66. *Sci. Rep.* **2015**, *5*, 16613. [CrossRef]
87. Lin, X.; Xiong, Y.; Dong, F. Sodium Alginate/UiO-66-NH_2 Nanocomposite for Phosphate Removal. *Nanomaterials* **2024**, *14*, 1176. [CrossRef] [PubMed]
88. Ghaffari, Y.; Gupta, N.K.; Bae, J.; Kim, K.S. Heterogeneous catalytic performance and stability of iron-loaded ZSM-5, zeolite-A, and silica for phenol degradation: A microscopic and spectroscopic approach. *Catalysts* **2019**, *9*, 859. [CrossRef]
89. Wu, W.; Yi, H.; Tang, X.; Kadja, G.T.M.; Zhao, S.; Zhang, Y.; Meng, F.; Sun, L.; Yao, Y.; Niu, Z.; et al. Review on Mn/zeolite: Active species, influence factors and application in air pollution control. *Chem. Eng. J.* **2024**, *497*, 154940. [CrossRef]
90. Hosseinzadeh, S.; Bonarrigo, G.; Verheust, Y.; Roccaro, P.; Van Hulle, S. Water reuse in closed hydroponic systems: Comparison of GAC adsorption, ion exchange and ozonation processes to treat recycled nutrient solution. *Aquac. Eng.* **2017**, *78*, 190–195. [CrossRef]
91. Hosseinzadeh, S.; Testai, D.; Bkheet, M.; De Graeve, J.; Roccaro, P.; Van Hulle, S. Degradation of root exudates in closed hydroponic systems using UV/H_2O_2: Kinetics and cost analysis. *Sci. Total Environ.* **2019**, *687*, 479–487. [CrossRef]
92. Shah, I.; Adnan, R.; Wan Ngah, W.S.; Mohamed, N. Iron impregnated activated carbon as an efficient adsorbent for the removal of methylene blue: Regeneration and kinetics studies. *PLoS ONE* **2015**, *10*, e0122603. [CrossRef]
93. Xu, Z.; Wan, Z.; Sun, Y.; Gao, B.; Hou, D.; Cao, X.; Komárek, M.; Ok, Y.S.; Tsang, D.C. Electroactive Fe-biochar for redox-related remediation of arsenic and chromium: Distinct redox nature with varying iron/carbon speciation. *J. Hazard. Mater.* **2022**, *430*, 128479. [CrossRef]
94. Okpara, E.C.; Quadri, T.W.; Ebenso, E.E.; Rowley-Neale, S.J.; Banks, C.E. Advancing water purification: The role of copper-carbon nanostructured heterojunctions in pollutant removal. *Coord. Chem. Rev.* **2025**, *542*, 216837. [CrossRef]
95. Alam, S.; Khan, M.S.; Umar, A.; Khattak, R.; Zekker, I.; Burlakovs, J.; Ghangrekar, M.M.; Bhowmick, G.D.; Kallistova, A.; Pimenov, N.; et al. Preparation of Pd–Ni nanoparticles supported on activated carbon for efficient removal of basic blue 3 from water. *Water* **2021**, *13*, 1211. [CrossRef]
96. Seo, W.H.; Kim, G.E.; Lee, J.Y.; Choi, S. Application of Iron-Modified Activated Carbon for Phosphate Removal in Aqueous Systems. *Appl. Sci.* **2025**, *15*, 5353. [CrossRef]
97. Król, M. Natural vs. Synthetic Zeolites. *Crystals* **2020**, *10*, 622. [CrossRef]
98. Wang, Y.; Wu, G.; Zhang, Y.; Su, Y.; Zhang, H. The deactivation mechanisms, regeneration methods and devices of activated carbon in applications. *J. Clean. Prod.* **2024**, *476*, 143751. [CrossRef]
99. Biliani, I.; Papadopoulou, E.; Zacharias, I. Evaluating Zeolites of Different Origin for Eutrophication Control of Freshwater Bodies. *Sustainability* **2025**, *17*, 7120. [CrossRef]

100. Mondal, M.; Biswas, B.; Garai, S.; Sarkar, S.; Banerjee, H.; Brahmachari, K.; Bandyopadhyay, P.K.; Maitra, S.; Brestic, M.; Skalicky, M.; et al. Zeolites Enhance Soil Health, Crop Productivity and Environmental Safety. *Agronomy* **2021**, *11*, 448. [CrossRef]
101. Saleem, J.; Khalid Baig Moghal, Z.; Tahir, F.; Al-Ansari, T.; Osman, A.I.; McKay, G. Life cycle assessment of high value activated carbon production based on mass and functional performance metrics. *Sci. Rep.* **2025**, *15*, 32797. [CrossRef]
102. Chen, X.; Xiao, G.; Li, T.; Wang, C.; Cui, Q.; Bao, X.; Yue, Y. Comparative environmental assessment of zeolites synthesized from chemicals and natural minerals. *Green Chem.* **2024**, *26*, 5273–5283. [CrossRef]

water

MDPI

Article

Combined Efficacy of Silver, Copper, and Hypochlorite Ions for Vector Control of Juvenile *Aedes aegypti* in Household Water Storage Containers

Sydney S. Turner [1,*], James A. Smith [2], Karin Brett [2], Patrick I. Hancock [3], Sophie L. Howle [2], Victoria Cecchetti [2], Lorin M. Bruno [2], Julia Davis [2] and Clay Ford [4]

1 Occoquan Watershed Monitoring Laboratory, Department of Civil and Environmental Engineering, Virginia Tech, 9408 Prince William St., Manassas, VA 20110, USA
2 Department of Civil and Environmental Engineering, University of Virginia, Charlottesville, VA 22904, USA; jas9e@virginia.edu (J.A.S.); karin.brett@duke.edu (K.B.); jd3td@virginia.edu (J.D.)
3 Institute on the Environment, University of Minnesota, Minneapolis, MN 55455, USA
4 Research Data Services, Shannon Library, University of Virginia, 160 McCormick Road, Charlottesville, VA 22904, USA; jcf2d@virginia.edu
* Correspondence: sydneyturner@vt.edu

Abstract: This study evaluates the larvicidal effects of three common water disinfectants, silver ($AgNO_3$), copper ($CuSO_4{\cdot}5H_2O$), and hypochlorite ($NaOCl$) ions. The treatments were combined at 40–50% of their recommended drinking water guidelines and tested against late first instar and third instar *Ae. aegypti* larvae. The findings demonstrate that the combined application of water disinfectants yields greater efficacy in suppressing the emergence of *Aedes aegypti* compared to the use of the individual disinfectants alone. The silver (Ag) and copper (Cu) combination treatment (40 ppb Ag + 600 ppb Cu) showed the greatest efficacy, achieving nearly complete inhibition of emergence of the older instar larvae (98.52% [96.50, 99.47]). All treatments demonstrated high efficacy against late 1st instar *Ae. aegypti* larvae, with the combined copper and chlorine (Cl) treatment yielding the lowest survival rates, though individual disinfectants also produced substantial mortality. The results of this study provide critical insights to inform the design and implementation of point-of-use water treatment technologies for household water storage containers that both ensure safe drinking water and also strategically target mosquito breeding within household storage containers, thus supporting integrated vector management approaches essential for controlling neglected tropical diseases.

Keywords: *Aedes aegypti*; household water storage; vector control management; water treatment

1. Introduction

Access to safe drinking water remains a global challenge, with over 2 billion people lacking safely managed drinking water services which contributes to an estimated 485,000 diarrheal deaths annually [1]. Simultaneously, the world faces a rising threat from vector-borne diseases (VBDs), such as dengue and Zika, affecting approximately 80% of the global population [2,3]. Mosquitoes are the most lethal disease vectors, responsible for nearly 700 million infections and over one million deaths each year [4,5]. Though transmitted through different mechanisms, both waterborne and vector-borne diseases are rooted in shared structural challenges: poor sanitation, poverty, and limited access to clean water [6,7]. In response to water insecurity, many communities rely on household water storage (HWS) [8,9], which, when improperly managed, can serve as both a breeding site for mosquitoes and a reservoir for waterborne pathogens [9–12], as illustrated in Figure 1. These overlapping vulnerabilities create an urgent need for integrated public health interventions that address water safety while simultaneously reducing vector breeding grounds, particularly in regions where household water storage is common.

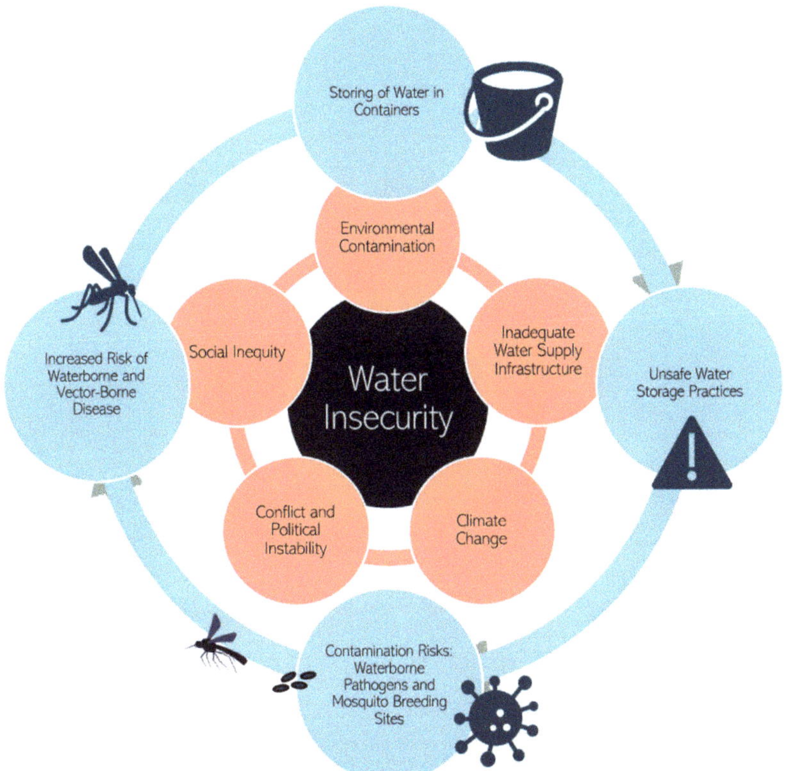

Figure 1. Water insecurity stems from a combination of factors that contribute to the lack of access to safe and clean water [13,14]. Orange circles indicate the primary drivers of water insecurity. The surrounding blue circles illustrate human coping behaviors in response to this insecurity, as well as the broader implications these behaviors carry for health and well-being. Key contributors include environmental contamination [15,16], inadequate water supply infrastructure [17], climate change [18,19], conflict and political instability [20,21], and social inequity, all limiting access to reliable and safe water sources [14,22,23]. When water is perceived as scarce or unreliable, individuals and communities often adopt water storage as a coping strategy which can lead to poor public health outcomes when improperly managed [8,9].

According to the World Health Organization's International Scheme to Evaluate Household Water Treatment Technologies, implementing household water treatment and safe storage (HWTS) can reduce diarrheal disease risk by up to 61% [24]. Furthermore, studies that have linked unsafe water storage practices with mosquito proliferation [6,25,26] provide plausible evidence that implementing appropriate HWTS practices can also contribute to the reduction in VBDs by reducing mosquito breeding sites. Point-of-use water treatment (POUWT) technologies provide a practical means of achieving these benefits at the household level. POUWT technologies refers to the treatment of water at the place where it is being used, commonly used for improving the quality of water for drinking and cooking purposes [24]. Since POUWT technologies and systems treat the water near the point of consumption, the risk of (re)contamination during storage and transport is reduced [27–29]. Design considerations for POUWT technologies are illustrated in Figure 2.

Figure 2. Key design considerations for point-of-use water treatment (POUWT) technologies. Effective POUWT systems must balance multiple factors, including: (1) effectiveness in improving public health outcomes by reducing exposure to waterborne pathogens; (2) safety standards, ensuring disinfection agents are safe for human consumption over extended periods; (3) cost and affordability for end-users; (4) durability and long-lasting performance; (5) ease of use and access, including low maintenance, minimal energy requirements, and compatibility with locally available materials; (6) social acceptability, reflecting cultural appropriateness and societal adoption.

A variety of POUWT technologies exist on the market with varying degrees of effectiveness in removing various contaminants and can be selected based on the following:

- Circumstance (e.g., local water source type, availability of replacement parts, ease of operation, cultural acceptance, and cost);
- Standards/regulatory frameworks (e.g., World Health Organization's Performance Criteria for Household Water Treatment Technologies [24]; National Sanitation Foundation and the American National Standards Institute standards [NSF/ANSI] [30]).

Many of these POUWT technologies treat HWS containers through the provision of an optimum dose of water disinfectants for the reduction in water pathogens originating from either the source water or conditions of storage [31,32]. Three common water disinfectants used in POUWT contexts include silver, copper, and chlorine. Chlorine is widely used because it is affordable, easy to apply, and highly effective against most waterborne pathogens, including bacteria and many viruses; however, chlorine is notably ineffective against certain protozoa, such as *Cryptosporidium parvum* oocysts and *Mycobacteria* species, and its use can result in objectionable taste and odor as well as the formation of disinfection by-products (DBPs) [29,33–41]. Silver has demonstrated broad biocidal activity against bacteria, viruses, and protozoa, including *Cryptosporidium*, and does not produce DBPs or alter water taste or odor [42–45]. Its primary drawback is the need for higher contact times to achieve inactivation [46]. Copper is effective against a wide spectrum of pathogens, such as *Escherichia coli*, *Legionella*, *Salmonella*, and protozoa like *Cryptosporidium* [47,48]. Although copper is generally less potent than silver at equivalent doses, higher allowable concentrations in drinking water standards enable its use at more practical levels. Moreover, copper tends to be significantly more affordable than silver.

Building on prior research on the larvicidal efficacy of individual water disinfectants silver, copper, and chlorine [10], this study investigates the effects of disinfectant combinations of silver (Ag), copper (Cu), and chlorine (Cl) for mosquito larval source management of *Aedes (Ae.) aegypti* and compares the performance of combined treatments to that of individual chemicals. The selection of silver, copper, and chlorine for this study was based on their established disinfection efficacies in water treatment processes and their common implementation in POUWT technologies and HWTS practices (Table 1 highlights some POUWT products on the global market). Thus, this study presented aims to investigate whether using these chemicals in combination can result in the inhibition of emergence of the disease vector *Ae. aegypti*. Our research addresses microbial safety and vector control in the same intervention, which is rarely considered in a combined framework. The findings bridge public health and vector control, thus expanding the potential application of results beyond traditional mosquito control research. To our knowledge, no prior studies have examined the simultaneous application of these three disinfectants for larval source management, representing a critical opportunity in both the vector control and household water treatment.

As there is limited research on multi-disinfectant strategies for larval control, this study seeks to address that gap. When using chemicals in tandem, there is a potential for greater disinfection efficiency while allowing for the application of lower concentrations of each chemical, resulting in reduced effects on taste and odor, minimized formation of DBPs when the disinfectants react with naturally present compounds in the water, avoidance of resistance forming by organisms/pathogens, shorter contact times, and cost savings [49,50]. Additionally, different chemical effects have been observed in the literature when water disinfectants have been used in combination. Some studies have observed synergistic effects, meaning that the interaction between two substances can produce a combined effect greater than the sum of their separate effects, for the utilization of water treatment chemicals for microbial disinfection efficiency [49,51–54].

Table 1. Examples of point-of-use water treatment (POUWT) technologies currently in use for improving drinking water quality and safety that utilize the silver, copper, or chlorine as disinfecting agents. Product descriptions include the manufacturer, active disinfectant(s), and evaluation status according to international standards such as the World Health Organization (WHO) performance criteria or National Sanitation Foundation (NSF) certification.

Product	Manufacturer	Disinfectant	Evaluation
Aquatabs® & Aquatabs Flo	Medentech Ltd. (Wexford, Ireland)	Chlorine in the form of sodium dichloroiso-cyanurate (NaDCC) as the active ingredient	Met World Health Organization's (WHO) performance criteria, receiving the designation of "targeted protection (bacteria and viruses only)"
H2gO Purifier	Aqua Research, LLC (Albuquerque, NM, USA)	Electrolytic chlorine generator: splits salt (NaCl) to chlorine (Cl_2) and hydrogen peroxide (H_2O_2)	Met WHO's performance criteria, receiving the designation of "targeted protection (bacteria and viruses only)"
H_2O ResQ Copper-Silver Ion Water Storage Treatment Drops	Legacy Food Storage (Salt Lake City, UT, USA)	Copper–silver ion formula	No third-party certifications (e.g., WHO or National Sanitation Foundation [NSF]) found.
MadiDrop®	Silivhere Technologies Inc. (Charlottesville, VA, USA)	Releases ionic silver	Certified for drinking water treatment by NSF International.
Oasis Water Purification Tablets	Hydrachem Ltd. (Billingshurst, England)	Chlorine in the form of sodium dichloroisocyanurate (NaDCC) as the active ingredient	Met WHO's performance criteria, receiving the designation of "targeted protection (bacteria and viruses only)"
P&G Purifier of Water Packets	Procter & Gamble (Cincinnati, OH, USA)	Calcium hypochlorite (chlorine disinfectant) + ferric sulfate (coagulant/flocculant)	Met WHO's performance criteria; provides comprehensive protection by combining coagulation, flocculation, and disinfection

2. Materials and Methods

This study follows protocol from previous work on the larvicidal efficacy of individual water disinfectants [10]. This study evaluates the performance of combined disinfectant treatments of silver, copper, and chlorine for the control of *Ae. aegypti* larvae in household water storage contexts; thus bioassay experiments were designed to simulate operational conditions relevant to point-of-use water treatment. The following sections describe the experimental procedures: culturing and rearing of *Ae. aegypti* in the laboratory (Section 2.1); test concentrations for treatment scenarios (Section 2.2); and evaluation of larvicidal efficacy (Section 2.3).

2.1. Culturing and Rearing

Ae. aegypti eggs were obtained from Benzon Research, Inc. (Carlisle, PA, USA). Eggs used in this study were 2–3 weeks old upon arrival. Rearing was conducted in the Water Quality Laboratory at the University of Virginia under a 12:12 h light–dark cycle. Environmental conditions were monitored using an Extech RHT20 Humidity and Temperature Datalogger (Teledyne FLIR LLC, Nashua, NH, USA). Eggs and larvae were maintained at 27.9 ± 0.2 °C (82.2 °F) in Sterlite plastic trays containing deionized (DI) water. A feeding regimen was followed using a ground larval diet composed of a 3:1 mixture of liver powder and brewer's yeast (MP Biomedicals™, Solon, OH, USA).

2.2. Test Concentrations for Treatment Scenarios

Ae. aegypti larvae were exposed to varying concentrations of disinfectants widely used in water treatment, particularly in POUWT applications in resource-limited settings: silver ions, copper ions, and hypochlorous acid. Disinfectant concentrations were selected based on established drinking water quality guidelines (DWQGs) from the WHO and

United States Environmental Protection Agency (US EPA), ensuring relevance to household water storage contexts. In the previous study [10], concentration ranges were designed relative to DWQG values for each chemical: low (~20–25%), mid (~40–50%), and high (~80–95%) of the guideline threshold (see Table 2). Testing across this range accounted for potential health, sensory, and cost considerations associated with higher disinfectant levels. Additionally, regulatory frameworks assess toxic exposures on a per-contaminant basis, with agencies such as the Agency for Toxic Substances and Disease Registry (ATSDR) and WHO evaluating silver and copper independently. Since no comprehensive studies have examined their combined effects in drinking water, caution is warranted when developing multi-metal treatments, even if each disinfectant is applied at concentrations below its individual safety threshold for human intake.

Table 2. Drinking water quality guidelines (DWQG) from the United States Environmental Protection Agency (US EPA) and World Health Organization (WHO) for disinfectants silver, copper, and chlorine alongside the low, medium, and high concentrations tested in larvicidal bioassays. μg/L = micrograms per liter; NTU = Nephelometric Turbidity Unit.

Disinfectant	Drinking Water Quality Guideline (DWQG)	Concentrations Tested (μg/L)		
		High (80–95% of DWQG)	Mid (40–50% of DWQG)	Low (20–25% of DWQG)
Silver (Ag):	US EPA and WHO: 100 μg/L	20	40	80
Copper (Cu):	US EPA: 1300 μg/L WHO: 2000 μg/L	300	600	1200
Free Chlorine (OCl^-/HOCl):	US EPA: 4000 μg/L WHO: 2000 μg/L free chlorine dose for clear water (<10 NTU) and 4000 μg/L for turbid water (≥10 NTU) for POUWT	500	1000	2000

This study specifically examines combinations at the mid-range concentration level, as illustrated in Table 3, as it represents a practical balance between safety, efficacy, and affordability. Selecting mid-range levels for combinations reduces the risk of compounding toxicity or triggering taste and odor concerns that can arise at higher doses, potential issues that affect both social acceptability and long-term adoption in POUWT applications.

Table 3. Concentrations of silver (Ag), copper (Cu), and free chlorine (Cl) tested individually and in combination in larvicidal bioassays with *Ae. aegypti*. Disinfectants were introduced as silver nitrate ($AgNO_3$), copper sulfate pentahydrate ($CuSO_4 \cdot 5H_2O$), and sodium hypochlorite (NaOCl). The tested concentrations included single-agent exposures (40 ppb Ag, 600 ppb Cu, or 1 ppm Cl) as well as combinations (600 ppb Cu + 40 ppb Ag; 1 ppm Cl + 40 ppb Ag; 1 ppm Cl + 600 ppb Cu). Units are expressed in parts per billion (ppb; equivalent to μg/L) for silver and copper, and parts per million (ppm; equivalent to mg/L) for chlorine.

Water Disinfectants	Silver (Ag) $AgNO_3$	Copper (Cu) $CuSO_4 \cdot 5H_2O$	Free Chlorine ($HOCl/OCl^-$) NaOCl
Silver	40 ppb Ag	600 ppb Cu + 40 ppb Ag	1 ppm Cl + 40 ppb Ag
Copper	-	600 ppb Cu	1 ppm Cl + 600 ppb Cu
Chlorine	-	-	1 ppm Cl

Preparing Stock Solutions

A 100 mMsilver stock solution was prepared with silver nitrate ($AgNO_3$; Artcraft Chemicals, CAS No. 7761-88-8, South Glens Falls, NY, USA) in deionized (DI) water. Serial dilutions (10 mg/L → 1 mg/L → 0.1 mg/L) were then performed. Silver concentrations were verified via inductively coupled plasma mass spectrometry (ICP-MS) using an Agilent 7900 ICP-MS instrument (Agilent Technologies, Santa Clara, CA, USA) following the US EPA Method 6020B [55]. ICP-MS samples were prepared by acidifying with 2% trace metal grade nitric acid (HNO_3; Fisher Chemical, Fair Lawn, NJ, USA).

A 100 mM stock solution of copper sulfate pentahydrate ($CuSO_4 \cdot 5H_2O$, Alfa Aesar, Thermo Fisher Scientific, CAS No. 7758-99-8, 99% purity, Waltham, MA, USA) was prepared in deionized (DI) water. This stock was sequentially diluted to create 10 mM, 1 mM and 10 µM solutions. Copper concentrations were validated via ICP-MS following the U.S. EPA Method 6020B [55], with each sample acidified to 2% HNO_3.

A sodium hypochlorite solution containing 10–15% available chlorine was obtained from Sigma-Aldrich (CAS No. 7681-52-9, St. Louis, MO, USA). This was initially diluted to prepare a 25 mg/L stock solution. From this stock, working solutions were freshly prepared immediately prior to each experimental trial. Free chlorine concentrations were quantified using the US EPA DPD Method 8021 (low-range), measured at baseline (prior to larval introduction), and subsequently at 4 and 8 h post-treatment, using a HACH DR6000 spectrophotometer.

2.3. *Survival Bioassays: Evaluation of Dose Response to Water Treatment Disinfectants*

For test concentrations and controls, experiments were carried out in triplicate. These trials were repeated on three separate days, each time using newly prepared test solutions and fresh batches of larvae. Larval survival and development were monitored daily under laboratory conditions. Mortality was assessed visually and by probing with a pipette tip; larvae were considered dead if they did not respond to gentle stimulation. Adult emergence were recorded to capture sub-lethal effects on larval development. These larval development outcomes were recorded for each container and analyzed statistically.

2.3.1. Experimental Setup

The laboratory procedures used in this study were adapted from the World Health Organization's guidelines for larvicide testing in both laboratory and field settings [56] and outlined in our previous study [10]. For each test concentration, 25 mosquito larvae were introduced into 200 mL of prepared solution contained in 250 mL glass beakers, with three replicates per treatment group. Control groups, also in triplicate, consisted of larvae placed in DI water. Transfers of larvae were performed using disposable pipettes. All beakers were maintained under standardized environmental conditions previously described. Two larval age groups were tested to simulate realistic household water storage scenarios. The first scenario modeled the introduction of disinfectants into water immediately after egg deposition, with newly hatched larvae exposed to treatment. The second scenario simulated older larvae entering a disinfected household container through the use of pre-contaminated source water. To reflect these conditions, both late first instar and late third instar larvae were used in experiments. For the older instar, larval survival, and adult emergence were monitored at 24-h intervals across both treated and control groups, continuing either until all mosquitoes had emerged in control groups or until day 16 post-exposure. Trials with <80% adult emergence in control groups were deemed invalid and repeated. For younger instar, mortality of the larvae were monitored across 3 days.

2.3.2. Data Analysis

Larvicidal efficacy was evaluated following WHO guidelines [56] and methods from Ngonzi et al. [57]. Data were pooled across replicates for each treatment. When adult emergence in controls ranged between 80–95%, mortality and emergence were corrected using Abbott's formula.

$$\text{Survival (\%)} = 100 - ((C - T)/C \times 100)$$

where C = percentage survival in the untreated control and T = percentage survival in the treated sample. Larvae that developed into successfully emerging adults was expressed in terms of emergence:

$$\text{Emergence (\%)} = 100 - ((C - T)/C \times 100)$$

where C = percentage emergence in the untreated control and T = percentage emergence in the treated sample. Inhibition of emergence (IE%) was calculated as:

$$\text{IE\%} = 100 - (T \times 100)/C$$

where T is emergence in the treatment and C is emergence in the control.

Statistical analyses were conducted in R (version 4.2.2, R Foundation for Statistical Computing, Vienna, Austria) using RStudio (version 2022.07.2 Build 576, RStudio, PBC, Boston, MA, USA). Descriptive statistics (mean, SD, SEM) were calculated, and probit regression was used to model the dose–response relationship between disinfectant concentration and larval survival or emergence. A mixed-effects probit model (via *glmer* from the *lme4* package) incorporated treatment, time (as a natural spline with 2 degrees of freedom), and their interaction as fixed effects, with each experiment treated as a random effect. Pairwise comparisons were performed using the *emmeans* package.

For emergence inhibition modeling, day 16 control values were derived from the emergence model, and 95% confidence intervals were estimated using 100 bootstrap simulations (*bootMer*). Probit curves show predicted probabilities of survival or emergence over time, with 95% confidence intervals. To evaluate model fit for the larval survival data, we conducted standard diagnostics for binomial generalized linear models using the DHARMa package.

To evaluate whether the effects of chlorine, silver, and copper in combination were synergistic, additive, or antagonistic, we compared the inhibition of emergence for third instar larvae or the mortality for the first instar larvae in combined treatments against the expected mortality predicted under the assumption of independent action. Expected mortality for additive effects was calculated using Bliss independence:

$$E_{AB} = E_A + E_B - (E_A \times E_B)$$

where E_A and E_B represent the fractional inhibition of emergence or mortalities (proportion dead) of larvae exposed to each disinfectant individually, and E_{AB} is the expected combined effect. If the observed mortality in the combined treatment was significantly greater than E_{AB}, we interpreted this as potential for synergistic effects taking place. If it was not significantly different from E_{AB}, it was considered additive. If it was lower, this suggested some potentially antagonistic effects.

3. Results and Discussion

In the following sections, we report results from tests of silver, copper, and chlorine on older juvenile *Ae. aegypti* (Section 3.1) and younger larvae (Section 3.2), followed by an analysis of chemical effects (Section 3.3).

It is important to note that our experiments were conducted in DI water, which provides a controlled baseline for assessing larvicidal effect. In real-world conditions, source waters often contain organic matter, suspended solids, and other constituents that exert disinfectant demand. These interactions can reduce the free concentrations of chlorine, silver, or copper through processes such as complexation with natural organic matter, precipitation with anions, adsorption to suspended solids or container surfaces, photodegradation, chlorine demand from inorganic species, and volatilization, thereby lowering efficacy.

3.1. Late Third Instar Experiments

This section presents outcomes from experiments on late 3rd instar *Ae. aegypti* exposed to silver and copper (Section 3.1.1), silver and chlorine (Section 3.1.2); copper and chlorine (Section 3.1.3) and copper, silver, and chlorine in combination (Section 3.1.4). Observed data results are reported as Percentage Mean ± SEM and model predictions are expressed as Predicted Probability Mean (95% Upper Confidence Interval, 95% Lower Confidence Interval).

Model diagnostics for older instars showed marginal departures from uniformity in two of the combined treatments (Cu + Ag, Ag + Cl; $p \leq 0.021$). Dispersion tests also indicated slight overdispersion in the Cl-only and Ag + Cl treatments, reflecting higher variability among replicates. No evidence of zero inflation was observed in these models. These minor deviations likely reflect biological variability across larval cohorts and chlorine volatility between experimental runs.

3.1.1. Silver + Copper Exposure to Older Instar

The 40 ppb Ag + 600 ppb Cu combination treatment (Ag + Cu) observed and model predicted data for day 16 are shown in Table 4. Figure 3 illustrates their alignment, supporting the use of model predictions moving forward. The observed data for the remainder of the treatments can be found in Supplementary Information.

Table 4. Observed (n = 4) and modeled data for the silver (Ag, 40 ppb), copper (Cu, 600 ppb), and their combination (Ag + Cu) treatment on day 16, corrected using Abbott's formula (1925). Observed values represent the mean, standard deviation (SD), and standard error of the mean (SEM). Modeled data are reported as predicted mean values with 95% confidence interval (CI). Outcomes include larval survival (%), adult emergence (Emerg, %), and inhibition of emergence (IE, %). Concentrations are reported in parts per billion (ppb), where 1 ppb = 1 μg/L.

Treatment	**Ag (40 ppb)**			**Cu (600 ppb)**			**Ag (40 ppb) + Cu (600 ppb)**			**Control**		
						Observed						
Variable	mean	sd	sem	mean	sd	sem	mean	sd	sem	mean	sd	sem
Survival (%)	11.23	7.74	3.87	29.19	14.78	7.39	3.25	4.11	2.06	91.33	1.72	0.86
Emerg (%)	9.94	8.80	4.40	22.00	13.90	6.95	2.52	4.10	2.05	88.33	5.03	2.52
IE (%)	89.01	9.47	4.74	74.47	17.08	8.54	97.29	4.38	2.19			
						Model						
Variable						Predicted mean [95% Confidence Interval]						
Survival (%)	10.00 [6.31, 14.68]			29.53 [23.09, 37.71]			3.53 [1.87, 5.33]			92.07 [86.79, 95.58]		
Emerg (%)	7.15 [3.90, 1 4.09]			22.47 [13.50, 33.15]			1.48 [0.39, 2.94]			91.81 [85.28,95.8]		
IE (%)	92.85 [88.24, 95.68]			77.53 [68.19, 85.34]			98.52 [96.50, 99.47]					

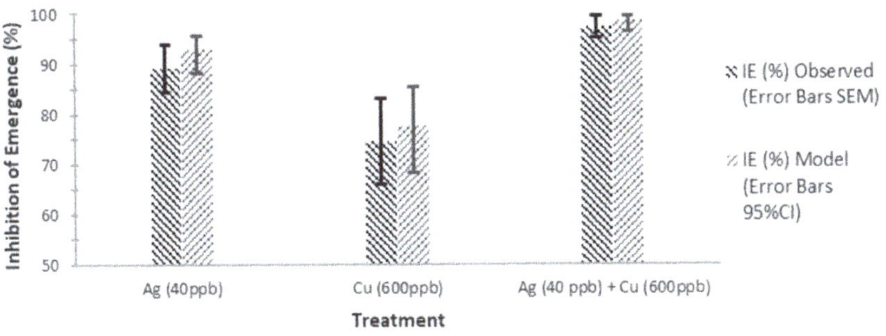

Figure 3. Comparison of observed and model values for inhibition of emergence (IE) for *Ae. aegypti* larvae treated with silver ions (40 ppb Ag) and copper ions (600 ppb Cu), relative to untreated controls, on day 16. Observed values are shown with error bars representing the standard error of the mean (SEM). Modeled values are presented with 95% confidence intervals (CI), which represent the range within which the true population mean is expected to fall with 95% confidence.

The Ag + Cu combo performed significantly differently from each of the chemicals separately in terms of survival ($p_{(Ag+Cu)\text{ vs. }Ag} = 0.002$, $p_{(Ag+Cu)\text{ vs. }Cu} < 0.001$) and predicting how many larva had emerged ($p_{(Ag+Cu)\text{ vs. }Ag} = 0.021$, $p_{(Ag+Cu)\text{ vs. }Cu} < 0.001$) by the end of the observational period. Additionally, the use of the chemicals together (Ag + Cu) resulted in less variability in outcome (inhibition of emergence) as illustrated in Figure 3. In the predicted survival curves illustrated in Figure 4, both copper and silver treatments initially reduced *Ae. aegypti* survival; however, copper effects began to plateau while silver toxicity remains relatively constant. This trend may indicate that larval susceptibility to copper diminishes with maturation, while silver maintains a sustained toxic effect throughout the developmental stages. Although we did not experimentally test mechanisms, previous work suggests several biological processes may contribute to the plateau in copper toxicity observed in older instars. For example, progressive cuticle thickening and sclerotization can reduce metal uptake across the integument [58,59] while studies in mosquitoes and aquatic invertebrates point to inducible detoxification pathways and acclimation responses that can mitigate copper stress [60–62].

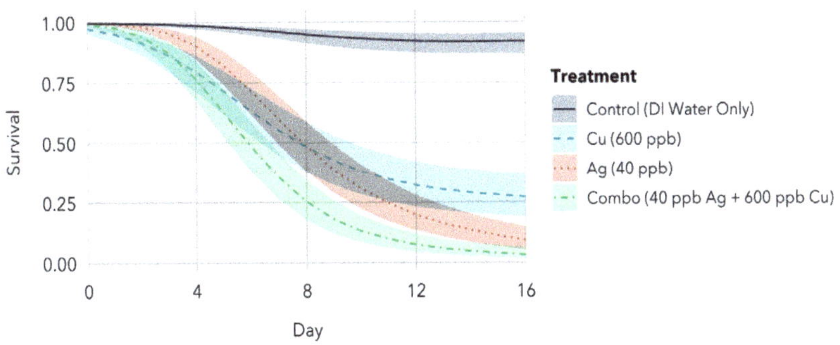

Figure 4. Predicted probabilities of survival for older instar *Ae. aegypti* larvae exposed to silver ions (40 ppb Ag), copper ions (600 ppb Cu), and their combination (40 ppb Ag + 600 ppb Cu) compared to untreated controls (larvae reared in deionized water without disinfectant) on day 16. Shaded areas represent the 95% confidence interval (CI). Concentrations are expressed as parts per billion (ppb, 1 ppb = 1 μg/L).

3.1.2. Silver + Chlorine Exposure to Older Instar

The 40 ppb Ag + 1 ppm free chlorine combination treatment (Ag + Cl) model predicted survival, emergence, and inhibition of emergence data for day 16 is reported in Table 5 and Figure 5 specifically illustrates modeled survival probabilities. Analysis of the model indicates that Ag + Cl performed significantly better than either Ag or Cl separately in terms of predicted survival ($p_{(Ag+Cl)\ vs.\ Ag} = <0.001$, $p_{(Ag+Cl)\ vs.\ Cl} = <0.001$). For the predicted probability that a larva has emerged, the Ag + Cl treatment performed better than the chlorine alone ($p_{(Ag+Cl)\ vs.\ Cl} = 0.002$), but was not significantly different from silver alone ($p_{(Ag+Cl)\ vs.\ Ag} = 0.154$, by the end of the observational period. The efficacy of the 40 ppb Ag treatment, $IE\%_{model} = 88.31$ [78.05, 94.66], was similar to that of the Ag + Cl treatment, $IE\%_{model} = 93.75$ [84.65, 97.76], $p_{(Ag+Cl)\ vs.\ Ag} = 0.154$).

Table 5. Predicted mean values with 95% confidence intervals (CI) for *Ae. aegypti* larvae exposed to silver ions (Ag, 40 ppb), free chlorine (Cl, 1 ppm), and their combination (Ag + Cl) relative to untreated controls reared in deionized water on day 16. Outcomes include survival (%), emergence (Emerg, %), and inhibition of emergence (IE, %). Data is corrected with Abbot's formula (1925). Concentrations are reported as parts per billion (ppb) and parts per million (ppm).

Treatment	**Ag (40 ppb)**	**Free Chlorine (1 ppm)**	**Ag (40 ppb) + Free Chlorine (1 ppm)**	**Control**
Variable		Predicted Mean [95% Confidence Interval]		
Survival (%)	17.30 [12.03, 23.11]	29.26 [23.25, 36.70]	7.38 [4.49, 11.36]	94.42 [90.99, 96.72]
Emerg (%)	11.69 [4.42, 24.67]	31.65 [21.95, 37.68]	6.25 [1.83, 15.35]	94.75 [85.62, 98.53]
IE (%)	88.31 [78.05, 94.66]	68.35 [51.93, 82.01]	93.75 [84.65, 97.76]	

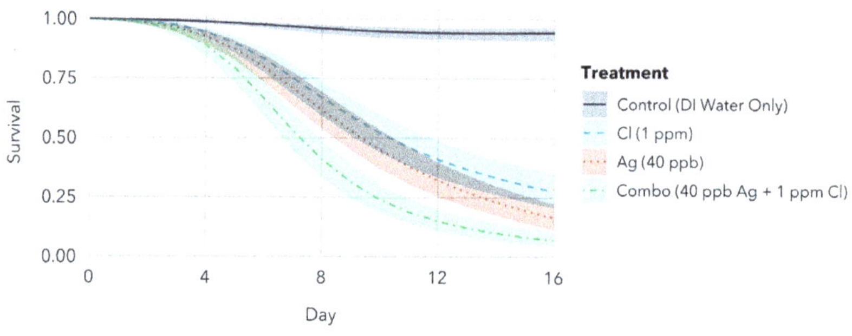

Figure 5. Predicted probabilities of survival of *Ae. aegypti* larvae exposed to silver ions (40 ppb Ag) and free chlorine (1 ppm free chlorine) compared to untreated controls (larvae reared in deionized water [DI] without disinfectant) on day 16. The shaded areas represent the 95% confidence interval (CI). Concentrations are expressed as parts per billion (ppb, 1 ppb = 1 µg/L) and parts per million (ppm, 1 ppm = 1 mg/L).

It is important to note that chlorine is known to degrade relatively quickly in stored water, especially in the presence of organic matter and high chlorine demand, which can limit the duration of larvicidal protection. This same issue has long been a concern in household water safety programs where recontamination of containers occurs once chlorine residuals fall below protective levels [27,63,64], meaning our laboratory estimates of chlorine's larvicidal potential may represent an upper bound of what is achievable in practice.

3.1.3. Chlorine + Copper Exposure to Older Instar

The 600 ppb Cu + 1 ppm free chlorine combination treatment (Cu + Cl) results for survival, emergence, and inhibition of emergence are shown in Table 6 for day 16. All treatments successfully decreased the larval population, resulting in the predicted probability of larval survival for 600 ppb Cu, 1 ppm Cl, and 600 ppb Cu + 1 ppm Cl was 19.87% (16.51, 24.48), 26.08% (*21.84, 31.84*), and 4.48% (3.19, 6.36) as compared to the predicted emergence of the controls at 93.12% (90.89, 94.90) on day 16.

Table 6. Predicted mean values with 95% confidence intervals (CI) for *Ae. aegypti* larvae exposed to copper ions (600 ppb Cu), free chlorine (1 ppm free chlorine), and their combination (Cu + Cl) treatments on day 16. Outcomes include survival (%), emergence (Emerg, %), and inhibition of emergence (IE, %). Data is corrected with Abbot's formula (1925).

Treatment	Cu (600 ppb)	Free Chlorine (1 ppm)	Ag (40 ppb) + Cl (1 ppm)	Control
Variable		Predicted Mean [95% Confidence Interval]		
Survival (%)	19.87 [16.51, 24.48]	26.08 [21.84, 31.84]	4.48 [3.19, 6.36]	91.59 [88.75, 93.85]
Emerg (%)	22.67 [19.19, 27.77]	15.75 [12.87, 19.09]	4.29 [2.66, 6.06]	93.12 [90.89, 94.90]
IE (%)	84.25 [81.08, 87.78]	77.33 [72.57, 81.12]	95.71 [94.25, 97.28]	

Figure 6 shows the predicted probability of larval survival during the experimental timeframe. The combination treatment showed similar toxicity to copper alone during the first five days of the observational period but demonstrated greater efficacy later, as copper's effects plateaued. This suggests that the chlorine in the combination enhanced overall toxicity, leading to increased mortality. The results of the model indicate that the Cu + Cl combo performed significantly differently from each of the chemicals separately in terms of survival ($p_{(Cu+Cl)\ vs.\ Cu}$ = <0.001, $p_{(Cu+Cl)\ vs.\ Cl}$ = <0.001) and predicting how many larva had emerged ($p_{(Cu+Cl)\ vs.\ Cu}$ = <0.001, $p_{(Cu+Cl)\ vs.\ Cl}$ = <0.001) by the end of the observational period.

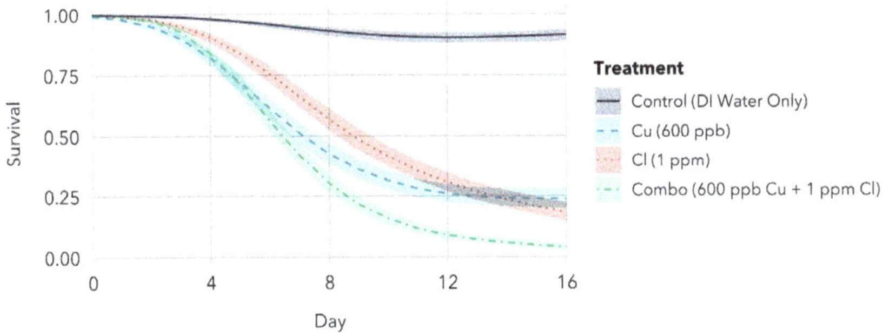

Figure 6. Predicted probabilities of survival of *Ae. aegypti* larvae exposed to copper ions (600 ppb Cu), free chlorine (1 ppm free chlorine), and their combination compared to untreated controls (larvae reared in deionized [DI] water without disinfectant). on day 16. The shaded areas represent the 95% confidence interval (CI). Concentrations are expressed as parts per billion (ppb) and parts per million (ppm).

3.1.4. Silver + Chlorine + Copper Exposure to Older Instar

The 40 ppb Ag + 600 ppb Cu + 1 ppm free chlorine combination treatment (Ag + Cu + Cl) predicted data for day 16 is illustrated in Figure 7. The combination of Ag + Cu + Cl demonstrated better performance compared to each chemical individually in terms of predicted survival ($p_{(Ag+Cu+Cl)\ vs.\ Ag}$ = 0.001, $p_{(Ag+Cu+Cl)\ vs.\ Cu}$ = <0.001, $p_{(Ag+Cu+Cl)\ vs.\ Cl}$ = <0.001) and predicting how many larva emerge ($p_{(Ag+Cu+Cl)\ vs.\ Ag}$ = 0.003, $p_{(Ag+Cu+Cl)\ vs.\ Cu}$ = 0.001, $p_{(Ag+Cu+Cl)\ vs.\ Cl}$ = <0.001) by the end of the observational period. The modeled IE% for treatments 40 ppb Ag, 600 ppb Cu and 1 ppm free chlorine were 89.61 [84.25, 94.24],

86.83 [81.06, 91.59], and 76.39 [69.13, 82.9], respectively. The Ag + Cu + Cl combo exhibited the highest IE% at 97.44 [94.77, 98.78] (Note: Experiments were conducted separately and do not share the same control group). The IE%, for the combined use of Ag + Cu + Cl resulted in reduced variability in outcomes/predicted outcomes comparably to the chemicals acting alone.

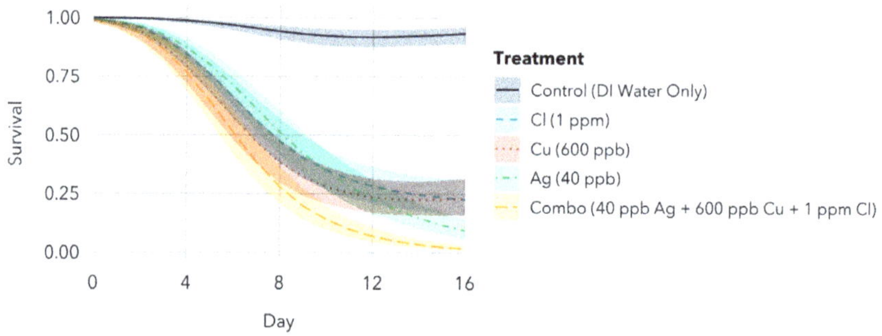

Figure 7. Predicted probabilities of survival of *Ae. aegypti* larvae exposed to copper ions (600 ppb Cu), free chlorine (1 ppm free chlorine), and silver ions (40 ppb Ag) and their combination (Ag + Cu + Cl) compared to untreated controls (larvae reared in deionized [DI] water without disinfectant) on day 16. The shaded areas represent the 95% confidence interval. Concentrations are expressed as parts per billion (ppb) and parts per million (ppm).

3.1.5. Comparing Efficacy of Water Disinfectants in Combination Exposure to Older Instar

Overall, all combination treatments demonstrated a high level of efficacy in inhibiting the emergence of the juvenile *Ae. aegypti* larvae and pupae into adult mosquitoes (as seen in Figure 8), and they outperformed the individual chemical treatments. All the combination models predicted viability for emergence, as the 95% confidence intervals did not ever reach 100% inhibition of emergence. When comparing combinations, the observed inhibition values had overlapping confidence intervals. Not only did the inclusion of a third disinfectant not yield a statistically distinct improvement over the dual combinations, but it was also not even the highest-performing treatment. From a practical perspective, this suggests that adding chlorine on top of silver and copper does not provide a substantial incremental benefit in larvicidal efficacy under the tested conditions. The added disinfectant simply increases cost, chemical complexity, and the potential risks of byproduct formation associated with introducing another agent. In practice, dual-agent strategies (e.g., Ag + Cu or Cu + Cl) may therefore strike the most efficient balance between efficacy and practicality.

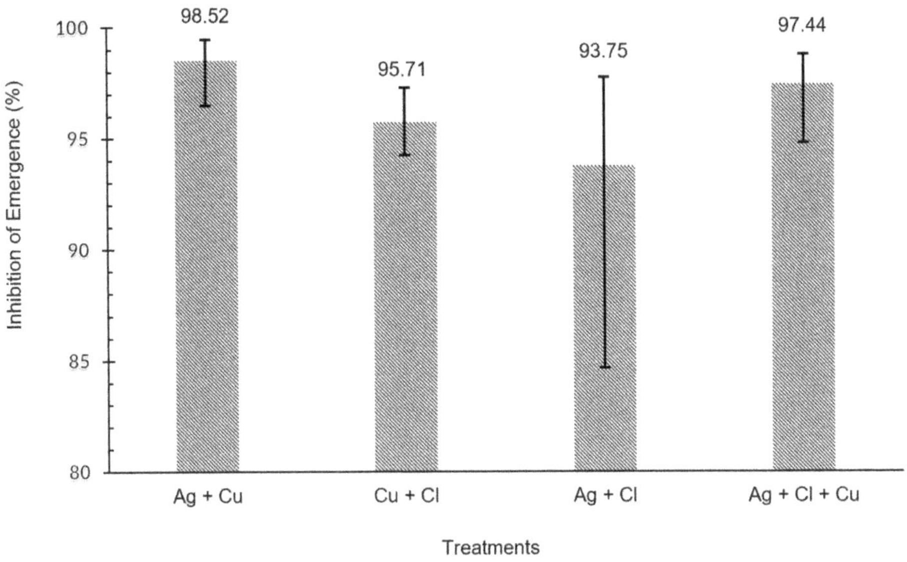

Figure 8. Comparing modeled inhibition of emergence (IE) values for combinations of silver ions (Ag, 40 ppb), copper ions (Cu, 600 ppb), free chlorine (free Cl, 1 ppm) treatments, and their combination (Ag + Cl + Cu) on older instar *Ae. aegypti* larvae in regard to untreated larvae on day 16. Error bars represent the 95% confidence interval. Experiments were conducted separately and do not share the same control group.

3.2. *Younger Instar Experiments*

This section presents the findings of larvicidal experiments conducted on late first instar *Aedes aegypti* using silver nitrate, copper sulfate pentahydrate, and sodium hypochlorite treatments. The results from the observed data are expressed as Percentage Mean ± SEM, while the model predicted data are expressed as Predicted Probability Mean (95% Upper Confidence Interval, 95% Lower Confidence Interval).

Model diagnostics for younger instars indicated overall good fit. Uniformity tests did not identify significant deviations. The copper–chlorine treatment exhibited an elevated dispersion ratio (>2) and evidence of zero inflation ($p = 0.008$), suggesting more replicate outcomes of complete survival or complete mortality than expected under a binomial distribution. These departures are likely attributable to biological variability and the strong efficacy of some treatments.

3.2.1. Silver + Copper Exposure to Younger Instar

The model-predicted survival outcomes for younger instar *Ae. aegypti* larvae are summarized in Table 7 for treatments with silver nitrate (40 ppb Ag), copper sulfate (600 ppb Cu), and their combination (Ag + Cu), compared with untreated controls maintained in deionized water. The table reports predicted mean survival values and 95% confidence intervals across 24-, 48-, and 72-h exposures, corrected using Abbott's formula (1925). After 72 h of treatment, survival rates were 22.67 ± 5.05% for 40 ppb Ag, 17.33 ± 4.29% for 600 ppb Cu, and 3.56 ± 1.18% for the Ag + Cu combination, as illustrated in Figure 9. The results of the model indicate that the Ag + Cu combo performed significantly differently from each of the chemicals separately ($p_{(Ag+Cu)\ vs.\ Ag}$ = <0.001, $p_{(Ag+Cu)\ vs.\ Cu}$ = <0.001). The difference between the Ag and Cu treatments were not significantly different ($p_{Cu\ vs.\ Ag} = 0.315$).

Table 7. Predicted probability of survival for younger instar *Ae. aegypti* larvae exposed to copper sulfate (600 ppb Cu), silver nitrate (40 ppb Ag), and their combination (Ag + Cu). Control larvae were reared in deionized (DI) water without disinfectant. Data is corrected with Abbot's formula (1925). Values are presented as predicted means with 95% confidence intervals (CI). Concentrations are expressed as parts per billion (ppb).

Treatment	**Ag (40 ppb)**	**Cu (600 ppb)**	**Ag (40 ppb) + Cu (600 ppb)**	**Control**
Time (h)	Predicted Mean [95% Confidence Interval]			
24	88.50 [82.32, 92.97]	84.20 [76.42, 90.07]	79.73 [70.86, 86.75]	97.58 [94.92, 98.96]
48	53.34 [43.08, 63.37]	34.49 [25.27, 44.74]	13.74 [8.28, 21.25]	90.60 [84.61, 94.66]
72	23.28 [15.69, 32.57]	17.14 [10.83, 25.41]	3.33 [1.35, 7.26]	82.97 [74.73, 89.25]

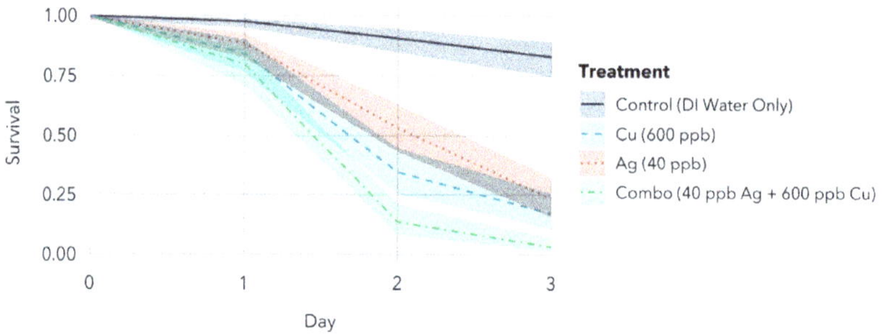

Figure 9. Predicted probabilities of survival of *Ae. aegypti* larvae exposed to silver ions (40 ppb Ag), copper ions (600 ppb Cu), and their combination (Ag + Cu) compared to control larvae reared in deionized (DI) water without disinfectant at 72 h. Shaded area represents the 95% confidence intervals of the probit regression model. Concentrations are expressed as parts per billion (ppb).

3.2.2. Silver + Chlorine Exposure to Younger Instar

The performance of silver individually in decreasing survival (40 ppb Ag: 30.22 ± 8.30%) was not as good as that of chlorine individually (1 ppm Cl: 8.44 ± 4.64%) or the combination of both chemicals together (40 ppb Ag + 1 ppm Cl: 5.78 ± 2.35%) after 72 h of exposure. Table 8 shows the predicted probability of survival data after exposure to the treatments for 24, 48, and 72 h. Figure 10 illustrates the predicted probability of survival throughout the duration of the experiment. Results of the model indicate that the Cl and Ag + Cl treatments perform better than the Ag alone treatment in terms of survival ($p_{(Ag+Cl)\text{ vs. }Ag}$ = <0.001, $p_{Ag\text{ vs. }Cl}$ = <0.001); however, the combination treatment was not significantly different from the Cl alone treatment ($p_{(Ag+Cl)\text{ vs. }Cl} = 0.446$).

Table 8. Predicted probabilities of survival for younger instar *Ae. aegypti* larvae exposed to silver nitrate (40 ppb Ag), sodium hypochlorite (1 ppm free chlorine) and their combination (Ag + Cl). Predictions are presented as mean values with 95% confidence intervals (CI), and data were corrected using Abbott's formula (1925).

Treatment	**Ag (40 ppb)**	**Cl (1 ppm)**	**Ag (40 ppb) + Cl (1 ppm)**	**Control**
Time (h)	Predicted Mean [95% Confidence Interval]			
24	83.81 [73.88, 90.89]	69.37 [56.61, 80.13]	64.79 [51.61, 76.40]	95.17 [90.17, 97.89]
48	46.51 [33.46, 59.95]	21.53 [12.91, 32.80]	16.64 [9.45, 26.65]	84.74 [74.89, 91.61]
72	30.31 [19.30, 43.48]	8.99 [4.26, 16.82]	6.22 [2.67, 12.68]	79.17 [67.55, 87.89]

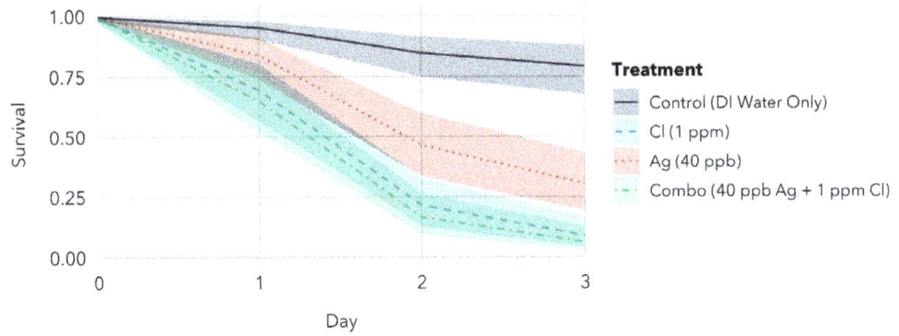

Figure 10. Predicted probabilities of survival of *Ae. aegypti* larvae in contact with silver ions (40 ppb Ag), free chlorine (1 ppm free chlorine), and their combination (Ag + Cl), compared to control larvae reared in deionized (DI) water without disinfectant at 72 h. Shaded area represents the 95% confidence interval of the probit regression model. Concentrations are expressed as parts per billion (ppb) and parts per million (ppm).

3.2.3. Chlorine + Copper Exposure to Younger Instar

The model predicted probability of survival data after exposure to the treatments for 24, 48, and 72 h is reported in Table 9 and Figure 11. After 72 h of exposure, all of the treatments were extremely effective in decreasing survival (600 ppb Cu: 10.67 ± 2.78%, 1 ppm Cl: 6.22 ± 4.95%, 600 ppb Cu + 1 ppm Cl: 1.33 ± 0.77%). Besides all treatments being significantly different from the controls, the only other statistical different treatments were Cu + Cl combo from the Cu alone treatment ($p_{(Cu+Cl)\ vs.\ Cu}$ = <0.001).

Table 9. Modeled predicted probability of survival for copper sulfate (600 ppb Cu) + sodium hypochlorite (1 ppm free chlorine) treatments for younger instar *Ae. aegypti*. Data is corrected with Abbot's formula (1925). Concentrations are expressed as parts per billion (ppb) and parts per million (ppm).

Treatment	Cu (600 ppb)	Cl (1 ppm)	Cu (600 ppb) + Cl (1 ppm)	Control
Time (h)	Predicted Mean [95% Confidence Interval]			
24	67.88 [61.89, 73.44]	68.53 [61.74, 74.74]	34.71 [28.60, 41.25]	94.54 [91.47, 96.67]
48	23.38 [18.62, 28.75]	11.32 [7.61, 16.19]	3.41 [2.08, 5.36]	80.37 [74.74, 85.17]
72	11.83 [7.98, 16.84]	6.18 [3.53, 10.18]	2.44 [1.01, 5.31]	75.04 [68.55, 80.73]

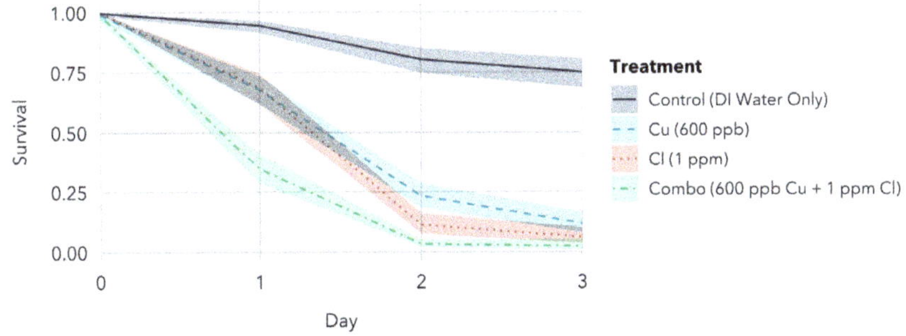

Figure 11. Predicted probabilities of survival of *Ae. aegypti* larvae in contact with copper (600 ppb Cu), free chlorine (1 ppm free chlorine), and a combination of the two disinfectants. Shaded area represents the 95% confidence interval of the probit regression model. Concentrations are expressed as parts per billion (ppb) and parts per million (ppm).

3.2.4. Comparing Efficacy of Water Disinfectants in Combination Exposure to Younger Instar

The modeled outcomes of the combined treatment are compared in Figure 12, depicting the predicted probability of survival for the various treatment combinations on day 16 of silver nitrate (40 ppb Ag), copper sulfate pentahydrate (600 ppb Cu) and sodium hypochlorite (1 ppm free chlorine). These experimental groups were not tested simultaneously and thus do not share the same controls. All treatments demonstrated a high level of efficacy in killing young *Ae. aegypti* larvae. Of the combinations, Cu + Cl appears to have had the most toxic effect on the younger instar larvae.

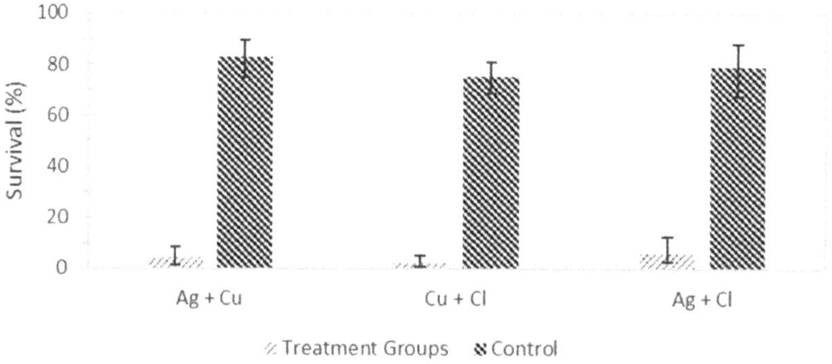

Figure 12. Comparison of modeled survival (%) for late first instar *Ae. aegypti* larvae combinations after 72 h of exposure to combinations of silver ions (40 ppb Ag), copper ions (600 ppb Cu) and free chlorine (1 ppm free chlorine). Results are presented relative to different control groups of larvae reared in deionized water without disinfectants.

Unlike the older instar larvae experiments, the combination treatments against younger instar larvae did not always outperform the individual chemical treatments. This pattern likely reflects stage-specific physiology, i.e., underdeveloped physical barriers and detox systems. Early instar *Ae. aegypti* larvae possess thinner cuticles and a relatively immature peritrophic matrix/gut epithelium, which increases permeability to the disinfectants in this study [58,59]. At the same time, their detoxification capacity and energetic reserves are limited [60,65], so a potent single disinfectant can already produce near-maximal inhibition, leaving little scope for combinations to show superior efficacy. As larvae mature, barrier and detoxification mechanisms strengthen, which can alter the relative performance of single versus combined treatments.

3.3. *Exploring Chemical Effects*

Bliss independence analysis across both late third instar (day 16 outcomes) and late first instar (72 h mortality) assays indicated that the chemical combinations behaved additively rather than synergistically. For Ag + Cu, observed inhibition exceeded Bliss predictions slightly (<1%), consistent with additivity. Both Ag + Cl and Cu + Cl combinations showed minor underperformance relative to Bliss expectations for both the older (Ag + Cl: −2.6%; Cu + Cl: −0.72%) and younger instar larvae (Ag + Cl: −2.3%; Cu + Cl: −2.0%), but these differences were small and occurred at already high efficacy levels (>95%), meaning they are not meaningfully antagonistic. This suggests a ceiling effect because efficacy was already near complete inhibition, there was little scope left for synergy to manifest. Taken together, the combinations produced strong larvicidal effects that were largely additive, with no clear evidence of true synergism under the tested conditions.

4. Conclusions

Improperly managed household water storage containers pose a dual threat as they can harbor disease-causing pathogens and serve as breeding grounds for mosquitoes, contributing to both waterborne and vector-borne disease transmission. This study evaluated the larvicidal potential of silver ions, copper ions, and hypochlorous acid against juvenile *Ae aegypti*. Across both late third-instar and late first-instar *Ae. aegypti* larvae, all three disinfectants produced strong larvicidal effects, with combinations generally enhancing efficacy relative to single disinfectants. For older instars, dual combinations (Ag + Cu, Ag + Cl, Cu + Cl) consistently outperformed individual disinfectants, achieving inhibition of emergence above 90% and reducing variability across replicates. Adding a third agent (Ag + Cu + Cl) yielded high inhibition but did not provide clear advantages beyond dual treatments. By contrast, younger instars were highly susceptible even to single disinfectants, most notably chlorine, reflecting their less developed cuticular and detoxification defenses. In these assays, combinations provided only modest improvements, and in some cases single disinfectant exposures were nearly as effective as multi-agent treatments. All together, these results highlight stage-specific differences in that combinations offer the greatest relative benefits in older, more resistant larvae, while early instars are broadly vulnerable to single chemical exposures. Bliss independence analysis confirmed that effects were primarily additive rather than synergistic, with near-maximal efficacy at higher doses creating a ceiling effect.

While this study demonstrates the larvicidal potential of chlorine, silver, and copper individually and in combination, translating these findings into public health practice requires consideration of real-world conditions. In household and community settings, water is often stored for days, during which disinfectant concentrations decline due to decay, container interactions, and recontamination risks. Because our experiments were conducted in DI water under controlled conditions, they likely represent an upper bound of achievable efficacy. Future field studies are therefore essential, particularly across a range of source waters with varying chemistry since parameters such as pH, alkalinity, organic matter, and container material strongly influence disinfectant stability, speciation, and persistence. Monitoring these factors in parallel with larvicidal efficacy will be critical to evaluating POUWT technologies under realistic conditions and ensuring their effectiveness for both mosquito control and safe water provision.

At the same time, the existence of point-of-use products already on the market provides a foundation, and our findings may help guide both the design of new technologies and the adaptation of existing ones to specific contexts. Cost and scalability of POUWT technologies remain central determinants of household feasibility, with chlorine-based systems benefiting from established global supply chains versus silver- and copper-based technologies which are less common and thus require additional infrastructure for production and distribution. Pricing volatility, availability of raw materials, and local manufacturing capacity may also shape feasibility, meaning cost-effectiveness will vary by setting. Acceptability of disinfectants is another critical factor as we have observed how taste and odor concerns often limit chlorine uptake, while silver and copper were generally more neutral but raised concerns related to safety, environmental accumulation, and affordability. Addressing these implementation challenges alongside navigating regulatory approval will be essential to determine whether disinfectant combinations can be advanced in POUWT technologies for improving household water safety and vector control.

Ultimately, this work provides a proof-of-concept that common disinfectants when combined can serve a dual role in safeguarding drinking water quality and disrupting mosquito breeding. While this work focused on *Ae. aegypti*, the same principles could extend to other container-breeding vectors such as *Culex* and *Anopheles*, which also pose significant public health threats. By bridging water treatment and vector control, such integrated strategies open new pathways for holistic household water management, particularly in resource-limited settings where innovative, scalable, and acceptable solutions are most urgently needed.

Supplementary Materials: The following supporting information can be downloaded at: https://www.mdpi.com/article/10.3390/w17182754/s1. Figure S1. Comparing observed and model values for inhibition of emergence for silver nitrate (40 ppb Ag) and sodium hypochlorite (1 ppm free chlorine) treatments on older instar *Ae. aegypti* larvae on day 16. Figure S2. Comparing observed and model values for inhibition of emergence for copper sulfate (600 ppb Cu) and sodium hypochlorite (1 ppm free chlorine) treated larvae in regards to untreated larvae on day 16. Figure S3. Comparing observed and model values for inhibition of emergence for silver nitrate (40 ppb Ag), copper sulfate (600 ppb Cu) and sodium hypochlorite (1 ppm free chlorine) treated older *Ae. aegypti* larvae in regards to untreated larvae on day 16. Figure S4. Comparing observed and model values for survival for copper sulfate (600 ppb Cu) and silver nitrate (40 ppb Ag) treated larvae in regards to untreated larvae at 72 h. Data is corrected with Abbot's formula (1925). Figure S5. Comparing observed and model values for survival for silver nitrate (40 ppb Ag) and sodium hypochlorite (1 ppm Cl) treated larvae in regards to untreated larvae at 72 h. Data is corrected with Abbot's formula (1925). Figure S6. Comparing observed and model values for survival for copper sulfate (600 ppb Cu) and sodium hypochlorite (1 ppm free chlorine) treated larvae in regards to untreated larvae at 72 h. Data is corrected with Abbot's formula (1925).

Author Contributions: Conceptualization, S.S.T. and J.A.S.; Data curation, S.S.T. and P.I.H.; Formal analysis, S.S.T., K.B., P.I.H., S.L.H., V.C., L.M.B., J.D. and C.F.; Funding acquisition, S.S.T.; Investigation, S.S.T., K.B., S.L.H., V.C., L.M.B. and J.D.; Methodology, S.S.T., J.A.S., P.I.H. and C.F.; Project administration, S.S.T.; Resources, S.S.T.; Supervision, S.S.T.; Validation, S.S.T.; Visualization, S.S.T. and P.I.H.; Writing—original draft, S.S.T.; Writing—review & editing, S.S.T. and J.A.S. All authors have read and agreed to the published version of the manuscript.

Funding: Sydney Turner received funding from the Graduate Assistance in Areas of National Need (GAAN) Program; American Association of University Women (AAUW); and the Metropolitan Washington Chapter of Achievement Rewards for College Scientists Foundation to complete her doctoral studies.

Data Availability Statement: The original contributions presented in this study are included in the article. Further inquiries can be directed to the corresponding author.

Acknowledgments: Thank you to Jose G. Juarez for the support and guidance throughout the study.

Conflicts of Interest: The authors declare no conflicts of interest.

References

1. World Health Organization (WHO). Drinking-Water. Available online: https://www.who.int/news-room/fact-sheets/detail/drinking-water (accessed on 30 June 2025).
2. World Health Organization (WHO). *Global Vector Control Response 2017–2030*; World Health Organization: Geneva, Switzerland, 2017; ISBN 978-92-4-151297-8.
3. Torto, B.; Tchouassi, D.P. Grand Challenges in Vector-Borne Disease Control Targeting Vectors. *Front. Trop. Dis.* **2021**, *1*, 635356. [CrossRef]
4. Chilakam, N.; Lakshminarayanan, V.; Keremutt, S.; Rajendran, A.; Thunga, G.; Poojari, P.G.; Rashid, M.; Mukherjee, N.; Bhattacharya, P.; John, D. Economic Burden of Mosquito-Borne Diseases in Low- and Middle-Income Countries: Protocol for a Systematic Review. *JMIR Res. Protoc.* **2023**, *12*, e50985. [CrossRef]
5. Chapter 2—Mosquito-Borne Diseases. In *Zika Virus Disease*; Qureshi, A.I., Ed.; Academic Press: Cambridge, MA, USA, 2018; pp. 27–45. ISBN 978-0-12-812365-2.
6. Overgaard, H.J.; Dada, N.; Lenhart, A.; Stenström, T.A.B.; Alexander, N. Integrated Disease Management: Arboviral Infections and Waterborne Diarrhoea. *Bull. World Health Organ.* **2021**, *99*, 583–592. [CrossRef] [PubMed]
7. Akanda, A.S.; Johnson, K.; Ginsberg, H.S.; Couret, J. Prioritizing Water Security in the Management of Vector-Borne Diseases: Lessons From Oaxaca, Mexico. *GeoHealth* **2020**, *4*, e2019GH000201. [CrossRef]

8. Venkataramanan, V.; Collins, S.M.; Clark, K.A.; Yeam, J.; Nowakowski, V.G.; Young, S.L. Coping Strategies for Individual and Household-Level Water Insecurity: A Systematic Review. *WIREs Water* **2020**, *7*, e1477. [CrossRef]
9. Barrera, R.; Avila, J.; González-Téllez, S. Unreliable Supply of Potable Water and Elevated *Aedes aegypti* Larval Indices: A Causal Relationship? *J. Am. Mosq. Control Assoc.* **1993**, *9*, 189–195.
10. Turner, S.S.; Smith, J.A.; Howle, S.L.; Hancock, P.I.; Brett, K.; Davis, J.; Bruno, L.M.; Cecchetti, V.; Ford, C. Analyzing the Efficacy of Water Treatment Disinfectants as Vector Control: The Larvicidal Effects of Silver Nitrate, Copper Sulfate Pentahydrate, and Sodium Hypochlorite on Juvenile *Aedes aegypti*. *Water* **2025**, *17*, 348. [CrossRef]
11. Padmanabha, H.; Soto, E.; Mosquera, M.; Lord, C.C.; Lounibos, L.P. Ecological Links between Water Storage Behaviors and *Aedes aegypti* Production: Implications for Dengue Vector Control in Variable Climates. *EcoHealth* **2010**, *7*, 78–90. [CrossRef]
12. Wright, J.; Gundry, S.; Conroy, R. Household Drinking Water in Developing Countries: A Systematic Review of Microbiological Contamination between Source and Point-of-Use. *Trop. Med. Int. Health TMIH* **2004**, *9*, 106–117. [CrossRef]
13. Hoekstra, A.Y.; Buurman, J.; Ginkel, K.C.H. van Urban Water Security: A Review. *Environ. Res. Lett.* **2018**, *13*, 053002. [CrossRef]
14. Deshpande, A.; Miller-Petrie, M.K.; Lindstedt, P.A.; Baumann, M.M.; Johnson, K.B.; Blacker, B.F.; Abbastabar, H.; Abd-Allah, F.; Abdelalim, A.; Abdollahpour, I.; et al. Mapping Geographical Inequalities in Access to Drinking Water and Sanitation Facilities in Low-Income and Middle-Income Countries, 2000–2017. *Lancet Glob. Health* **2020**, *8*, e1162–e1185. [CrossRef] [PubMed]
15. Delpla, I.; Jung, A.-V.; Baures, E.; Clement, M.; Thomas, O. Impacts of Climate Change on Surface Water Quality in Relation to Drinking Water Production. *Environ. Int.* **2009**, *35*, 1225–1233. [CrossRef]
16. Falkenmark, M. The Greatest Water Problem: The Inability to Link Environmental Security, Water Security and Food Security. *Int. J. Water Resour. Dev.* **2001**, *17*, 539–554. [CrossRef]
17. United Nations-Water Summary Progress Update 2021: SDG 6—Water and Sanitation for All. Available online: https://www.unwater.org/publications/summary-progress-update-2021-sdg-6-water-and-sanitation-all (accessed on 10 May 2024).
18. Stoler, J.; Brewis, A.; Kangmennang, J.; Keough, S.B.; Pearson, A.L.; Rosinger, A.Y.; Stauber, C.; Stevenson, E.G. Connecting the Dots between Climate Change, Household Water Insecurity, and Migration. *Curr. Opin. Environ. Sustain.* **2021**, *51*, 36–41. [CrossRef]
19. Gosling, S.N.; Arnell, N.W. A Global Assessment of the Impact of Climate Change on Water Scarcity. *Clim. Change* **2016**, *134*, 371–385. [CrossRef]
20. Levy, B.S. Increasing Risks for Armed Conflict: Climate Change, Food and Water Insecurity, and Forced Displacement. *Int. J. Health Serv. Plan. Adm. Eval.* **2019**, *49*, 682–691. [CrossRef]
21. Hidalgo, J.P.; Boelens, R.; Vos, J. De-Colonizing Water. Dispossession, Water Insecurity, and Indigenous Claims for Resources, Authority, and Territory. *Water Hist.* **2017**, *9*, 67–85. [CrossRef]
22. Deitz, S.; Meehan, K. Plumbing Poverty: Mapping Hot Spots of Racial and Geographic Inequality in U.S. Household Water Insecurity. *Ann. Am. Assoc. Geogr.* **2019**, *109*, 1092–1109. [CrossRef]
23. Ezbakhe, F.; Giné-Garriga, R.; Pérez-Foguet, A. Leaving No One behind: Evaluating Access to Water, Sanitation and Hygiene for Vulnerable and Marginalized Groups. *Sci. Total Environ.* **2019**, *683*, 537–546. [CrossRef] [PubMed]
24. World Health Organization (WHO). *Results of Round II of the WHO Household Water Treatment Evaluation Scheme*; World Health Organization: Geneva, Switzerland, 2019; ISBN 978-92-4-151603-7.
25. Pinchoff, J.; Silva, M.; Spielman, K.; Hutchinson, P. Use of Effective Lids Reduces Presence of Mosquito Larvae in Household Water Storage Containers in Urban and Peri-Urban Zika Risk Areas of Guatemala, Honduras, and El Salvador. *Parasit. Vectors* **2021**, *14*, 167. [CrossRef] [PubMed]
26. Vannavong, N.; Seidu, R.; Stenström, T.-A.; Dada, N.; Overgaard, H.J. Effects of Socio-Demographic Characteristics and Household Water Management on *Aedes aegypti* Production in Suburban and Rural Villages in Laos and Thailand. *Parasit. Vectors* **2017**, *10*, 170. [CrossRef]
27. Ehdaie, B.; Rento, C.T.; Son, V.; Turner, S.S.; Samie, A.; Dillingham, R.A.; Smith, J.A. Evaluation of a Silver-Embedded Ceramic Tablet as a Primary and Secondary Point-of-Use Water Purification Technology in Limpopo Province, S. Africa. *PLoS ONE* **2017**, *12*, e0169502. [CrossRef]
28. Fewtrell, L.; Kaufmann, R.B.; Kay, D.; Enanoria, W.; Haller, L.; Colford, J.M. Water, Sanitation, and Hygiene Interventions to Reduce Diarrhoea in Less Developed Countries: A Systematic Review and Meta-Analysis. *Lancet Infect. Dis.* **2005**, *5*, 42–52. [CrossRef] [PubMed]
29. World Health Organization. *Managing Water in the Home Accelerated Health Gains from Improved Water Supply*; Sobsey, M.D., World Health Organization, Eds.; World Health Organization: Geneva, Switzerland, 2002.
30. National Sanitation Foundation. NSF Standards for Water Treatment Systems. Available online: https://www.nsf.org/consumer-resources/articles/standards-water-treatment-systems (accessed on 15 August 2025).
31. Pooi, C.K.; Ng, H.Y. Review of Low-Cost Point-of-Use Water Treatment Systems for Developing Communities. *Npj Clean Water* **2018**, *1*, 11. [CrossRef]

32. Clasen, T.F.; Alexander, K.T.; Sinclair, D.; Boisson, S.; Peletz, R.; Chang, H.H.; Majorin, F.; Cairncross, S. Interventions to Improve Water Quality for Preventing Diarrhoea. *Cochrane Database Syst. Rev.* **2015**, *2015*, CD004794. [CrossRef]
33. Nakada, L.Y.K.; Franco, R.M.B.; da Silva Fiuza, V.R.; dos Santos, L.U.; Branco, N.; Guimarães, J.R. Pre-Ozonation of Source Water: Assessment of Efficacy against Giardia Duodenalis Cysts and Effects on Natural Organic Matter. *Chemosphere* **2019**, *214*, 764–770. [CrossRef]
34. Rose, J.B.; Huffman, D.E.; Gennaccaro, A. Risk and Control of Waterborne Cryptosporidiosis. *FEMS Microbiol. Rev.* **2002**, *26*, 113–123. [CrossRef] [PubMed]
35. Korich, D.G.; Mead, J.R.; Madore, M.S.; Sinclair, N.A.; Sterling, C.R. Effects of Ozone, Chlorine Dioxide, Chlorine, and Monochloramine on Cryptosporidium Parvum Oocyst Viability. *Appl. Environ. Microbiol.* **1990**, *56*, 1423–1428. [CrossRef]
36. Sobsey, M.D. Inactivation of Health-Related Microorganisms in Water by Disinfection Processes. *Water Sci. Technol.* **1989**, *21*, 179–195. [CrossRef]
37. Mostafa, S.; Camacho-Galván, M.N.; Castelán-Martínez, H.E.; Galdos-Balzategui, A.; Reygadas, F. A Hybrid Centralized-Point-of-Use Drinking Water Treatment System in a Rural Community in Chiapas, Mexico. *Environ. Eng. Sci.* **2021**, *38*, 418–429. [CrossRef]
38. Bruchet, A.; Duguet, J.P. Role of Oxidants and Disinfectants on the Removal, Masking and Generation of Tastes and Odours. *Water Sci. Technol.* **2004**, *49*, 297–306. [CrossRef] [PubMed]
39. Li, X.-F.; Mitch, W.A. Drinking Water Disinfection Byproducts (DBPs) and Human Health Effects: Multidisciplinary Challenges and Opportunities. *Environ. Sci. Technol.* **2018**, *52*, 1681–1689. [CrossRef] [PubMed]
40. Wang, X.; Mao, Y.; Tang, S.; Yang, H.; Xie, Y.F. Disinfection Byproducts in Drinking Water and Regulatory Compliance: A Critical Review. *Front. Environ. Sci. Eng.* **2015**, *9*, 3–15. [CrossRef]
41. Estrella-You, A.; Harris, J.; Singh, R.; Smith, J. Inactivation of Waterborne Pathogens by Copper and Silver Ions, Free Chlorine, and N-Chloramines in Point-of-Use Technology: A Review. In *Water Purification: Processes, Applications and Health Effects*; LeBlanc, P., Ed.; Water Purification; Nova Science Publishers, Inc.: Hauppauge, NY, USA, 2022; pp. 1–88. ISBN 978-1-68507-622-1.
42. Singh, R.; Rento, C.; Son, V.; Turner, S.; Smith, J.A. Optimization of Silver Ion Release from Silver-Ceramic Porous Media for Household Level Water Purification. *Water* **2019**, *11*, 816. [CrossRef]
43. Pathak, S.P.; Gopal, K. Evaluation of Bactericidal Efficacy of Silver Ions on Escherichia Coli for Drinking Water Disinfection. *Environ. Sci. Pollut. Res.* **2012**, *19*, 2285–2290. [CrossRef] [PubMed]
44. Swathy, J.R.; Sankar, M.U.; Chaudhary, A.; Aigal, S.; Anshup; Pradeep, T. Antimicrobial Silver: An Unprecedented Anion Effect. *Sci. Rep.* **2014**, *4*, 7161. [CrossRef]
45. Cameron, P.; Gaiser, B.K.; Bhandari, B.; Bartley, P.M.; Katzer, F.; Bridle, H. Silver Nanoparticles Decrease the Viability of Cryptosporidium Parvum Oocysts. *Appl. Environ. Microbiol.* **2016**, *82*, 431–437. [CrossRef]
46. World Health Organization. *World Health Organization Guidelines for Drinking-Water Quality*, 4th ed.; Incorporating the 1st Addendum; World Health Organization: Geneva, Switzerland, 2017; ISBN 978-92-4-154995-0.
47. Sudha, V.B.P.; Ganesan, S.; Pazhani, G.P.; Ramamurthy, T.; Nair, G.B.; Venkatasubramanian, P. Storing Drinking-Water in Copper Pots Kills Contaminating Diarrhoeagenic Bacteria. *J. Health Popul. Nutr.* **2012**, *30*, 17–21. [CrossRef]
48. Vincent, M.; Hartemann, P.; Engels-Deutsch, M. Antimicrobial Applications of Copper. *Int. J. Hyg. Environ. Health* **2016**, *219*, 585–591. [CrossRef]
49. Patil, R.A.; Ahmad, D.; Kausley, S.B.; Balkunde, P.L.; Malhotra, C.P. A Compact Point-of-Use Water Purification Cartridge for Household Use in Developing Countries. *J. Water Health* **2014**, *13*, 91–102. [CrossRef]
50. Lambert, R.J.W.; Johnston, M.D.; Hanlon, G.W.; Denyer, S.P. Theory of Antimicrobial Combinations: Biocide Mixtures—Synergy or Addition? *J. Appl. Microbiol.* **2003**, *94*, 747–759. [CrossRef] [PubMed]
51. Estrella-You, A.; Smith, J.A. Synergistic Bacterial Inactivation by Silver Ions and Free Chlorine in Natural Waters. *J. Environ. Eng.* **2022**, *148*, 04022072. [CrossRef]
52. Soliman, M.Y.M.; Medema, G.; Bonilla, B.E.; Brouns, S.J.J.; van Halem, D. Inactivation of RNA and DNA Viruses in Water by Copper and Silver Ions and Their Synergistic Effect. *Water Res. X* **2020**, *9*, 100077. [CrossRef] [PubMed]
53. Landeen, L.K.; Yahya, M.T.; Gerba, C.P. Efficacy of Copper and Silver Ions and Reduced Levels of Free Chlorine in Inactivation of Legionella Pneumophila. *Appl. Environ. Microbiol.* **1989**, *55*, 3045–3050. [CrossRef]
54. Estrella-You, A.; Duti, I.J.; Luo, Q.; Harris, J.D.; Letteri, R.A.; Smith, J.A. N-Chloramine Functionalized Polymer Gels for Point-of-Use Water Disinfection. *Water* **2024**, *16*, 3128. [CrossRef]
55. United States Environmental Protection Agency (EPA) EPA Method 6020B (SW-846): Inductively Coupled Plasma—Mass Spectrometry. Available online: https://www.epa.gov/esam/epa-method-6020b-sw-846-inductively-coupled-plasma-mass-spectrometry (accessed on 17 August 2025).
56. World Health Organization. *World Health Organization Guidelines for Laboratory and Field Testing of Mosquito Larvicides*; World Health Organization: Geneva, Switzerland, 2005.

57. Ngonzi, A.J.; Muyaga, L.L.; Ngowo, H.; Urio, N.; Vianney, J.-M.; Lwetoijera, D.W. Susceptibility Status of Major Malaria Vectors to Novaluron, an Insect Growth Regulator South-Eastern Tanzania. *Pan Afr. Med. J.* **2022**, *41*, 273. [CrossRef]
58. Jacobs, E.; Chrissian, C.; Rankin-Turner, S.; Wear, M.; Camacho, E.; Broderick, N.A.; McMeniman, C.J.; Stark, R.E.; Casadevall, A. Cuticular Profiling of Insecticide Resistant *Aedes aegypti*. *Sci. Rep.* **2023**, *13*, 10154. [CrossRef] [PubMed]
59. Ren, Y.; Li, Y.; Ju, Y.; Zhang, W.; Wang, Y. Insect Cuticle and Insecticide Development. *Arch. Insect Biochem. Physiol.* **2023**, *114*, e22057. [CrossRef]
60. Mireji, P.O.; Keating, J.; Hassanali, A.; Impoinvil, D.E.; Mbogo, C.M.; Njeri, M.; Nyambaka, H.; Kenya, E.; Githure, J.I.; Beier, J.C. Expression of Metallothionein and α-Tubulin in Heavy Metal-Tolerant Anopheles Gambiae Sensu Stricto (Diptera: Culicidae). *Ecotoxicol. Environ. Saf.* **2010**, *73*, 46–50. [CrossRef]
61. Chain, F.J.J.; Finlayson, S.; Crease, T.; Cristescu, M. Variation in Transcriptional Responses to Copper Exposure across *Daphnia PULEX* Lineages. *Aquat. Toxicol.* **2019**, *210*, 85–97. [CrossRef]
62. Bossuyt, B.T.A.; Janssen, C.R. Acclimation of Daphnia Magna to Environmentally Realistic Copper Concentrations. *Comp. Biochem. Physiol. Toxicol. Pharmacol. CBP* **2003**, *136*, 253–264. [CrossRef] [PubMed]
63. Sobsey, M.D.; Stauber, C.E.; Casanova, L.M.; Brown, J.M.; Elliott, M.A. Point of Use Household Drinking Water Filtration: A Practical, Effective Solution for Providing Sustained Access to Safe Drinking Water in the Developing World. *Environ. Sci. Technol.* **2008**, *42*, 4261–4267. [CrossRef] [PubMed]
64. Lantagne, D.S.; Quick, R.; Mintz, E.D. *Household Water Treatment and Safe Storage Options in Developing Countries: A Review of Current Implementation Practices*; Woodrow Wilson International Center for Scholars: Washington, DC, USA, 2010.
65. Perez, M.H.; Noriega, F.G. *Aedes aegypti* Pharate 1st Instar Quiescence Affects Larval Fitness and Metal Tolerance. *J. Insect Physiol.* **2012**, *58*, 824–829. [CrossRef] [PubMed]

BioTech

MDPI

Article

Bioremediation Potential of a Non-Axenic Cyanobacterium *Synechococcus* sp. for Municipal Wastewater Treatment in the Peruvian Amazon: Growth Kinetics, Ammonium Removal, and Biochemical Characterization Within a Circular Bioeconomy Framework

Remy G. Cabezudo [1,*], Juan C. Castro [2,3,*], Carlos G. Castro [2], Hicler N. Rodriguez [2,3], Gabriela L. García [2], Paul M. Vizcarra [4], Carmen Ruiz-Huamán [1] and Marianela Cobos [2,3,*]

1 Environmental Engineering Career, Engineering Faculty, Universidad Peruana de Ciencias Aplicadas (UPC), Lima 15023, Peru; pcigcrui@upc.edu.pe
2 Specialized Unit of Biotechnology Research Laboratory (UELIB), Natural Resources Research Center of UNAP (CIRNA), Universidad Nacional de la Amazonía Peruana (UNAP), Iquitos 16001, Peru; castrocoboscarlosgilbertox13@gmail.com (C.G.C.); hicler.rodriguez@unapiquitos.edu.pe (H.N.R.); 2160831@unapiquitos.edu.pe (G.L.G.)
3 Academic Department of Biomedical Sciences and Biotechnology, Faculty of Biological Sciences, Universidad Nacional de la Amazonía Peruana (UNAP), Iquitos 16001, Peru
4 Specialized Unit of Soils Research Laboratory (UELIS), Natural Resources Research Center of UNAP (CIRNA), Universidad Nacional de la Amazonía Peruana (UNAP), Iquitos 16001, Peru; paul.vizcarra@unapiquitos.edu.pe
* Correspondence: u201916614@upc.edu.pe (R.G.C.); juan.castro@unapiquitos.edu.pe (J.C.C.); marianela.cobos@unapiquitos.edu.pe (M.C.)

Abstract: Effective wastewater management is critical for mitigating environmental and health impacts in ecologically sensitive regions like the Peruvian Amazon, where rapid urbanization has led to increased discharge of nutrient-rich effluents into freshwater systems. Conventional treatment methods often fail to address nutrient imbalances while generating secondary pollutants. This study aims to evaluate the bioremediation potential of a non-axenic cyanobacterium, *Synechococcus* sp., isolated from the Amazon Basin, for municipal wastewater treatment within a circular bioeconomy framework. The strain was cultivated in different concentrations of municipal wastewater (25%, 50%, 75%, 100%) from Moronacocha Lake in the Peruvian Amazon to assess growth kinetics, ammonium removal efficiency, and biochemical composition. The cyanobacterium exhibited optimal performance in 25% wastewater, achieving the highest specific growth rate (22.8×10^{-2} $\mu \cdot day^{-1}$) and biomass increase (393.2%), exceeding even the standard BG-11 medium. This treatment also demonstrated exceptional ammonium removal efficiency (95.4%) and enhanced phycocyanin production (33.6 µg/mg, 56% higher than the control). As wastewater concentration increased, both growth parameters and removal efficiency progressively declined. Biochemical analysis revealed that higher wastewater concentrations resulted in decreased protein content and increased lipid accumulation in the biomass. These findings demonstrate the dual potential of *Synechococcus* sp. for effective wastewater remediation and production of valuable biomass with modifiable biochemical characteristics, offering a sustainable approach for wastewater management in the Peruvian Amazon region.

Keywords: biodegradation; circular bioeconomy; environmental remediation; phycocyanin; wastewater

https://doi.org/10.3390/biotech14020036

Key Contribution: This study reveals that a non-axenic *Synechococcus* sp. strain can effectively remediate municipal wastewater while simultaneously producing valuable biomass with tailored biochemical properties, providing a sustainable solution for the Peruvian Amazon. The exceptional performance at 25% wastewater dilution, with ammonium removal efficiency of 95.4% and enhanced phycocyanin production, demonstrates the dual-purpose potential of this approach. By utilizing a locally isolated cyanobacterial strain adapted to tropical conditions, this research establishes a practical framework for integrating wastewater treatment with resource recovery. The findings contribute significantly to developing circular bioeconomy models for regions lacking conventional infrastructure, where biological resources can transform environmental challenges into economic opportunities while addressing critical water quality issues in sensitive Amazonian ecosystems.

1. Introduction

Effective wastewater management is critical for mitigating the environmental and health impacts of urbanization, particularly in ecologically sensitive regions, such as the Peruvian Amazon. Rapid population growth and inadequate infrastructure have exacerbated the discharge of nutrient-rich effluents into freshwater systems, contributing to eutrophication and loss of biodiversity [1,2]. Conventional wastewater treatments, such as anaerobic oxidation and chemical precipitation [3,4], often fail to address nitrogen and phosphorus imbalances while generating secondary pollutants [3,5,6]. This scenario has spurred interest in biotechnological approaches that align with circular bioeconomy principles by simultaneously addressing pollution control and resource recovery [7].

Phototrophic microorganisms, particularly cyanobacteria, have emerged as promising candidates for wastewater bioremediation due to their ability to assimilate nutrients while producing valuable biomass under relatively low-cost conditions [8–10]. Among cyanobacteria, members of the genus *Synechococcus* are particularly noteworthy for their metabolic versatility, environmental tolerance, and potential for industrial applications [11,12]. These unicellular organisms can effectively remove nitrogen and phosphorus from wastewater through direct uptake mechanisms, while simultaneously harvesting light energy to generate biomass rich in high-value compounds, including proteins, lipids, carbohydrates, and bioactive pigments [13,14].

The application of cyanobacteria for wastewater treatment represents a sustainable approach that transforms the conventional linear model of resource consumption into a circular system, where waste streams become valuable inputs for biomass production [15]. This paradigm shift aligns with the principles of a circular bioeconomy [7] and is particularly relevant for developing regions like the Peruvian Amazon. Here, the scarcity of conventional infrastructure is counterbalanced by the abundant sunlight and warm temperatures in the region, which create favorable conditions for phototrophic bioremediation systems.

Despite the growing body of research in cyanobacterial bioremediation, much of the existing literature has predominantly focused on axenic strains cultivated under controlled laboratory conditions. This limitation presents a significant gap in our understanding of the behavior and advantages of non-axenic cultures that are adapted to local environmental conditions, which may be more representative of practical applications in natural systems. Furthermore, the characterization of biomass composition in relation to wastewater cultivation is another area that remains underexplored. Comprehensive analyses are crucial, as the nutrient profiles of wastewater can significantly influence the metabolic pathways of cyanobacteria. Understanding these dynamics is essential for tailoring cyanobacterial biomass for downstream applications, including biofuel production, biofertilizers, and other bioproducts in a circular bioeconomy context [15,16].

The Peruvian Amazon region presents unique challenges for wastewater management due to its distinctive climatic conditions, remoteness, and sensitive ecosystems. Conventional treatment technologies often prove economically unfeasible or environmentally unsuitable for this region, creating an urgent need for alternative approaches that are both effective and contextually appropriate. The integration of cyanobacterial bioremediation systems could potentially address these challenges while supporting local economic development through the production of valuable biomass.

This study aims to evaluate the bioremediation potential of a non-axenic *Synechococcus* sp. strain isolated from the Peruvian Amazon for treating municipal wastewater at varying concentrations. Specifically, we investigate the growth profile, ammonium removal efficiency, biochemical composition, and pigment production of the cyanobacterium under different wastewater treatments. The findings are discussed within a circular bioeconomy framework that considers both environmental remediation and resource recovery, with particular attention to the contextual realities of the Peruvian Amazon region.

2. Materials and Methods

2.1. Study Area and Sample Collection

Municipal wastewater samples were collected from the primary effluent discharge point of Moronacocha Lake (Iquitos, Loreto Region, northeastern Peru; 3°44′41.89″ S, 73°15′56.65″ W), a tropical freshwater system that receives untreated domestic waste from an urban population of 479,866 (Figure S1). The sampling site was selected due to its role as a critical point of entry for municipal wastewater into the lake, with an estimated maximum daily discharge of 1000 m^3. Notably, approximately 40% of households in the adjacent urban areas lack formal wastewater infrastructure, contributing to the direct effluent released into the lake.

For the experimental assay, 10 L of a complex wastewater sample was collected from the effluent in pre-cleaned, acid-washed, high-density polyethylene containers at intervals of four hours (0, 4, 8, 12, 16, 20, 24, 28, 32, and 36 h) to ensure representative sampling. The samples were immediately placed on ice in the dark and transported to the laboratory within 30 min to minimize biological activity and photodegradation [17].

2.2. Sample Processing

Upon arrival, the wastewater samples were filtered through Whatman filter paper (grade 1) to remove the suspended solids. This step ensured particle-free effluent for downstream cyanobacterial inoculation, while preserving the dissolved nutrients critical for growth analysis.

2.3. Cyanobacterial Strain and Cultivation

The non-axenic cyanobacterium *Synechococcus* sp., sourced from the Peruvian Amazon Native Microalgae Culture Collection (maintained by UELIB-CIRNA, Iquitos, Peru), was selected for its documented adaptability to tropical freshwater ecosystems [18]. Stock cultures were maintained in sterile BG-11 medium (pH 7.4 ± 0.2) under controlled laboratory conditions for 4 weeks before experimentation.

2.4. Growth Conditions and Scaling

Cultures were incubated in a climate-controlled growth chamber (25.27 ± 0.06 °C) under a 12:12 h light–dark photoperiod. Illumination was provided by 50-W LED panels (Wellmax®, Shanghai, China) with Samsung LED chips, calibrated to deliver a photosynthetic photon flux density of 265 ± 10 μmol photons m^{-2} s^{-1} at the culture surface. To ensure homogeneous growth and gas exchange, the cultures were continuously aerated and agitated at 150 rpm using an orbital shaker (Thermo Scientific MaxQ 6000, Waltham, MA, USA).

A semi-continuous cultivation strategy was employed to scale biomass production: the initial 100 mL cultures were aseptically transferred to 500 mL and subsequently to 1 L vessels in a stepwise manner, with transfers timed to coincide with late-exponential growth phases (optical density at 730 nm [OD_{730}] ≈ 0.8–1.2). This scaling protocol minimized physiological stress while ensuring a sufficient cyanobacterial inoculum for subsequent bioremediation experiments.

2.5. Experimental Treatments

All bioremediation assays were conducted under the standardized culture conditions described previously (25.27 ± 0.06 °C, 12:12 h light–dark cycle, 265 ± 10 μmol photons m^{-2} s^{-1} continuous aeration, and agitation at 150 rpm) [6,19]. Four experimental groups (wastewater at 25, 50, 75, and 100%) and their respective control groups (without the addition of the cyanobacterial inoculum) were established in triplicate. Ultrapure water was used to dilute the wastewater to 25, 50, and 75%. A BG-11 medium control was also included, which provided a baseline for growth under optimized synthetic conditions.

Erlenmeyer flasks containing 200 mL of BG-11 medium or wastewater at four concentrations were inoculated with 2 mL of mid-exponential growth phase *Synechococcus* sp. culture (equivalent to 6.7 mg of dry cyanobacterial biomass). Daily monitoring over the 7-day experimental period tracked dry cyanobacterial biomass accumulation (μg/mL of culture) and ammonium concentration in the cultures (except in the culture with BG-11 medium)

2.6. Growth Kinetics

Cyanobacterial growth was monitored at 24 h intervals over the experimental period. One-milliliter culture aliquots were sampled and centrifuged (10,000× g for 5 min at 10 °C). The supernatants were transferred to 1.5 mL microtubes and used to determine the concentration of ammonium. Cyanobacterial biomass obtained after centrifugation was dried at 60 °C for 24 h, weighed, and recorded. Measurements were performed in triplicate to ensure reproducibility and expressed as micrograms of cyanobacterial dried biomass by milliliter of culture (μg of cdb/mL). Specific growth rates and percentage of cdb increases were estimated according to [20] using Equations (1) and (2), where X_0 and X_7 represent the cdb/mL at days 0 and 7, respectively:

$$\text{Specific growth rate } (\mu \cdot day^{-1}) = \frac{l_n\left(\frac{X_7}{X_0}\right)}{7} \quad (1)$$

$$\text{Percentage of cdb increase } (\%) = \frac{X_7 - X_0}{7} \times 100 \quad (2)$$

2.7. Ammonium Removal Efficiency

Ammonium concentration was evaluated in each experimental group and their respective control groups at the start (day 0) and every 24 h for seven days. Ammonium concentration was determined via the indophenol blue direct method [21,22], where ammonium ions react with hypochlorite and phenol to form a blue-green chromophore detectable at 640 nm. A standard calibration curve was constructed using analytical grade (>99.9% purity) ammonium chloride (NH_4Cl). The calibration curve was prepared using at least six concentration points, spanning the expected range of sample concentrations. The linearity of the calibration curve was validated with correlation coefficient (R^2) exceeding

0.995, ensuring reliable quantification (Figure S2). Quality control standards were analyzed every 10 samples to verify analytical accuracy throughout the analysis period.

The ammonium removal efficiency (RE) was determined using Equation (3), where S_0 and S_7 represent the ammonium concentrations in $\mu g \cdot mL^{-1}$ on days 0 and 7, respectively:

$$RE\ (\%) = \frac{S_0 - S_7}{S_0} \times 100 \quad (3)$$

2.8. Biochemical Characterization

At the end of the bioremediation experiments (day 7), the total cyanobacterial biomass was harvested by centrifugation and dried, as previously described. From five to 10 mg of cyanobacterial dried biomass (cdb) were used for biochemical analysis.

Ash, lipids, carbohydrates, and proteins were quantified according to the methods described by Cobos et al. [23]. Standard calibration curves for carbohydrate and protein quantification were constructed using analytical-grade (>99.9% purity) glucose and bovine serum albumin (Figure S2).

The photosynthetic pigment content (chlorophyll-a, carotenoids, and phycobiliproteins) was quantified spectrophotometrically using a modified protocol adapted from Cobos et al. [18]. Absorption spectra were recorded using a NanoDrop™ 2000 spectrophotometer (Thermo Fisher Scientific, Waltham, MA, USA) at wavelengths corresponding to pigment-specific absorption maxima: 440 nm and 680 nm for chlorophyll-a, 490 nm for carotenoids (combined β-carotene and xanthophyll absorption), 620 nm (c-phycocyanin, CPC), and 652 nm (allophycocyanin, APC) for phycobiliproteins. Phycocyanin concentrations were calculated using empirically derived Equations (4) and (5) to correct for the spectral overlap between CPC and APC, as defined by Bennett and Bogorad [24]:

$$CPC\ (mg\ mL^{-1}) = \frac{A_{620} - 0.474 \times A_{652}}{5.34} \quad (4)$$

$$APC\ (mg\ mL^{-1}) = \frac{A_{652} - 0.208 \times A_{620}}{5.09} \quad (5)$$

To verify and quantify carotenoids in the cyanobacterial biomass, we conducted spectrophotometric analysis using a chloroform solution of analytical-grade β-carotene (C4582, Sigma-Aldrich (St. Louis, MO, USA), >95% purity) at a concentration of 50 mg/mL as a reference standard (Figure S3). The absorption spectra of the extracted carotenoids from experimental samples were compared against this standard to confirm their presence and relative abundance.

For phycocyanin characterization, we performed partial purification of these pigment-proteins from 10 g of fresh cyanobacterial biomass harvested during mid-exponential growth phase. The purification process followed the protocol described by Patel et al. [25], which involved ammonium sulfate precipitation followed by dialysis (Figure S4). This procedure allowed us to isolate phycocyanins from the cellular matrix while maintaining their structural integrity for subsequent spectral analysis. The partially purified phycocyanin exhibited the characteristic absorption peaks, confirming the presence of these valuable pigment-proteins in the *Synechococcus* sp. culture.

2.9. Statistical Analysis

Statistical analyses were conducted to evaluate the differences in growth kinetics, ammonium removal efficiency, and biochemical composition across the experimental groups. Normality and homogeneity of variances were first assessed using the Anderson-Darling test ($\alpha = 0.05$) for all variables.

A two-way ANOVA (factors: treatment × time) with Tukey's post hoc test was applied to analyze differences in *Synechococcus* sp. growth profile over the 7-day experimental period. This model assesses both temporal dynamics and treatment-specific effects.

For endpoint measurements (day 7), parametric data meeting normality assumptions were analyzed using one-way analysis of variance (ANOVA) followed by Tukey's multi-comparison test. Non-parametric datasets were evaluated using the Kruskal–Wallis test with Dunn's post hoc correction. Pairwise comparisons between specific treatments were conducted using the Mann–Whitney U test with Bonferroni adjustment for family wise error rate control.

All analyses were performed using GraphPad Prism 10.0 (GraphPad Software, San Diego, CA, USA), with statistical significance defined at $\alpha = 0.05$. Results are reported as the average ± SD unless otherwise noted.

3. Results and Discussion

3.1. Growth Kinetics

The growth in *Synechococcus* sp. was successfully established across all tested wastewater concentrations, with quantifiable differences in growth kinetics and biomass accumulation between treatments (Figure 1). The specific growth rate ($\mu \cdot day^{-1}$) and percentage increase in biomass concentration revealed notable patterns that provide insight into the optimal conditions for cyanobacterial cultivation in municipal wastewater from the Peruvian Amazon region. *Synechococcus* sp. exhibited its highest specific growth rate (22.8×10^{-2} $\mu \cdot day^{-1}$) and greatest biomass increase (393.2%) in 25% wastewater, surpassing even the performance in the standard BG-11 medium (21.4×10^{-2} $\mu \cdot day^{-1}$, 347.9% increase).

A clear inverse relationship was observed between wastewater concentration and growth parameters. As wastewater concentration increased from 25% to 100%, both specific growth rate and percent biomass increase progressively declined. The 50% wastewater treatment maintained a growth rate (22.0×10^{-2} $\mu \cdot day^{-1}$) and biomass increase (365.7%) comparable to the BG-11 control, while the 75% and 100% treatments showed substantially reduced performance (17.6×10^{-2} $\mu \cdot day^{-1}$ with 243.2% increase, and 13.0×10^{-2} $\mu \cdot day^{-1}$ with 148.9% increase, respectively). This pattern suggests that while wastewater from Moronacoha Lake provides essential nutrients for cyanobacterial growth, certain components may exert concentration-dependent inhibitory effects, as similarly observed by several investigations [9,19,26,27]. However, these findings have important implications for practical applications, as they suggest that a fed-batch or semi-continuous cultivation system with appropriate dilution could optimize biomass productivity while simultaneously treating wastewater, as was demonstrated in recent publications [28–32].

Another potential strategy for practical implementation could involve a multi-stage cultivation approach as suggested by Sutherland et al. [33], where initial treatment uses diluted wastewater to maximize biomass productivity, followed by subsequent stages with progressively higher wastewater concentrations to achieve comprehensive nutrient removal while maintaining reasonable growth rates.

But it is important to take into account that the relationship between nutrient stoichiometry and biomass productivity has significant implications for large-scale applications. Anaerobically digested wastewater, characterized by high ammonium and orthophosphate content, could also serve as a cost-effective growth medium for *Synechococcus* cultivation when maintained below inhibitory thresholds [34]. Furthermore, wastewater nutrient ratios influence not only growth rates but also cellular stoichiometry, which directly impacts the nutrient cycling capacity of the cyanobacteria and bioremediation efficiency [35].

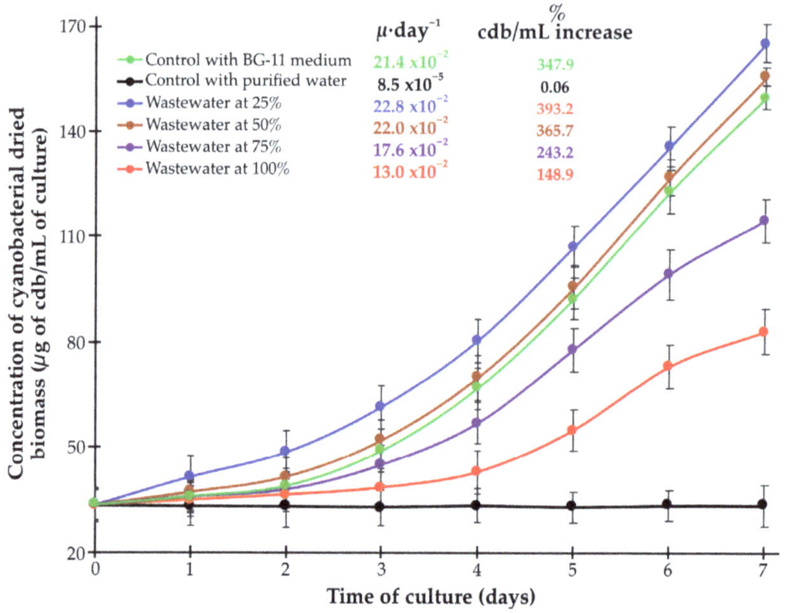

Figure 1. Growth kinetics of *Synechococcus* sp. cultivated in different concentrations of municipal wastewater over a 7-day period. The graph displays the concentration of dry cyanobacterial biomass (μg/mL of culture) for each treatment condition. Specific growth rates ($\mu \cdot day^{-1}$) and percentage of cyanobacterial dried biomass increase are indicated in the legend for each treatment. The highest growth performance was observed in 25% wastewater (22.8×10^{-2} $\mu \cdot day^{-1}$, 393.2% increase), exceeding even the standard BG-11 medium (21.4×10^{-2} $\mu \cdot day^{-1}$, 347.9% increase). Growth parameters progressively decreased with increasing wastewater concentration, while the purified water control showed minimal growth (8.5×10^{-5} $\mu \cdot day^{-1}$, 0.06% increase). Error bars represent standard deviation (n = 3).

3.2. Ammonium Removal Efficiency

The bioremediation capacity of *Synechococcus* sp. was evaluated by quantifying its ammonium removal efficiency in municipal wastewater across different dilution levels and corresponding initial ammonium concentrations (Figure 2). The cyanobacterial cultures exhibited consistent ammonium reduction throughout the 7-day experimental period, while control cultures maintained stable levels, confirming that the observed removal was biologically mediated.

In 25% wastewater treatment (initial ammonium = 162.3 ± 9.3 μg/mL), *Synechococcus* sp. demonstrated exceptional performance with a removal efficiency of 95.4 ± 2.8%, achieving near-complete ammonium removal (Figure 2A). This high efficiency coincided with optimal growth rates and biomass production (Figure 1), suggesting favorable conditions for nutrient assimilation and metabolic activity. At higher wastewater concentrations, removal efficiency declined proportionally. In 50% wastewater (324.3 ± 9.7 μg/mL), efficiency decreased to 39.1 ± 2.5% (Figure 2B). Further reductions occurred in 75% (14.6 ± 1.1%) and 100% of wastewater (11.7 ± 2.3%; Figure 2C,D), relating to growth inhibition patterns (Figure 1). This inverse relationship between wastewater concentration and, therefore, ammonium concentration and removal efficiency parallels the growth inhibition pattern observed in Figure 1, providing strong evidence that optimal bioremediation performance occurs under diluted conditions.

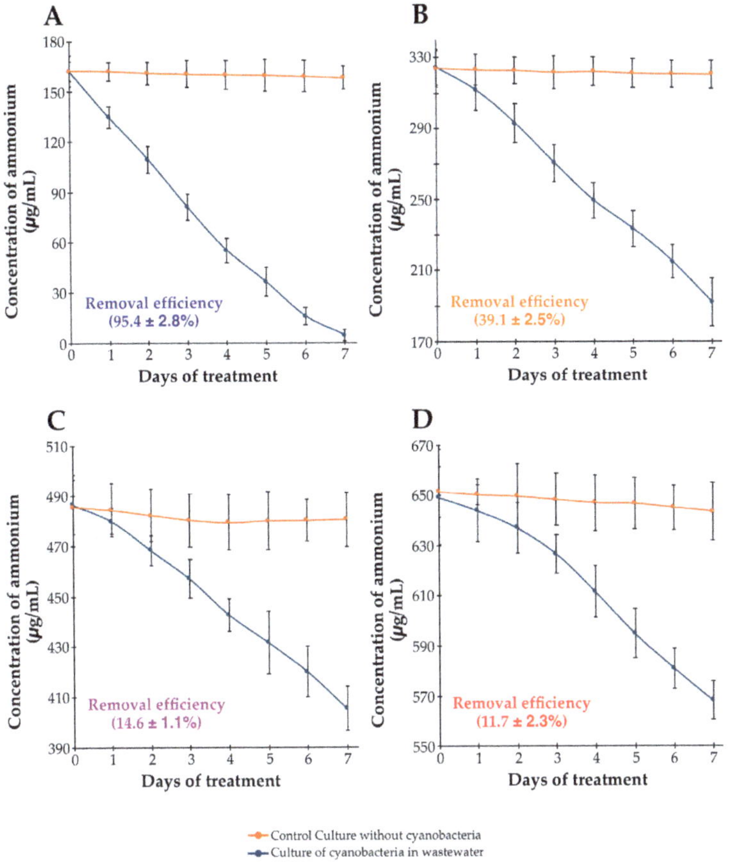

Figure 2. Ammonium removal by *Synechococcus* sp. in different wastewater concentrations over a 7-day treatment period. (**A**) 25% wastewater with initial ammonium concentration ~160 μg/mL, resulting in 95.4 ± 2.8% removal efficiency; (**B**) 50% wastewater with initial concentration ~330 μg/mL, resulting in 39.1 ± 2.5% removal efficiency; (**C**) 75% wastewater with initial concentration ~490 μg/mL, resulting in 14.6 ± 1.1% removal efficiency; (**D**) 100% wastewater with initial concentration ~650 μg/mL, resulting in 11.7 ± 2.3% removal efficiency. Red lines represent control cultures without cyanobacteria; blue lines represent cultures with cyanobacteria. Error bars represent standard deviation (n = 3). Note the inverse relationship between wastewater concentration and removal efficiency, with optimal performance at the 25% wastewater dilution.

The removal of ammonium in non-axenic cyanobacterial cultures appears to be governed by a complex interplay of direct assimilation and bacterial nitrification processes. Cyanobacteria are known to directly assimilate ammonium into their biomass, a process that has been observed to function in synergy with nitrifying bacteria that oxidize ammonium to nitrite and nitrate [36,37]. It was demonstrated that microbial cultures in municipal wastewater can exhibit high rates of both ammonium assimilation and nitrification, emphasizing the potential role of associated bacteria in these mixed communities [36].

The reduced efficiency of ammonium removal at higher substrate concentrations may be attributed to substrate inhibition and physiological stress imposed by co-occurring wastewater pollutants. Under high ammonium conditions, cyanobacteria often exhibit altered photophysiological behavior to mitigate the risk of photodamage by modulating their light-harvesting capacities and protective pigment composition [38]. When cyanobacteria are supplied with ammonium as the primary nitrogen source, there is a pronounced shift in metabolic pathways geared toward photoprotection, including modifications in pigment synthesis and energy dissipation mechanisms. Such changes can be directly correlated with declines in growth rates, as the reallocative stress responses impede the normal assimilation processes [38].

Additionally, the interplay between ammonium concentration and other wastewater components further complicates the metabolic balance within the culture. It was observed that ammonia-rich environments could trigger significant changes in the nitrogen sensing and assimilation pathways at the proteomic level, affecting both biomass production and nutrient removal efficiency [39]. This aligns with observations where higher loads of ammonium, possibly compounded by other contaminants, result in altered biochemical composition and pigment production, ultimately corroborating the hypothesis of substrate inhibition and stress-induced metabolic adjustments [38,39].

Together, these studies illustrate that while simultaneous direct assimilation and nitrification play crucial roles in ammonium removal within non-axenic cyanobacterial systems, environmental factors and the intricacies of intracellular metabolic regulation under high-ammonium conditions can lead to reduced treatment efficiency. The cumulative evidence emphasizes the need for a balanced nutrient load to maintain optimal metabolic function and highlights the complex dynamics that govern wastewater treatment processes in phototrophic cultures.

3.3. Biochemical Characterization

The wastewater composition significantly influenced the proximate biochemical profile of the harvested *Synechococcus* sp. biomass (Figure 3). The protein content, which constituted the largest fraction of biomass across all treatments, exhibited a notable decrease with increasing wastewater concentration, from 530.1 μg/mg in 25% wastewater to 373.3 μg/mg in 100% wastewater. This trend may reflect cellular responses to stress conditions at higher wastewater concentrations, as protein synthesis is often downregulated under various environmental stressors [40,41].

In contrast, lipid content displayed a marked increase with increasing wastewater concentration, reaching its maximum of 295.7 μg/mg in 100% wastewater, a 38% increase compared to the BG-11 control (214.5 μg/mg). This enhanced lipid accumulation represents a potentially valuable outcome from a biorefinery perspective, as lipid-rich biomass has applications in biofuel production and high-value nutraceuticals [23,42]. Carbohydrate content showed more complex variation, initially decreasing in 25% wastewater compared to the BG-11 control, but then increasing at higher wastewater concentrations. This non-linear response suggests complex metabolic adaptations to the changing nutrient profiles and potential stressors in the wastewater treatments. The inverse relationship between protein and lipid content aligns with observations by other researchers [2,43–47], who described the plasticity in the metabolism of proteins, lipids, and carbohydrates under nutrient stress and changing environmental conditions in various cyanobacterial and microalgal species.

The ash content observed in the dried biomass of cyanobacteria exhibited a significant increase correlated with wastewater concentration, rising from 43.6 μg/mg in the BG-11 control to 103.4 μg/mg in a 100% wastewater condition. This phenomenon can be attributed to the accumulation of inorganic ions derived from the wastewater matrix. In this context, the ash content in cyanobacteria cultured in different concentrations of wastewater can vary significantly due to the nutrient composition and concentration of the wastewater. Cyanobacteria, being efficient in nutrient uptake, can alter their biochemical composition, including ash content, based on the availability of nutrients and heavy metal content in the wastewater. Studies indicate that the concentration of wastewater influences the growth and biochemical composition of cyanobacteria, which in turn affects the ash content [48–50].

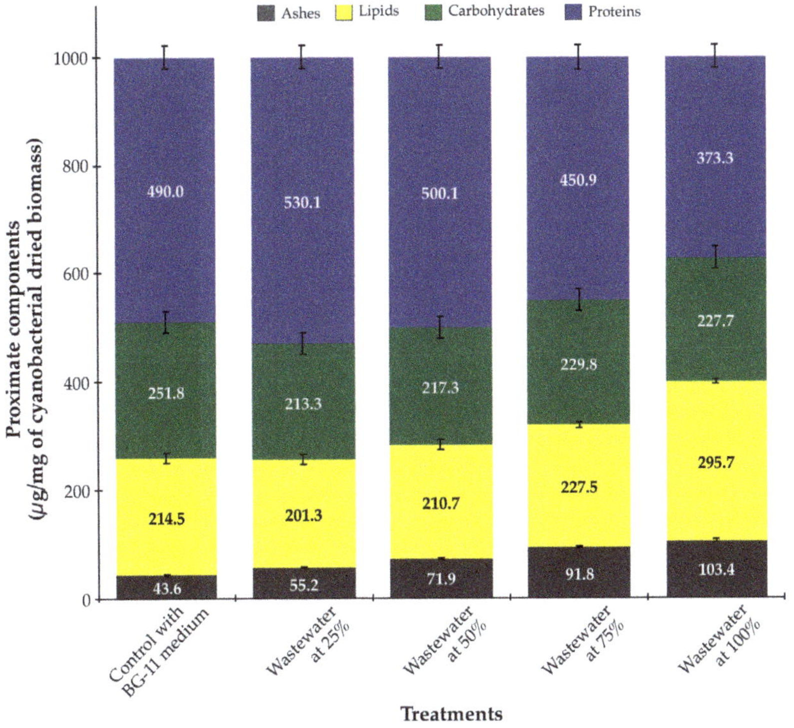

Figure 3. Proximate biochemical composition of *Synechococcus* sp. dried biomass cultivated in different wastewater concentrations. The stacked bars show the content of proteins (blue), carbohydrates (green), lipids (yellow), and ashes (gray) expressed as μg/mg of cyanobacterial dried biomass. Error bars represent standard deviation (n = 3). Note the inverse relationship between protein and lipid content with increasing wastewater concentration, with protein content decreasing and lipid content increasing at higher wastewater concentrations.

The spectrophotometric analysis and quantification of photosynthetic pigments revealed significant adaptations in the *Synechococcus* sp. pigment apparatus in response to wastewater cultivation (Figures 4 and S4). The most striking finding was the enhanced production of phycocyanin, a high-value blue protein pigment with applications in food coloring, cosmetics, and pharmaceuticals [51], in the 25% wastewater treatment, which reached 33.6 μg/mg, representing a 56% increase compared to the BG-11 control (21.5 μg/mg). The absorbance spectrum (Figures 4A and S4) corroborated the phycocyanin quantification results, with the 25% wastewater treatment exhibiting higher absorption peaks in the characteristic phycocyanin regions (550–650 nm). The visual color differences in the microtubes with extracted phycocyanins further confirmed these spectroscopic findings, with more intense blue coloration in the 25% wastewater culture compared to other treatments.

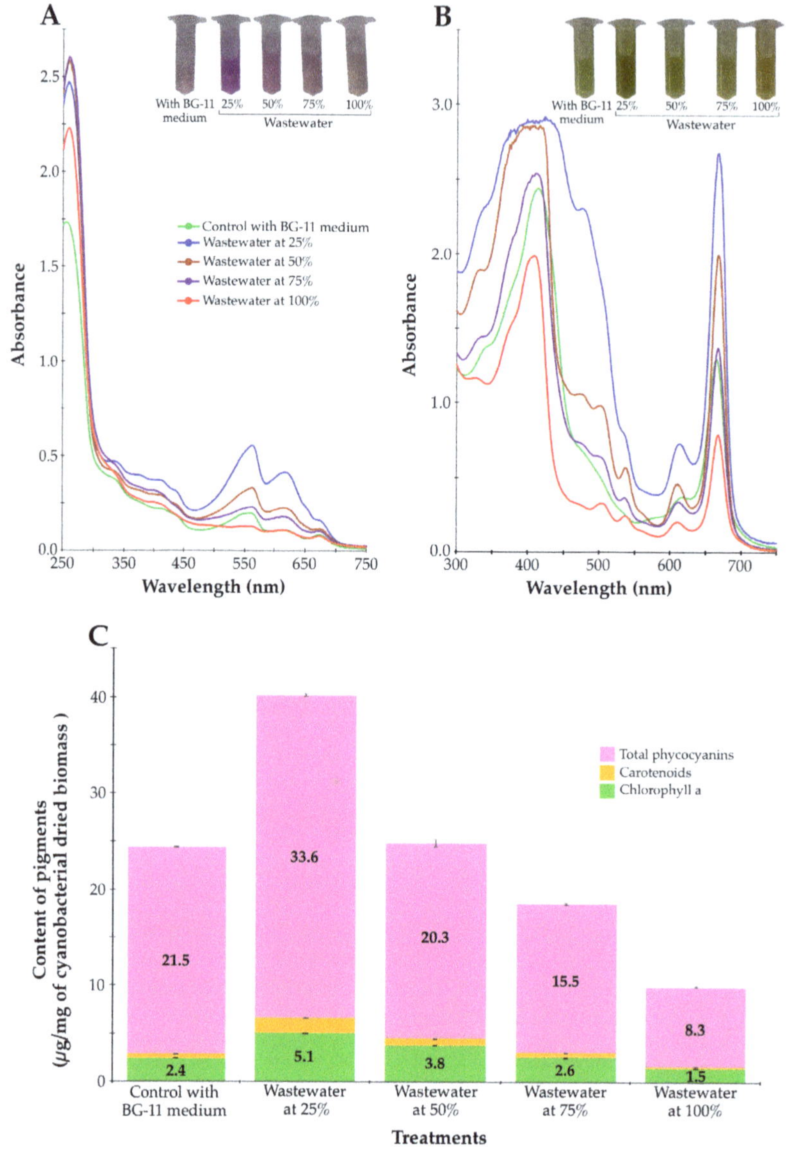

Figure 4. Spectral characteristics and pigment composition of *Synechococcus* sp. cultivated in different wastewater concentrations. (**A**) UV-visible absorption spectra (250–750 nm) with inset showing culture tube coloration across treatments. (**B**) Visible light absorption spectra (300–750 nm) highlighting peaks associated with photosynthetic pigments. (**C**) Content of photosynthetic pigments including total phycocyanins (purple), carotenoids (orange), and chlorophyll a (green) expressed as µg/mg of cyanobacterial dried biomass. Error bars represent standard deviation (n = 3). Note the enhanced phycocyanin production in 25% wastewater treatment compared to other conditions.

However, all pigments (phycocyanins, carotenoids, and chlorophyll a) showed progressive decreases as wastewater concentration increased beyond 25%, reaching minimum values in the 100% wastewater treatment (Figure 4A–C). Other investigations have also shown that the concentration of phycocyanin and other photosynthetic pigments in cyanobacteria cultures can vary based on the concentration of wastewater used in the culture medium. This variation is influenced by the availability of nutrients, particularly nitrogen and phosphorus, which are abundant in wastewater. For instance, in cultures of *Synechococcus* sp., phycocyanin content was found to vary according to nitrogen levels, while carotenoid content remained stable regardless of wastewater composition [13]. Similarly, the presence of nutrients such as ammonium and phosphate in swine wastewater supports the growth and influences the phycocyanin production in *Thermosynechococcus* sp. CL-1 [52].

The observed pattern indicates that moderate dilution of wastewater (25%) creates optimal conditions for the biosynthesis of pigments in the cyanobacteria *Synechococcus* sp. At this dilution level, the balance of nutrients appears to favor pigment production, enhancing the metabolic pathways associated with pigment biosynthesis. However, as the concentration of wastewater increases beyond this moderate level, the cyanobacteria may experience stress conditions that lead to a redirection of metabolic resources. Elevated concentrations of certain toxic substances or nutrient imbalances can trigger defensive strategies in cyanobacteria, resulting in reduced pigment synthesis. This shift in metabolic focus is well-documented; for instance, Allen and Smith noted that cyanobacteria under nitrogen-stressed conditions may tap into pigment reserves and reutilize nitrogen-rich compounds to sustain growth, thereby compromising pigment production [53].

Additionally, toxic byproducts typically present in higher concentrations of wastewater can inhibit the metabolic pathways involved in pigment biosynthesis. Under high-stress conditions, such as those induced by nutrient limitations or toxic ion exposure, cyanobacteria prioritize growth and survival over the synthesis of secondary metabolites, including pigments [53]. Furthermore, the iron homeostasis and nutrient regulation in cyanobacterial systems are critical, indicating that disturbances in nutrient availability could further detract from optimal pigment production [54]. Thus, while moderate wastewater concentration fosters pigment biosynthesis through nutrient abundance, elevated levels may detrimentally affect cellular processes, illustrating a complex interplay between environmental stressors and metabolic allocation in cyanobacteria.

4. Conclusions

This study demonstrates the effective bioremediation potential of a non-axenic *Synechococcus* sp. strain for treating municipal wastewater from the Peruvian Amazon, while simultaneously producing valuable biomass with modifiable biochemical characteristics. The cyanobacterium exhibited robust growth across all wastewater concentrations tested, with performance in diluted wastewater (25%) exceeding even the standard BG-11 medium. This finding is particularly significant as it indicates that municipal wastewater can serve not merely as a substrate for remediation but as an optimized growth medium when appropriately diluted.

The substantial ammonium removal capacity demonstrated across various initial concentrations highlights the potential of this *Synechococcus* strain for addressing one of the primary nutrient concerns in municipal wastewater. Complete removal at lower concentrations and significant reductions at higher concentrations suggest that a multistage treatment approach could maximize remediation efficiency while producing biomass with different characteristics at each stage.

The modulation of biochemical composition in response to different wastewater concentrations presents opportunities for tailored biomass production targeted toward specific applications. The enhanced lipid accumulation at higher wastewater concentrations could be particularly valuable for biofuel applications, while the increased phycocyanin production in 25% wastewater opens possibilities for high-value pigment extraction within an integrated biorefinery concept.

The non-axenic nature of the cyanobacterial culture likely contributed to its robust performance across different wastewater conditions, suggesting that engineered consortia may offer advantages over axenic cultures for real-world applications. This approach aligns with emerging understanding that microbial community interactions can enhance resilience and functional performance in bioremediation systems.

Several limitations of this study should be acknowledged. The 7-day experimental period, while sufficient to demonstrate bioremediation potential, may not fully capture long-term performance and stability. Additionally, while ammonium removal was comprehensively evaluated, other important wastewater parameters such as phosphorus, organic carbon, and potential micropollutants warrant further investigation.

Future research should explore continuous or semi-continuous operational modes that better approximate real-world treatment scenarios. The optimization of hydraulic and solid retention times, along with investigations into seasonal variations and scaling considerations, would provide valuable insights for practical implementation. Furthermore, detailed economic and life cycle analyses would help quantify the sustainability benefits of the proposed approach within a circular bioeconomy framework.

In conclusion, this study provides compelling evidence for the potential of *Synechococcus* sp. to serve as a cornerstone organism in developing integrated wastewater treatment systems that simultaneously address environmental challenges while generating valuable bioproducts in the Peruvian Amazon region. The findings contribute to the growing body of knowledge supporting the transition from linear to circular approaches in wastewater management, particularly in regions where conventional infrastructure is limited but biological resources are abundant.

Supplementary Materials: The following supporting information can be downloaded at: https://www.mdpi.com/article/10.3390/biotech14020036/s1, Figure S1. Geographic location and aerial views of the municipal wastewater discharge point in Moronacocha Lake, Iquitos, Peru. Top: Location of Iquitos city in northeastern Peru and its position relative to the Amazon River. Middle: Satellite image of the urban area showing the wastewater discharge point (03°44′41.89″ S, 73°15′56.65″ W). Bottom: Detailed aerial view and ground-level photograph of the discharge infrastructure where wastewater samples were collected for this study; Figure S2. Standard calibration curves for analytical measurements used in this study. (A) Ammonium chloride calibration curve (indophenol blue method at 640 nm) showing linear relationship (R^2 = 0.9995) used for quantifying ammonium in wastewater treatments; (B) Bovine serum albumin calibration curve (R^2 = 0.9979) used for protein content determination in cyanobacterial biomass; (C) Glucose calibration curve (R^2 = 0.9978) used for carbohydrate content determination in cyanobacterial biomass. Each calibration curve demonstrates high linearity with correlation coefficients exceeding 0.995, ensuring reliable quantification of respective compounds. Linear regression equations and correlation coefficients (R^2) are shown for each calibration curve; Figure S3. Absorption spectrum of carotenoids extracted from *Synechococcus* sp. biomass. The spectrum shows a characteristic absorbance pattern of carotenoids with an absorption maximum around 330–340 nm and multiple smaller peaks corresponding to vibrational fine structure. This reference spectrum was used for identification and relative quantification of carotenoid content in cyanobacterial biomass grown under different wastewater treatments; Figure S4. Absorption spectra of phycobiliproteins extracted from *Synechococcus* sp. biomass. (A) Full spectral scan (200–800 nm) of partially purified phycocyanin showing characteristic peaks in the 500–650 nm region; (B) Detailed absorption profile of phycobiliproteins in the visible region (450–700 nm) with characteristic absorption maxima at 565 nm and 620 nm, corresponding to phycoerythrin and phycocyanin, respectively. These spectra were used for pigment identification and quantification in cyanobacterial biomass grown under different wastewater treatments.

Author Contributions: Conceptualization, Funding acquisition, Investigation, Project administration, Resources, Supervision, Formal analysis, Writing—review and editing, M.C. and J.C.C.; Investigation,

Methodology, Software, Validation, Writing—original draft, R.G.C.; Conceptualization, Data curation, Investigation, Writing—review and editing, C.R.-H.; Methodology, Software, Writing—review and editing, C.G.C.; Methodology, Writing—review and editing, Formal analysis, H.N.R.; Methodology, Validation, Writing—review and editing, G.L.G.; Methodology, Validation, Writing—review and editing, P.M.V. All authors have read and agreed to the published version of the manuscript.

Funding: This research was funded by two institutional grants: (1) Universidad Nacional de la Amazonía Peruana (UNAP) under Grant No. 0449-2024-UNAP (Rectoral Resolution), awarded through the Basic Research Project "Bioremediation of Wastewater with a Microalgal Consortium and Production of Biofertilizers: A Circular Bioeconomy Approach in Loreto."; and (2) Universidad Peruana de Ciencias Aplicadas (UPC) under Grant No. UPC-EXPOST-2025-1. The authors gratefully acknowledge the financial support provided by both institutions, which made this research possible.

Institutional Review Board Statement: Not applicable.

Informed Consent Statement: Not applicable.

Data Availability Statement: All relevant data are presented within the manuscript and Supplementary Materials. Additional raw data and detailed experimental protocols are available upon request from the corresponding author.

Acknowledgments: The authors gratefully acknowledge the support provided by the Universidad Nacional de la Amazonía Peruana (UNAP) through the Specialized Unit of Biotechnology Research Laboratory (UELIB), Natural Resources Research Center of UNAP (CIRNA) for providing access to laboratory facilities, specialized equipment, and technical support. We thank the laboratory technicians (Elsa Maveth Mendoza, Julio Rodrigo Vásquez, Maria José Panduro, Oscar Alexander Saldaña, Sergio Antonio Valles) and administrative staff for their valuable assistance throughout the experimental work. We also extend our appreciation to the Dirección de Investigación de la Universidad Peruana de Ciencias Aplicadas (UPC) for providing complementary infrastructure and administrative support that facilitated the completion of this research through the incentive UPC-EXPOST-2025-1. The dedication and professionalism of the technical and administrative personnel from both universities significantly contributed to the success of this study.

Conflicts of Interest: The authors declare no conflicts of interest.

References

1. UNESCO. *Water and Climate Change*; The United Nations World Water Development Report; UNESCO: Paris, France, 2020; ISBN 978-92-3-100371-4.
2. Sisman-Aydin, G.; Simsek, K. Municipal Wastewater Effects on the Performance of Nutrient Removal, and Lipid, Carbohydrate, and Protein Productivity of Blue-Green Algae *Chroococcus turgidus*. *Sustainability* **2022**, *14*, 17021. [CrossRef]
3. Verma, A.; Sharma, A.; Kumar, R.; Sharma, P. Nitrate Contamination in Groundwater and Associated Health Risk Assessment for Indo-Gangetic Plain, India. *Groundw. Sustain. Dev.* **2023**, *23*, 100978. [CrossRef]
4. Wang, Q.; Wang, X.; Hong, Y.; Liu, X.; Zhao, G.; Zhang, H.; Zhai, Q. Microalgae Cultivation in Domestic Wastewater for Wastewater Treatment and High Value-Added Production: Species Selection and Comparison. *Biochem. Eng. J.* **2022**, *185*, 108493. [CrossRef]
5. Ye, S.; Gao, L.; Zhao, J.; An, M.; Wu, H.; Li, M. Simultaneous Wastewater Treatment and Lipid Production by *Scenedesmus* sp. HXY2. *Bioresour. Technol.* **2020**, *302*, 122903. [CrossRef]
6. Wu, K.; Atasoy, M.; Zweers, H.; Rijnaarts, H.; Langenhoff, A.; Fernandes, T.V. Impact of Wastewater Characteristics on the Removal of Organic Micropollutants by *Chlorella sorokiniana*. *J. Hazard. Mater.* **2023**, *453*, 131451. [CrossRef] [PubMed]
7. Haider, M.N.; Liu, C.-G.; Tabish, T.A.; Balakrishnan, D.; Show, P.-L.; Qattan, S.Y.A.; Gull, M.; Mehmood, M.A. Resource Recovery of the Wastewater-Derived Nutrients into Algal Biomass Followed by Its Cascading Processing to Multiple Products in a Circular Bioeconomy Paradigm. *Fermentation* **2022**, *8*, 650. [CrossRef]
8. Aditya, L.; Mahlia, T.M.I.; Nguyen, L.N.; Vu, H.P.; Nghiem, L.D. Microalgae-Bacteria Consortium for Wastewater Treatment and Biomass Production. *Sci. Total Environ.* **2022**, *838*, 155871. [CrossRef]
9. Janpum, C.; Pombubpa, N.; Monshupanee, T.; Incharoensakdi, A.; In-Na, P. Advancement on Mixed Microalgal-Bacterial Cultivation Systems for Nitrogen and Phosphorus Recoveries from Wastewater to Promote Sustainable Bioeconomy. *J. Biotechnol.* **2022**, *360*, 198–210. [CrossRef]

10. Goyal, S.; Dhanker, R.; Hussain, T.; Ferreira, A.; Gouveia, L.; Kumar, K.; Mohamed, H.I. Modern Advancement in Biotechnological Applications for Wastewater Treatment through Microalgae: A Review. *Water. Air Soil Pollut.* **2023**, *234*, 417. [CrossRef]
11. Pishbin, M.; Sarrafzadeh, M.-H.; Faramarzi, M.A. Nitrate and Phosphate Removal Efficiency of *Synechococcus elongatus* Under Mixotrophic and Heterotrophic Conditions for Wastewater Treatment. *Iran. J. Sci. Technol. Trans. Civ. Eng.* **2021**, *45*, 1831–1843. [CrossRef]
12. De Farias Silva, C.E.; Bertucco, A.; Vieira, R.C.; Abud, A.K.D.S.; Silva Almeida, F.B.P.D. *Synechococcus* PCC 7002 to Produce a Carbohydrate-Rich Biomass Treating Urban Wastewater. *Biofuels* **2022**, *13*, 551–558. [CrossRef]
13. Bellver, M.; Ruales, E.; Díez-Montero, R.; Escolà Casas, M.; Matamoros, V.; Ferrer, I. Natural Pigments and Biogas Recovery from Cyanobacteria Grown in Treated Wastewater. Fate of Organic Microcontaminants. *Water Res.* **2025**, *273*, 123005. [CrossRef] [PubMed]
14. Mathimani, T.; Alshiekheid, M.A.; Sabour, A.; Le, T.; Xia, C. Appraising the Phycoremediation Potential of Cyanobacterial Strains *Phormidium* and *Oscillatoria* for Nutrient Removal from Textile Wastewater (TWW) and Synchronized Biodiesel Production from TWW-Tolerant Biomass. *Environ. Res.* **2024**, *241*, 117628. [CrossRef] [PubMed]
15. Prabha, S.; Vijay, A.K.; Paul, R.R.; George, B. Cyanobacterial Biorefinery: Towards Economic Feasibility through the Maximum Valorization of Biomass. *Sci. Total Environ.* **2022**, *814*, 152795. [CrossRef]
16. Satya, A.D.M.; Cheah, W.Y.; Yazdi, S.K.; Cheng, Y.-S.; Khoo, K.S.; Vo, D.-V.N.; Bui, X.D.; Vithanage, M.; Show, P.L. Progress on Microalgae Cultivation in Wastewater for Bioremediation and Circular Bioeconomy. *Environ. Res.* **2023**, *218*, 114948. [CrossRef]
17. Ummalyma, S.B.; Singh, A. Biomass Production and Phycoremediation of Microalgae Cultivated in Polluted River Water. *Bioresour. Technol.* **2022**, *351*, 126948. [CrossRef]
18. Cobos, M.; Condori, R.C.; Grandez, M.A.; Estela, S.L.; Del Aguila, M.T.; Castro, C.G.; Rodríguez, H.N.; Vargas, J.A.; Tresierra, A.B.; Barriga, L.A.; et al. Genomic Analysis and Biochemical Profiling of an Unaxenic Strain of *Synechococcus* sp. Isolated from the Peruvian Amazon Basin Region. *Front. Genet.* **2022**, *13*, 973324. [CrossRef]
19. Sousa, J.F.; Amaro, H.M.; Ribeirinho-Soares, S.; Esteves, A.F.; Salgado, E.M.; Nunes, O.C.; Pires, J.C.M. Native Microalgae-Bacteria Consortia: A Sustainable Approach for Effective Urban Wastewater Bioremediation and Disinfection. *Microorganisms* **2024**, *12*, 1421. [CrossRef]
20. Richmond, A.; Hu, Q. (Eds.) *Handbook of Microalgae Culture: Applied Phycology and Biotechnology*, 2nd ed.; Wiley Blackwell: Hoboken, NJ, USA, 2013; ISBN 978-0-470-67389-8.
21. American Public Health Association. *Standard Methods for the Examination of Water and Wastewater*; American Public Health Association: Washington, DC, USA, 2017; ISBN 978-0-87553-287-5.
22. Huizenga, J.R.; Teelken, A.W.; Tangerman, A.; De Jager, A.E.J.; Gips, C.H.; Jansen, P.L.M. Determination of Ammonia in Cerebrospinal Fluid Using the Indophenol Direct Method. *Mol. Chem. Neuropathol.* **1998**, *34*, 169–177. [CrossRef]
23. Cobos, M.; Paredes, J.D.; Maddox, J.D.; Vargas-Arana, G.; Flores, L.; Aguilar, C.P.; Marapara, J.L.; Castro, J.C. Isolation and Characterization of Native Microalgae from the Peruvian Amazon with Potential for Biodiesel Production. *Energies* **2017**, *10*, 224. [CrossRef]
24. Bennett, A.; Bogorad, L. Complementary Chromatic Adaptation in a Filamentous Blue-Green Alga. *J. Cell Biol.* **1973**, *58*, 419–435. [CrossRef] [PubMed]
25. Patel, A.; Mishra, S.; Pawar, R.; Ghosh, P.K. Purification and Characterization of C-Phycocyanin from Cyanobacterial Species of Marine and Freshwater Habitat. *Protein Expr. Purif.* **2005**, *40*, 248–255. [CrossRef]
26. Agawin, N.S.R.; Duarte, C.M.; Agustí, S. Response of Mediterranean *Synechococcus* Growth and Loss Rates to Experimental Nutrient Inputs. *Mar. Ecol. Prog. Ser.* **2000**, *206*, 97–106. [CrossRef]
27. Korosh, T.C.; Dutcher, A.; Pfleger, B.F.; McMahon, K.D. Inhibition of Cyanobacterial Growth on a Municipal Wastewater Sidestream Is Impacted by Temperature. *mSphere* **2018**, *3*, e00538-17. [CrossRef]
28. Solís-Salinas, C.E.; Patlán-Juárez, G.; Okoye, P.U.; Guillén-Garcés, A.; Sebastian, P.J.; Arias, D.M. Long-Term Semi-Continuous Production of Carbohydrate-Enriched Microalgae Biomass Cultivated in Low-Loaded Domestic Wastewater. *Sci. Total Environ.* **2021**, *798*, 149227. [CrossRef]
29. Bresaola, M.D.; Morocho-Jácome, A.L.; Matsudo, M.C.; de Carvalho, J.C.M. Semi-Continuous Process as a Promising Technique in *Ankistrodesmus braunii* Cultivation in Photobioreactor. *J. Appl. Phycol.* **2019**, *31*, 2197–2205. [CrossRef]
30. Álvarez, X.; Otero, A. Nutrient Removal from the Centrate of Anaerobic Digestion of High Ammonium Industrial Wastewater by a Semi-Continuous Culture of *Arthrospira* sp. and *Nostoc* sp. PCC 7413. *J. Appl. Phycol.* **2020**, *32*, 2785–2794. [CrossRef]
31. Markou, G. Fed-Batch Cultivation of *Arthrospira* and *Chlorella* in Ammonia-Rich Wastewater: Optimization of Nutrient Removal and Biomass Production. *Bioresour. Technol.* **2015**, *193*, 35–41. [CrossRef] [PubMed]
32. García-Cañedo, J.C.; Cristiani-Urbina, E.; Flores-Ortiz, C.M.; Ponce-Noyola, T.; Esparza-García, F.; Cañizares-Villanueva, R.O. Batch and Fed-Batch Culture of *Scenedesmus incrassatulus*: Effect over Biomass, Carotenoid Profile and Concentration, Photosynthetic Efficiency and Non-Photochemical Quenching. *Algal Res.* **2016**, *13*, 41–52. [CrossRef]

33. Sutherland, D.L.; Park, J.; Heubeck, S.; Ralph, P.J.; Craggs, R.J. Size Matters—Microalgae Production and Nutrient Removal in Wastewater Treatment High Rate Algal Ponds of Three Different Sizes. *Algal Res.* **2020**, *45*, 101734. [CrossRef]
34. Dutcher, A. Optimization of Growth Conditions for the Model Bacterium *Synechococcus* sp. PCC 7002 for Chemical Production Using Wastewater-Based Media. Master's Thesis, University of Wisconsin-Madison, Madison, WI, USA, 2016.
35. Garcia, N.S.; Bonachela, J.A.; Martiny, A.C. Interactions between Growth-Dependent Changes in Cell Size, Nutrient Supply and Cellular Elemental Stoichiometry of Marine *Synechococcus*. *ISME J.* **2016**, *10*, 2715–2724. [CrossRef]
36. Leong, W.H.; Kiatkittipong, K.; Kiatkittipong, W.; Cheng, Y.W.; Lam, M.K.; Shamsuddin, R.; Mohamad, M.; Lim, J.W. Comparative Performances of Microalgal-Bacterial Co-Cultivation to Bioremediate Synthetic and Municipal Wastewaters Whilst Producing Biodiesel Sustainably. *Processes* **2020**, *8*, 1427. [CrossRef]
37. Stauch-White, K.; Srinivasan, V.N.; Camilla Kuo-Dahab, W.; Park, C.; Butler, C.S. The Role of Inorganic Nitrogen in Successful Formation of Granular Biofilms for Wastewater Treatment That Support Cyanobacteria and Bacteria. *AMB Express* **2017**, *7*, 146. [CrossRef] [PubMed]
38. Yang, X.; Dong, W.; Liu, L.; Bi, Y.; Xu, W.; Wang, X. Uncovering the Differential Growth of *Microcystis aeruginosa* Cultivated under Nitrate and Ammonium from a Photophysiological Perspective. *ACS EST Water* **2023**, *3*, 1161–1171. [CrossRef]
39. Patel, A.K.; Huang, E.L.; Low-Décarie, E.; Lefsrud, M.G. Comparative Shotgun Proteomic Analysis of Wastewater-Cultured Microalgae: Nitrogen Sensing and Carbon Fixation for Growth and Nutrient Removal in *Chlamydomonas reinhardtii*. *J. Proteome Res.* **2015**, *14*, 3051–3067. [CrossRef]
40. Anand, J.; Arumugam, M. Enhanced Lipid Accumulation and Biomass Yield of *Scenedesmus quadricauda* Under Nitrogen Starved Condition. *Bioresour. Technol.* **2015**, *188*, 190–194. [CrossRef]
41. Breuer, G.; Lamers, P.P.; Martens, D.E.; Draaisma, R.B.; Wijffels, R.H. The Impact of Nitrogen Starvation on the Dynamics of Triacylglycerol Accumulation in Nine Microalgae Strains. *Bioresour. Technol.* **2012**, *124*, 217–226. [CrossRef]
42. Arguelles, E.D.; Laurena, A.C.; Monsalud, R.G.; Martinez-Goss, M.R. Fatty Acid Profile and Fuel-Derived Physico-Chemical Properties of Biodiesel Obtained from an Indigenous Green Microalga, *Desmodesmus* sp. (I-AU1), as Potential Source of Renewable Lipid and High Quality Biodiesel. *J. Appl. Phycol.* **2018**, *30*, 411–419. [CrossRef]
43. Rodolfi, L.; Chini Zittelli, G.; Bassi, N.; Padovani, G.; Biondi, N.; Bonini, G.; Tredici, M.R. Microalgae for Oil: Strain Selection, Induction of Lipid Synthesis and Outdoor Mass Cultivation in a Low-Cost Photobioreactor. *Biotechnol. Bioeng.* **2009**, *102*, 100–112. [CrossRef]
44. Castielli, O.; De la Cerda, B.; Navarro, J.A.; Hervás, M.; De la Rosa, M.A. Proteomic Analyses of the Response of Cyanobacteria to Different Stress Conditions. *FEBS Lett.* **2009**, *583*, 1753–1758. [CrossRef]
45. Xiong, W.; Cano, M.; Wang, B.; Douchi, D.; Yu, J. The Plasticity of Cyanobacterial Carbon Metabolism. *Curr. Opin. Chem. Biol.* **2017**, *41*, 12–19. [CrossRef] [PubMed]
46. Yalcin, D. Growth, Lipid Content, and Fatty Acid Profile of Freshwater Cyanobacteria *Dolichospermum affine* (Lemmermann) Wacklin, Hoffmann, & Komárek by Using Modified Nutrient Media. *Aquac. Int.* **2020**, *28*, 1371–1388. [CrossRef]
47. Ciebiada, M.; Kubiak, K.; Daroch, M. Modifying the Cyanobacterial Metabolism as a Key to Efficient Biopolymer Production in Photosynthetic Microorganisms. *Int. J. Mol. Sci.* **2020**, *21*, 7204. [CrossRef] [PubMed]
48. Singh, D.V.; Upadhyay, A.K.; Singh, R.; Singh, D.P. Eco-Friendly and Eco Technological Approaches in Treatment of Wastewater by Different Algae and Cyanobacteria. In *Algae and Sustainable Technologies*; CRC Press: Boca Raton, FL, USA, 2020; ISBN 978-1-003-00191-1.
49. Sánchez-Bayo, A.; Morales, V.; Rodríguez, R.; Vicente, G.; Bautista, L.F. Cultivation of Microalgae and Cyanobacteria: Effect of Operating Conditions on Growth and Biomass Composition. *Molecules* **2020**, *25*, 2834. [CrossRef]
50. El-Bestawy, E. Treatment of Mixed Domestic–Industrial Wastewater Using Cyanobacteria. *J. Ind. Microbiol. Biotechnol.* **2008**, *35*, 1503–1516. [CrossRef]
51. Eriksen, N.T. Production of Phycocyanin—A Pigment with Applications in Biology, Biotechnology, Foods and Medicine. *Appl. Microbiol. Biotechnol.* **2008**, *80*, 1–14. [CrossRef] [PubMed]
52. Winayu, B.N.R.; Ho, J.-Y.; Hsueh, H.-T.; Chu, H. Multifunctional *Thermosynechococcus* sp. CL-1 Cultivation in Swine Wastewater for Nutrients Utilization, CO_2 Fixation, and C-Phycocyanin Production. *J. Taiwan Inst. Chem. Eng.* **2025**, *166*, 105046. [CrossRef]
53. Erratt, K.J.; Creed, I.F.; Trick, C.G. Comparative Effects of Ammonium, Nitrate and Urea on Growth and Photosynthetic Efficiency of Three Bloom-Forming Cyanobacteria. *Freshw. Biol.* **2018**, *63*, 626–638. [CrossRef]
54. González, A.; Bes, M.T.; Valladares, A.; Peleato, M.L.; Fillat, M.F. FurA Is the Master Regulator of Iron Homeostasis and Modulates the Expression of Tetrapyrrole Biosynthesis Genes in *Anabaena* sp. PCC 7120. *Environ. Microbiol.* **2012**, *14*, 3175–3187. [CrossRef]

molecules

MDPI

Article

Synthesis of Amorphous MnFe@SBA Composites for Efficient Adsorptive Removal of Pb(II) and Sb(V) from Aqueous Solution

Zhou Shi, Aogui Zhu, Fan Chen, Yishu Cai * and Lin Deng *

Hunan Engineering Research Center of Water Security Technology and Application, College of Civil Engineering, Hunan University, Changsha 410082, China; shiz61@hnu.edu.cn (Z.S.); zhuaogui1@hnu.edu.cn (A.Z.); chenfan@cmie.cn (F.C.)

* Correspondence: yishu.cai@hnu.edu.cn (Y.C.); lindeng@hnu.edu.cn (L.D.)

Abstract: The extensive release of water contaminated with lead (Pb(II)) and antimony (Sb(V)) constitutes a serious threat to the human living environment and public health, necessitating immediate attention. In this study, a novel MnFe@SBA composite was synthesized using the hydrothermal method through the in situ growth of $MnFe_2O_4$ on SBA-15. The MnFe@SBA exhibits an amorphous structure with a high specific surface area of 405.9 m^2/g and pore sizes ranging from 2 to 10 nm. Adsorption experiments demonstrated that MnFe@SBA removed over 99% of Pb(II) and 80% of Sb(V) within 120 min at initial concentrations of 10 mg/L, whereas both $MnFe_2O_4$ and SBA-15 exhibited poor adsorption capacities. Additionally, the MnFe@SBA displayed excellent tolerance towards coexisting cations, including Na^+, K^+, Mg^{2+}, Ca^{2+}, Zn^{2+}, Ni^{2+}, and Cd^{2+}, as well as anions such as Cl^-, NO_3^-, CO_3^{2-}, and PO_4^{3-}. The adsorption behavior of Pb(II) onto MnFe@SBA was satisfactorily described by the pseudo-second-order kinetic model and the Freundlich isotherm, while the adsorption of Sb(V) was well-fitted by the pseudo-second-order kinetic model and the Langmuir isotherm. At 318 K, the maximum adsorption capacities of MnFe@SBA for Pb(II) and Sb(V) were determined to be 329.86 mg/g and 260.40 mg/g, respectively. Mechanistic studies indicated that the adsorption of Pb(II) and Sb(V) onto MnFe@SBA involved two primary steps: electrostatic attraction and complexation. In conclusion, the MnFe@SBA is anticipated to serve as an ideal candidate for efficient removal of Pb(II) and Sb(V) from contaminated water.

Keywords: SBA-15; $MnFe_2O_4$; Pb(II); Sb(V); adsorption mechanisms

1. Introduction

Lead (Pb(II)) is a stable heavy metal existing in the environment and is highly toxic to the human body even at very low concentrations [1,2]. Thus, the World Health Organization (WHO) sets a guideline limit of 10 μg/L for lead in drinking water [3]. Lead can affect the hematopoietic system by inhibiting the synthesis of hemoglobin [4], leading to anemia, and lead poisoning can have serious adverse effects on the nervous system, liver, kidney, and reproductive system [5]. Over the past half-century, more than 800,000 tons of lead have been released worldwide from the battery manufacturing, glass manufacturing, and cosmetics industries [6]. Antimony is another toxic nonessential element, existing primarily in the forms of Sb(III) and Sb(V). Its destructive effects on cellular enzyme activity have serious implications for human health [7]. As its use in the industrial production of flame retardants, batteries, catalysts, and other applications has increased, the concentration of antimony in water has escalated rapidly [8]. The discharge of antimony not only contaminates surface water but also affects sediments, groundwater aquifers, and soil [9]. Furthermore, Sb(III) is readily oxidized to Sb(V) under natural environmental conditions. These environmental and health threats make developing effective methods for removing Pb(II) and Sb(V) from water is an urgent priority.

Commonly used techniques for removing heavy metal ions from aqueous phases include chemical precipitation [10], electrochemical recovery [11], ion exchange [12], membrane separation [13], and adsorption [14]. Adsorption is extensively applied due to its simplicity, energy efficiency, environmental friendliness, and reusability [15]. Graphene oxide [16], chitosan [17], metal–organic frameworks [18], carbon nanotubes [19], and other adsorbents have been studied to remove heavy metal ions from water and wastewater.

SBA-15 is a mesoporous silicon-based material with controllable pore size and high specific surface area, making it suitable to be used as a carrier for adsorbents and catalysts [20,21]. The presence of hydroxyl groups on the inner surface and channels facilitates the adsorption of heavy metal ions by functional nanoparticles immobilized on SBA-15 [22]. Fereshte et al. successfully loaded Fe_3O_4 onto mesoporous SBA-15 to prevent nanoparticle agglomeration, resulting in an adsorbent with abundant guanidine and −OH groups, exhibiting maximum adsorption capacities of 344.82 mg/g for Pb(II) and 303.03 mg/g for Cu(II) [23]. Xing et al. incorporated bimetallic Fe/Ni into the six-hole structure of SBA-15 to achieve a remarkable removal efficiency of 97.62% for Cr(VI) [24]. He et al. uniformly dispersed Cu^0 onto SBA-15, and the obtained Cu^0-SBA-15 composite displayed impressive adsorption capacities toward gaseous iodine (954 mg/g) and liquid iodine (842 mg/g) [25].

The recyclability of magnetic materials from aqueous solutions using external magnets makes them highly promising for environmental remediation [26]. Cobalt ferrite ($CoFe_2O_4$) and manganese ferrite ($MnFe_2O_4$) are two typical magnetic nanomaterials commonly applied in catalysis due to their exceptional stability, moderate magnetic saturation, and convenient separation properties [27]. Nevertheless, due to their magnetic properties, nanoparticles of $CoFe_2O_4$ and $MnFe_2O_4$ tend to agglomerate during synthesis, leading to reduced specific surface areas and inadequate adsorption capacities [28]. The incorporation of magnetic nanoparticles into the pore structures of a mesoporous matrix can effectively mitigate agglomeration, thereby enhancing the adsorption capacity toward pollutants. Ren et al. synthesized porous spinel $MnFe_2O_4$ via sol–gel method, and the resulting adsorbent effectively removed Pb(II) and Cu(II) by forming complexes with carboxyl and hydroxyl groups on the surface of $MnFe_2O_4$ [29]. Verma et al. loaded $MnFe_2O_4$ onto graphene oxide through the hydrothermal method, and the as-synthesized composite achieved a remarkable removal efficiency of 99.99% for Pb(II). In the adsorption process, hydroxyl and carboxyl groups are the main functions [30]. Yang et al. constructed a Fe–Mn bimetallic composite, which demonstrated an impressive adsorption capacity of 95 mg/g for Sb(V), and the main mechanism of adsorption is the inner sphere complexation between $Sb(OH)_6^-$ and the exposed hydroxyl groups [31]. Although significant progress has been made in utilizing ferrite materials as adsorbents for the removal of heavy metal ions, limited research has been conducted on the construction of $MnFe_2O_4$ loaded SBA composite for adsorptive removal Pb(II) and Sb(V).

In this study, a $MnFe_2O_4$-loaded SBA composite (MnFe@SBA) with a high specific surface area and excellent adsorption capacity was prepared for the first time using a hydrothermal method. The morphology, crystal structure, and physicochemical properties of the MnFe@SBA were characterized using scanning electron microscopy (SEM), X-ray powder diffraction (XRD), Brunauer–Emmett–Teller (BET) analysis, X-ray photoelectron spectroscopy (XPS), and Fourier-transform infrared spectroscopy (FTIR). The effects of factors such as initial solution pH, adsorbent dosage, coexisting ions, and temperature on the adsorption of Pb(II) and Sb(V) were examined. The kinetics, isotherms, thermodynamics, and mechanisms of Pb(II) and Sb(V) adsorption on MnFe@SBA were also investigated.

2. Results and Discussion

2.1. Characterizations

The crystal structures of $MnFe_2O_4$, SBA-15, and MnFe@SBA-0.15 (denoted as MnFe@SBA for short) were determined based on their XRD patterns. As shown in Figure 1, the pure $MnFe_2O_4$ displays five distinct diffraction peaks at 2θ values of 29.59°, 34.88°, 42.40°, 56.01°, and 61.51°, corresponding to (202), (311), (400), (333), and (440) planes of standard $MnFe_2O_4$ (PDF#38-0430) [32]. SBA-15 exhibits a broad and faint peak within the 2θ range of 20° to 30°, which is consistent with the characteristic peak observed for amorphous silica [33]. Interstingly, no discernible peaks are observed in the XRD patterns of MnFe@SBA, indicating a significantly low degree of crystallization and confirming the amorphous nature of the as-synthesized composite. Moreover, we find that MnFe@SBA exhibits almost no magnetic characteristics, possibly due to the preparation of the material at lower temperatures.

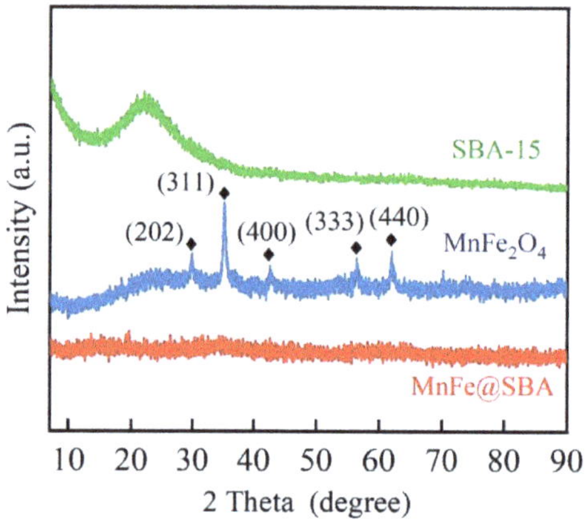

Figure 1. XRD patterns of SBA-15, $MnFe_2O_4$, and MnFe@SBA.

Figure 2a–c illustrates the surface morphologies of $MnFe_2O_4$, SBA-15, and MnFe@SBA. It is evident that SBA-15 exhibits an ordered channel structure formed by the accumulation of small tubes with lengths ranging from 1 to 2 μm [24]. $MnFe_2O_4$ showcases aggregated spherical nanoparticles measuring approximately 50 nm in size (Figure 2b) [34]. In contrast, the as-synthesized MnFe@SBA displays smaller nanoparticles, with the disappearance of the channel structure of SBA (Figure 2c). From Figure 2d, the elements of Si, O, Mn, and Fe uniformly distribute throughout the whole sample, and the relative mass ratios of O, Fe, Mn, and Si in MnFe@SBA are 41.4%, 29.6%, 14.2%, and 14.8%, respectively (Figure 2e).

To compare the pore structures of SBA-15, $MnFe_2O_4$, and MnFe@SBA, N_2 adsorption/desorption isotherms were performed, as shown in Figure 3. SBA-15 shows a typical type-IV isotherm with an H1 hysteresis loop at relative pressures ranging from 0.6 to 0.9 (Figure 3a), indicating its classification as a mesoporous material with a pore diameter of 8 nm, as observed from the inset figure [35]. As illustrated in Figure 3b, $MnFe_2O_4$ displays a type-III isotherm accompanied by an H3 hysteresis loop, suggesting enhanced interparticle interactions compared to those between nanoparticles and pollutants. In other words, this observation confirms the propensity of $MnFe_2O_4$ nanoparticles to agglomerate due to the existence of strong interactions. The average pore diameter of $MnFe_2O_4$ ranges from 20 to 30 nm. The as-synthesized MnFe@SBA composite demonstrates a type-IV isotherm with a noticeable H1 hysteresis loop observed between 0.5 and 0.9 (Figure 3c), indicating its mesoporous structure. The pore size distribution curves reveal that the pore diameter of MnFe@SBA falls within the range of 2–10 nm. At the same time, Table S1 further confirms the mesoporous structure of MnFe@SBA. The calculated specific surface area (S_{BET}) values follow the order: SBA-15 (537.3 m^2/g) > MnFe@SBA (405.9 m^2/g) > $MnFe_2O_4$ (67.9 m^2/g). Despite some pores in/on SBA-15 being occupied by $MnFe_2O_4$ nanoparticles during the synthesis process, MnFe@SBA maintains a significantly higher specific surface area than $MnFe_2O_4$, facilitating enhanced exposure of adsorption sites for pollutants.

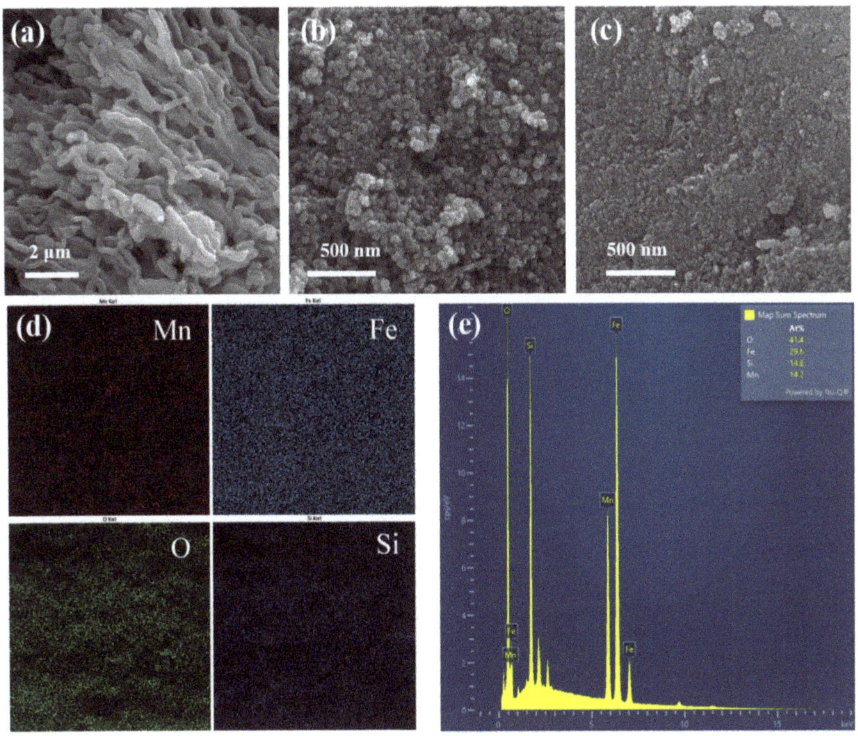

Figure 2. SEM images of (**a**) SBA-15, (**b**) $MnFe_2O_4$, and (**c**) MnFe@SBA; (**d**) EDX elemental mapping images and (**e**) corresponding map sum spectrum of MnFe@SBA.

The surface functional groups on $MnFe_2O_4$, SBA-15, and MnFe@SBA were identified via FTIR spectroscopy, as depicted in Figure 3d. A distinctive peak at a wavenumber of 525 cm^{-1} in the FTIR spectrum of $MnFe_2O_4$ is indicative of the presence of Mn–O and Fe–O bonds [36]. In the case of SBA-15, the peaks observed at 1065 and 803 cm^{-1} correspond to characteristic Si–O–Si peaks [33]. The FTIR spectrum of MnFe@SBA demonstrates these aforementioned characteristic peaks, confirming the successful loading of $MnFe_2O_4$ onto SBA-15. Additionally, a broad absorption band at 3424 cm^{-1} appears, corresponding to the O–H stretching vibration. The band at 1634 cm^{-1} can be assigned to Si–OH vibrations.

To distinguish the difference before and after SBA-15 loading, the chemical compositions and elemental valence states of $MnFe_2O_4$ and MnFe@SBA were further analyzed using XPS. In Figure 4a, the high-resolution O 1s spectrum of MnFe@SBA can be resolved into three individual peaks at binding energies of 530.4, 531.6, and 533.1 eV, which can be assigned to the lattice oxygen (O^{2-}), hydroxyl functional groups (–OH), and adsorbed water (H_2O) [37], with relative contents of 9.55%, 53.65%, 36.80%, respectively. The incorporation of $MnFe_2O_4$ with SBA-15 results in the decrease in O^{2-} content from 95.17% to 9.55%, along with the enhanced content in –OH to 53.65%. The increased amount of surface –OH could be beneficial for metal ions adsorption. The high-resolution Fe 2p spectrum (Figure 4b) exhibits fitted peaks at 711.3 and 713.9 eV, corresponding to the Fe(II) and Fe(III) orbitals in the form of their respective spin-orbit splitting components, Fe(II)-2p3/2 orbital and Fe(III)-2p3/2 orbital, as well as peaks at 724.6 and 727.6 eV attributed to the Fe(II) and Fe(III) orbitals in terms of their respective spin-orbit splitting components, Fe(II)-2p1/2 orbital and Fe(III)-2p1/2 orbital. Additionally, an oscillating satellite peak is also detectable. The relative proportions of Fe(II) and Fe(III) in $MnFe_2O_4$, as presented in Table S2, are 42.42% and 20.19%, respectively. In the MnFe@SBA sample, the proportion of Fe(III) increased to 47.55%, while the proportion of Fe(II) decreased to 26.04%. The deconvoluted Mn 2p spectrum in Figure 4c shows peaks at 641.5 and 652.7 eV, corresponding to Mn(II), as well as peaks at 643.2 and 654.5 eV, indicating the presence of Mn(IV) [38,39]. Additionally, a satellite peak is observed at 646.5 eV, which is usually caused by electron transitions related to the metal's d orbitals. As listed in Table S2, the Mn(II) contents in $MnFe_2O_4$ and MnFe@SBA are determined to be 58.68% and 12.32%, respectively, while approximately 50% of Mn(IV) is detected in MnFe@SBA. These findings confirm the successful loading of $MnFe_2O_4$ onto SBA-15 and the predominant form of Mn(IV) in MnFe@SBA.

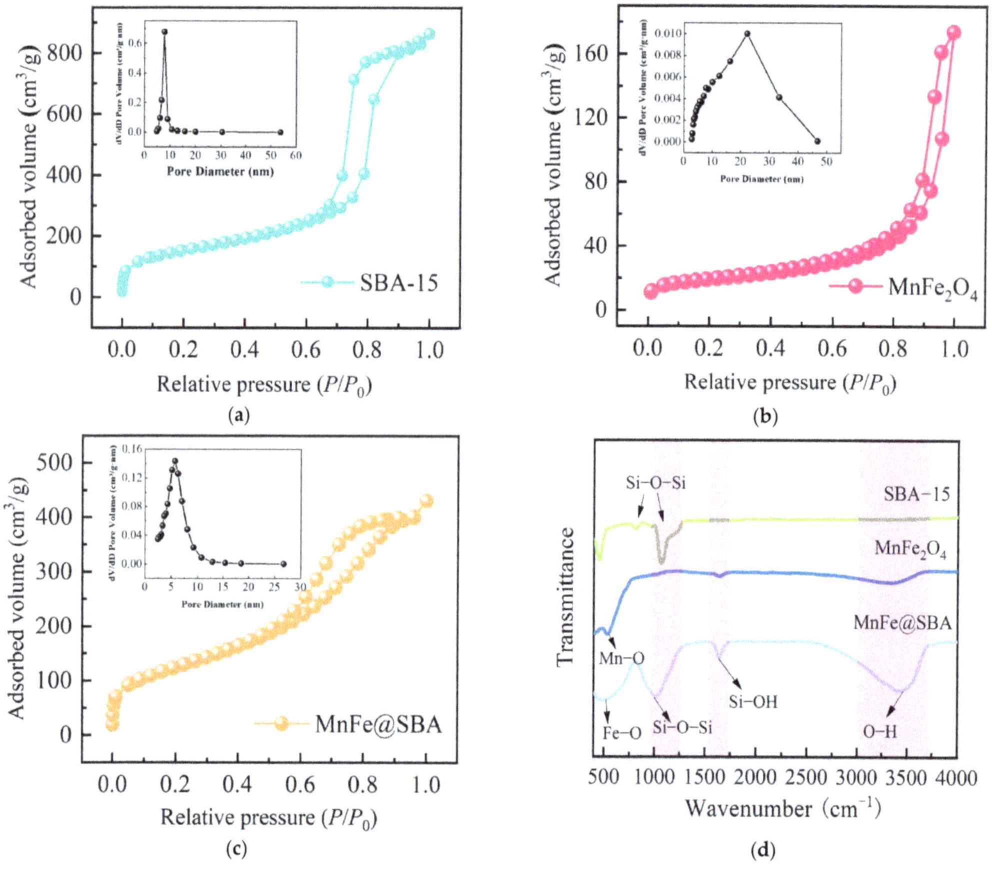

Figure 3. Nitrogen adsorption–desorption isotherms and pore size distributions of (**a**) SBA-15, (**b**) $MnFe_2O_4$, and (**c**) MnFe@SBA; (**d**) FT-IR spectrum of SBA-15, $MnFe_2O_4$, and MnFe@SBA.

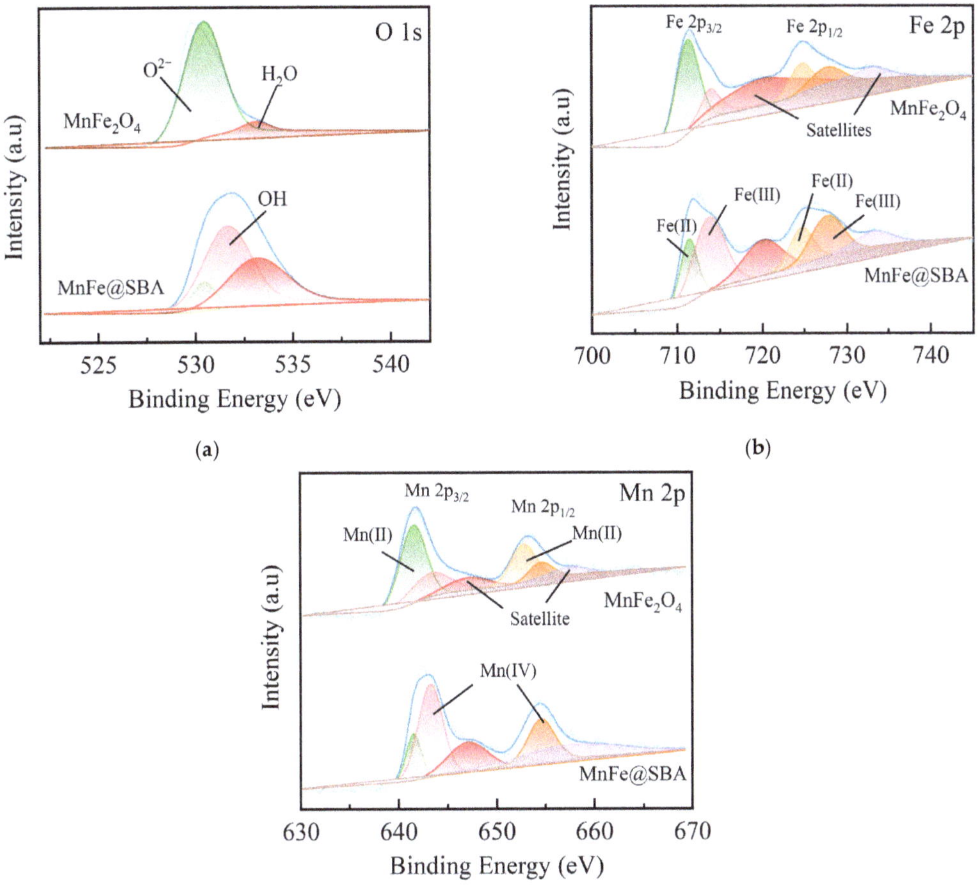

Figure 4. High-resolution XPS spectra: (**a**) O 1s; (**b**) Fe 2p; (**c**) Mn 2p for $MnFe_2O_4$ and MnFe@SBA.

2.2. Adsorption Performance

2.2.1. Effect of Doping Amount of SBA-15

The impact of SBA-15 loading amounts on the adsorption performance of MnFe@SBA-*x* for Pb(II) and Sb(V) was investigated. As shown in Figure 5a, both $MnFe_2O_4$ and SBA-15 exhibited negligible removal efficiencies, suggesting their limited adsorption capacity towards Pb(II). The mass loading of SBA-15 significantly influenced the adsorption capacity of MnFe@SBA-*x*. With increasing loading amount from 0.025 to 0.15 g, the removal efficiency of Pb(II) increased from 20% to 100% within 60 min. In Figure 5b, the as-synthesized MnFe@SBA-0.15 achieved the highest adsorption, exceeding 80% for Sb(V) in 120 min. The enhanced adsorption capacity observed upon incorporating SBA-15 can be attributed to the promotion of grain size reduction in the MnFe@SBA composite [40]. However, excessive loading of SBA-15 (0.2 g) resulted in a reduction in removal efficiency to approximately 30% due to the limited adsorption capacity of SBA-15. Therefore, MnFe@SBA-0.15 was selected for subsequent adsorption experiments.

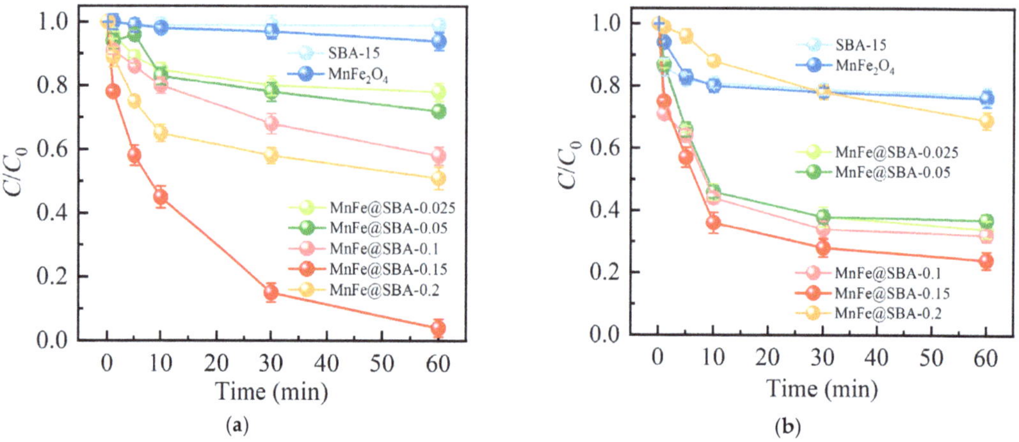

Figure 5. Effect of SBA-15 doping amount on the removal of (**a**) Pb(II) and (**b**) Sb(V).

2.2.2. Effect of Solution pH and Coexisting Cations

The initial solution pH was found to be a crucial factor influencing the effectiveness of adsorption, which can be explained by its effects on both the surface charge of the adsorbent and the speciation of metal ions in the solution [41]. Below pH 6.0, monovalent ionic Pb(II) is predominant in the solution, and small amounts of $Pb(OH)^+$ and $Pb(OH)_2$ are present at a solution pH of 6–8. $Pb(OH)_2$ is the major species when the solution pH ranges from 8 to 11, and $Pb(OH)_3^-$ is present at pH above 8 [42]. In this study, an initial solution pH range of 2–6 was studied to examine how it affected the adsorption of Pb(II) by MnFe@SBA (Figure 6a). The adsorption was found to be negligible within the pH range of 2.0 to 3.0 due to the pronounced acidity. The removal efficiency significantly improved to 63.8% when increasing the solution pH to 4. The zeta potential for MnFe@SBA at different pH values was also recorded in Figure 6c. The surface charge of MnFe@SBA was negative. The poor removal of Pb(II) at extremely acidic conditions was mainly attributed to the competitive effects caused by predominant H^+ ions competing with Pb(II) [25]. As the solution pH increased from 3 to 11, the potential dropped from −24.2 to −51.2 mV. Thus, the electrostatic attraction force existing between the negatively charged MnFe@SBA and positively charged Pb(II) improved, facilitating the adsorption of Pb(II).

The influence of initial solution pH (2–9) on the removal of Sb(V) was studied, as shown in Figure 6b. Obviously, acidic conditions are more favorable for Sb(V) adsorption. The removal showed a decreasing trend with a gradual increase in solution pH from 3 to 9. This phenomenon may be related to the fact that, at pH higher than 3, $Sb(OH)_6^-$ becomes the predominant species in solution, resulting in an increased electrostatic repulsion force between the negatively charged MnFe@SBA and $Sb(OH)_6^-$, thus inhibiting the adsorption.

Various metal ions and inorganic anions coexist in natural waters. In order to investigate the effectiveness of MnFe@SBA in practical applications, the effects of prevalent metal ions (Na^+, K^+, Mg^{2+}, and Ca^{2+}), as well as divalent heavy metal ions (Zn^{2+}, Ni^{2+}, and Cd^{2+}) on the adsorption of Pb(II) were examined as shown in Figure S2a–g. Additionally, the influences of commonly existing inorganic anions (Cl^-, NO_3^-, CO_3^{2-}, and PO_4^{3-}) on the adsorption of Sb(V) were studied, as shown in Figure S3a–d. It can be observed that Na^+, K^+, Mg^{2+}, and Ca^{2+} at concentrations of 50 and 100 mg/L had minimal impact on Pb(II) removal. And 100 mg/L Cd^{2+}, Ni^{2+}, and Zn^{2+} decreased the Pb(II) adsorption from 100% to 80.2%, 82.5%, and 83.2%, respectively [43]. Figure S3a–d demonstrates that high concentrations of Cl^-, NO_3^-, CO_3^{2-}, and PO_4^{3-} had little effect on the adsorption of Sb(V) by MnFe@SBA. These findings confirm the strong interference resistance of MnFe@SBA in practical application.

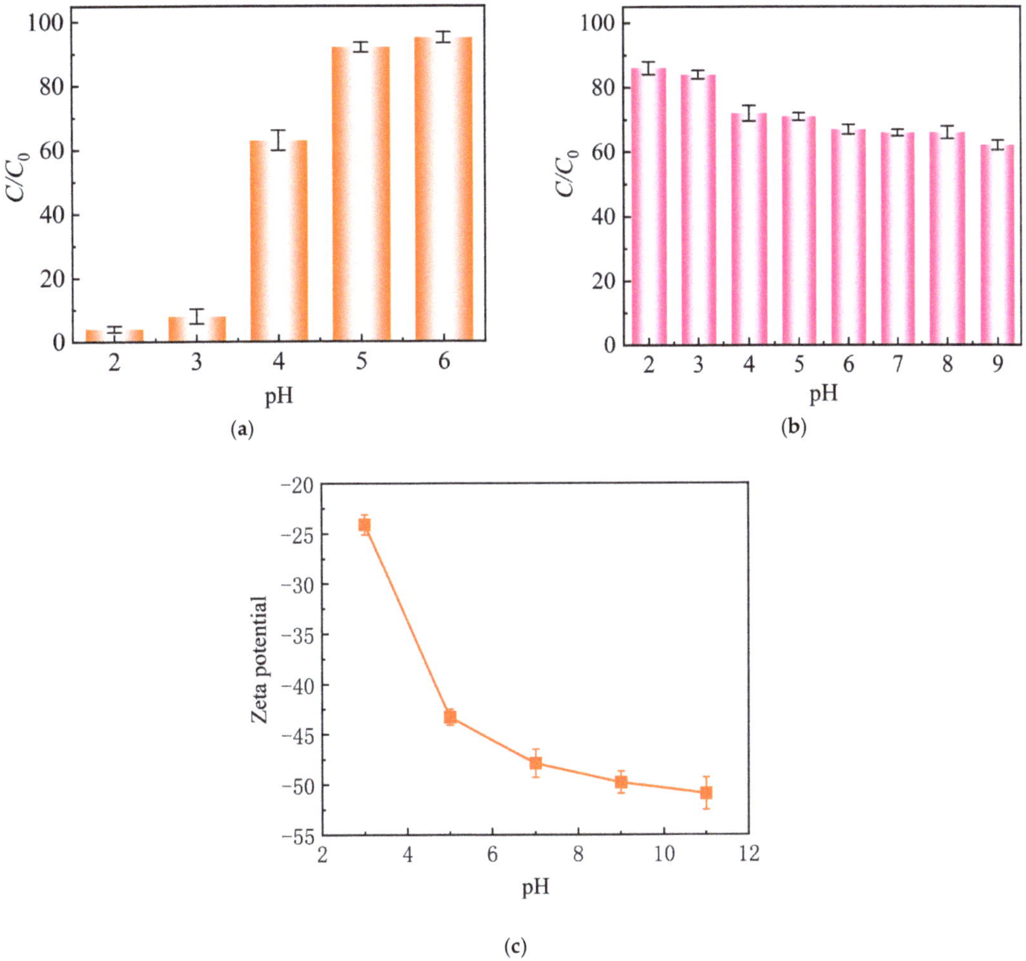

Figure 6. Effect of initial solution pH on removal of (**a**) Pb(II) and (**b**) Sb(V) by MnFe@SBA. (**c**) Zeta potential of MnFe@SBA at pH ranging from 3 to 11.

2.2.3. Adsorption Kinetics

The adsorption kinetics of Pb(II) and Sb(V) onto MnFe@SBA-15 were studied at initial concentrations of 5, 10, and 20 mg/L, as shown in Figure 7. A rapid adsorption process occurred within the initial 30 min, with over 80% of the Pb(II) being adsorbed by the adsorbent. Subsequently, the adsorption process gradually approached equilibrium. In the initial stage, abundant vacant sites were available, leading to a high rate of adsorption. As the reaction time exceeded 30 min, the adsorption rate decreased due to a reduced number of available sites for Pb(II) uptake. Once all the sites were occupied, the adsorption reached a dynamic equilibrium. Similarly, MnFe@SBA achieved more than 70% adsorption of Sb(V) within the initial 60 min. The adsorption behavior of Pb(II) and Sb(V) onto MnFe@SBA was modeled using pseudo-first-order and pseudo-second-order kinetic models, represented by the nonlinear Equations (1) and (2) [44]:

$$q_t = q_e\left(1 - e^{-k_1 t}\right) \quad (1)$$

$$q_t = \frac{k_2 q_e^2 t}{1 + k_2 q_e t} \quad (2)$$

where q_e and q_t are the adsorption capacity at equilibrium and at time t (min), respectively; k_1 (1/min) is the pseudo-first-order rate constant, and k_2 (g/(mg min)) is the pseudo-second-order rate constant. The nonlinear fitting parameters of the pseudo-first-order and pseudo-second-order kinetic models are summarized in Table 1. Based on the higher coefficients of determination (R^2), the adsorption behavior of Pb(II) and Sb(V) onto MnFe@SBA was better described by the pseudo-second-order kinetic model. Chemisorption was considered as the primary adsorption mechanism. However, the high R^2 values obtained from the pseudo-first-order kinetic model suggested that physical adsorption could not be entirely disregarded, and it might also contribute to the overall adsorption [45].

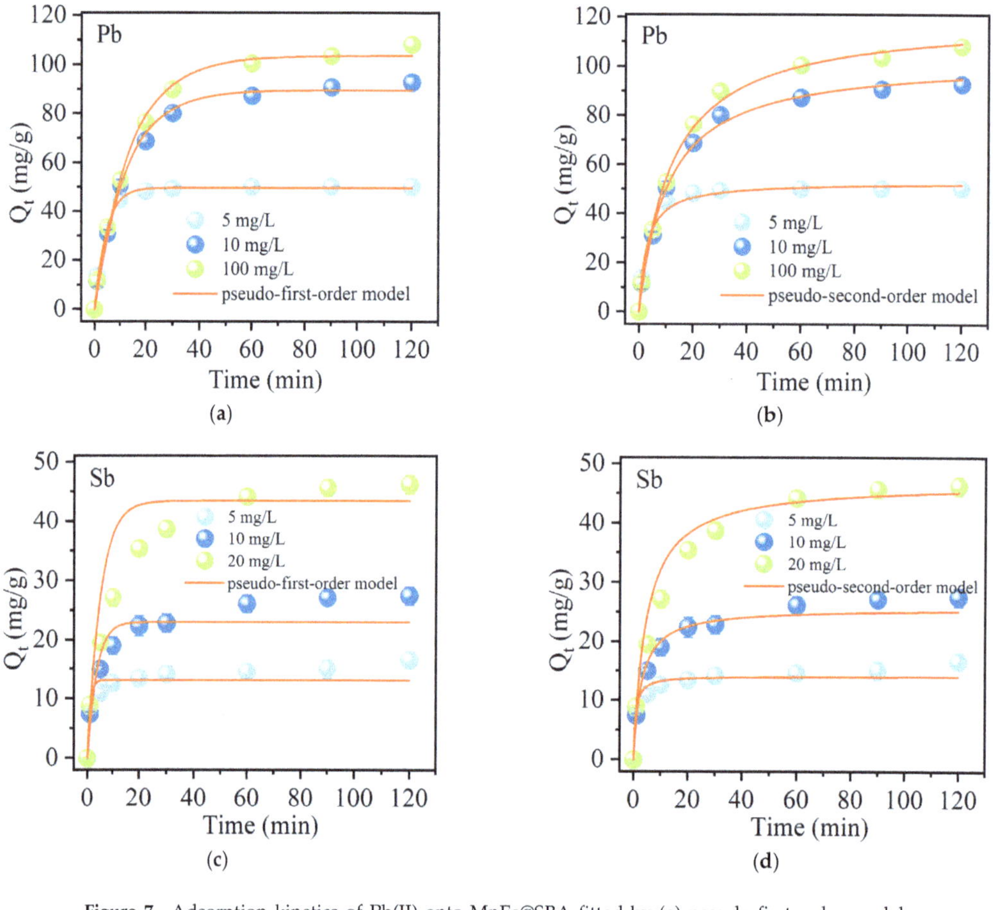

Figure 7. Adsorption kinetics of Pb(II) onto MnFe@SBA fitted by (**a**) pseudo-first-order model and (**b**) pseudo-second-order model; adsorption kinetics of Sb(V) onto MnFe@SBA fitted by (**c**) pseudo-first-order model and (**d**) pseudo-second-order model.

Table 1. Parameters of pseudo-first-order and pseudo-second-order kinetic models for the adsorption of Pb(II) and Sb(V) onto MnFe@SBA.

	C0 (mg/L)	Pseudo-First-Order Model			Pseudo-Second-Order Model		
		K1 (L/min)	Qe (mg/g)	R2	K2 (g/(mg min))	Qe (mg/g)	R2
Pb(II)	5	0.22611	49.68	0.9983	0.00663	53.07	0.9983
	10	0.08088	89.61	0.9983	0.00092	103.27	0.9986
	20	0.07198	103.58	0.9982	0.00069	120.08	0.9988
Sb(V)	5	1.00037	13.15	0.9928	0.09955	14.15	0.9969
	10	0.28702	23.02	0.9680	0.01387	25.69	0.9920
	20	0.20328	43.47	0.9765	0.00474	46.84	0.9937

2.2.4. Adsorption Isotherms

To gain a comprehensive understanding of the surface characteristics of the adsorbent, adsorption isotherm studies were conducted at initial concentrations of Pb(II) and Sb(V) ranging from 10 to 400 mg/L, as well as temperatures of 298, 308, and 318 K. The experimental data were fitted using two well-established models: Langmuir and Freundlich isotherms [46]. The Langmuir model assumes a uniform adsorbent surface with no interaction between adjacent sites where ions are adsorbed, thus categorizing the process as monolayer adsorption [47]. On the other hand, the Freundlich model describes the complex nature of multiphase and multilayer adsorption on various types of adsorbents. These models can be represented mathematically as follows [48]:

$$q_e = \frac{q_m k_L C_e}{1 + k_L C_e} \tag{3}$$

$$q_e = k_F C_e^{\frac{1}{n}} \tag{4}$$

where q_e (mg/g) is the equilibrium adsorption capacity, q_m (mg/g) is the maximum adsorption capacity, C_e (mg/L) is the equilibrium concentration, k_L (L/g) is the Langmuir constant associated with adsorption free energy, k_F ((mg g^{-1})(L mg^{-1})$^{1/n}$) is the Freundlich constant related to adsorption capacity, and n is the adsorption energy correlation constant. As depicted in Figure 8, the adsorption of Pb(II) and Sb(V) onto MnFe@SBA exhibited enhanced favorability at elevated temperatures, thereby suggesting an endothermic nature of the adsorption process. According to Table 2, the adsorption of Pb(II) was better fitted by the Freundlich model, as evidenced by higher R^2 values, implying that multilayer adsorption was primarily involved in the adsorption of Pb(II) onto MnFe@SBA. Moreover, a value less than 0.5 for parameter $1/n$ indicated a facile occurrence of adsorption [49]. Conversely, the adsorption behavior of Sb(V) on MnFe@SBA aligned more closely with the Langmuir model, with R^2 values higher than 0.99 at temperatures of 298, 308, and 318 K. The result suggested that monolayered adsorption primarily characterized the interaction between Sb(V) and MnFe@SBA. The maximum adsorption capacities of Sb(V) by MnFe@SBA were found to be 211.49, 241.57, and 260.40 mg/g, respectively.

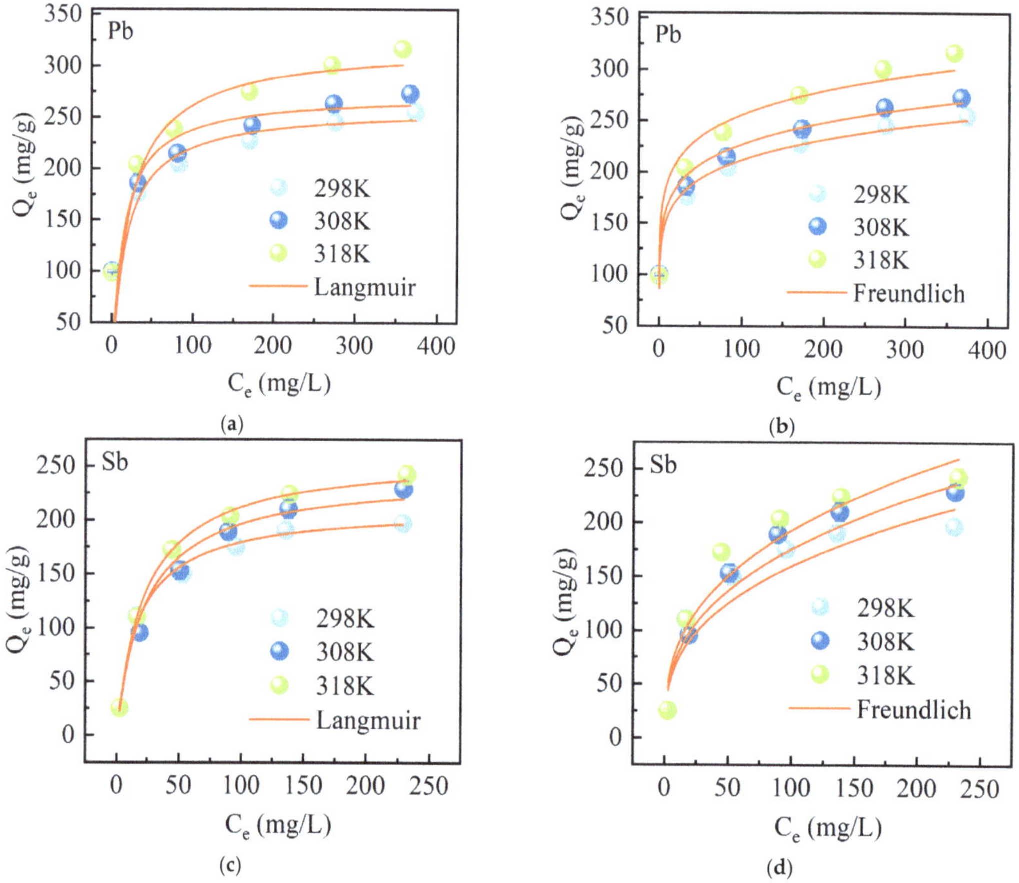

Figure 8. Adsorption isotherms of Pb(II) and Sb(V) on MnFe@SBA at temperatures of 298, 308, and 318 K: (**a**,**c**) Langmuir; (**b**,**d**) Freundlich.

Table 2. Parameters of Langmuir and Freundlich isotherms models for the adsorption of Pb(II) and Sb(V) onto MnFe@SBA.

	T (K)	**Langmuir**			**Freundlich**		
		K_L (L/min)	**Q_m (mg/g)**	**R^2**	**1/*n***	**K_F [(mg/g)(L/mg)1/n]**	**R^2**
Pb(II)	298	0.0569	258.77	0.852	0.130	116.41	0.977
	308	0.0691	271.85	0.885	0.129	125.19	0.991
	318	0.0447	329.86	0.872	0.126	143.09	0.974
Sb(V)	298	0.0562	211.49	0.998	0.349	31.95	0.863
	308	0.0434	241.57	0.993	0.357	33.87	0.856
	318	0.0438	260.40	0.999	0.356	37.45	0.823

2.2.5. Adsorption Mechanisms Investigation

To gain a deeper understanding of the adsorption mechanisms of Pb(II) and Sb(V) onto MnFe@SBA, XPS analysis was conducted to investigate the surface composition and chemical states of both the fresh and used adsorbent, as illustrated in Figure 9. In the survey spectra (Figure 9a), a distinct peak corresponding to Pb appeared at a binding energy of approximately 140 eV following the adsorption of Pb(II). After the adsorption of Sb(V), there was an evident overlap between O 1s and Sb 3d peaks. These results confirmed the successful adsorption of Pb(II) and Sb(V). The high-resolution O 1s spectrum of the fresh and used MnFe@SBA is presented in Figure 9b. According to Table S3, the adsorption of Pb(II) led to an increase in the proportion of lattice oxygen (O^{2-}) from 9.55% to 45.80%, indicating the formation of Pb-O through the interaction between Pb(II) and –OH groups. Consequently, the proportion of –OH dropped from 53.65% to 47.14%, suggesting that –OH on MnFe@SBA played a crucial role in Pb(II) adsorption [50]. Due to the energy overlap between Sb 3d and O 1s, there was a shift in the peak positions of O^{2-}, –OH, and H_2O, leading to an overlap between the –OH peak and Sb $3d_{5/2}$ [51]. The deconvolution analysis of the Sb $3d_{3/2}$ peak revealed a single peak at 540.66 eV corresponding to Sb(V), indicating that the adsorption process did not alter the valence state of Sb(V) and was thus straightforward. The relative proportion of Sb $3d_{3/2}$ was 18.66%, and the proportion of Sb $3d_{5/2}$ was estimated to be 69.2% of Sb $3d_{3/2}$ or a calculated value of 12.91%. The initial ratio of –OH to O^{2-} in the fresh adsorbent was 2.35, which subsequently declined to 1.15 after the adsorption of Sb(V), indicating the involvement of –OH.

In the high-resolution Mn 2p XPS spectrum (Figure 9c), the relative proportion of Mn(II) peak at 641.5 and 652.7 eV increased from 12.43% to 28.89% upon Pb(II) adsorption. Concurrently, the proportion of the Mn(IV) peak at 643.2 and 654.5 eV decreased from 50% to 44.24%. This phenomenon was likely due to the release of Mn(IV) from the interlayer, which subsequently re-adsorbed onto the surface as Mn(II) [52]. Analogous alterations were observed in the XPS spectrum of the Sb(V) adsorbed MnFe@SBA, suggesting that Mn(IV) was similarly released from the interlayer and adsorbed on the surface as Mn(II). This process enhanced the zeta potential and facilitated the formation of a Mn–Sb–O complex via coordination with Sb(V) [53].

As illustrated in Figure 9d, following the adsorption of Pb(II) and Sb(V), the relative proportion of Fe(II) in the adsorbent increased while the proportion of Fe(III) decreased. The result strongly suggested the occurrence of electron transfer and redox reactions throughout the process. Figure 9e represents the high-resolution XPS spectrum of Pb 4f, which can be deconvoluted into two individual peaks at 137.77 and 142.60 eV, corresponding to Pb $4f_{7/2}$ and Pb $4f_{5/2}$, respectively. Further, the peaks at 138.50 and 143.20 eV are indicative of Pb=O and Pb−O species [54]. These findings can be attributed to the complex interaction between MnFe@SBA and Pb(II).

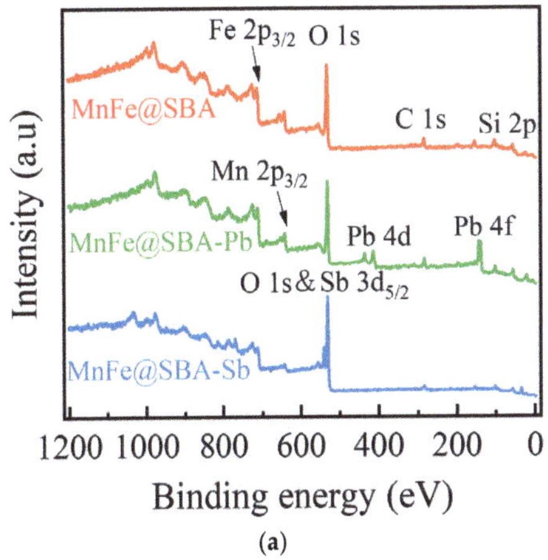

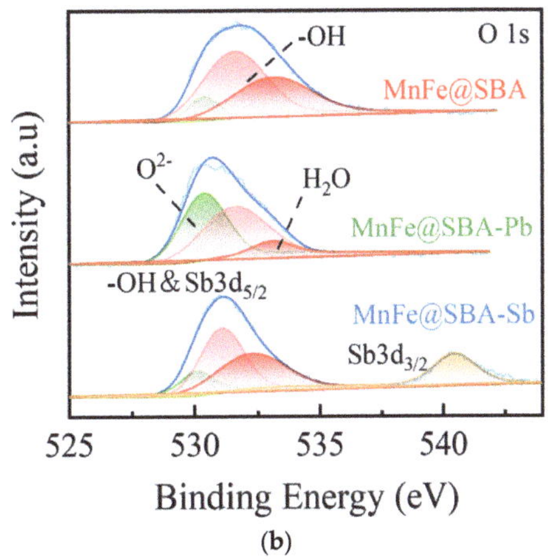

Figure 9. *Cont.*

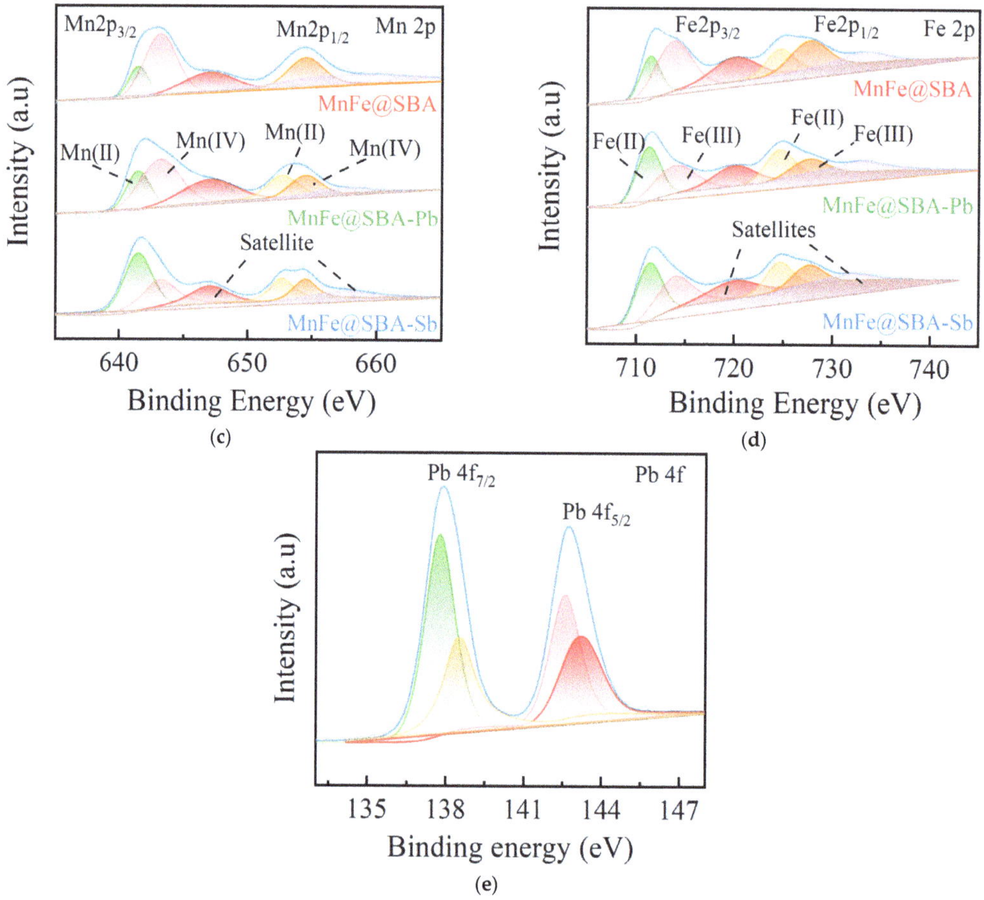

Figure 9. (**a**) XPS survey spectra for the fresh MnFe@SBA, Pb(II) adsorbed MnFe@SBA (MnFe@SBA-Pb), and Sb(V) adsorbed MnFe@SBA (MnFe@SBA-Sb). High-resolution XPS spectra for the fresh MnFe@SBA, MnFe@SBA-Pb, and MnFe@SBA-Sb: (**b**) O 1s, (**c**) Mn 2p and (**d**) Fe 2p. (**e**) High-resolution XPS Pb 4f spectra for MnFe@SBA-Pb.

In summary, the adsorption mechanisms of Pb(II) and Sb(V) onto MnFe@SBA were found to exhibit similarities and consist of two steps, as illustrated in Figure 10. Initially, an electrostatic attraction occurred between MnFe@SBA and Pb(II)/Sb(V), which played a significant role in contributing to the rapid adsorption process. Subsequently, –OH groups on MnFe@SBA formed the inner-sphere complexes with Pb(II). In the case of Sb(V) adsorption, MnFe@SBA enhanced the surface zeta potential via the interlayer release and re-adsorption of Mn(IV), thereby attracting Sb(V) and forming Sb–O–Mn complexes.

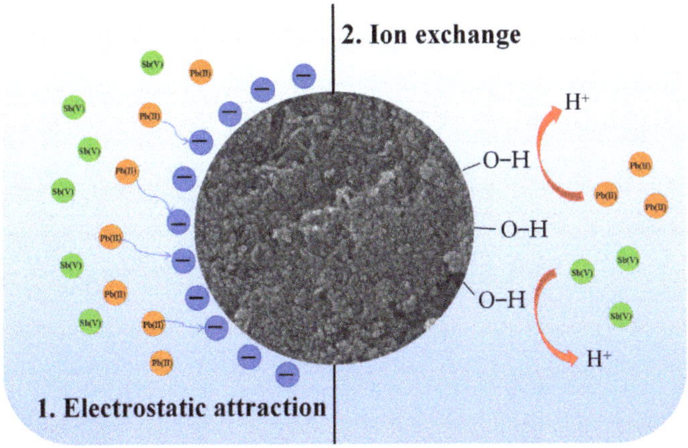

Figure 10. Diagram of the potential mechanisms of Sb(V) and Pb(II) adsorption by MnFe@SBA.

3. Chemicals and Methods

3.1. Chemicals

SBA-15 was purchased from Jiangsu Xianfeng Nanomaterial Technology Co., Ltd. (Nanjing, China). Tetrahydrate manganese chloride ($MnCl_2 \cdot 4H_2O$), sodium hydroxide (NaOH), and hexahydrate iron chloride ($FeCl_3 \cdot 6H_2O$) were obtained from Shanghai Guoyao Chemical Reagent Co., Ltd. (Shanghai, China). All chemicals were of analytical purity and were used as received without further purification. All solutions were prepared using ultrapure water produced using an ultrapure water purification system (Veolia Water Solutions and Technologies, Singapore).

3.2. Synthesis of MnFe@SBA

MnFe@SBA was synthesized using a simple hydrothermal method, as shown in Figure S2. First, 2 mmol of $MnCl_2 \cdot 4H_2O$ and 4 mmol of $FeCl_3 \cdot 6H_2O$ were dissolved in a beaker to form 100 mL of mixed solution. Subsequently, 10 mL of NaOH solution was added, and the resulting solution was stirred continuously for 10 min. SBA-15 with mass loads from 0.025 to 0.2 g was then added. After being agitated for 10 min, the solution was placed into a Teflon-lined stainless-steel autoclave and heated to and held at a temperature of 100 °C for 24 h. The autoclave was then cooled naturally to room temperature. The precipitate was collected, washed with deionized water, and dried in an oven at 60 °C for 12 h. The product is referred to hereinafter as MnFe@SBA-x, where x denotes the loading amount of SBA-15, and the loading capacity of $MnFe_2O_4$ is maintained at 0.46 g. The pure $MnFe_2O_4$ was synthesized as well without the addition of SBA-15.

3.3. Characterizations

The surface morphologies of the samples were observed using a scanning electron microscope (MIRA3 LMH, TESCAN Group a.s., Brno, Czech). Elemental analyses of the adsorbents were performed using energy-dispersive X-ray spectrometry (EDS, One Max 20, UK). X-ray diffraction (XRD) analysis was performed using an X-ray diffractometer (Rigaku SmartLab SE, Akishima, Japan). Fourier-transform infrared spectroscopy (FTIR) was performed using an analytical FTIR spectrometer (Thermo Fisher Scientific Nicolet iS50, Waltham, MA, USA). The BET surface areas and pore size distribution of the N_2 adsorption–desorption isotherms were measured using an analytical instrument (Micromeritics ASAP 2460, Norcross, GA, USA). X-ray photoelectron spectroscopy (XPS) spectra were characterized by an XPS spectrometer (AXIS SUPRA+, Shimadzu, Kyoto, Japan). The pH values of the solutions were measured using a pHs-3C digital pH meter (Shanghai Precision Scientific Instruments Co., Ltd., Shanghai, China). The zeta potentials of the adsorbents were determined using a Zetasizer Nano ZS90 (Malvern Instruments, Malvern, UK).

3.4. Batch Adsorption Experiments

Typically, at room temperature, 10 and 30 mg of MnFe@SBA were dispersed in 100 mL of Pb(II) (10 mg/L) and Sb(V) solutions (10 mg/L), respectively. At preset time intervals spanning 0 to 120 min, 0.5 mL of water sample was extracted and filtered using a syringe filter membrane (0.22 μm) immediately. The concentration of residue Pb(II) and Sb(V) ions in the solution was determined using atomic absorption spectrophotometry (AA-6300; Shimadzu, Kyoto, Japan). The removal efficiency and adsorption capacity were computed using the equations of $Q_e = \frac{(C_0 - C_e) \times V}{m}$ and $R = \frac{C_0 - C_e}{C_0} \times 100\%$, respectively [55], where C_0 and C_e (mg/L) are the initial and equilibrium concentrations in the solution, respectively; V (L) is the solution volume; m (g) is the adsorbent mass; Q_e (mg/g) is the adsorption capacity at equilibrium; and R (%) is the removal efficiency. To examine the impact of pH on the adsorption, a solution containing 10 mg/L Pb(II) or Sb(V) was adjusted to the desired pH range using 1 M HCl and NaOH. The interference of commonly coexisting cations (Na^+, K^+, Ca^{2+}, Mg^{2+}, Cd^{2+}, Ni^{2+} and Zn^{2+}) and anions (Cl^-, NO_3^-, CO_3^{2-}, and PO_4^{3-}) on the adsorption was also investigated.

To study the adsorption kinetics, 10 and 30 mg of MnFe@SBA were added to Pb(II) and Sb(V) solutions with initial concentrations of 5, 10, and 20 mg/L at a constant pH of 6, respectively. To study the adsorption isotherms, 10 and 30 mg of MnFe@SBA were added to Pb(II) and Sb(V) solutions with initial concentrations ranging from 10 to 400 mg/L at a pH of 6, respectively. The adsorption of Pb(II) and Sb(V) using MnFe@SBA at temperatures of 25, 35, and 45 °C were studied to understand the adsorption thermodynamics.

4. Conclusions

In this work, a novel amorphous MnFe@SBA composite was synthesized through the in situ growth of $MnFe_2O_4$ on SBA-15 to effectively remove heavy metals (Pb(II) and Sb(V)) from aqueous solutions. Both $MnFe_2O_4$ and SBA-15 demonstrated limited adsorption capacities for Pb(II) and Sb(V). However, by optimizing the material synthesis and adsorption conditions, the maximum adsorption capacities of MnFe@SBA for Pb(II) and Sb(V) at 318 K were significantly enhanced to 329.86 mg/g and 260.40 mg/g, respectively. The adsorption process for both Pb(II) and Sb(V) by MnFe@SBA was found to be endothermic and spontaneous. The adsorption of Pb(II) onto MnFe@SBA was well-fitted by the pseudo-second-order kinetic and Freundlich models, while the adsorption of Sb(V) was well-described by the pseudo-second-order kinetic and Langmuir models. Mechanistic studies indicated that the adsorption of Pb(II) and Sb(V) onto MnFe@SBA involved two primary steps: electrostatic attraction and complexation. Initially, rapid adsorption occurred due to the electrostatic attraction between MnFe@SBA and metal ions. Subsequently, –OH groups on MnFe@SBA formed inner-sphere complexes with Pb(II). For Sb(V) adsorption, MnFe@SBA enhanced its surface zeta potential through the interlayer release and re-adsorption of Mn(IV), facilitating the formation of Sb–O–Mn complexes.

Supplementary Materials: The following supporting information can be downloaded at: https://www.mdpi.com/article/10.3390/molecules30030679/s1, Figure S1: Synthesis procedure for amorphous MnFe@SBA; Figure S2: Effects of (a) Na^+, (b) K^+, (c) Ca^{2+}, (d) Mg^{2+}, (e) Cd^{2+}, (f) Ni^{2+}, (g) Zn^{2+} on the removal of Pb(II) by MnFe@SBA; Figure S3: Effects of (a) Cl^-, (b) NO_3^-, (c) CO_3^{2-}, (d) PO_4^{3-} on the removal of Sb(V) by MnFe@SBA; Table S1: Textural properties of the SBA-15, $MnFe_2O_4$, and MnFe@SBA composites; Table S2: Peaks information of O 1s, Fe 2p, and Mn 2p; Table S3: Peaks information of O 1s, Fe 2p, and Mn 2p.

Author Contributions: Conceptualization, Z.S. and A.Z.; methodology, Z.S.; software, F.C.; validation, Z.S., A.Z. and F.C.; formal analysis, A.Z. and Y.C.; investigation, Z.S. and A.Z.; resources, L.D.; data curation, A.Z.; writing—original draft preparation, A.Z.; writing—review and editing, Z.S. and L.D.; visualization, Y.C.; supervision, L.D.; project administration, L.D.; funding acquisition, L.D. All authors have read and agreed to the published version of the manuscript.

Funding: This research received no external funding.

Institutional Review Board Statement: Not applicable.

Informed Consent Statement: Not applicable.

Data Availability Statement: Data is contained within the article and Supplementary Materials.

Conflicts of Interest: The authors declare no conflict of interest.

References

1. Saif, S.; Zaidi, A.; Khan, M.S.; Rizvi, A. Metal-Legume-Microbe Interactions: Toxicity and Remediation. In *Microbes for Legume Improvement*; Zaidi, A., Khan, M.S., Musarrat, J., Eds.; Springer International Publishing: Cham, Switzerland, 2017; pp. 367–385.
2. Cao, Y.; Hu, X.; Zhu, C.; Zhou, S.; Li, R.; Shi, H.; Miao, S.; Vakili, M.; Wang, W.; Qi, D. Sulfhydryl functionalized covalent organic framework as an efficient adsorbent for selective Pb (II) removal. *Colloids Surf. A Physicochem. Eng. Asp.* **2020**, *600*, 125004. [CrossRef]
3. Wajima, T. Preparation of Adsorbent with Lead Removal Ability Using Sulfur-Impregnation. *Adv. Mater. Res.* **2013**, *831*, 253–257. [CrossRef]
4. Hasan, R.; Chong, C.C.; Bukhari, S.N.; Jusoh, R.; Setiabudi, H.D. Effective removal of Pb(II) by low-cost fibrous silica KCC-1 synthesized from silica-rich rice husk ash. *J. Ind. Eng. Chem.* **2019**, *75*, 262–270. [CrossRef]
5. Shukla, V.; Shukla, P.; Tiwari, A. Lead poisoning. *Indian J. Med. Spec.* **2018**, *9*, 146–149. [CrossRef]
6. Yang, Q.; Li, Z.; Lu, X.; Duan, Q.; Huang, L.; Bi, J. A review of soil heavy metal pollution from industrial and agricultural regions in China: Pollution and risk assessment. *Sci. Total Environ.* **2018**, *642*, 690–700. [CrossRef] [PubMed]
7. Teng, F.; Zhang, Y.; Wang, D.; Shen, M.; Hu, D. Iron-modified rice husk hydrochar and its immobilization effect for Pb and Sb in contaminated soil. *J. Hazard. Mater.* **2020**, *398*, 122977. [CrossRef]
8. Feng, X.; Yan, R.; Zhang, Q.; Wan, Q.; Hagio, T.; Ichino, R.; Kong, L.; Cao, X.; Li, L. Nano ferric oxide adsorbents with self-acidification effect for efficient adsorption of Sb(V). *Chemosphere* **2021**, *272*, 129933. [CrossRef] [PubMed]
9. Guo, F.; Wu, F.C.; Yu, F.; Bai, Y.C.; Fu, Z.Y.; Zhu, Y.R.; Guo, W.J. Fate and removal of antimony in response to stringent control activities after a mine tailing spill. *Sci. Total Environ.* **2019**, *693*, 133604. [CrossRef] [PubMed]
10. Zhang, P.; Ouyang, S.; Li, P.; Sun, Z.; Ding, N.; Huang, Y. Ultrahigh removal performance of lead from wastewater by tricalcium aluminate via precipitation combining flocculation with amorphous aluminum. *J. Clean. Prod.* **2020**, *246*, 118728. [CrossRef]
11. Choumane, R.; Peulon, S. Development of an efficient electrochemical process for removing and separating soluble Pb(II) in aqueous solutions in presence of other heavy metals: Studies of key parameters. *Chem. Eng. J.* **2021**, *423*, 130161. [CrossRef]
12. Rong, Y.; Yan, W.; Wang, Z.; Hao, X.; Guan, G. Rapid and selective removal of Pb ions by electroactive titanium Dioxide/Polyaniline ion exchange film. *Sep. Purif. Technol.* **2023**, *312*, 123386. [CrossRef]
13. Xu, Z.; Gu, S.; Rana, D.; Matsuura, T.; Lan, C.Q. Chemical precipitation enabled UF and MF filtration for lead removal. *J. Water Process Eng.* **2021**, *41*, 101987. [CrossRef]
14. Huang, Z.; Xiong, C.; Ying, L.; Wang, W.; Wang, S.; Ding, J.; Lu, J. A post-functional Ti-based MOFs composite for selective removal of Pb (II) from water. *J. Hazard. Mater.* **2022**, *432*, 128700. [CrossRef]
15. Liu, T.; Lawluvy, Y.; Shi, Y.; Ighalo, J.O.; He, Y.; Zhang, Y.; Yap, P.-S. Adsorption of cadmium and lead from aqueous solution using modified biochar: A review. *J. Environ. Chem. Eng.* **2022**, *10*, 106502. [CrossRef]
16. Kumar, R.; Bhattacharya, S.; Sharma, P. Novel insights into adsorption of heavy metal ions using magnetic graphene composites. *J. Environ. Chem. Eng.* **2021**, *9*, 106212. [CrossRef]
17. Huang, Y.; Zheng, H.; Hu, X.; Wu, Y.; Tang, X.; He, Q.; Peng, S. Enhanced selective adsorption of lead(II) from complex wastewater by DTPA functionalized chitosan-coated magnetic silica nanoparticles based on anion-synergism. *J. Hazard. Mater.* **2022**, *422*, 126856. [CrossRef]
18. Abdel-Magied, A.F.; Abdelhamid, H.N.; Ashour, R.M.; Fu, L.; Dowaidar, M.; Xia, W.; Forsberg, K. Magnetic metal-organic frameworks for efficient removal of cadmium(II), and lead(II) from aqueous solution. *J. Environ. Chem. Eng.* **2022**, *10*, 107467. [CrossRef]

19. Abdulkareem, A.S.; Hamzat, W.A.; Tijani, J.O.; Egbosiuba, T.C.; Mustapha, S.; Abubakre, O.K.; Okafor, B.O.; Babayemi, A.K. Isotherm, kinetics, thermodynamics and mechanism of metal ions adsorption from electroplating wastewater using treated and functionalized carbon nanotubes. *J. Environ. Chem. Eng.* **2023**, *11*, 109180. [CrossRef]
20. Eren, Z.S.; Tunçer, S.; Gezer, G.; Yildirim, L.T.; Banerjee, S.; Yilmaz, A. Improved solubility of celecoxib by inclusion in SBA-15 mesoporous silica: Drug loading in different solvents and release. *Microporous Mesoporous Mater.* **2016**, *235*, 211–223. [CrossRef]
21. Li, G.; Wang, B.; Sun, Q.; Xu, W.Q.; Han, Y. Adsorption of lead ion on amino-functionalized fly-ash-based SBA-15 mesoporous molecular sieves prepared via two-step hydrothermal method. *Microporous Mesoporous Mater.* **2017**, *252*, 105–115. [CrossRef]
22. Li, Y.; Zhong, N.; Cheong, L.Z.; Huang, J.; Chen, H.; Lin, S. Immobilization of Candida antarctica Lipase B onto organically-modified SBA-15 for efficient production of soybean-based mono and diacylglycerols. *Int. J. Biol. Macromol.* **2018**, *120 Pt A*, 886–895. [CrossRef]
23. Hassanzadeh-Afruzi, F.; Esmailzadeh, F.; Asgharnasl, S.; Ganjali, F.; Taheri-Ledari, R.; Maleki, A. Efficient removal of Pb(II)/Cu(II) from aqueous samples by a guanidine-functionalized SBA-15/Fe3O4. *Sep. Purif. Technol.* **2022**, *291*, 120956. [CrossRef]
24. Xing, X.; Ren, X.; Alharbi, N.S.; Chen, C. Efficient adsorption and reduction of Cr(VI) from aqueous solution by Santa Barbara Amorphous-15 (SBA-15) supported Fe/Ni bimetallic nanoparticles. *J. Colloid. Interface Sci.* **2023**, *629 Pt A*, 744–754. [CrossRef]
25. Zhang, H.; Wang, J.; Liu, T.; Zhang, M.; Hao, L.; Phoutthavong, T.; Liang, P. Cu-Zn oxides nanoparticles supported on SBA-15 zeolite as a novel adsorbent for simultaneous removal of H2S and Hg0 in natural gas. *Chem. Eng. J.* **2021**, *426*, 131286. [CrossRef]
26. Yao, J.; Deng, Y.; Pan, S.; Korna, R.; Wen, J.; Yuan, N.; Wang, K.; Li, H.; Yang, Y. The difference in the adsorption mechanisms of magnetic ferrites modified carbon nanotubes. *J. Hazard. Mater.* **2021**, *415*, 125551. [CrossRef] [PubMed]
27. Cao, X.; Liu, H.; Wan, B.; Zhang, C.; Yao, S. Effective removal of Cd(II) from aqueous solution by MnFe2O4 composite modified by surfactant. *Desalination Water Treat.* **2022**, *262*, 168–179. [CrossRef]
28. Liu, J.; Ren, S.; Cao, J.; Tsang, D.C.W.; Beiyuan, J.; Peng, Y.; Fang, F.; She, J.; Yin, M.; Shen, N.; et al. Highly efficient removal of thallium in wastewater by MnFe(2)O(4)-biochar composite. *J. Hazard. Mater.* **2021**, *401*, 123311. [CrossRef] [PubMed]
29. Ren, Y.; Li, N.; Feng, J.; Luan, T.; Wen, Q.; Li, Z.; Zhang, M. Adsorption of Pb(II) and Cu(II) from aqueous solution on magnetic porous ferrospinel MnFe2O4. *J. Colloid Interface Sci.* **2012**, *367*, 415–421. [CrossRef]
30. Verma, M.; Kumar, A.; Singh, K.P.; Kumar, R.; Kumar, V.; Srivastava, C.M.; Rawat, V.; Rao, G.; Kumari, S.; Sharma, P.; et al. Graphene oxide-manganese ferrite (GO-MnFe2O4) nanocomposite: One-pot hydrothermal synthesis and its use for adsorptive removal of Pb2+ ions from aqueous medium. *J. Mol. Liq.* **2020**, *315*, 113769. [CrossRef]
31. Yang, K.; Li, C.; Wang, X.; Liu, Y.; Li, Y.; Zhou, C.; Wang, Z.; Zhou, J.; Cao, Z.; Xu, X. Functional adjustment and cyclic application of Fe–Mn bimetal composite for Sb(V) removal: Transformation of iron oxides forms and a stable regeneration method. *J. Clean. Prod.* **2020**, *266*, 122007. [CrossRef]
32. Ding, W.; Zheng, H.; Sun, Y.; Zhao, Z.; Zheng, X.; Wu, Y.; Xiao, W. Activation of MnFe(2)O(4) by sulfite for fast and efficient removal of arsenic(III) at circumneutral pH: Involvement of Mn(III). *J. Hazard. Mater.* **2021**, *403*, 123623. [CrossRef]
33. Cui, J.; Wang, Q.; Gao, J.; Guo, Y.; Cheng, F. The selective adsorption of rare earth elements by modified coal fly ash based SBA-15. *Chin. J. Chem. Eng.* **2022**, *47*, 155–164. [CrossRef]
34. Shen, Y.; Fu, F.; Tang, B. Ethylenediamine-functionalized MnFe2O4 @ferrihydrite as a magnetic adsorbent for removal of Cr(VI): Adsorption and mechanism studies. *J. Environ. Chem. Eng.* **2023**, *11*, 110230. [CrossRef]
35. Betiha, M.A.; Moustafa, Y.M.; El-Shahat, M.F.; Rafik, E. Polyvinylpyrrolidone-Aminopropyl-SBA-15 schiff Base hybrid for efficient removal of divalent heavy metal cations from wastewater. *J. Hazard. Mater.* **2020**, *397*, 122675. [CrossRef] [PubMed]
36. Wen, Z.; Xi, J.; Lu, J.; Zhang, Y.; Cheng, G.; Zhang, Y.; Chen, R. Porous biochar-supported MnFe(2)O(4) magnetic nanocomposite as an excellent adsorbent for simultaneous and effective removal of organic/inorganic arsenic from water. *J. Hazard. Mater.* **2021**, *411*, 124909. [CrossRef] [PubMed]
37. Wang, J.; Liu, Y.; Yin, W.; Cao, Y.; Hou, J.; Wang, S.; Wang, X. Solvent-induced facile synthesis of MnFe2O4 and the As(V) removal mechanism study. *J. Mol. Liq.* **2023**, *371*, 120845. [CrossRef]
38. Bai, M.; Chai, Y.; Chen, A.; Yuan, J.; Shang, C.; Peng, L.; Peng, C. Enhancing cadmium removal efficiency through spinel ferrites modified biochar derived from agricultural waste straw. *J. Environ. Chem. Eng.* **2023**, *11*, 109027. [CrossRef]
39. Huang, W.-H.; Wu, R.-M.; Chang, J.-S.; Juang, S.-Y.; Lee, D.-J. Manganese ferrite modified agricultural waste-derived biochars for copper ions adsorption. *Bioresour. Technol.* **2023**, *367*, 128303. [CrossRef] [PubMed]
40. Huang, L.; Zhang, H.; Zeng, T.; Chen, J.; Song, S. Synergistically enhanced heterogeneous activation of persulfate for aqueous carbamazepine degradation using Fe3O4@SBA-15. *Sci. Total Environ.* **2021**, *760*, 144027. [CrossRef] [PubMed]
41. Xia, S.; Deng, L.; Liu, X.; Yang, L.; Yang, X.; Shi, Z.; Pei, Y. Fabrication of magnetic nickel incorporated carbon nanofibers for superfast adsorption of sulfadiazine: Performance and mechanisms exploration. *J. Hazard. Mater.* **2022**, *423 Pt B*, 127219. [CrossRef]
42. Kavand, M.; Eslami, P.; Razeh, L. The adsorption of cadmium and lead ions from the synthesis wastewater with the activated carbon: Optimization of the single and binary systems. *J. Water Process Eng.* **2020**, *34*, 101151. [CrossRef]

43. Chen, Q.; Tang, Z.; Li, H.; Wu, M.; Zhao, Q.; Pan, B. An electron-scale comparative study on the adsorption of six divalent heavy metal cations on MnFe2O4@CAC hybrid: Experimental and DFT investigations. *Chem. Eng. J.* **2020**, *381*, 122656. [CrossRef]
44. Foroutan, R.; Peighambardoust, S.J.; Latifi, P.; Ahmadi, A.; Alizadeh, M.; Ramavandi, B. Carbon nanotubes/β-cyclodextrin/MnFe2O4 as a magnetic nanocomposite powder for tetracycline antibiotic decontamination from different aqueous environments. *J. Environ. Chem. Eng.* **2021**, *9*, 106344. [CrossRef]
45. Huang, X.; Kong, L.; Huang, S.; Liu, M.; Li, L. Synthesis of novel magnetic sulfur-doped Fe3O4 nanoparticles for efficient removal of Pb(II). *Sci. China Chem.* **2017**, *61*, 164–171. [CrossRef]
46. Medina, R.P.; Nadres, E.T.; Ballesteros, F.C.; Rodrigues, D.F. Incorporation of graphene oxide into a chitosan–poly(acrylic acid) porous polymer nanocomposite for enhanced lead adsorption. *Environ. Sci. Nano* **2016**, *3*, 638–646. [CrossRef]
47. Chen, Y.; Tang, J.; Wang, S.; Zhang, L. Ninhydrin-functionalized chitosan for selective removal of Pb(II) ions: Characterization and adsorption performance. *Int. J. Biol. Macromol.* **2021**, *177*, 29–39. [CrossRef]
48. Ivanets, A.; Prozorovich, V.; Kouznetsova, T.; Dontsova, T.; Yanushevska, O.; Hosseini-Bandegharaei, A.; Srivastava, V.; Sillanpaa, M. Effect of Mg(2+) ions on competitive metal ions adsorption/desorption on magnesium ferrite: Mechanism, reusability and stability studies. *J. Hazard. Mater.* **2021**, *411*, 124902. [CrossRef] [PubMed]
49. Liu, S.; Huang, J.; Zhang, W.; Shi, L.; Yi, K.; Zhang, C.; Pang, H.; Li, J.; Li, S. Investigation of the adsorption behavior of Pb(II) onto natural-aged microplastics as affected by salt ions. *J. Hazard. Mater.* **2022**, *431*, 128643. [CrossRef] [PubMed]
50. Yan, X.; Fei, Y.; Zhong, L.; Wei, W. Arsenic stabilization performance of a novel starch-modified Fe-Mn binary oxide colloid. *Sci. Total Environ.* **2020**, *707*, 136064. [CrossRef] [PubMed]
51. Xue, C.; Zhang, Q.; Li, W.; Ownes, G.; Chen, Z. Removal mechanism of Sb(V) by a hybrid ZIF-8@FeNPs and used for treatment of mining wastewater. *Chem. Eng. J.* **2023**, *451*, 138691. [CrossRef]
52. Xue, Y.; Teng, W.; Chen, Y.; Ma, Q.; Chen, X.; Sun, Y.; Fan, J.; Qiu, Y.; Fu, R. Amorphous Mn-La oxides immobilized on carbon sphere for efficient removal of As(V), Cd(II), and Pb(II): Co-adsorption and roles of Mn species. *Chem. Eng. J.* **2022**, *429*, 132262. [CrossRef]
53. Liu, R.; Xu, W.; He, Z.; Lan, H.; Liu, H.; Qu, J.; Prasai, T. Adsorption of antimony(V) onto Mn(II)-enriched surfaces of manganese-oxide and FeMn binary oxide. *Chemosphere* **2015**, *138*, 616–624. [CrossRef] [PubMed]
54. Deng, S.; Ren, B.; Hou, B.; Deng, X.; Deng, R.; Zhu, G.; Cheng, S. Adsorption of Sb(III) and Pb(II) in wastewater by magnetic γ-Fe2O3-loaded sludge biochar: Performance and mechanisms. *Chemosphere* **2024**, *349*, 140914. [CrossRef] [PubMed]
55. Fu, Q.; Hu, J.; Zhou, S. Adsorption properties of graphene oxide@sodium alginate composite for Pb(II). *J. Iran. Chem. Soc.* **2023**, *21*, 399–407. [CrossRef]

water

MDPI

Article

Analyzing the Efficacy of Water Treatment Disinfectants as Vector Control: The Larvicidal Effects of Silver Nitrate, Copper Sulfate Pentahydrate, and Sodium Hypochlorite on Juvenile *Aedes aegypti*

Sydney S. Turner [1,*], James A. Smith [2], Sophie L. Howle [2], Patrick I. Hancock [3], Karin Brett [2], Julia Davis [2], Lorin M. Bruno [2], Victoria Cecchetti [2] and Clay Ford [4]

1 Occoquan Watershed Monitoring Laboratory, Department of Civil and Environmental Engineering, Virginia Tech, 9408 Prince William St., Manassas, VA 20110, USA
2 Department of Civil and Environmental Engineering, University of Virginia, Charlottesville, VA 22904, USA; jas9e@virginia.edu (J.A.S.)
3 Institute on the Environment, University of Minnesota, Minneapolis, MN 55455, USA
4 Research Data Services, Shannon Library, University of Virginia, 160 McCormick Road, Charlottesville, VA 22904, USA
* Correspondence: sydneyturner@vt.edu

Abstract: For communities without access to uninterrupted, piped water, household water storage (HWS) practices can lead to adverse public health outcomes caused by water degradation and mosquito proliferation. With over 700,000 deaths caused by vector-borne diseases annually, the objective of this study was to determine whether water disinfectants, at concentrations deemed safe for human consumption and beneficial for water treatment, are effective in reducing the emergence of adult mosquitoes that transmit disease. Laboratory bioassays, designed to resemble the context of treating HWS containers, were conducted to assess the larvicidal effects of chemicals at concentrations below regulatory limits for drinking water: silver (20, 40, 80 μg/L Ag), copper (300, 600, 1200 μg/L Cu), and chlorine (500, 1000, 2000 ug/L free chlorine). The water disinfectants demonstrated the ability to significantly reduce the population of juvenile *Ae. aegypti*. Sodium hypochlorite was found to be the most effective in decreasing the survival rate of late first instar larvae, while silver nitrate exhibited the highest effectiveness in inhibiting the emergence of late third instar larvae. Ultimately, this study highlights the potential of an integrated approach to Water, Sanitation, and Health (WASH) solutions with vector control management.

Keywords: *Aedes aegypti*; household water storage; vector control management; water treatment

1. Introduction

Eighty percent of the world's human population is at risk of contracting one or more vector-borne diseases (VBDs), which are caused by the spread of viruses, bacteria, and parasites from infected animals to a human [1,2]. These diseases account for more than 17% of the global burden of infectious diseases. With rising global temperatures driven by climate change, the risk of VBDs spreading to new regions continues to grow [3–5]. Many VBDs fall under the category of neglected tropical diseases (NTDs) due to their disproportionate impact on underserved or marginalized populations, particularly communities lacking access to basic healthcare, clean water, and sanitation [6,7]. In the absence of accessible vaccines and treatments for many VBDs, vector control management is essential to reducing the risk of disease transmission by controlling vector populations or disrupting human–vector contact [8–11]. Effective vector control programs require a deep understanding of the infected animals responsible for spreading disease, which are often blood-sucking insects [2], with mosquitoes being the most prolific [12]. One of the two most abundant mosquito species is the *Aedes (Stegomyia) aegypti* (Linneaus), responsible for transmitting several arboviral diseases such as Chikungunya, Dengue, Yellow fever, and Zika [2,13,14]. Table 1 highlights key public health data for each of these diseases.

Table 1. Public health information regarding VBDs transmitted by *Aedes aegypti.*

Vector-Borne Disease	Common Symptoms	Global Burden	Location of Prevalence	Vaccine for Prevention
Chikungunya (CHIKV)	Fever, severe joint pain, joint swelling, muscle pain [15]	Since 2023, 5+ million cases have been reported in last 15 yrs [16]	Detected in >100 countries as of 2021, circulating mainly in Africa, Asia, South America, and regions of the Pacific Ocean [17].	IXCHIQ (manufactured by Valenva) [16]
Dengue	High fever (40 °C/104 °F), severe headache, pain behind the eyes, muscle and joint pains, nausea, vomiting, swollen glands, rash [18]	In 2019, age-standardized incidence rate (ASIR) (A statistical measurement that compares the number of new cases of a disease in a population to a standard population. It's used to compare health metrics across populations with different age distributions.) estimated to be 7.40 per 1000 [19]. An estimated 100–400 million infections occur each year [18].	Endemic in >100 countries in WHO regions of Africa, the Americas, the Eastern Mediterranean, Southeast Asia and the Western Pacific. The Americas, Southeast Asia, and Western Pacific regions are most significantly affected [18].	Dengvaxia® (CYD-TDV), developed by Sanofi Pasteur, Qdenga® (TAK-003), developed by Takeda [18].
Yellow Fever	Fever, muscle pain, headache, loss of appetite, nausea or vomiting, jaundice, dark urine, abdominal pain [20]	In 2018, estimated 109,000 severe infections and 51,000 deaths in Africa and South America [21].	Thirty-four countries in Africa and thirteen countries in Central and South America are either endemic for, or have regions that are endemic for, yellow fever as of 2023 [20].	YF-VAX®, manufactured by Sanofi Pasteur [22]
Zika virus (ZIKV)	Fever, rash, headache, joint pain, conjunctivitis (red eye), muscle pains. Virus can be passed through sex and from a pregnant woman to her fetus [23]	In 2019, ASIR estimated to be 3.44 per 100,000 [24]	Eighty-nine countries and territories have documented evidence of current or previous transmission as of February 2022, circulating primarily in the Americas, South Asia, and the Pacific Islands [24,25].	No [24].

Understanding the life cycle of the *Ae. aegypti* (depicted in Figure 1), as well as its behaviors and habitats, is crucial for developing effective strategies to prevent and control the spread of these VBDs [9,26–28]. Typically living in tropical and subtropical regions, *Ae. aegypti* are found in close association with humans [29]. Jansen et al. cites urbanization, socioeconomic factors, building design and constructions features, the quality of water supply and management, and the quality of other public health infrastructure services as significant influences shaping the geographic spread of *Ae. aegypti* [29]. The following behavioral traits have been observed of adult female *Ae. aegypti* mosquitoes [9,29–32]:

1. They prefer to bite hosts during the daytime, with peak activity occurring between dawn and dusk.
2. They feed on humans relative to other vertebrate species, obtaining bloodmeals which provide the necessary nutrients required for a female mosquito's egg production and reproduction.
3. They feed on several hosts within one reproductive cycle which increases the potential for the transmission of disease.
4. They prefer to lay eggs in manmade or artificial containers (e.g., household water storage containers, plant pots, tires, etc.), which is why they are commonly referred to as a container-breeding species.

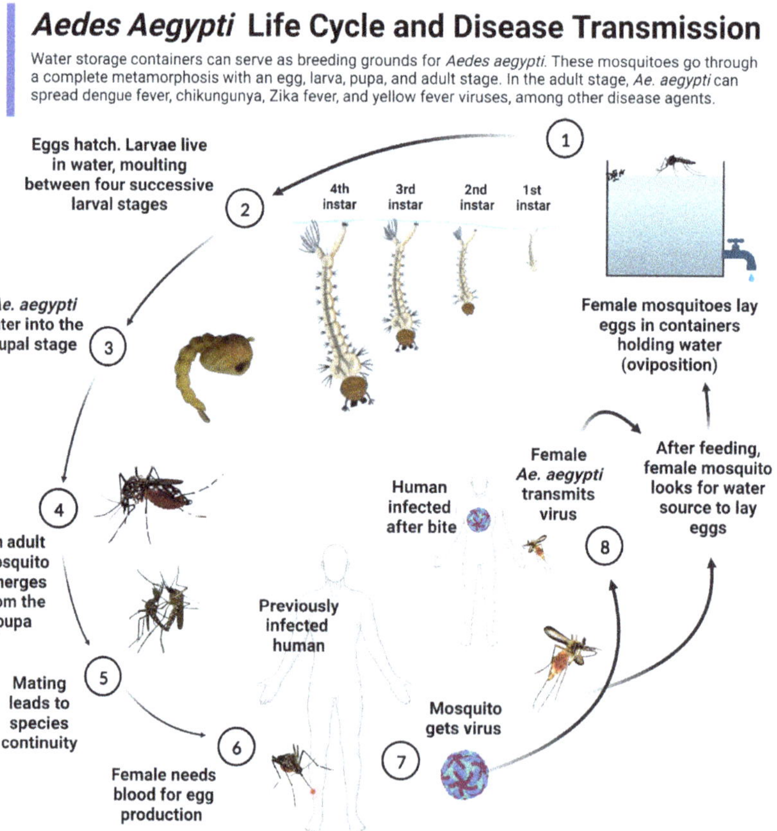

Figure 1. Life cycle of *Aedes aegypti* and its role in disease transmission. Figure created with Biorender.com.

Ae. aegypti eggs can hatch within minutes after coming into contact with water. Eggs also have been observed to remain viable for over a year in dry conditions (i.e., desiccation-resistant) [33]. As a result, water storage practices play a crucial role in controlling their breeding.

Vector control management targets different life stages, including (1) larval and pupal control, involving the stages at which the living organism is typically aquatic, and (2) adult control, involving the stage which is responsible for transmitting diseases. Figure 2 depicts a range of vector control management strategies aimed at different stages of the *Ae. aegypti* life cycle. These strategies include *chemical* methods (spraying chemicals, introducing chemicals to water containers), *biological* methods (release of transgenic vectors with reduced capacity to transmit disease and/or reproduce), and *environmental* approaches (environmental modification, environmental manipulation) [34,35]. When these methods are employed in combination, it constitutes an *integrated vector management approach* [35,36]. The control of *Ae. aegypti* has primarily relied on the use of chemicals, particularly pyrethroids [37]. This has led to the onset of mass insect resistance [38–40] through the following four primary mechanisms: behavioral resistance, reduced penetration/cuticular resistance, metabolic detoxification, and target site resistance [41]. Thus, while chemicals serve as a useful tool, they need to be used in moderation within targeted approaches that apply lesser quantities of chemicals to delay or reduce the rate at which insects build resistance. Insecticides and larvicides, even at sublethal doses, have been shown to affect the different mechanisms associated with *Ae. aegypti's* ability to reproduce such as their fecundity [42–44], egg hatching [44], immature development [44–46], adult longevity [47], sex proportion [48], adult size [48], and blood feeding [49,50]. One such targeted approach to applying chemicals, as a component in an integrated management plan, would be to treat water storage containers in or around the home.

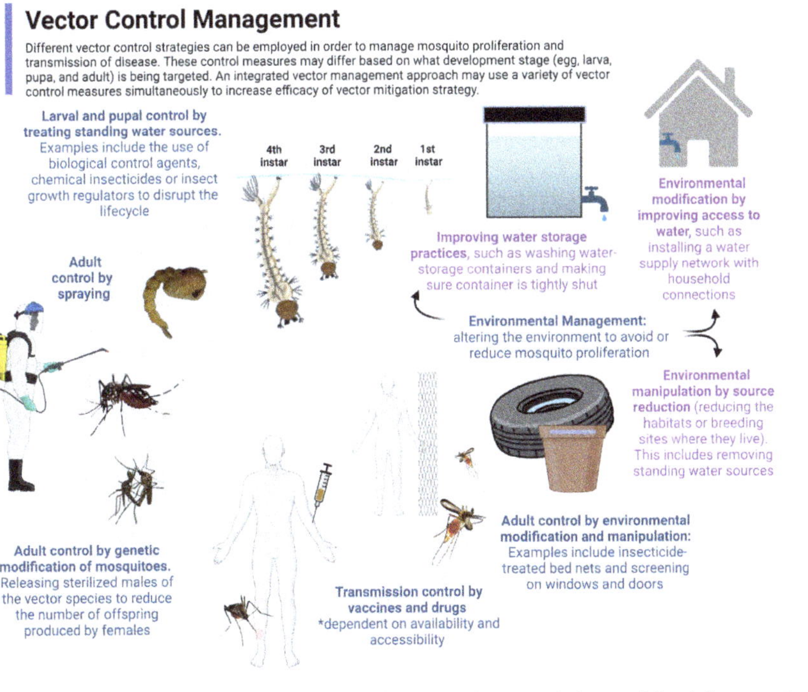

Figure 2. Vector control strategies that target different developmental phases of the *Aedes aegypti*. Vector control examples (biological, chemical, environmental management, and integrated practices) [35,51,52]. Figure created with Biorender.com.

In 2020, around one in four people lacked safely managed drinking water in their homes [53]. For regions experiencing water shortages and water stress, particularly within communities without access to uninterrupted and reliable piped water connections [54,55], water storage within or around the home is a common coping strategy [56,57]; however, unsafe water storage practices, such as those illustrated in Figure 3, creates ideal breeding grounds for mosquitoes, leading to their proliferation [29,30,58–62] and consequently increasing the risk of contracting VBDs [60]. For this reason, it is important to implement proper water storage and hygiene practices, such as covering and/or treating containers, to reduce the risk of disease transmission [61]. Chemicals commonly used in HWS containers include temephos [63–68], diflubenzuron [69–74], pyriproxyfen [75–80], Bacillus thuringiensis israelensis (Bti) [81–83], Spinosad DT [69,84,85], and various plant oils [86,87]. The *Ae. aegypti*, among other species, have demonstrated resistance to many of these common insecticides currently employed in HWS containers, such as temephos, within the context of large vector-borne disease mitigation efforts [39,64,88–92]. Thus, in this study, we assess the efficacy of water disinfectants to mitigate larval growth, reduce adult mosquito emergence, and ensure water being stored is safe for human consumption. Since clean water access and VBD control are often interconnected, this approach can inform the design of interventions for water storage containers, such as improved point-of-use water treatment (POUWT) technologies that can address both issues simultaneously.

Figure 3. Water storage containers, such as buckets, drums, or jars, are commonly used to store and access water for daily use, particularly in settings with limited access to continuous water supply. Poor water storage practices can result in standing water, providing ideal conditions for mosquito-breeding sites. Addressing these issues typically falls under the umbrella of environmental management in integrated vector management. Photos captured by Sydney Turner in in Dzimali, South Africa, March of 2020, of water storage containers surrounding two different households. Standard sizes of water storage containers surrounding homes in these images typically range from 50 to 200 L, reflecting common household and community water storage practices in many regions.

This paper examines the larvicidal effects of water disinfectants commonly used in point-of-use water treatment (POUWT) technologies, on juvenile *Ae. aegypti* at concentrations within drinking water quality guidelines established by the U.S. Environmental Protection Agency (EPA) [93] and the World Health Organization (WHO) [94–96]. The study evaluates the effectiveness of sodium hypochlorite, silver nitrate, and copper sulfate pentahydrate for mosquito control while ensuring treated water remains safe for human consumption. The findings provide insights for practical, integrated public health strategies that address both water quality and VBDs, enabling communities, governments, and organizations to implement effective, sustainable, and context-specific solutions. The study did find that elongated exposure to silver nitrate, copper sulfate pentahydrate, and sodium hypochlorite at these low concentrations negatively impacted juvenile *Ae. aegypti* development. While the goal was to achieve 100% inhibition of emergence, the study was constrained by the drinking water regulations, thus not permitting the use of chemicals at concentrations deemed unsafe for human consumption.

2. Materials and Methods

To analyze the efficacy of water disinfectants for mosquito larval source management of *Ae. aegypti*, the bioassay experiments in this study were designed to resemble the context of treating household water storage containers. This section describes the methods used to culture and rear *Ae. aegypti* in the laboratory (Section 2.1), select disinfectants for the bioassays and determine the test concentrations (Section 2.2), and test silver nitrate, copper sulfate pentahydrate, and sodium hypochlorite against *Ae. aegypti* larvae (Section 2.3).

2.1. Culturing and Rearing

Ae. aegypti eggs were obtained commercially from Benzon Research, Inc. (Carlisle, PA, USA). The colony, derived from the USDA "Gainesville" strain, has been continuously colonized at Benzon Research since 1994. Eggs procured from Benzon were 2–3 weeks old. The mosquitoes were reared in the Water Quality Laboratory at the University of Virginia on a 12:12 h light–dark cycle. The Extech RHT20 Humidity and Temperature Datalogger (Teledyne FLIR LLC, Nashua, NH, USA) was used to monitor the environmental conditions. The *Ae. aegypti* eggs and larvae were cultured at 27.9 ± 0.2 (82.2 °F) in Sterlite plastic trays ($35.6 \times 27.9 \times 8.3$ cm) containing deionized (DI) water. The larvae were fed daily with ground larval food, a 3:1 mixture of liver powder–brewer's yeast (MP Biomedicals™, Solon, Ohio, USA). Five grams of this mixture was added to 400 mL water. The DI water was deoxygenated by adding 1/8 oz of the food slurry to the rearing trays. Twenty-four hours later, eggs attached to strips of paper were submerged into the trays. Larvae of an intended instar were collected for each experiment. No food was added to rearing trays on the day of the hatch. The larvae were fed 0.25, 0.5, and 1 oz on the first three days post hatching, respectively. After day 3, larvae feed between 1 and 1.5 oz/day until pupation.

2.2. Water Treatment Disinfectants

Ae. aegypti larvae were assessed against varying concentrations of silver nitrate, copper sulfate pentahydrate, and sodium hypochlorite. These disinfectants were selected for this study due to their widespread use in water treatment processes, particularly in point-of-use water treatment (POUWT) technologies commonly employed in resource-limited settings [97]. Chlorine, particularly in the form of sodium or calcium hypochlorite, is one of the most widely used disinfectants in water treatment processes worldwide due to its strong oxidizing properties which allow it to eliminate a broad range of common water-borne pathogens such as *Escherichia coli*, *Legionella*, *Salmonella* Typhi, *Shigella*, adenoviruses, norovirus, and rotaviruses [97–99]. While cost effectiveness and availability make chlorine an attractive option for water treatment interventions, the taste and odor of chlorine has made it less culturally relevant in certain contexts [100,101], as well as its production of harmful disinfection byproducts at high concentrations [102]. Thus, silver and copper have gained traction as effective alternatives in water treatment applications [97,103–113]. Silver has become increasingly prominent as a disinfecting agent in water purification systems because of its high efficacy against killing Gram-negative bacteria at very low concentrations, as well as its minimal impact on the taste or odor of treated water [94,114,115]. While copper has been observed in many studies to be less potent than silver at similar concentrations, it has been demonstrated as effective against both Gram-negative and Gram-positive bacteria, copper's higher drinking water quality guideline allows for greater flexibility in dosing and copper is approximately ten times less expensive than silver [97,113,116]. Established microbial efficacy ranges reported in previous laboratory and field studies for chlorine, silver, and copper provided the rationale for testing these water disinfectants at similar ranges for larvicidal efficacy.

The concentrations of disinfectants used in this study were selected based on established drinking water quality guidelines (DWQGs) set by public health agencies: the WHO [94–96] and the US EPA [93]. The DQWG values, reported in Table 2, served as the reference points for the creation of a range of concentrations to test each disinfectant that would maintain safety for human consumption when applied in the context of household water storage containers. For free chlorine, 2 mg/L was selected as the upper boundary condition as it aligns with the WHO guideline for POUWT contexts [117]. The WHO DWQG was chosen over the EPA's maximum contaminant level goal (MCLG: the level of a contaminant in drinking water below which there is no known or expected risk to health) of 4 mg/L [93] as it is more conservative, ensuring broader applicability across diverse global contexts, including taste and odor thresholds. The EPA's MCLG of 1.3 mg/L for copper [93] was selected as the upper boundary instead of the higher 2 mg/L guideline set by the WHO [95]. For silver, both the EPA and WHO DWQG align on a threshold of 100 μg/L, which was used as the upper boundary [93,96].

Table 2. Drinking water quality guidelines for disinfectants (silver, copper, and chlorine) and the low, medium, and high concentrations tested in larvicidal bioassays.

Disinfectant	Drinking Water Quality Guideline (DWQG)	Concentrations Tested (μg/L)		
		High (80–95% of DWQG)	Mid (40–50% of DWQG)	Low (20–25% of DWQG)
Silver (Ag): $AgNO_3$	EPA and WHO: 100 μg/L	20	40	80
Copper (Cu): $CuSO_4.5H_2O$	EPA: 1300 μg/L WHO: 2000 μg/L	300	600	1200
Chlorine (OCl^-/HOCl): NaOCl	EPA: 4000 μg/L(EPA) WHO: 2000 μg/L free chlorine dose for clear water (<10 NTU) and 4000 μg/L for turbid water (≥10 NTU) for POUWT	500	1000	2000

To avoid potential health effects associated with concentrations exceeding recommended DWQG for consumption (e.g., high levels of chlorine may cause eye and nose irritation and stomach discomfort) and to address social acceptability factors (e.g., chlorine taste thresholds in different populations, cost implications), a range of concentrations were tested, none of which were directly at the uppermost boundary identified from reviewing DWQG set by public health organizations. The concentration ranges selected for this study represent high-, mid-, and low-range dosing for water treatment. A mid-range concentration was chosen at approximately 40–50% of the DWQG. The lower range was determined by halving the selected mid-range value, which resulted in about 20–25% of the DWQG value, while the higher range was set at double the mid-range value, making it 80–95% of the DWQG value of the specified disinfectant. For example, the DWQG set by the United States EPA and WHO for silver is 100 μg/L and the concentrations tested in the present study represent 20%, 40%, and 80% of that DWQG. This approach aligns with the design and operational practices of POUWT technologies, which aim to dose disinfectants at levels below the maximum DWQG to ensure safety while simultaneously producing sufficient disinfectant to effectively target microbial disinfection of waterborne pathogens.

2.2.1. Silver Nitrate

The stock solution was made by dissolving 16.99 g of $AgNO_3$ powder (Artcraft Chemicals, CAS No. 7761-88-8, South Glens Falls, NY, USA) into 1000 mL of DI water. One milliliter of this 100 mM $AgNO_3$ stock solution (10.79 g of silver in 1000 mL) was added to 1077 mL of DI water to make a 10 mg/L silver solution. Serial dilutions of the stock solution (10 mg/L to 1 mg/L to 0.1 mg/L) were performed to make test concentrations of 80 μg/L (high dose), 40 μg/L (mid dose), and 20 μg/L (low dose). Nominal silver concentrations were confirmed by inductively coupled plasma mass spectrometric analysis (ICP-MS) using the Agilent 7900 ICP-MS instrument (Agilent, Santa Clara, CA, USA). Samples for the ICP-MS were prepared by adding 2% trace metal grade nitric acid (HNO_3, Fisher Chemical, Fair Lawn, NJ, USA).

2.2.2. Copper Sulfate Pentahydrate

The 100 mM stock solution was prepared by dissolving 25.22 g of copper sulfate pentahydrate ($CuSO_4.5H_2O$, Alfa Aesar, Thermo Fisher Scientific, CAS No. 7758-99-8, purity: 99%, Waltham, MA, USA) into 1000 mL of deonized (DI) water. The 100 mM was diluted to 10 mM by transferring 10 mL of the stock solution into a 100 mL volumetric flask and filling to the mark with DI water. A 1 mM solution was prepared by transferring 10 mL of the 10 mM solution into another 100 mL volumetric flask and diluted with DI water. For a 10 μM solution of copper sulfate pentahydrate (equivalent to 2497 μg/L), 1 mL of the 1 mM solution was transferred into a 1000 mL volumetric flask and diluted to the mark with DI water. Further serial dilutions were performed to make test concentrations of 1200 μg/L (high dose), 600 μg/L (mid dose), and 300 μg/L (low dose).The copper level was confirmed using the ICP-MS. Samples for the ICP-MS were prepared by adding 2% HNO_3.

2.2.3. Sodium Hypochlorite

Free chlorine is an indicator of the overall chlorine concentration in the water available for disinfection and lethality. Sodium hypochlorite (NaOCl) is a powerful oxidizing agent that can react with water to release free chlorine. Free chlorine is in a solution as hypochlorous acid (HOCl) or hypochlorite ion $(OCl)^-$. When sodium hypochlorite is added to water, the hydrolysis reaction occurs. These processes are represented in the following equations:

$$NaOCl \rightleftharpoons Na^+ + OCl^- \quad (1)$$

$$HOCl \rightleftharpoons H^+ + OCl^- \quad (2)$$

The dissociation of HOCl in water has an equilibrium constant, K_a, much smaller than 1, indicating only a small fraction of HOCl dissociates into H^+ and OCl^-, making it a weak acid.

$$K_a = \frac{[H^+][OCL^-]}{[HOCl]} = 3.5 \times 10^{-8} \quad (3)$$

where $[H^+]$ is the concentration of hydrogen ions, $[OCl^-]$ is the concentration of hypochlorite ions, and [HOCl] is the concentration of undissociated hypochlorous acid. Depending on factors such as pH, temperature, and pressure, the ratio of OCl^- and HOCl will change. For instance, higher temperatures might slightly increase the dissociation of HOCl in water. At lower pH, HOCl dominates over OCl^- and is more effective as a disinfectant. Depending on the conditions, the resulting mixture of free chlorine, OCl^-, and HOCl reacts with and disinfects microorganisms present in the water, making it safe for consumption or other uses.

The stock solution of sodium hypochlorite with 10–15% available chlorine was procured from Sigma-Aldrich (CAS No. 7681-52-9, St. Louis, MO, USA). Serial dilutions were performed to create a new stock solution at a concentration of 25 mg/L. Further serial dilutions to make free chlorine test concentrations of 0.5 mg/L (low dose), 1 mg/L (mid dose), and 2 mg/L (high dose) were performed directly before the start of an experiment. Concentrations of free chlorine were measured directly, utilizing the USEPA DPD Method 8021 for low-range free chlorine, before and after the addition of larvae, as well as at 4 and 8 h, using a HACH DR6000 spectrophotometer.

2.3. Survival Bioassays: Evaluation of Dose Response to Water Treatment Disinfectants

The main objective of the survival bioassays was to determine the optimum application of common water disinfectants silver, copper, and chlorine against juvenile *Ae. aegypti* to create recommendations for water treatment within household water storage containers for the purpose of vector control. Each experiment was conducted with three replicates of each test concentration and control. Each set of experiments was performed in triplicate on different days, using fresh solutions and batches of larvae each time to counter confounders in the bioassay. The resulting data, which included counts of larvae, pupae, and adult mosquitoes in each container, were subject to statistical analysis.

2.3.1. Experimental Setup

Experimental methods for the laboratory study were influenced primarily by *WHO's Guidelines for laboratory and field testing of mosquito larvicides* [118]. Per test concentration, 25 larvae were placed in aliquots of 200 mL prepared solution within a 250 mL beaker (3 replicates) for bioassays. Control beakers (three replicates) were filled with DI water. The larvae were transferred to test beakers by means of disposable transfer pipettes. The test beakers were held at the same environmental conditions described in Section 2.1: 27.9 ± 0.2 °C, photoperiod of 12L:12D.

Larval activity (surviving larvae, the emerging pupae, and adult mosquitoes) was recorded in both treated and control beakers. Observations were made every 24 h until completion of adult emergence in all treatment groups or until the experiment reached day 16 post exposure. Test and control beakers were covered with netting to prevent successfully emerged adults from escaping. The number of successfully emerged adults were confirmed by observations of the empty pupal cases left in the solution in the instance that an adult mosquito may have escaped. The water disinfectant's inhibitory effect on mosquito emergence was indicated by the observation of adult mosquitoes that failed to fully separate from their pupal cases and the presence of moribund or dead larvae and pupae. Larvae were considered moribund or dead when they did not move when their aquatic environment was disturbed and, furthermore, could not be induced to move even after being probed. Larvae matching this description were counted as dead larvae for calculating survival. To calculate mortality, any larvae that successfully completed their metamorphosis into adult mosquitoes were considered alive for the duration of the experiment.

This study examined two distinct larval stages: late third instar larvae and late first instar larvae. By testing these stages, the study aimed to illustrate two potential scenarios within the context of household water storage:

1. A mosquito deposits eggs in a water storage container, resulting in newly hatched larvae being exposed to a freshly applied disinfectant. This scenario evaluates whether

a POUWT technology introduced into the water storage container can effectively reduce the survival and emergence of newly hatched larvae.

2. Larvae are already present in the source water supplying the household water storage (HWS) system, allowing them to develop further before encountering the water storage container and disinfectant.

The older instar experiments took place over 16 days (when the control larvae reached full emergence and larvae in the treatment groups either emerged, died, or were moribund). If adult emergence in the control was less than 80%, the test was discarded and repeated. For the younger instar experiments, larvicidal data were collected for 3 days.

The older larvae were fed after observations on day 4, and the younger instar larvae were fed after observations on day 2. Two drops of the slurry were added per test beaker at each feeding, which occurred every other day until the experiment was terminated. While delaying the feeding, as described above, is suboptimal for typical lab-grown *Ae. aegypti* growth rates as observed in the literature, this particular food regiment was chosen for this particular study for two main reasons: (1) to simulate the scenario when the water within the household water storage container has been disinfected/safe to drink, and (2) to decrease the confounding variable of the water treatment disinfectant interacting with the food slurry (e.g., volatility of the chlorine, interactions between silver or copper ions and brewer's yeast or liver powder that could affect toxicity).

2.3.2. Data Analysis

Larvicidal activity was calculated using the WHO's bioassay protocol [118] and methods established in Ngonzi et al. [119]. Data from all replicates in an experiment were pooled for analysis. Where the emergence was between 80% and 95%, mortality was calculated using Abbott's (1925) formula [120].

$$\text{Survival (\%)} = 100 - ((C - T)/C * 100) \quad (4)$$

where C = percentage survival in the untreated control and T = percentage survival in the treated sample. Larvae that developed into successfully emerging adults was expressed in terms of emergence:

$$\text{Emergence (\%)} = 100 - ((C - T)/C * 100) \quad (5)$$

where C = percentage emergence in the untreated control and T = percentage emergence in the treated sample. Inhibition of emergence (IE) was calculated on the basis of the number of larvae exposed. IE% is calculated using the following formula:

$$\text{Inhibition of Emergence (IE\%)} = 100 - (T * 100)/C \quad (6)$$

where T = percentage emergence in treated batches and C = percentage emergence in the control. Abbot's correction was also applied when appropriate according to the WHO guidelines for laboratory testing of mosquito larvicides [118].

All statistical analyses were performed using R (version 4.2.2) and RStudio (2022.07.2 Build 576). To assess the dose response of silver, copper, and chlorine on the number of larvae, pupae, and emerged adult mosquitoes, the sample mean, standard deviation (SD), and standard error of mean (SEM) were calculated on the observed data.

Probit analysis, a method first published in *Science* by Chester Ittner Bliss in 1934 [121] and then popularized by the work of Finney [122], is commonly used in toxicology in order to analyze the relationship between a dose and a response [123]. To investigate the impact of the treatments on the probability that a larva survives and/or has emerged, a mixed-effect probit regression model was fit to the data in R. The response variable was the classification of percent survival or percent emergence, while the independent variables were time (days), dosage, and their interaction. A natural spline with two degrees of freedom was applied to time to account for non-linearity in the rate of change per day. Each experiment was treated as a random effect. The function *glmer* (which stands for "generalized linear mixed effects regression"), from the *lme4* package, was employed to produce predictive models for each of the treatments. The function *emmeans* (which stands for "estimated marginal means") was used to perform pairwise comparisons of the estimated means.

For the inhibition of emergence predictive model, day 16 post exposure values for the control were obtained from the emergence model. Bootstrapping was used to generate a confidence interval using the upper and lower limits of the 95% confidence interval for emergence. The *bootMer* (which stands for "bootstrap for mixed effects models") function simulated 100 samples based on the bounds of the model.

The probit regression curves presented in the analysis represent the probability of the binary response variable taking on the value of 1 ("alive" or "has emerged") as a function of the predictor variables with a 95% confidence interval. Thus, the probit regression model predicts the probability of survival or emergence for the juvenile *Ae. aegypti* over the course of the experiment. Within the analysis in this paper, extrapolation of the inhibition of emergence model beyond the experimental range of this study was performed to estimate the conditions required to achieve a predicted 100% inhibition of emergence by day 16. This estimation was used to assess whether the required concentration would remain within drinking water quality guidelines. Further experimentation would be necessary to validate these predictions and confirm whether 100% inhibition can be achieved under these conditions.

3. Results and Discussion

Within this study, testing larvae at different life stages allowed the researchers to understand if susceptibility of the larvae to the water disinfectants' effects was dependent on the age of the larvae as reflected in previous studies [124–126]. For older instar experiments, survival and emergence data are presented over a 16-day post exposure period. For the younger larvae, results reflect survival over the course of 72 h of exposure to the treatments.

3.1. Late Third Instar Experiments

This section reports the survival and emergence results for the experiments conducted on late third instar *Ae. aegypti* utilizing silver nitrate, copper sulfate pentahydrate, and sodium hypochlorite treatments. Observed data (summarized in Table 3) serve as input for probit regression models. Observed data results are expressed in terms of Percentage Mean ± SEM, and the model data results (summarized in Table 4) are expressed in terms of Predicted Probability Mean [Upper 95% Confidence Interval, Lower 95% Confidence Interval].

3.1.1. Silver Nitrate Exposure to Older Instar

Introducing silver nitrate into the aquatic environment of juvenile *Ae. aegypti* negatively impacted their growth and development at all concentrations tested (20, 40, and 80 μg/L Ag) when compared to the growth and development of the control larvae by the end of the experimental period ($p_{\text{control}} < 0.001$). Within the environmental conditions tested, silver nitrate treatment shows great potential to inhibit emergence. The data reflected a dose–response relationship: 80 μg/L achieved the highest inhibition of emergence of 88.66 ± 5.70% on day 16 of exposure, followed by 40 μg/L at 81.56 ± 3.39% and 20 μg/L at 72.40 ± 4.13%. Emergence and survival values are comparable, suggesting that most of the larvae that survived the treatment emerged into adult mosquitoes during the experimental period.

Table 3. Observed data for silver nitrate ($AgNO_3$) treatments of 20, 40, and 80 µg/L; copper sulfate pentahydrate ($CuSO_4.5H_2O$) treatments of 300, 600, and 1200 µg/L; and for sodium hypochlorite (NaOCl) treatments of 0.5, 1, and 2 mg/L free chlorine (OCl^-/HOCl). Standard deviation (St. dev) and standard error of mean (SEM) is presented with the mean percentage survival, emergence percentage, and inhibition of emergence (IE) percentage of older instar *Ae. aegypti* larvae on day 16 post exposure. Data are corrected with Abbot's formula (1925).

Observed Data												
Silver Treatment	20 µg/L Ag			40 µg/L Ag			80 µg/L Ag			Control		
Variable	Mean	St. Dev	SEM	Mean	St. Dev	SEM	Mean	St. Dev	SEM	Mean	St. Dev	SEM
Survival (%)	26.85	6.25	3.61	17.09	4.58	2.64	10.75	8.64	4.99	91.11	0.77	0.44
Emergence (%)	25.01	6.44	3.72	16.69	5.24	3.02	10.29	8.94	5.16	90.67	1.33	0.77
IE (%)	72.40	7.16	4.13	81.56	5.86	3.39	88.66	9.87	5.70			
Copper Treatment	300 µg/L Cu			600 µg/L Cu			1200 µg/L Cu			Control		
Variable	Mean	St. Dev	SEM	Mean	St. Dev	SEM	Mean	St. Dev	SEM	Mean	St. Dev	SEM
Survival (%)	42.55	7.13	4.12	25.80	10.91	6.30	11.92	7.50	4.33	89.78	0.77	0.44
Emergence (%)	40.30	5.38	3.11	24.88	9.94	5.74	11.44	7.51	4.34	89.33	0.00	0.00
IE (%)	54.89	6.02	3.48	72.15	11.12	6.42	87.19	8.41	4.86			
Chlorine Treatment	500 µg/L Free Cl			1000 µg/L Free Cl			2000 µg/L Free Cl			Control		
Variable	Mean	St. Dev	SEM	Mean	St. Dev	SEM	Mean	St. Dev	SEM	Mean	St. Dev	SEM
Survival (%)	58.40	21.88	12.63	24.75	12.85	7.42	17.39	14.70	8.49	90.67	2.67	1.54
Emergence (%)	56.53	20.75	11.98	24.98	12.99	7.50	17.59	14.95	8.63	89.78	2.78	1.60
IE (%)	37.44	21.35	12.32	72.46	13.77	7.95	80.65	16.04	9.26			

Table 4. Predicted mean probability of survival and predicted probability of emergence of older instar *Aedes aegypti* on day 16 after contact with silver nitrate ($AgNO_3$) treatments of 20, 40, and 80 µg/L; copper sulfate pentahydrate ($CuSO_4.5H_2O$) treatments of 300, 600, and 1200 µg/L; and for sodium hypochlorite (NaOCl) treatments of 500, 1000, and 2000 µg/L free chlorine (OCl^-/HOCl). Modeled inhibition of emergence (IE) percentage produced with the probit regression model for emergence. Data are corrected with Abbot's formula (1925).

Probit Regression Model				
Silver Treatment	**20 µg/L Ag**	**40 µg/L Ag**	**80 µg/L Ag**	**Control**
Survival (%)	23.57 [18.10, 28.57]	15.91 [11.49, 20.01]	11.21 [8.25, 14.29]	91.39 [87.75, 94.15]
Emergence (%)	21.04 [15.48, 29.37]	14.10 [8.79, 20.55]	8.20 [4.81, 11.96]	92.84 [88.22, 95.93]
IE (%)	78.96 [71.87, 85.19]	85.90 [78.13, 90.51]	91.80 [86.72, 94.84]	
Copper Treatment	**300 µg/L Cu**	**600 µg/L Cu**	**1200 µg/L Cu**	**Control**
Survival (%)	42.56 [34.02, 53.57]	27.50 [19.17, 37.88]	12.52 [7.05, 20.42]	90.52 [84.18, 94.76]
Emergence (%)	42.65 [36.46, 49.15]	24.43 [18.71, 30.17]	11.77 [7.96, 15.68]	93.36 [90.36, 95.58]
IE (%)	57.35 [49.48, 64.47]	75.57 [70.91, 82.82]	88.23 [84.47, 91.90]	
Chlorine Treatment	**500 µg/L Free Cl**	**1000 µg/L Free Cl**	**2000 µg/L Free Cl**	**Control**
Survival (%)	57.53 [8.14, 17.24]	19.77 [14.32, 26.56]	12.19 [7.64, 18.30]	91.63 [86.98, 94.90]
Emergence (%)	58.34 [45.76, 69.53]	21.57 [14.00, 34.51]	13.73 [7.55, 21.99]	92.85 [86.58, 96.58]
IE (%)	41.66 [31.12, 54.55]	78.43 [67.94, 86.38]	86.27 [79.95, 92.99]	

Figure 4 depicts the experimental data for survival with curves fitted by probit regression overlaid, illustrating the relationship between the model and observed experimental data. The probit regression curves for probability of survival across the different treatments are plotted together on a single graph in Figure 5. The model predicted that 11.21% [8.25, 14.29] of the juvenile *Ae. aegypti* treated with 80 μg/L survived by day 16, followed by 15.91% [11.49, 20.01] at 40 μg/L, and 23.57% [18.10, 28.57] at 20 μg/L.

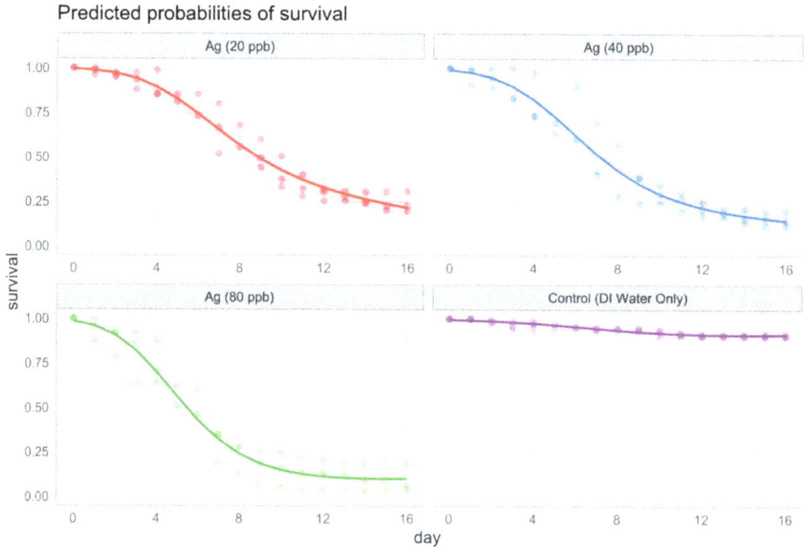

Figure 4. The predicted probabilities of survival of older instar *Ae. aegypti* during the experimental period for larvae treated with silver nitrate (20, 40, and 80 μg/L) and controls. The shaded areas represent the 95% confidence interval.

As seen in Figure 5, there is a clear indication that the final proportion of individuals that have emerged into adult mosquitoes by day 16 in the treatment groups is significantly lower than that of the control ($p_{Cntrl–20Ag} \leq 0.001$, $p_{Cntrl–40Ag} \leq 0.001$, $p_{Cntrl–80Ag} \leq 0.001$). The visual interpretation of the model shows separation between all treatment groups from the control by day 7, with the first four days of the experiment characterized by minimal emergence in the controls. While the predicted probability of emergence for the controls by day 16 was 92.84% [88.22, 95.91], the emergence predicted for larvae in water containing 20, 40, and 80 μg/L Ag was 21.04% [15.48, 29.37], 14.10% [8.79, 20.55], and 8.20% [4.81, 11.96], respectively. In the model, the 20 μg/L and 40 μg/L Ag treatments were found to be significantly different [$p_{40Ag–20Ag} = 0.032$] for the predicted probability that a larva has emerged by day 16.

While the ultimate objective of vector control is to achieve a 100% inhibition of emergence (IE% = 100%), excessive use of disinfectants beyond what is necessary to achieve this goal may lead to diminishing returns in terms of cost-effectiveness and human health; thus, it is important to balance the concentration of the chemical used with the desired outcome. Depicted in Figure 6, extrapolation of the IE% model outside of this study's experimental range suggests that 100% inhibition of emergence on day 16 would occur at a concentration of 117.32. μg/L Ag, which is slightly greater than the drinking water quality guideline.

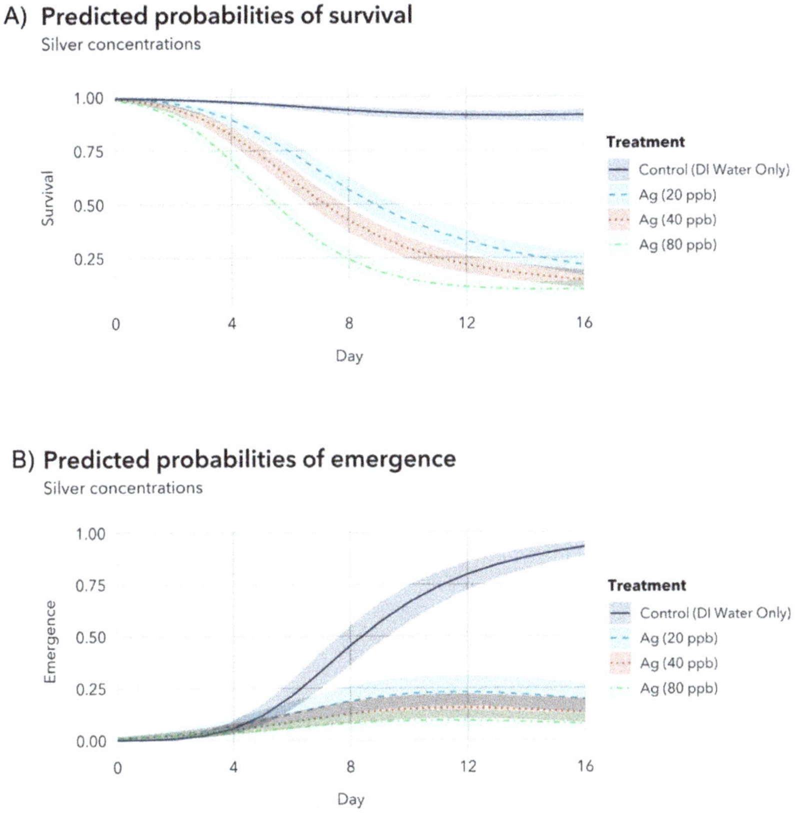

Figure 5. Predicted probabilities of (**A**) survival of *Ae. aegypti* larvae exposed to 20, 40, and 80 μg/L of silver nitrate and (**B**) emergence of *Ae. aegypti* larvae into adult mosquitoes following exposure to these varying concentrations of silver nitrate. Shaded areas indicate the 95% confidence interval.

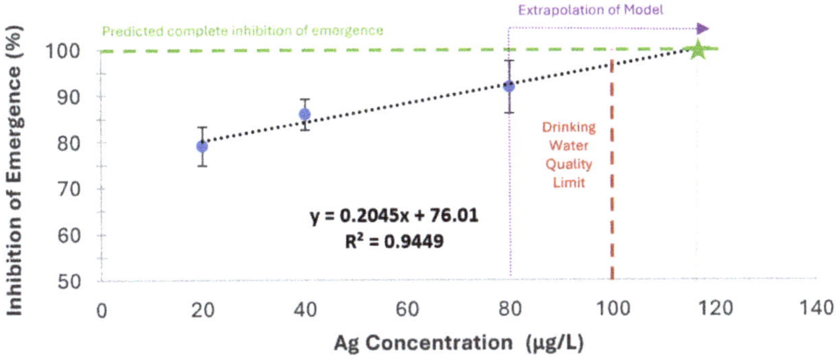

Figure 6. The modeled inhibition of emergence of *Ae. aegypti* larvae on day 16 of exposure to silver nitrate, with concentrations of 20, 40, and 80 μg/L Ag. Model error bars represent a 95% confidence interval.

To validate this prediction, more experiments would need to be conducted within a wider range of silver concentrations, including those concentrations not safe for human consumption, to validate whether linear regression is the most appropriate model for this relationship between concentration of the larvicide and percent inhibition of emergence.

Prolonged exposure of the larvae to higher concentrations of the water disinfectant may lead to compound effects (e.g., threshold toxicity concentration reached for the larvae or cumulative developmental effects over the duration of the experiment) that may cause the relationship to be more non-linear in fit.

3.1.2. Copper Sulfate Pentahydrate Exposure to Older Instar

All copper concentrations tested (300, 600, and 1200 μg/L Cu) resulted in decreased survival and emergence of juvenile *Ae. aegypti* compared to the control group by the end of the experimental period ($p_{\text{control}} < 0.001$), as reported in Table 3. The control group exhibited a survival and emergence of roughly 90%, while those that came into contact with 300 μg/L Cu had a survival and emergence of only 42.55 ± 4.12% and 40.30 ± 5.38%, respectively. Increasing the dose to 1200 μg/L resulted in a more drastic reduction in both the emergence and survival of the *Ae. aegypti*, with on average 11.44 ± 4.34% having emerged by day 16 and 11.92 ± 4.33% probability of survival. A dose response is reflected within the inhibition of emergence data, with 300, 600, and 1200 μg/L Cu, resulting in IE% on day 16 of 54.89 ± 3.48%, 72.15 ± 6.42%, and 87.19 ± 4.86%, respectively.

The probit model for the predicted probability of survival of juvenile *Ae. aegypti* in the presence of different concentrations of copper sulfate pentahydrate and the predicted probability that a larva has emerged is illustrated in Figure 7. When compared to the control group on day 16, the predicted survival and emergence of *Ae. aegypti* exposed to copper treatments was significantly lower at all concentrations tested ($p_{\text{cntrl–300Cu}} \leq 0.001$, $p_{\text{cntrl–600Cu}} \leq 0.001$, $p_{\text{cntrl–1200Cu}} \leq 0.001$). As was observed with silver nitrate, a dose response is evident in emergence with copper sulfate ($p_{\text{600Cu–300Cu}} \leq 0.001$, $p_{\text{1200Cu–600Cu}} \leq 0.001$, $p_{\text{1200Cu–300Cu}} \leq 0.001$).

The probit regression model was applied to create a model for inhibition of emergence, depicted in Figure 8. At the conclusion of the experimental period, this model predicts that 1200 μg/L concentration results in an 88.23% [84.47, 91.90] inhibition of emergence, which is roughly 1.5× more effective than the 300 μg/L treatment (IE%: 57.35% [49.48, 64.47]); however, the concentration is four times greater. From the linear regression model which establishes a relationship between concentration of copper and percentage inhibition of emergence, and through extrapolation, it is predicted that 100% inhibition of emergence could potentially occur at roughly 1500 μg/L Cu. This concentration is over the drinking water quality standard set by the EPA, but does not surpass the guideline set by the WHO. The linear regression line appears to underestimate the inhibition of emergence potential of the 600 μg/L Cu treatment, which may indicate that there is a non-linear relationship between the concentration of the water disinfectant and percent inhibition of emergence at these low concentrations after a prolonged exposure to the treatment.

3.1.3. Sodium Hypochlorite Exposure to Older Instar

The results indicate that all concentrations of free chlorine tested (500, 1000, and 2000 μg/L) adversely affected the development and survival of the *Ae. aegypti* when compared to the control group ($p_{\text{Cntrl–500Cl}} \leq 0.001$, $p_{\text{Cntrl–1000Cl}} \leq 0.001$, $p_{\text{Cntrl–2000Cl}} \leq 0.001$), as reported in Table 3. The control group survival and emergence were both near 90% by day 16, while the treatment groups of 500, 1000, and 2000 μg/L free chlorine caused survival to drop down to 58.40 ± 12.63%, 24.75 ± 7.42%, and 17.39 ± 8.49%, respectively. Inhibition of emergence for free chlorine concentrations on day 16 for 500, 1000, and 2000 μg/L were 37.44 ± 12.32%, 72.46 ± 7.95%, and 80.65 ± 9.26, respectively.

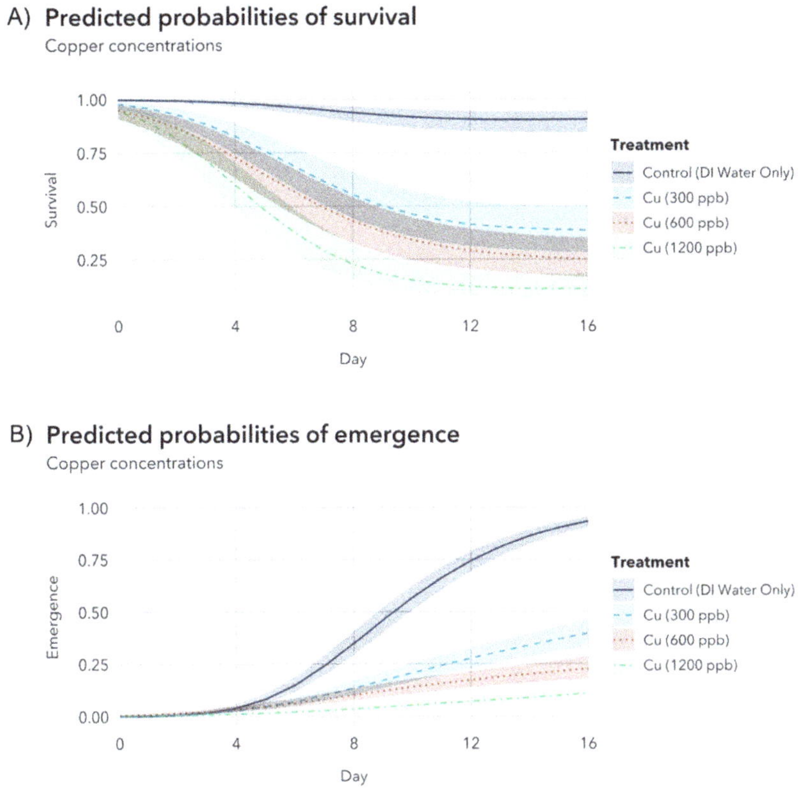

Figure 7. Predicted probabilities (**A**) of survival and (**B**) that *Ae. aegypti* larvae have emerged into adult mosquitoes after being exposed to copper sulfate pentahydrate at concentrations of 300, 600, and 1200 µg/L. Shaded areas indicate the 95% confidence interval.

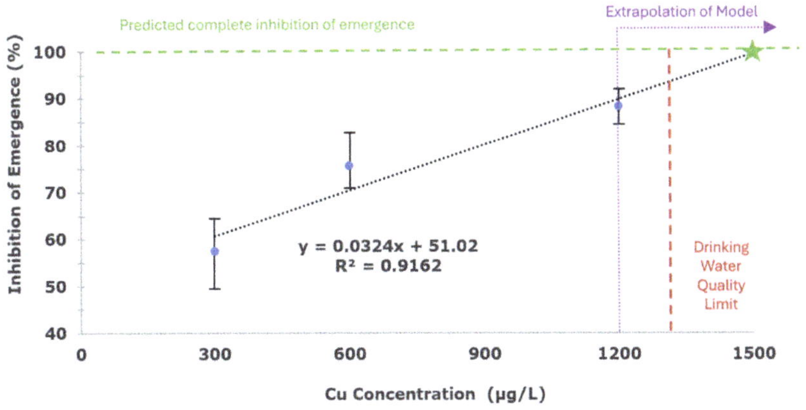

Figure 8. The modeled inhibition of emergence of *Ae. aegypti* larvae on day 16 of exposure to copper sulfate pentahydrate, with concentrations of 300, 600, and 1200 µg/L Cu. Observed error bars represent a 95% confidence interval.

The probit model that predicts the survival of juvenile *Ae. aegypti* after they have encountered a single dose of free chlorine of concentration 500, 1000, and 2000 µg/L is illustrated in Figure 9. Similar to the response observed with the other two water disinfectants, a dose response was evident with free chlorine, with predicted survival probabilities of 57.53% [8.14, 17.24], 19.77% [14.32, 26.56], and 12.19% [7.64, 18.30] at the end of the observational period for 500, 1000, and 2000 µg/L, respectively. Similarly, the chlorine treatment decreased the larvae's ability to reach adulthood ($p_{Cntrl\text{–}500Cl} \leq 0.001$, $p_{Cntrl\text{–}1000Cl} \leq 0.001$, $p_{Cntrl\text{–}2000Cl} \leq 0.001$). As the concentration of free chlorine increased, the predicted probabilities that a larva has emerged decreased: 58.34% [45.76, 69.53] for 500 µg/L, 21.57% [14.00, 34.51] for 1000 µg/L, and 13.73 [7.55, 21.99] for 2000 µg/L. These results suggest that chlorine has a dose-dependent effect on the development of *Ae. aegypti* ($p_{500Cl\text{–}1000Cl} \leq 0.001$, $p_{2000Cl\text{–}1000Cl} = 0.008$, $p_{2000Cl\text{–}500Cl} \leq 0.001$). Depicted in Figure 9, the model also portrays that while a single dose of 500 µg/L free chlorine reduced emergence, the treatments of 1000 and 2000 µg/L were roughly twice as effective.

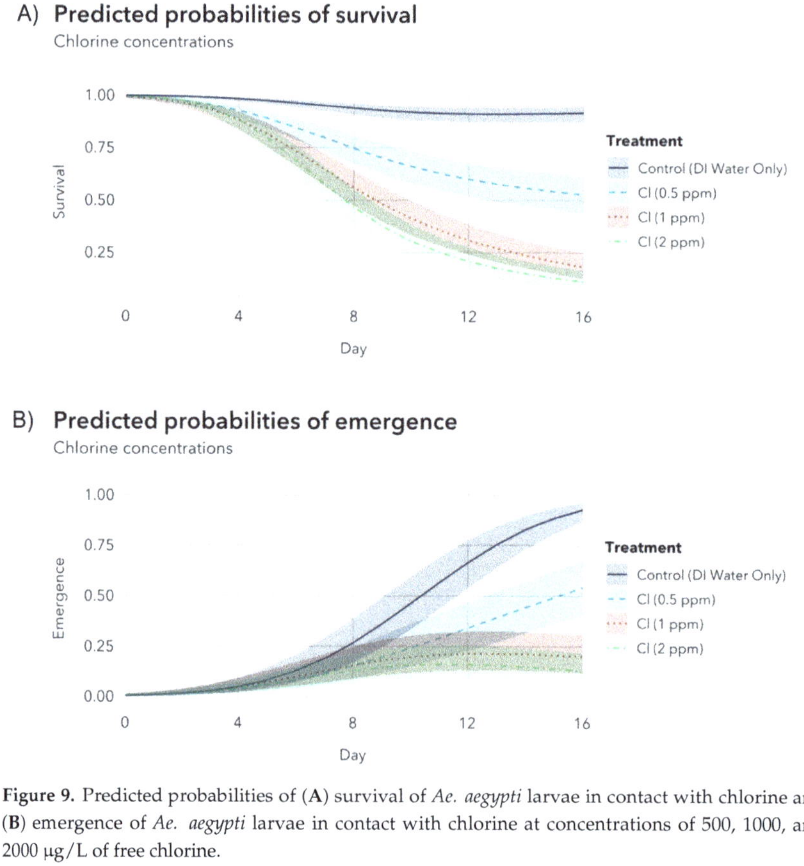

Figure 9. Predicted probabilities of (**A**) survival of *Ae. aegypti* larvae in contact with chlorine and (**B**) emergence of *Ae. aegypti* larvae in contact with chlorine at concentrations of 500, 1000, and 2000 µg/L of free chlorine.

When considering the other treatments tested in this study (silver and copper), there is greater variability in the chlorine data for larval emergence, as indicated by the larger standard deviation in the observed data and the wider confidence interval in the model. Three possible explanations for this variation include the following:

1. The volatility of chlorine, which makes it less stable compared to silver and copper during long exposure periods, especially in elevated temperatures that can increase chlorine evaporation;

2. Differences in the rate of free chlorine consumption within each beaker, potentially caused by larvae, external contamination sources, or variations in pH and temperature;
3. The ratio of hypochlorous acid (HOCl) to hypochlorite ion (OCl^-), as HOCl is the more effective disinfectant of the two forms and may also exhibit stronger larvicidal effects (ratio is influenced by the pH of the solution; higher pH values correspond to a greater concentration of OCl^-).

These factors lead to different contact times and forms of chlorine exposure for the larvae, depending on the specific environmental conditions in each beaker.

The relationship between free chlorine concentration and the inhibition of *Ae. aegypti* emergence is shown in Figure 10. Data extrapolation suggests that a free chlorine concentration of approximately 2.3 mg/L would completely inhibit larval emergence under the tested environmental conditions. The linear model for inhibition of emergence underestimates the potential for the 1 mg/L free chlorine treatment, indicating that prolonged larval exposure may not follow a strictly linear relationship.

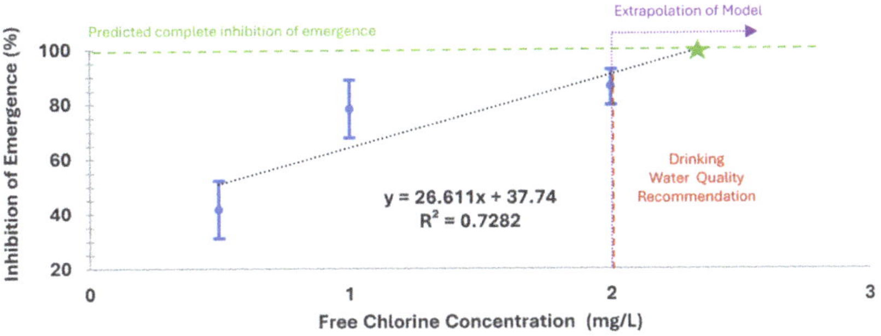

Figure 10. The modeled inhibition of emergence of *Ae. aegypti* larvae on day 16 of exposure to copper sulfate, with concentrations of 0.5, 1, and 2 mg/L free chlorine. Observed error bars represent a 95% confidence interval.

3.2. *Younger Instar Experiments*

This section reports the results for the experiments conducted on late first instar *Ae. aegypti* utilizing silver nitrate, copper sulfate, and sodium hypochlorite treatments. Each of the following sections will start with observed data (summarized in Table 5), which will serve as input for probit regression models (represented in Table 6). Significant differences will then be examined through analysis of the models. Observed data results are expressed in terms of Percentage Mean ± SEM, and the model data results are expressed in terms of *Predicted Probability Mean (UCI_95%, LCI_95%)*.

3.2.1. Silver Nitrate Exposure to Younger Instar

Silver nitrate was effective against first instar *Ae. aegpyti*, contributing to the observed 57.73 ± 7.10%, 36.44 ± 11.23%, and 7.95 ± 6.41% survival of the larvae after 72 h of exposure to 20, 40, and 80 μg/L treatments, respectively. Table 5 presents the observed survival of larvae after 24, 48, and 72 h exposure to the silver nitrate treatments. The probit regression model was built using the observed data as the input. Table 6 and Figure 11 present the predicted probability of survival models. By 72 h, each of the treatments performed significantly differently from each other ($p_{40Ag\text{–}20Ag} \leq 0.001$, $p_{80Ag\text{–}40Ag} \leq 0.001$, $p_{80Ag\text{–}20Ag} \leq 0.001$), with the highest concentration of silver nitrate (80 μg/L) leading to the greatest mortality.

Table 5. Observed data for silver nitrate ($AgNO_3$) treatments of 20, 40, and 80 μg/L, copper sulfate pentahydrate ($CuSO_4.5H_2O$) treatments of 300, 600, and 1200 μg/L, and sodium hypochlorite (NaOCl) treatments of 500, 1000, and 2000 μg/L. Standard deviation (St. dev) and standard error of mean (SEM) is presented with the mean percentage survival of younger instar *Ae. aegypti* larvae after 24, 48, and 72 h of exposure. Data are corrected with Abbot's formula (1925).

Silver Treatment	20 μg/L Ag			40 μg/L Ag			80 μg/L Ag			Control		
Time (h)	Mean	St. Dev	SEM	Mean	St. Dev	SEM	Mean	St. Dev	SEM	Mean	St. Dev	SEM
24	92.78	4.47	2.58	89.70	1.92	1.11	86.06	7.46	4.31	99.11	1.54	0.89
48	81.13	11.95	6.90	65.55	2.36	1.36	54.11	15.69	9.06	94.22	2.04	1.18
72	57.73	12.30	7.10	36.44	19.45	11.23	7.95	11.10	6.41	82.22	2.04	1.18
Copper Treatment	300 μg/L Cu			600 μg/L Cu			1200 μg/L Cu			Control		
Time (h)	Mean	St. Dev	SEM	Mean	St. Dev	SEM	Mean	St. Dev	SEM	Mean	St. Dev	SEM
24	82.59	14.96	8.64	79.47	14.77	8.53	68.31	24.64	14.23	99.11	1.54	0.89
48	57.24	25.06	14.47	49.21	25. 64	14.80	32.41	33.08	19.10	94.22	2.04	1.18
72	21.85	23.74	13.71	17.71	12.43	7.18	12.81	15.23	8.79	82.22	2.04	1.18
Chlorine Treatment	500 μg/L Free Cl			1000 μg/L Free Cl			2000 μg/L Free Cl			Control		
Time (h)	Mean	St. Dev	SEM	Mean	St. Dev	SEM	Mean	St. Dev	SEM	Mean	St. Dev	SEM
24	76.67	6.55	3.78	72.56	12.53	7.23	36.91	11.13	6.43	93.78	5.39	3.11
48	25.67	3.79	2.19	14.72	17.20	9.93	6.13	5.36	3.10	81.33	4.00	2.31
72	15.40	1.20	0.69	8.60	11.93	6.89	0.00	0.00	0.00	75.11	3.36	1.94

Table 6. Predicted mean probability of survival (%) of younger instar *Aedes aegypti* after 24, 48, and 72 h of exposure to silver nitrate treatments of 20 μg/L, 40 μg/L, and 80 μg/L; copper sulfate pentahydrate treatments of 300 μg/L, 600 μg/L, and 1200 μg/L; and sodium hypochlorite (NaOCl) treatments of 500, 1000, and 2000 μg/L. Data corrected by Abbot's formula (1925).

Silver Treatment	20 μg/L Ag	40 μg/L Ag	80 μg/L Ag	Control
t = 24 h	94.75 [91.50, 96.92]	91.51 [87.19, 94.63]	89.85 [85.30, 93.27]	99.24 [97.52, 99.81]
t = 48 h	79.20 [72.29, 84.97]	63.89 [55.72, 71.47]	50.09 [41.91, 58.26]	94.24 [90.01, 96.91]
t = 72 h	58.66 [49.92, 66.99]	36.87 [28.85, 45.52]	8.38 [4.90, 13.45]	82.43 [75.48, 87.98]
Copper Treatment	300 μg/L Cu	600 μg/L Cu	1200 μg/L Cu	Control
t = 24 h	88.71 [75.01, 95.97]	85.46 [69.89, 94.42]	74.39 [54.91, 88.25]	99.42 [96.88, 99.93]
t = 48 h	53.40 [32.80, 73.11]	44.86 [25.47, 65.60]	26.94 [12.57, 46.72]	95.02 [85.93, 98.66]
t = 72 h	20.67 [8.66, 39.21]	16.82 [6.59, 33.90]	11.38 [3.92, 25.69]	83.63 [66.76, 93.64]
Chlorine Treatment	500 μg/L Free Cl	1000 μg/L Free Cl	2000 μg/L Free Cl	Control
t = 24 h	77.29 [71.79, 82.13]	72.55 [66.53, 77.97]	41.02 [35.19, 47.05]	94.55 [91.65, 96.59]
t = 48 h	25.37 [20.28, 31.05]	14.57 [10.53, 19.56]	3.34 [2.02, 5.29]	80.39 [75.26, 84.82]
t = 72 h	15.24 [11.04, 20.38]	8.92 [5.75, 13.23]	0.80 [0.18, 2.87]	75.44 [69.54, 80.66]

3.2.2. Copper Sulfate Pentahydrate Exposure to Younger Instar

The results suggest that copper sulfate pentahydrate was effective in controlling first instar *Ae. aegypti*. The larvae's survival rates after exposure to 300, 600, and 1200 μg/L treatments for 72 h were 21.85 ± 13.71%, 17.71 ± 7.18%, and 12.81 ± 8.79%, respectively. The predicted probability of survival, presented in Figure 12, for first instar *Ae. aegypti* exposed to 300, 600, and 1200 μg/L treatments for 72 h were 20.67% [8.66, 39.21], 16.82% [6.59, 33.90], and 11.38% [3.92, 25.69], respectively. By the end of the experimental period, all treatments performed similarly in terms of producing larvicidal effects ($p_{CuTreatments–Control} \leq 0.001$; $p_{600Cu–300Cu} = 0.744$, $p_{1200Cu–600Cu} = 0.430$, $p_{1200Cu–300Cu} = 0.122$).

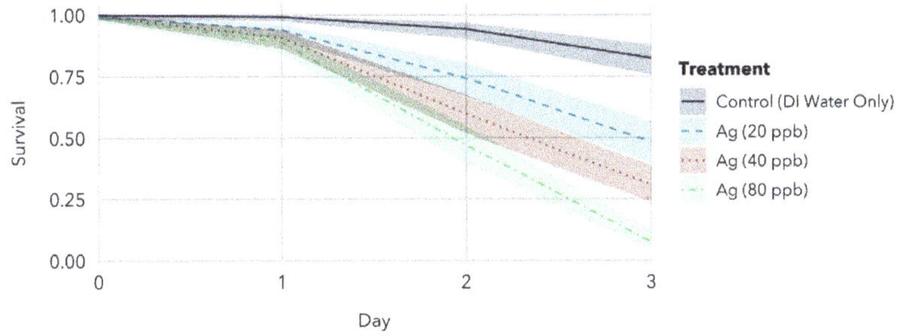

Figure 11. Predicted probabilities of survival of *Ae. aegypti* larvae in contact with silver nitrate concentrations of 20, 40, and 80 μg/L Ag.

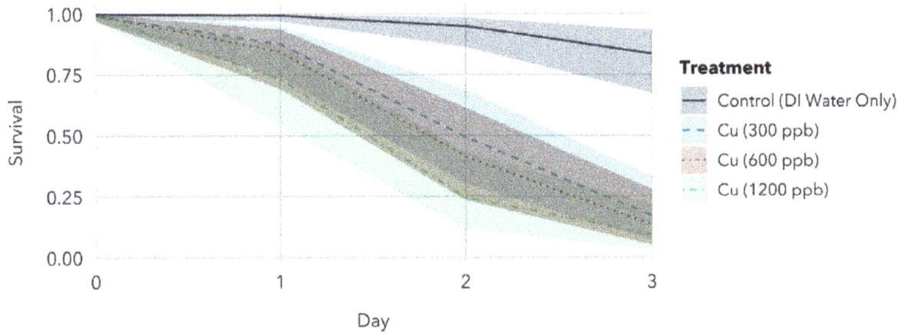

Figure 12. Predicted probabilities of survival of *Ae. aegypti* larvae in contact with copper sulfate pentahydrate concentrations of 300, 600, and 1200 μg/L Cu.

3.2.3. Sodium Hypochlorite Exposure to Younger Instar

This study finds that sodium hypochlorite shows great efficacy in managing young *Ae. aegypti* larvae. After being subjected to treatments of 500, 1000, and 2000 μg/L for 72 h, the larvae exhibited survival rates of 15.40 ± 0.69%, 8.60 ± 6.89%, and 0 ± 0%, respectively. The survival rates of the larvae following exposure to the sodium hypochlorite treatments for 24, 48, and 72 h are shown in Table 5. As illustrated in Figure 13, the first instar *Ae. aegypti* that were exposed to 500, 1000, and 2000 μg/L free Cl treatments for 72 h have a predicted probability of survival of 15.24% [11.04, 20.38], 8.92% [5.75, 13.23], and 0.80% [0.18, 2.87], respectively. While the 500 μg/L and 1000 μg/L free chlorine treatments were not statistically significant on day 3 ($p_{1000Cl–500Cl} = 0.161$), both treatments were statistically different from the better performance of the 2000 μg/L treatment ($p_{(2000Cl–500Cl)} \leq 0.001$; $p_{(2000Cl–1000Cl)} \leq 0.001$).

3.3. *Comparing and Contextualizing the Results*

To compare differences across water disinfectants, Figure 14 provides a summary of the model generated IE% data for older instar on day 16. Based on the model output, it was determined that within the low dosage range, the treatment of 20 μg/L Ag was notably more effective than both the 300 μg/L Cu treatment and the 500 μg/L free Cl treatment. The interpretation of the model data also demonstrated that as the concentrations of the water disinfectants increased into the mid-range and high range, the differences in efficacy between the various treatments became less pronounced. The analysis also considered the statistical significance across the treatments in order to determine if significant dose responses were occurring. The key findings of this analysis are summarized below for each of the three disinfectants for the late third instar *Ae. aegypti*:

- Silver nitrate treatments: The 80 μg/L Ag treatment was significantly different from the 20 μg/L treatment, but not significantly different from the 40 μg/L treatment.
- Copper sulfate pentahydrate treatments: The model found that all copper treatments were statistically significant from each other.
- Free chlorine treatments: The 500 μg/L free chlorine treatment was statistically significant from the 1000 μg/L and 2000 μg/L treatments; however, the 1000 μg/L and 2000 μg/L treatments were not statistically different from each other.

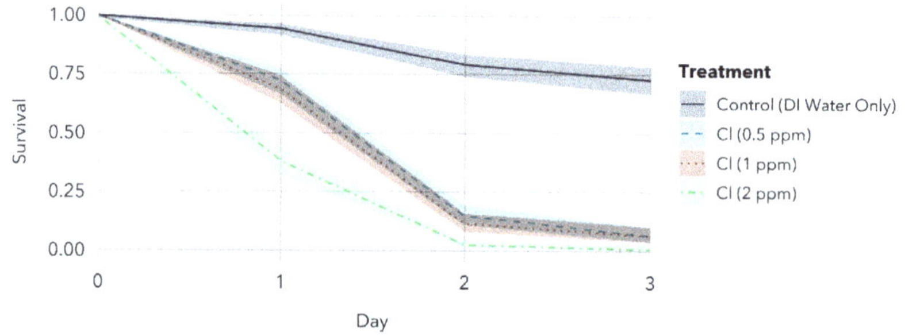

Figure 13. Predicted probabilities of survival of *Ae. aegypti* larvae in contact with chlorine at concentrations of 500, 1000, and 2000 μg/L of free chlorine.

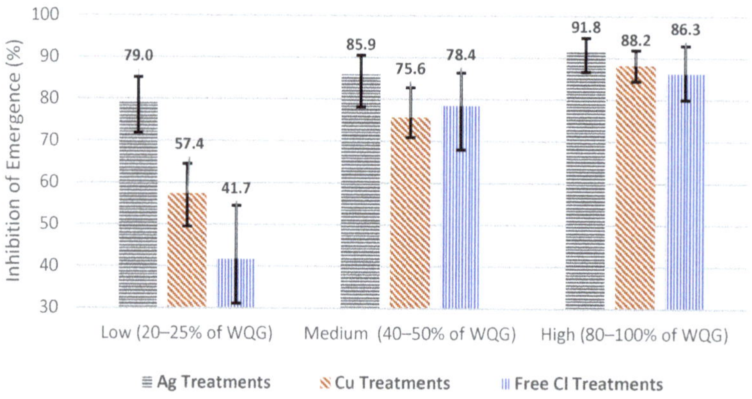

Figure 14. Summary of predicted probabilities of inhibition of emergence results regarding the efficacy of water disinfectants as larval control for older instar *Ae. aegypti*. IE% calculated from the model generated data from day 16. Error bars represent the 95% confidence interval.

While the study design revealed that the first instar larvae were highly vulnerable to all of the water disinfectants, the treatment that yielded the best results was 2000 μg/L Cl, with no larvae surviving the 72 h exposure period. Figure 15 displays the summary of the predicted probability of survival of the late first instar *Ae. aegypti* after 72 h of exposure to the water disinfectants. The relatively large error bars representing the 95% confidence interval in Figure 15 reflect the biological variability inherent in working with live mosquito larvae, particularly at concentrations that require extended time to achieve lethal effects. Each batch of *Ae. aegypti* mosquitoes may exhibit slight differences in susceptibility, as evidenced by variability seen even in the controls. This can be expected at such young and vulnerable stages and may suggest that *Ae. aegypti* egg batches tested were not equally fit or there were minor inconsistencies in environmental conditions or handling.

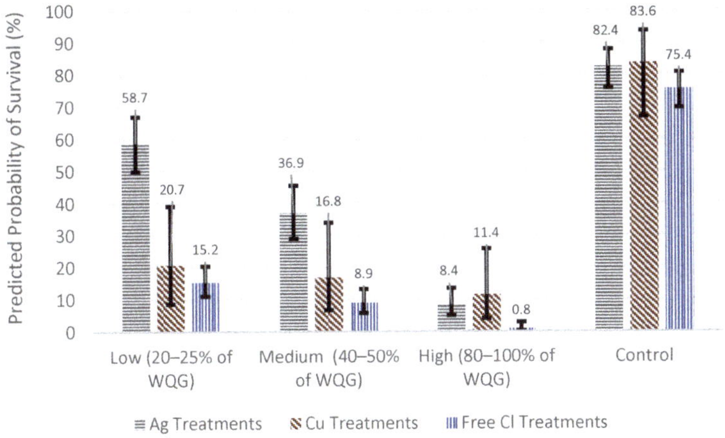

Figure 15. Summary of predicted probability of survival at 72 h regarding the efficacy of water disinfectants as larval control for younger instar *Ae. aegypti*. Error bars represent the 95% confidence interval.

The results of this study are consistent with previous research demonstrating that small aquatic invertebrates (relevant to the egg and larval stages of mosquito development) experience toxicity to metals silver and copper [127–138] and to various formulations of chlorine [139–142].

The effectiveness of these water disinfectants as vector control agents is highly dependent on contact time with the larvae and the developmental stage at which the larvae encounter the disinfectant. This has been observed in other studies examining mosquito larvae development in metal stressed environments [127,128,135] and chlorinated environments [139,140]. For example, a study utilizing household bleach with 5.25% sodium hypochlorite as the active ingredient found that the lethal dose for first instar *Ae. aegypti* larvae in the presence of food was 16 mg/L, whereas the lethal dose for third and fourth instar larvae jumped up to 250 mg/L [141], illustrating the significant impact of instar age on treatment efficacy. Building on the concept that age at which mosquito larvae come into contact with a disinfectant matters, a study that dosed first instar *Ae. aegypti* larvae with a copper sulfate treatment of 1250 µg/L observed that most mortality occurred during the first instar stage. Larvae that reached the second instar were generally able to progress to the pupal stage, suggesting mosquitoes are particularly sensitive to metal exposure as first instar. The findings also suggest the concept that surviving larvae rapidly develop tolerance to metal exposure [127]. In another study examining the effects of copper sulfate on third instar *Ae. aegypti*, researchers found that exposure to the disinfectant at concentrations of 1500 and 15,000 µg/L reduced the survival of *Ae. aegypti* during the immature stages, specifically citing that the development of *Ae. aegypti* was impaired both in the transition from larvae to pupae and from pupae to adult; however, in that study, the larvae were exposed to copper sulfate for only 24 h, which may explain why larvicidal effects were not observed at concentrations below 1500 µg/L [134]. In contrast, the larvicidal effects presented in this paper reflect the cumulative effects of prolonged exposure to the disinfectant as the larvae are in contact with the disinfectant during the duration of the observation period.

Many prior studies investigating the effects of silver, copper, and chlorine on mosquitoes did not specifically focus on the application of these chemicals in treating water storage containers for both vector control and drinking water purposes; thus, higher concentrations of the chemicals were often used in these studies yielding increases in larvicidal efficacy in shorter exposure periods [130,131,138,141]. These studies typically aimed to determine LC_{50} or LC_{90} values, which represent the concentration of a chemical required to kill 50% or 90% of the exposed population, respectively, within a period of 24 to 48 h. In contrast, the experiments presented in this paper involved observing older instar larvae over a 16-day exposure period and younger instar larvae over a 72 h exposure period. For example, one study found that the LC_{50} for copper sulfate at 24 h was 33 mg/L for late third instar *Ae. Aegypti*, which is over 25 times that of EPA's drinking water standard for copper. Within the same study, researchers observed 7% mortality at 3200 μg/L and 1% mortality at 320 μg/L after 24 h [132]. For comparison, in the experiments described in this paper, we observed a 6.6 ± 4.0% and 5.8 ± 3.2% mortality at 24 h for third instar *Ae. aegypti* exposed to 1200 μg/L and 300 μg/L, respectively. The discrepancy in the mortality outcomes between the two studies can likely be attributed to differences in environmental conditions, such as the use of tap water versus deionized (DI) water; *Ae. aegypti* strains; culturing methods; and feeding protocols.

Studies considering these water disinfectants have also tested various mosquito species such as the *Ae. albopictus, Anopheles (An.) subpictus, An. quadrimaculatus, An. farauti, Culex (Cx.) quinquefasciatus*, and *Cx. pipiens* [128,130,131,133,136–138,140,142]. Among the studies considering other species, some included test concentrations of the disinfectants at levels considered safe for human consumption. In one study, first instar larvae of *Ae. albopictus*, Cx. quinquefasciatus, and *Cx. pipiens* were exposed to copper sulfate at concentrations between 1100 and 10,000 μg/L, with *Cx. pipiens* found to be the most sensitive to the treatments, highlighting that different mosquito species exhibit varying levels of sensitivity to the same treatments. Focusing on the most affected species, *Cx. pipiens*, at 3300 μg/L, 50% of larvae died within 24 h, whereas no significant mortality was observed at 1100 μg/L with the same contact time; however, after 72 h of exposure, approximately 50% mortality was recorded at 1100 μg/L, and 80% mortality was observed at 3300 μg/L [136]. Comparing these results with our study, at 24 h, approximately 30% mortality was observed in first instar *Ae. aegypti*, and at 72 h, roughly 87% mortality was recorded at 1200 μg/L.

The mechanisms of action underlying the developmental effects of silver, copper, and chlorine treatments on mosquito larvae reveal a complex interplay of physiological and biochemical disruptions, with many details still largely understudied. Sodium hypochlorite treatments have been observed to significantly prolong the larval development period, with adverse effects on the integument (the outer covering of the larvae) and abnormalities in the siphon (the breathing tube). Many treated larvae exhibited an inability to shed their exoskeleton, ultimately failing to complete the metamorphosis process, suggesting that hypochlorite disrupts key developmental processes essential for molting and metamorphosis [140]. For copper, one proposed mechanism of action involves its negative impact on the larval gut microbiota, with some studies suggesting that this can lead to gut dysfunction, impairing nutrient absorption. This, in turn, could reduce the amount of energy available for larval development, including molt, metamorphosis, and adult development [113,114,120,121]. Other mechanisms of action for the effects of copper on mosquito larvae include direct toxicity to the larvae, interference with physiological processes such as respiration or ion transport, or disruption of key enzymes or proteins involved in development [143]. The mechanisms of action underlying the effects of silver nitrate on mosquito larvae are not well understood. While silver's antimicrobial properties suggest potential disruptions in cellular processes, specific studies detailing its effects on mosquito larval development are lacking.

4. Conclusions

The presence of water storage containers near and within households provides an opportunity for mosquitoes to breed in close proximity to humans, thus increasing the risk for the spread of vector-borne diseases. The increase in cases and burden of mosquito-borne diseases has been exacerbated by the emergence of resistance to common chemical interventions, such as temephos, currently utilized in large global mosquito mitigation efforts. As a result, there is a need for innovative and creative approaches to manage and control mosquito populations, including the treatment of water storage containers. This research seeks to provide communities, organizations, and governments with guidance regarding chemical water treatment alternatives that may serve as viable options for addressing both vector control and water treatment management. The findings of this laboratory study suggest that using water disinfectants in water storage containers can serve two purposes simultaneously—making the water safe to drink and preventing the proliferation of disease-carrying mosquitoes.

In this study, alternative methods for treating water storage containers were examined, focusing on the use of silver nitrate, copper sulfate pentahydrate, and sodium hypochlorite. These treatments, tested at concentrations deemed safe for human consumption, were assessed based on their ability to reduce the survival of juvenile *Ae. aegypti* and prevent the emergence of larvae into adult mosquitoes capable of disease transmission. The findings demonstrate that low concentrations of commonly used water disinfectants exhibit significant larvicidal effects, offering a viable option for treating water storage containers. While the ideal vector control management approach would involve rapidly and completely eliminating all larvae, this study represents an initial step toward understanding the potential for water treatment disinfectants as a viable solution. Future work could explore the use of these treatments in tandem with other strategies to achieve the ultimate goal of comprehensive larval elimination.

By addressing two important issues simultaneously, public health projects that incorporate this approach may be more appealing to funding agencies and donors who prioritize a holistic approach to health and environmental concerns. Additionally, the use of water treatment chemicals for vector control can be seen as a cost-effective strategy for controlling disease transmission, as it targets the mosquito larvae at the source, which may be more efficient than other control methods. Other necessary ingredients for a successful integrated vector management approach include integrated vector and disease surveillance, education to promote community awareness, social mobilization (e.g., commitment from governments and community engagement), and a multisectoral approach.

Author Contributions: Conceptualization, S.S.T. and J.A.S.; methodology, S.S.T., J.A.S., P.I.H. and C.F.; formal analysis, S.S.T., P.I.H., C.F., S.L.H., K.B., J.D., L.M.B. and V.C.; investigation, S.S.T., S.L.H., K.B., J.D., L.M.B. and V.C.; data curation, S.S.T. and P.I.H.; visualization, S.S.T. and P.I.H.; project administration, S.S.T.; writing—original draft preparation, S.S.T.; writing—review and editing, J.A.S., S.L.H., K.B., J.D., L.M.B. and V.C. All authors have read and agreed to the published version of the manuscript.

Funding: This research received no external funding.

Data Availability Statement: All data and models used appear in the published article.

Acknowledgments: Thank you to Jose G. Juarez for the support and guidance throughout the study. Thank you to Jamie Harris and Ana Estrella You for the experimental assistance. Sydney Turner received funding from the Graduate Assistance in Areas of National Need (GAAN) Program; American Association of University Women (AAUW); and the Metropolitan Washington Chapter of Achievement Rewards for College Scientists Foundation to complete her doctoral studies.

Conflicts of Interest: The authors declare no conflicts of interest.

References

1. World Health Organization (WHO). *Global Vector Control Response 2017–2030*; World Health Organization: Geneva, Switzerland, 2017; ISBN 978-92-4-151297-8.
2. World Health Organization (WHO). Vector-Borne Diseases. Available online: https://www.who.int/news-room/fact-sheets/detail/vector-borne-diseases (accessed on 3 March 2022).
3. Rocklöv, J.; Dubrow, R. Climate Change: An Enduring Challenge for Vector-Borne Disease Prevention and Control. *Nat. Immunol.* **2020**, *21*, 479–483. [CrossRef]
4. El-Sayed, A.; Kamel, M. Climatic Changes and Their Role in Emergence and Re-Emergence of Diseases. *Environ. Sci. Pollut. Res.* **2020**, *27*, 22336–22352. [CrossRef] [PubMed]
5. Githeko, A.K.; Lindsay, S.W.; Confalonieri, U.E.; Patz, J.A. Climate Change and Vector-Borne Diseases: A Regional Analysis. *Bull. World Health Organ.* **2000**, *78*, 1136–1147.
6. Boisson, S.; Wohlgemuth, L.; Yajima, A.; Peralta, G.; Obiageli, N.; Matendechero, S.; Baayenda, G.; Seife, F.; Hamilton, H.; Chase, C.; et al. Building on a Decade of Progress in Water, Sanitation and Hygiene to Control, Eliminate and Eradicate Neglected Tropical Diseases. *Trans. R. Soc. Trop. Med. Hyg.* **2021**, *115*, 185–187. [CrossRef] [PubMed]
7. Tidman, R.; Abela-Ridder, B.; de Castañeda, R.R. The Impact of Climate Change on Neglected Tropical Diseases: A Systematic Review. *Trans. R. Soc. Trop. Med. Hyg.* **2021**, *115*, 147–168. [CrossRef] [PubMed]
8. Fazeli-Dinan, M.; Nikookar, S.H.; Azarnoosh, M.; Jafari, A.; Daneshpour, E.; Enayati, A.; Zaim, M. An Overview of Different Control Methods of Invasive Aedes Aegypti and Aedes Albopictus. *J. Maz. Univ. Med. Sci.* **2024**, *34*, 260–286.
9. Facchinelli, L.; Badolo, A.; McCall, P.J. Biology and Behaviour of Aedes Aegypti in the Human Environment: Opportunities for Vector Control of Arbovirus Transmission. *Viruses* **2023**, *15*, 636. [CrossRef]
10. Wilson, A.L.; Courtenay, O.; Kelly-Hope, L.A.; Scott, T.W.; Takken, W.; Torr, S.J.; Lindsay, S.W. The Importance of Vector Control for the Control and Elimination of Vector-Borne Diseases. *PLoS Negl. Trop. Dis.* **2020**, *14*, e0007831. [CrossRef]
11. World Health Organization (WHO). *Dengue Guidelines, for Diagnosis, Treatment, Prevention and Control*; World Health Organization: Geneva, Switzerland, 2009; ISBN 978-92-4-154787-1.
12. de Almeida, J.P.; Aguiar, E.R.; Armache, J.N.; Olmo, R.P.; Marques, J.T. The Virome of Vector Mosquitoes. *Curr. Opin. Virol.* **2021**, *49*, 7–12. [CrossRef]
13. European Centre for Disease Prevention and Control Aedes Aegypti-Factsheet for Experts. Available online: https://www.ecdc.europa.eu/en/disease-vectors/facts/mosquito-factsheets/aedes-aegypti (accessed on 6 February 2023).
14. Leta, S.; Beyene, T.J.; De Clercq, E.M.; Amenu, K.; Revie, C.W.; Kraemer, M.U.G. Global Risk Mapping for Major Diseases Transmitted by Aedes Aegypti and Aedes Albopictus. *Int. J. Infect. Dis. IJID Off. Publ. Int. Soc. Infect. Dis.* **2018**, *67*, 25–35. [CrossRef]
15. World Health Organization (WHO). Chikungunya Fact Sheet. Available online: https://www.who.int/news-room/fact-sheets/detail/chikungunya (accessed on 13 May 2024).
16. U.S. Food and Drug Administration FDA Approves First Vaccine to Prevent Disease Caused by Chikungunya Virus. Available online: https://www.fda.gov/news-events/press-announcements/fda-approves-first-vaccine-prevent-disease-caused-chikungunya-virus (accessed on 13 May 2024).
17. Costa, L.B.; Barreto, F.K.d.A.; Barreto, M.C.A.; dos Santos, T.H.P.; de Andrade, M.d.M.O.; Farias, L.A.B.G.; de Freitas, A.R.R.; Martinez, M.J.; Cavalcanti, L.P.d.G. Epidemiology and Economic Burden of Chikungunya: A Systematic Literature Review. *Trop. Med. Infect. Dis.* **2023**, *8*, 301. [CrossRef]
18. World Health Organization Dengue and Severe Dengue. Available online: https://www.who.int/news-room/fact-sheets/detail/dengue-and-severe-dengue (accessed on 15 May 2024).
19. Yang, X.; Quam, M.B.M.; Zhang, T.; Sang, S. Global Burden for Dengue and the Evolving Pattern in the Past 30 Years. *J. Travel. Med.* **2021**, *28*, taab146. [CrossRef]
20. World Health Organization. Yellow Fever. Available online: https://www.who.int/news-room/fact-sheets/detail/yellow-fever (accessed on 15 May 2024).
21. Gaythorpe, K.A.; Hamlet, A.; Jean, K.; Garkauskas Ramos, D.; Cibrelus, L.; Garske, T.; Ferguson, N. The Global Burden of Yellow Fever. *eLife* **2021**, *10*, e64670. [CrossRef]
22. United States Food and Drug Administration (FDA) Yellow Fever Vaccine: YF-Vax. Available online: https://www.fda.gov/vaccines-blood-biologics/vaccines/yf-vax (accessed on 15 May 2024).
23. Centers for Disease Control and Prevention (CDC). What We Know About Zika. Available online: https://www.cdc.gov/zika/about/index.html (accessed on 15 May 2024).
24. Guo, Z.; Jing, W.; Liu, J.; Liu, M. The Global Trends and Regional Differences in Incidence of Zika Virus Infection and Implications for Zika Virus Infection Prevention. *PLoS Negl. Trop. Dis.* **2022**, *16*, e0010812. [CrossRef]

25. World Health Organization. Countries and Territories with Current or Previous Zika Virus Transmission (Data as of Feb 2022). Available online: https://www.who.int/publications/m/item/countries-and-territories-with-current-or-previous-zika-virus-transmission (accessed on 13 May 2024).
26. McGregor, B.L.; Connelly, C.R. A Review of the Control of Aedes Aegypti (Diptera: Culicidae) in the Continental United States. *J. Med. Entomol.* **2021**, *58*, 10–25. [CrossRef]
27. Roiz, D.; Wilson, A.L.; Scott, T.W.; Fonseca, D.M.; Jourdain, F.; Müller, P.; Velayudhan, R.; Corbel, V. Integrated Aedes Management for the Control of Aedes-Borne Diseases. *PLoS Negl. Trop. Dis.* **2018**, *12*, e0006845. [CrossRef]
28. Smith, D.L.; Perkins, T.A.; Reiner, R.C.; Barker, C.M.; Niu, T.; Chaves, L.F.; Ellis, A.M.; George, D.B.; Le Menach, A.; Pulliam, J.R.C.; et al. Recasting the Theory of Mosquito-Borne Pathogen Transmission Dynamics and Control. *Trans. R. Soc. Trop. Med. Hyg.* **2014**, *108*, 185–197. [CrossRef]
29. Jansen, C.C.; Beebe, N.W. The Dengue Vector *Aedes Aegypti*: What Comes Next. *Microbes Infect.* **2010**, *12*, 272–279. [CrossRef]
30. Carvalho, F.D.; Moreira, L.A. Why Is Aedes Aegypti Linnaeus so Successful as a Species? *Neotrop. Entomol.* **2017**, *46*, 243–255. [CrossRef]
31. Pruszynski, C.A.; Stenn, T.; Acevedo, C.; Leal, A.L.; Burkett-Cadena, N.D. Human Blood Feeding by Aedes Aegypti (Diptera: Culicidae) in the Florida Keys and a Review of the Literature. *J. Med. Entomol.* **2020**, *57*, 1640–1647. [CrossRef]
32. Christophers, S.R. *Aedes Aegypti: The Yellow Fever Mosquito: Its Life History, Bionomics and Structure*; Cambridge University Press: Cambridge, MA, USA, 1960; ISBN 978-0-521-04638-1.
33. Faull, K.J.; Williams, C.R. Intraspecific Variation in Desiccation Survival Time of Aedes Aegypti (L.) Mosquito Eggs of Australian Origin. *J. Vector Ecol.* **2015**, *40*, 292–300. [CrossRef] [PubMed]
34. Lima, E.P.; Goulart, M.O.F.; Rolim Neto, M.L. Meta-Analysis of Studies on Chemical, Physical and Biological Agents in the Control of Aedes Aegypti. *BMC Public Health* **2015**, *15*, 858. [CrossRef] [PubMed]
35. Erlanger, T.E.; Keiser, J.; Utzinger, J. Effect of Dengue Vector Control Interventions on Entomological Parameters in Developing Countries: A Systematic Review and Meta-Analysis. *Med. Vet. Entomol.* **2008**, *22*, 203–221. [CrossRef]
36. Barrera, R.; Harris, A.; Hemme, R.R.; Felix, G.; Nazario, N.; Muñoz-Jordan, J.L.; Rodriguez, D.; Miranda, J.; Soto, E.; Martinez, S.; et al. Citywide Control of Aedes Aegypti (Diptera: Culicidae) during the 2016 Zika Epidemic by Integrating Community Awareness, Education, Source Reduction, Larvicides, and Mass Mosquito Trapping. *J. Med. Entomol.* **2019**, *56*, 1033–1046. [CrossRef]
37. van den Berg, H.; da Silva Bezerra, H.S.; Al-Eryani, S.; Chanda, E.; Nagpal, B.N.; Knox, T.B.; Velayudhan, R.; Yadav, R.S. Recent Trends in Global Insecticide Use for Disease Vector Control and Potential Implications for Resistance Management. *Sci. Rep.* **2021**, *11*, 23867. [CrossRef]
38. Metcalf, R.L. Insect Resistance to Insecticides. *Pestic. Sci.* **1989**, *26*, 333–358. [CrossRef]
39. Dusfour, I.; Vontas, J.; David, J.-P.; Weetman, D.; Fonseca, D.M.; Corbel, V.; Raghavendra, K.; Coulibaly, M.B.; Martins, A.J.; Kasai, S.; et al. Management of Insecticide Resistance in the Major Aedes Vectors of Arboviruses: Advances and Challenges. *PLoS Negl. Trop. Dis.* **2019**, *13*, e0007615. [CrossRef]
40. Guedes, R.N.C.; Beins, K.; Navarro Costa, D.; Coelho, G.E.; Bezerra, H.S.d.S. Patterns of Insecticide Resistance in Aedes Aegypti: Meta-Analyses of Surveys in Latin America and the Caribbean. *Pest Manag. Sci.* **2020**, *76*, 2144–2157. [CrossRef]
41. Bharati, M.; Saha, D. Insecticide Resistance Status and Biochemical Mechanisms Involved in Aedes Mosquitoes: A Scoping Review. *Asian Pac. J. Trop. Med.* **2021**, *14*, 52. [CrossRef]
42. Antonio, G.E.; Sánchez, D.; Williams, T.; Marina, C.F. Paradoxical Effects of Sublethal Exposure to the Naturally Derived Insecticide Spinosad in the Dengue Vector Mosquito, Aedes Aegypti. *Pest Manag. Sci.* **2009**, *65*, 323–326. [CrossRef]
43. Sulaiman, S.; Fadhlina, K.; Hidayatulfathi, O. Evaluation of Pyrethrin Formulations on Dengue/Dengue Haemorrhagic Fever Vectors in the Laboratory and Sublethal Effects. *J. Arthropod-Borne Dis.* **2007**, *1*, 1–6.
44. Shaalan, E.A.-S.; Canyon, D.V.; Younes, M.W.F.; Abdel-Wahab, H.; Mansour, A.-H. Effects of Sub Lethal Concentrations of Synthetic Insecticides and Callitris Glauscophylla Extracts on the Development of Aedes Aegypti. *J. Vector Ecol.* **2005**, *30*, 295–298. [PubMed]
45. Silva, I.M.; Martins, G.F.; Melo, C.R.; Santana, A.S.; Faro, R.R.; Blank, A.F.; Alves, P.B.; Picanço, M.C.; Cristaldo, P.F.; Araújo, A.P.A.; et al. Alternative Control of Aedes Aegypti Resistant to Pyrethroids: Lethal and Sublethal Effects of Monoterpene Bioinsecticides. *Pest Manag. Sci.* **2018**, *74*, 1001–1012. [CrossRef]
46. Salokhe, S.G.; Mukherjee, S.N.; Deshpande, S.G.; Ghule, V.P.; Mathad, J.R. Effect of Sub-Lethal Concentrations of Insect Growth Regulator, Lufenuron on Larval Growth and Development of Aedes Aegypti. *Curr. Sci.* **2010**, *99*, 1256–1259.
47. da Silva, J.J.; Mendes, J.; Lomônaco, C. Effects of Sublethal Concentrations of Diflubenzuron and Methoprene on Aedes Aegypti (Diptera: Culicidae) Fitness. *Int. J. Trop. Insect Sci.* **2009**, *29*, 17–23. [CrossRef]
48. Aldridge, R.L.; Alto, B.W.; Connelly, C.R.; Okech, B.; Siegfried, B.; Linthicum, K.J. Lethal and Sublethal Concentrations of Formulated Larvicides Against Susceptible Aedes Aegypti. *J. Am. Mosq. Control Assoc.* **2022**, *38*, 250–260. [CrossRef]

49. Belinato, T.A.; Valle, D. The Impact of Selection with Diflubenzuron, a Chitin Synthesis Inhibitor, on the Fitness of Two Brazilian Aedes Aegypti Field Populations. *PLoS ONE* **2015**, *10*, e0130719. [CrossRef]
50. Liu, W.; Todd, R.G.; Gerberg, E.J. Effect of Three Pyrethroids on Blood Feeding and Fecundity of Aedes Aegypti. *J. Am. Mosq. Control Assoc.* **1986**, *2*, 310–313.
51. Agyemang-Badu, S.Y.; Awuah, E.; Oduro-Kwarteng, S.; Dzamesi, J.Y.W.; Dom, N.C.; Kanno, G.G. Environmental Management and Sanitation as a Malaria Vector Control Strategy: A Qualitative Cross-Sectional Study Among Stakeholders, Sunyani Municipality, Ghana. *Environ. Health Insights* **2023**, *17*, 11786302221146890. [CrossRef]
52. Keiser, J.; Singer, B.H.; Utzinger, J. Reducing the Burden of Malaria in Different Eco-Epidemiological Settings with Environmental Management: A Systematic Review. *Lancet Infect. Dis.* **2005**, *5*, 695–708. [CrossRef]
53. United Nations Children's Fund (UNICEF). 2.1 Billion People Lack Safe Drinking Water at Home, More than Twice as Many Lack Safe Sanitation. Available online: https://www.who.int/news/item/12-07-2017-2-1-billion-people-lack-safe-drinking-water-at-home-more-than-twice-as-many-lack-safe-sanitation (accessed on 2 May 2022).
54. United Nations-Water Summary Progress Update 2021: SDG 6—Water and Sanitation for All. Available online: https://www.unwater.org/publications/summary-progress-update-2021-sdg-6-water-and-sanitation-all (accessed on 10 May 2024).
55. Deshpande, A.; Miller-Petrie, M.K.; Lindstedt, P.A.; Baumann, M.M.; Johnson, K.B.; Blacker, B.F.; Abbastabar, H.; Abd-Allah, F.; Abdelalim, A.; Abdollahpour, I.; et al. Mapping Geographical Inequalities in Access to Drinking Water and Sanitation Facilities in Low-Income and Middle-Income Countries, 2000–2017. *Lancet Glob. Health* **2020**, *8*, e1162–e1185. [CrossRef]
56. Venkataramanan, V.; Collins, S.M.; Clark, K.A.; Yeam, J.; Nowakowski, V.G.; Young, S.L. Coping Strategies for Individual and Household-Level Water Insecurity: A Systematic Review. *WIREs Water* **2020**, *7*, e1477. [CrossRef]
57. Barrera, R.; Avila, J.; González-Téllez, S. Unreliable Supply of Potable Water and Elevated Aedes Aegypti Larval Indices: A Causal Relationship? *J. Am. Mosq. Control Assoc.* **1993**, *9*, 189–195. [PubMed]
58. Pinchoff, J.; Silva, M.; Spielman, K.; Hutchinson, P. Use of Effective Lids Reduces Presence of Mosquito Larvae in Household Water Storage Containers in Urban and Peri-Urban Zika Risk Areas of Guatemala, Honduras, and El Salvador. *Parasites Vectors* **2021**, *14*, 167. [CrossRef]
59. Akanda, A.S.; Johnson, K.; Ginsberg, H.S.; Couret, J. Prioritizing Water Security in the Management of Vector-Borne Diseases: Lessons From Oaxaca, Mexico. *GeoHealth* **2020**, *4*, e2019GH000201. [CrossRef]
60. Vannavong, N.; Seidu, R.; Stenström, T.-A.; Dada, N.; Overgaard, H.J. Effects of Socio-Demographic Characteristics and Household Water Management on Aedes Aegypti Production in Suburban and Rural Villages in Laos and Thailand. *Parasites Vectors* **2017**, *10*, 170. [CrossRef]
61. Quintero, J.; Brochero, H.; Manrique-Saide, P.; Barrera-Pérez, M.; Basso, C.; Romero, S.; Caprara, A.; De Lima Cunha, J.C.; Beltrán-Ayala, E.; Mitchell-Foster, K.; et al. Ecological, Biological and Social Dimensions of Dengue Vector Breeding in Five Urban Settings of Latin America: A Multi-Country Study. *BMC Infect. Dis.* **2014**, *14*, 38. [CrossRef]
62. Overgaard, H.J.; Alexander, N.; Mátiz, M.I.; Jaramillo, J.F.; Olano, V.A.; Vargas, S.; Sarmiento, D.; Lenhart, A.; Seidu, R.; Stenström, T.A. Diarrhea and Dengue Control in Rural Primary Schools in Colombia: Study Protocol for a Randomized Controlled Trial. *Trials* **2012**, *13*, 182. [CrossRef]
63. George, L.; Lenhart, A.; Toledo, J.; Lazaro, A.; Han, W.W.; Velayudhan, R.; Ranzinger, S.R.; Horstick, O. Community-Effectiveness of Temephos for Dengue Vector Control: A Systematic Literature Review. *PLoS Negl. Trop. Dis.* **2015**, *9*, e0004006. [CrossRef]
64. Garelli, F.M.; Espinosa, M.O.; Weinberg, D.; Trinelli, M.A.; Gürtler, R.E. Water Use Practices Limit the Effectiveness of a Temephos-Based Aedes Aegypti Larval Control Program in Northern Argentina. *PLoS Negl. Trop. Dis.* **2011**, *5*, e991. [CrossRef]
65. Tawatsin, A.; Thavara, U.; Chompoosri, J.; Bhakdeenuan, P.; Asavadachanukorn, P. Larvicidal Efficacy of New Formulations of Temephos in Non-Woven Sachets against Larvae of Aedes Aegypti (L.) (Diptera: Culicidae) in Water-Storage Containers. *Southeast Asian J. Trop. Med. Public Health* **2007**, *38*, 641–645. [PubMed]
66. Chen, C.D.; Nazni, W.A.; Lee, H.L.; Sofian-Azirun, M. Weekly Variation on Susceptibility Status of Aedes Mosquitoes against Temephos in Selangor, Malaysia. *Trop. Biomed.* **2005**, *22*, 195–206. [PubMed]
67. Pinheiro, V.C.S.; Tadei, W.P. Evaluation of the Residual Effect of Temephos on Aedes Aegypti (Diptera, Culicidae) Larvae in Artificial Containers in Manaus, Amazonas State, Brazil. *Cad. Saude Publica* **2002**, *18*, 1529–1536. [CrossRef] [PubMed]
68. Laws, E.R.; Sedlak, V.A.; Miles, J.W.; Joseph, C.R.; Lacomba, J.R.; Diaz Rivera, A. Field Study of the Safety of Abate for Treating Potable Water and Observations on the Effectiveness of a Control Programme Involving Both Abate and Malathion. *Bull. World Health Organ.* **1968**, *38*, 439–445.
69. Marcombe, S.; Chonephetsarath, S.; Thammavong, P.; Brey, P.T. Alternative Insecticides for Larval Control of the Dengue Vector Aedes Aegypti in Lao PDR: Insecticide Resistance and Semi-Field Trial Study. *Parasites Vectors* **2018**, *11*, 616. [CrossRef]
70. Bellinato, D.F.; Viana-Medeiros, P.F.; Araújo, S.C.; Martins, A.J.; Lima, J.B.P.; Valle, D. Resistance Status to the Insecticides Temephos, Deltamethrin, and Diflubenzuron in Brazilian Aedes Aegypti Populations. *BioMed Res. Int.* **2016**, *2016*, 8603263. [CrossRef]

71. Lau, K.W.; Chen, C.D.; Lee, H.L.; Norma-Rashid, Y.; Sofian-Azirun, M. Evaluation of Insect Growth Regulators Against Field-Collected Aedes Aegypti and Aedes Albopictus (Diptera: Culicidae) from Malaysia. *J. Med. Entomol.* **2015**, *52*, 199–206. [CrossRef]
72. Neto, L.; Pamplona, L.; José Soares Pontes, R.; Oliveira Lima, J.W. Influence of Water Replacement on Diflubenzuron Duration Effect in the Control of Aedes Aegypti in Simulated Field Conditions, in Northeastern Brazil. *J. Health Biol. Sci.* **2013**, *1*, 21–26. [CrossRef]
73. World Health Organization (WHO). *Diflubenzuron in Drinking-Water: Use for Vector Control in Drinking-Water Sources and Containers Background Document for Development of WHO Guidelines for Drinking-Water Quality*; World Health Organization: Geneva, Switzerland, 2008.
74. Thavara, U.; Tawatsin, A.; Chansang, C.; Asavadachanukorn, P.; Zaim, M.; Mulla, M.S. Simulated Field Evaluation of the Efficacy of Two Formulations of Diflubenzuron, a Chitin Synthesis Inhibitor against Larvae of Aedes Aegypti (L.) (Diptera: Culicidae) in Water-Storage Containers. *Southeast Asian J. Trop. Med. Public Health* **2007**, *38*, 269–275.
75. Juarez, J.G.; Garcia-Luna, S.M.; Roundy, C.M.; Branca, A.; Banfield, M.G.; Hamer, G.L. Susceptibility of South Texas Aedes Aegypti to Pyriproxyfen. *Insects* **2021**, *12*, 460. [CrossRef]
76. Hustedt, J.C.; Boyce, R.; Bradley, J.; Hii, J.; Alexander, N. Use of Pyriproxyfen in Control of Aedes Mosquitoes: A Systematic Review. *PLoS Negl. Trop. Dis.* **2020**, *14*, e0008205. [CrossRef] [PubMed]
77. Oo, S.Z.M.; Thaung, S.; Maung, Y.N.M.; Aye, K.M.; Aung, Z.Z.; Thu, H.M.; Thant, K.Z.; Minakawa, N. Effectiveness of a Novel Long-Lasting Pyriproxyfen Larvicide (SumiLarv®2MR) against Aedes Mosquitoes in Schools in Yangon, Myanmar. *Parasites Vectors* **2018**, *11*, 16. [CrossRef] [PubMed]
78. Maoz, D.; Ward, T.; Samuel, M.; Müller, P.; Runge-Ranzinger, S.; Toledo, J.; Boyce, R.; Velayudhan, R.; Horstick, O. Community Effectiveness of Pyriproxyfen as a Dengue Vector Control Method: A Systematic Review. *PLoS Negl. Trop. Dis.* **2017**, *11*, e0005651. [CrossRef] [PubMed]
79. Seng, C.M.; Setha, T.; Nealon, J.; Socheat, D.; Nathan, M.B. Six Months of Aedes Aegypti Control with a Novel Controlled-Release Formulation of Pyriproxyfen in Domestic Water Storage Containers in Cambodia. *Southeast Asian J. Trop. Med. Public Health* **2008**, *39*, 822–826.
80. Sihuincha, M.; Zamora-Perea, E.; Orellana-Rios, W.; Stancil, J.D.; López-Sifuentes, V.; Vidal-Oré, C.; Devine, G.J. Potential Use of Pyriproxyfen for Control of Aedes Aegypti (Diptera: Culicidae) in Iquitos, Perú. *J. Med. Entomol.* **2005**, *42*, 620–630. [CrossRef]
81. Ritchie, S.A.; Rapley, L.P.; Benjamin, S. Bacillus Thuringiensis Var. Israelensis (Bti) Provides Residual Control of Aedes Aegypti in Small Containers. *Am. J. Trop. Med. Hyg.* **2010**, *82*, 1053–1059. [CrossRef] [PubMed]
82. Setha, T.; Chantha, N.; Socheat, D. Efficacy of Bacillus Thuringiensis Israelensis, VectoBac WG and DT, Formulations against Dengue Mosquito Vectors in Cement Potable Water Jars in Cambodia. *Southeast Asian J. Trop. Med. Public Health* **2007**, *38*, 261–268.
83. Mulla, M.S.; Thavara, U.; Tawatsin, A.; Chompoosri, J. Procedures for the Evaluation of Field Efficacy of Slow-Release Formulations of Larvicides against Aedes Aegypti in Water-Storage Containers. *J. Am. Mosq. Control Assoc.* **2004**, *20*, 64–73.
84. Marcombe, S.; Darriet, F.; Agnew, P.; Etienne, M.; Yp-Tcha, M.-M.; Yébakima, A.; Corbel, V. Field Efficacy of New Larvicide Products for Control of Multi-Resistant Aedes Aegypti Populations in Martinique (French West Indies). *Am. J. Trop. Med. Hyg.* **2011**, *84*, 118–126. [CrossRef]
85. World Health Organization. *Spinosad DT in Drinking-Water: Use for Vector Control in Drinking-Water Sources and Containers Background Document for Development of WHO Guidelines for Drinking-Water Quality*; World Health Organization: Geneva, Switzerland, 2010.
86. Njoroge, T.M.; Berenbaum, M.R. Laboratory Evaluation of Larvicidal and Oviposition Deterrent Properties of Edible Plant Oils for Potential Management of Aedes Aegypti (Diptera: Culicidae) in Drinking Water Containers. *J. Med. Entomol.* **2019**, *56*, 1055–1063. [CrossRef]
87. Amer, A.; Mehlhorn, H. Larvicidal Effects of Various Essential Oils against Aedes, Anopheles, and Culex Larvae (Diptera, Culicidae). *Parasitol. Res.* **2006**, *99*, 466–472. [CrossRef] [PubMed]
88. Vontas, J.; Kioulos, E.; Pavlidi, N.; Morou, E.; della Torre, A.; Ranson, H. Insecticide Resistance in the Major Dengue Vectors Aedes Albopictus and Aedes Aegypti. *Pestic. Biochem. Physiol.* **2012**, *104*, 126–131. [CrossRef]
89. Saeung, M.; Ngoen-Klan, R.; Thanispong, K.; Muenworn, V.; Bangs, M.J.; Chareonviriyaphap, T. Susceptibility of Aedes Aegypti and Aedes Albopictus (Diptera: Culicidae) to Temephos in Thailand and Surrounding Countries. *J. Med. Entomol.* **2020**, *57*, 1207–1220. [CrossRef]
90. Valle, D.; Bellinato, D.F.; Viana-Medeiros, P.F.; Lima, J.B.P.; Martins, A.d.J. Resistance to Temephos and Deltamethrin in Aedes Aegypti from Brazil between 1985 and 2017. *Mem. Inst. Oswaldo Cruz* **2019**, *114*, e180544. [CrossRef]
91. Khan, H.A.A.; Akram, W. Resistance Status to Deltamethrin, Permethrin, and Temephos Along With Preliminary Resistance Mechanism in Aedes Aegypti (Diptera: Culicidae) From Punjab, Pakistan. *J. Med. Entomol.* **2019**, *56*, 1304–1311. [CrossRef]
92. Morales, D.; Ponce, P.; Cevallos, V.; Espinosa, P.; Vaca, D.; Quezada, W. Resistance Status of Aedes Aegypti to Deltamethrin, Malathion, and Temephos in Ecuador. *J. Am. Mosq. Control Assoc.* **2019**, *35*, 113–122. [CrossRef]

93. United States Environmental Protection Agency (EPA) National Primary Drinking Water Regulations. Available online: https://www.epa.gov/ground-water-and-drinking-water/national-primary-drinking-water-regulations (accessed on 5 January 2025).
94. World Health Organization (WHO). *Alternative Drinking-Water Disinfectants: Bromine, Iodine and Silver*; World Health Organization (WHO): Geneva, Switzerland, 2018; ISBN 978-92-4-151369-2.
95. World Health Organization (WHO). Copper in Drinking-Water: Background Document for Development of WHO Guidelines for Drinking-Water Quality. Available online: https://cdn.who.int/media/docs/default-source/wash-documents/wash-chemicals/copper.pdf?sfvrsn=194c0f12_4s://www.who.int/teams/environment-climate-change-and-health/water-sanitation-and-health/chemical-hazards-in-drinking-water/copper (accessed on 12 April 2023).
96. World Health Organization (WHO). Silver in Drinking-Water: Background Document for Development of WHO Guidelines for Drinking-Water Quality. Available online: https://cdn.who.int/media/docs/default-source/wash-documents/wash-chemicals/silver.pdf?sfvrsn=195cf8b3_4 (accessed on 12 April 2023).
97. Estrella-You, A.; Harris, J.; Singh, R.; Smith, J. Inactivation of Waterborne Pathogens by Copper and Silver Ions, Free Chlorine, and N-Chloramines in Point-of-Use Technology: A Review. In *Water Purification: Processes, Applications and Health Effects*; LeBlanc, P., Ed.; Water Purification; Nova Science Publishers, Inc.: Hauppauge, NY, USA, 2022; pp. 1–88, ISBN 978-1-68507-622-1.
98. Arnold, B.F.; Colford, J.M. Treating Water with Chlorine at Point-of-Use to Improve Water Quality and Reduce Child Diarrhea in Developing Countries: A Systematic Review and Meta-Analysis. *Am. J. Trop. Med. Hyg.* **2007**, *76*, 354–364. [CrossRef]
99. Collivignarelli, M.C.; Abbà, A.; Benigna, I.; Sorlini, S.; Torretta, V. Overview of the Main Disinfection Processes for Wastewater and Drinking Water Treatment Plants. *Sustainability* **2018**, *10*, 86. [CrossRef]
100. Crider, Y.; Sultana, S.; Unicomb, L.; Davis, J.; Luby, S.P.; Pickering, A.J. Can You Taste It? Taste Detection and Acceptability Thresholds for Chlorine Residual in Drinking Water in Dhaka, Bangladesh. *Sci. Total Environ.* **2018**, *613–614*, 840–846. [CrossRef]
101. Albert, J.; Luoto, J.; Levine, D. End-User Preferences for and Performance of Competing POU Water Treatment Technologies among the Rural Poor of Kenya. *Environ. Sci. Technol.* **2010**, *44*, 4426–4432. [CrossRef]
102. Kim, J.; Chung, Y.; Shin, D.; Kim, M.; Lee, Y.; Lim, Y.; Lee, D. Chlorination By-Products in Surface Water Treatment Process. *Desalination* **2003**, *151*, 1–9. [CrossRef]
103. Thurman, R.B.; Gerba, C.P.; Bitton, G. The Molecular Mechanisms of Copper and Silver Ion Disinfection of Bacteria and Viruses. *Crit. Rev. Environ. Control* **1989**, *18*, 295–315. [CrossRef]
104. Ngwenya, N.; Ncube, E.J.; Parsons, J. Recent Advances in Drinking Water Disinfection: Successes and Challenges. In *Reviews of Environmental Contamination and Toxicology*; Whitacre, D.M., Ed.; Springer: New York, NY, USA, 2013; pp. 111–170, ISBN 978-1-4614-4717-7.
105. Lucier, K.J.; Dickson-Anderson, S.E.; Schuster-Wallace, C.J. Effectiveness of Silver and Copper Infused Ceramic Drinking Water Filters in Reducing Microbiological Contaminants. *J. Water Supply Res. Technol.-AQUA* **2017**, *66*, 528–536. [CrossRef]
106. Pooi, C.K.; Ng, H.Y. Review of Low-Cost Point-of-Use Water Treatment Systems for Developing Communities. *Npj Clean Water* **2018**, *1*, 1–8. [CrossRef]
107. Singh, R.; Edokpayi, J.N.; Odiyo, J.O.; Smith, J.A. *E. Coli* Inactivation by Metals and Effects of Changes in Water Chemistry. *J. Environ. Eng.* **2019**, *145*, 04018136. [CrossRef]
108. Ehdaie, B.; Su, Y.-H.; Swami, N.S.; Smith, J.A. Protozoa and Virus Disinfection by Silver- and Copper-Embedded Ceramic Tablets for Water Purification. *J. Environ. Eng.* **2020**, *146*, 04020015. [CrossRef]
109. Ndebele, N.; Edokpayi, J.N.; Odiyo, J.O.; Smith, J.A. Field Investigation and Economic Benefit of a Novel Method of Silver Application to Ceramic Water Filters for Point-Of-Use Water Treatment in Low-Income Settings. *Water* **2021**, *13*, 285. [CrossRef]
110. Estrella-You, A.; Smith, J.A. Synergistic Bacterial Inactivation by Silver Ions and Free Chlorine in Natural Waters. *J. Environ. Eng.* **2022**, *148*, 04022072. [CrossRef]
111. Hill, C.L.; Harris, J.D.; Turner, S.S.; Wason, K.L.; Gaylord, A.P.; Hatley, M.G.; Hardcastle, L.T.; Roberts, I.T.; You, J.Y.; Renneker, K.O.; et al. Field and Laboratory Assessment of a New Electrolytic Point-of-Use Water Treatment Technology. *Water* **2022**, *14*, 1077. [CrossRef]
112. Alherek, M.; Basu, O.D. Impact of Low Levels of Silver, Zinc and Copper Nanoparticles on Bacterial Removal and Potential Synergy in Water Treatment Applications. *J. Chem. Technol. Biotechnol.* **2023**, *98*, 1137–1146. [CrossRef]
113. Harris, J.D.; Homola, L.M.J.; Estrella-You, A.; Smith, J.A. Enhancing Microbial Disinfection in Household Water Treatment by Combining a Silver–Ceramic Tablet with Copper and Chlorine Technologies. *J. Environ. Eng.* **2024**, *150*, 04024026. [CrossRef]
114. Silvestry-Rodriguez, N.; Sicairos-Ruelas, E.E.; Gerba, C.P.; Bright, K.R. Silver as a Disinfectant. In *Reviews of Environmental Contamination and Toxicology*; Springer: New York, NY, USA, 2007; pp. 23–45, ISBN 978-0-387-69163-3.
115. Ehdaie, B.; Rento, C.T.; Son, V.; Turner, S.S.; Samie, A.; Dillingham, R.A.; Smith, J.A. Evaluation of a Silver-Embedded Ceramic Tablet as a Primary and Secondary Point-of-Use Water Purification Technology in Limpopo Province, S. Africa. *PLoS ONE* **2017**, *12*, e0169502. [CrossRef] [PubMed]
116. Armstrong, A.M.; Sobsey, M.D.; Casanova, L.M. Disinfection of Escherichia Coli and Pseudomonas Aeruginosa by Copper in Water. *J. Water Health* **2016**, *14*, 424–432. [CrossRef] [PubMed]

117. World Health Organization. *Guidelines for Drinking-Water Quality, 4th Edition, Incorporating the 1st Addendum*; World Health Organization: Geneva, Switzerland, 2017; ISBN 978-92-4-154995-0.
118. World Health Organization. *Guidelines for Laboratory and Field Testing of Mosquito Larvicides*; World Health Organization: Geneva, Switzerland, 2005.
119. Ngonzi, A.J.; Muyaga, L.L.; Ngowo, H.; Urio, N.; Vianney, J.-M.; Lwetoijera, D.W. Susceptibility Status of Major Malaria Vectors to Novaluron, an Insect Growth Regulator South-Eastern Tanzania. *Pan Afr. Med. J.* **2022**, *41*, 273. [CrossRef]
120. Abbott, W.S. A Method of Computing the Effectiveness of an Insecticide. *J. Econ. Entomol.* **1925**, *18*, 265–267. [CrossRef]
121. Bliss, C.I. The Method of Probits. *Science* **1934**, *79*, 38–39. [CrossRef]
122. Finney, D.J. *Probit Analysis*; Cambridge University Press: Cambridge, MA, USA; References-Scientific Research Publishing: Wuhan, China, 1971; 333p. Available online: https://www.scirp.org/(S(i43dyn45teexjx455qlt3d2q))/reference/ReferencesPapers.aspx?ReferenceID=1493267 (accessed on 4 April 2023).
123. Postelnicu, T. Probit Analysis. In *International Encyclopedia of Statistical Science*; Lovric, M., Ed.; Springer: Berlin/Heidelberg, Germany, 2011; pp. 1128–1131, ISBN 978-3-642-04898-2.
124. Ong, S.-Q.; Jaal, Z. Larval Age and Nutrition Affect the Susceptibility of Culex Quinquefasciatus (Diptera: Culicidae) to Temephos. *J. Insect Sci.* **2018**, *18*, 38. [CrossRef]
125. Nartey, R.; Owusu-Dabo, E.; Kruppa, T.; Baffour-Awuah, S.; Annan, A.; Oppong, S.; Becker, N.; Obiri-Danso, K. Use of Bacillus Thuringiensis Var Israelensis as a Viable Option in an Integrated Malaria Vector Control Programme in the Kumasi Metropolis, Ghana. *Parasites Vectors* **2013**, *6*, 116. [CrossRef]
126. Patil, C.D.; Borase, H.P.; Patil, S.V.; Salunkhe, R.B.; Salunke, B.K. Larvicidal Activity of Silver Nanoparticles Synthesized Using Pergularia Daemia Plant Latex against Aedes Aegypti and Anopheles Stephensi and Nontarget Fish Poecillia Reticulata. *Parasitol. Res.* **2012**, *111*, 555–562. [CrossRef]
127. Perez, M.H.; Noriega, F.G. Aedes Aegypti Pharate 1st Instar Quiescence Affects Larval Fitness and Metal Tolerance. *J. Insect Physiol.* **2012**, *58*, 824–829. [CrossRef]
128. Miah, M.A. Effect of Copper Sulphate Pentahydrate on Mosquito Larval Aedes Aegypti, Culex Quinquefasciatus, and Anopheles Quadrimaculatus in Laboratory and under Semi-Field Conditions. *J. Fla. Mosq. Control. Assoc.* **2021**, *68*, 79–85. [CrossRef]
129. Ratte, H.T. Bioaccumulation and Toxicity of Silver Compounds: A Review. *Environ. Toxicol. Chem.* **1999**, *18*, 89–108. [CrossRef]
130. Shanmugasundaram, T.; Balagurunathan, R. Mosquito Larvicidal Activity of Silver Nanoparticles Synthesised Using Actinobacterium, Streptomyces Sp. M25 against Anopheles Subpictus, Culex Quinquefasciatus and Aedes Aegypti. *J. Parasites Dis. Off. Organ Indian Soc. Parasitol.* **2015**, *39*, 677–684. [CrossRef] [PubMed]
131. Velayutham, K.; Ramanibai, R.; Umadevi, M. Green Synthesis of Silver Nanoparticles Using Manihot Esculenta Leaves against Aedes Aegypti and Culex Quinquefasciatus. *J. Basic Appl. Zool.* **2016**, *74*, 37–40. [CrossRef]
132. Rayms-Keller, A.; Olson, K.E.; McGaw, M.; Oray, C.; Carlson, J.O.; Beaty, B.J. Effect of Heavy Metals onAedes Aegypti (Diptera:Culicidae) Larvae. *Ecotoxicol. Environ. Saf.* **1998**, *39*, 41–47. [CrossRef]
133. Reza, M.; Ilmiawati, C. Laboratory Testing of Very Low-Copper-Treated Water to Prolong Pupation and Emerging Time of Mosquito Larvae: An Alternative Method to Delay Mosquito Breeding Capability. *PLoS ONE* **2020**, *15*, e0226859. [CrossRef]
134. Miranda, F.R.; Fernandes, K.M.; Farder-Gomes, C.F.; Bernardes, R.C.; de Oliveira, A.H.; Arthidoro de Castro, M.B.; Dourado, L.A.; Oliveira, L.L.; Martins, G.F.; Serrão, J.E. Exposure to Copper Sulfate Impairs Survival, Post-Embryonic Midgut Development and Reproduction in Aedes Aegypti. *Infect. Genet. Evol.* **2022**, *97*, 105185. [CrossRef]
135. Neff, E.; Dharmarajan, G. The Direct and Indirect Effects of Copper on Vector-Borne Disease Dynamics. *Environ. Pollut.* **2021**, *269*, 116213. [CrossRef]
136. Reza, M.; Yamamoto, D.S.; Matsuoka, H. Larvicidal and Ovipositional Preference Test of Copper Solution for Mosquitoes. *Med. Entomol. Zool.* **2014**, *65*, 147–150. [CrossRef]
137. El-Sheikh, E.-S.M.Y.; Fouda, M.A.; Hassan, M.I.; Abd-Elghaphar, A.-E.A.; Hasaballah, A.I. Toxicological Effects of Some Heavy Metal Ions on Culex Pipiens L. (Diptera: Culicidae). *Egypt. Acad. J. Biol. Sci. F Toxicol. Pest Control* **2010**, *2*, 63–76. [CrossRef]
138. Salunkhe, R.B.; Patil, S.V.; Patil, C.D.; Salunke, B.K. Larvicidal Potential of Silver Nanoparticles Synthesized Using Fungus Cochliobolus Lunatus against Aedes Aegypti (Linnaeus, 1762) and Anopheles Stephensi Liston (Diptera; Culicidae). *Parasitol. Res.* **2011**, *109*, 823–831. [CrossRef]
139. Sherman, C.; Fernandez, E.A.; Chan, A.S.; Lozano, R.C.; Leontsini, E.; Winch, P.J. La Untadita: A Procedure for Maintaining Washbasins and Drums Free of Aedes Aegypti Based on Modification of Existing Practices. *Am. J. Trop. Med. Hyg.* **1998**, *58*, 257–262. [CrossRef]
140. Shahen, M.; El-Wahsh, H.; Ramadan, H.; Hegazi, M.; Al-Sharkawi, I.; Seif, A. A Comparison Of The Toxicity Of Calcium And Sodium Hypochlorite Against Culex Pipiens (Diptera: Culicidae) Larvae. *Environ. Sci. Curr. Res.* **2020**, *3*, 1–9. [CrossRef]
141. Barrera, R.; Amador, M.; Clark, G.G. The Use of Household Bleach to Control Aedes Aegypti. *J. Am. Mosq. Control Assoc.* **2004**, *20*, 444–448.

142. Brown, M.D.; Walker, D.O.; Hendrikz, J.K.; Cabral, C.P.; Araujo, D.B.; Ribeiro, Z.M.; Kay, B.H. Chlorine Tolerance of Mesocyclops (Cyclopoida: Cyclopidae) Copepods and Three Container-Breeding Species of Mosquitoes. *Environ. Entomol.* **1994**, *23*, 1245–1249. [CrossRef]
143. Muttkowski, R.A. Copper: Its Occurrence and Rôle in Insects and Other Animals. *Trans. Am. Microsc. Soc.* **1921**, *40*, 144–157. [CrossRef]

molecules

MDPI

Article

Preparation and Screening of SRB Gel Particles Used for Deep Purification of Acid Mine Drainage

Chunpeng Leng [1,2,3], Xi He [2,4], Yukuo Liu [2,4], Lifeng Shi [2,4], Fuping Li [1,3,*], Hao Wang [1,2,3,*], Cong Zhao [2,4], Siyu Yi [2,4] and Lei Yu [2,4]

1 College of Mining Engineering, North China University of Science and Technology, Tangshan 063210, China; lengchunpeng@ncst.edu.cn
2 Key Laboratory of Bioelectrochemical Water Pollution Control Technology in Tangshan City, North China University of Science and Technology, Tangshan 063210, China; chris20010926@163.com (L.Y.)
3 Hebei Industrial Technology Institute of Mine Ecological Remediation, Tangshan 063210, China
4 College of Civil and Architectural Engineering, North China University of Science and Technology, Tangshan 063210, China
* Correspondence: lifuping@ncst.edu.cn (F.L.); wanghao@ncst.edu.cn (H.W.)

Abstract: The progressive decline of the coal industry necessitates the development of effective treatment solutions for acid mine drainage (AMD), which is characterized by high acidity and elevated concentrations of heavy metals. This study proposes an innovative approach leveraging sulfate-reducing bacteria (SRB) acclimated to contaminated anaerobic environments. The research focused on elucidating the physiological characteristics and optimal growth conditions of SRB, particularly in relation to the pH level and temperature. The experimental findings reveal that the SRB exhibited a sulfate removal rate of 88.86% at an optimal temperature of 30 °C. Additionally, SRB gel particles were formulated using sodium alginate (SA) and carboxymethyl cellulose (CMC), and their performance was assessed under specific conditions (pH = 6, C/S = 1.5, T = 30 °C, CMC = 4.5%, BSNa = 0.4 mol/L, and cross-linking time = 9 h). Under these conditions, the SRB gel particles demonstrated an enhanced sulfate removal efficiency of 91.6%. Thermal analysis via differential scanning calorimetry (DSC) and thermogravimetric analysis (TGA) provided further insights into the stability and properties of the SRB gel spheres. The findings underscore the potential of SRB-based bioremediation as a sustainable and efficient method for AMD treatment, offering a novel and environmentally friendly solution to mitigating the adverse effects of environmental contamination.

Keywords: acid mine drainage; sulfate reducing bacteria; sulfate root removal rate; gel particles

1. Introduction

In addition to the decline in the coal industry's backward production capacity, a significant number of coal mines have closed. The closure of coal mines has resulted in a number of environmental issues, with acid mine drainage (AMD) being a significant concern [1–3]. It is imperative that the issue of effective management of AMD be addressed with the utmost urgency.

AMD is characterized by strong acidity, high hardness, and a high concentration of heavy metals [4–6]. Such contamination will not only affect surface water quality but will also have an impact on groundwater resources. As a consequence of the substantial influx of groundwater into the goaf water, precious water resources are transformed into mine wastewater, thereby resulting in a significant waste of groundwater resources [7–9].

Sulfate-reducing bacteria (SRB) are a class of anaerobic microorganisms that metabolize by using sulfate (SO_4^{2-}) as an electron acceptor. These bacteria are able to survive in anoxic or anaerobic environments and obtain energy by reducing sulfate to hydrogen sulfide (H_2S). The wide distribution of sulfate-reducing bacteria in nature and their ecological and industrial importance have made them a focus of research. Recent studies have demonstrated that sulfate-reducing bacteria (SRB) can effectively treat acid mine water through inoculation, with the resulting treatment being referred to as AMD [10–16]. In the absence of oxygen, SRB reduce the sulfate in AMD to hydrogen sulfide, thereby reducing the sulfate content. Hydrogen sulfide continues to react with the metal ions in wastewater to form metal sulfide precipitation, which then removes the metal ions in AMD while producing alkaline substances to increase the pH level [17–20]. The creation of an optimal microenvironment for SRB and the immobilization of SRB gel particles are beneficial for enhancing the efficacy and stability of the treatment process [21–23].

In this study, wastewater samples from an area affected by AMD and the pretreatment process were collected. These samples were then used for the enrichment, culture, separation, and screening of SRB. A highly efficient strain was developed for the treatment of sulfate in AMD. Sodium alginate (SA) was employed to immobilize the SRB, while carboxymethyl cellulose (CMC) was used to enhance the mechanical properties of SA. This resulted in the formation of SRB gel particles through a cross-linking and embedding method. Furthermore, the physicochemical properties, electrical conductivity, water loss rate, and sulfate removal rate of the gel particles were investigated in order to provide experimental guidance for the advanced treatment of AMD.

2. Results and Discussion

This section may be divided into subheadings. The text should provide a succinct and accurate account of the experimental outcomes, their interpretation, and the conclusions that can be drawn from the experiment.

2.1. Preliminary Screening of Sulfate-Reducing Bacteria

The preliminary screening results, as indicated in Table 1 and Figure 1, indicate that the media of strains 1, 6, 9, and 10 exhibited the fastest blackening, accompanied by a notable increase in pH at the end of the culture period. This suggests that these four strains exhibited a pronounced ability to reduce sulfate, and they were therefore selected for subsequent experiments. The four strains in the aforementioned preliminary screening were designated as SRB-1, SRB-6, SRB-9, and SRB-10.

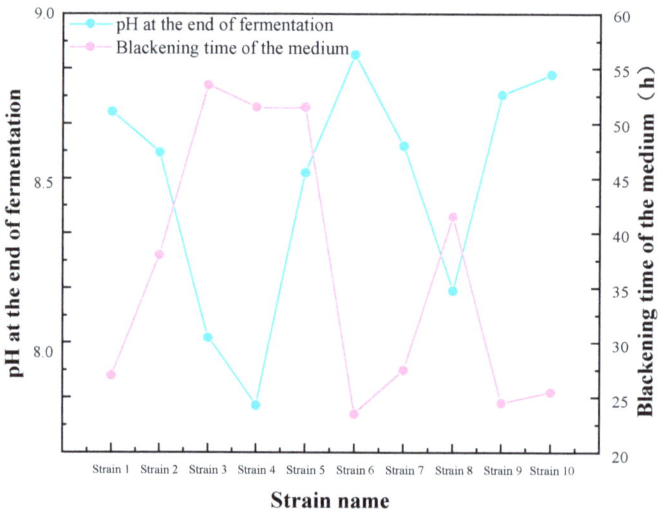

Figure 1. Preliminary screening and pH variations of sulfate-reducing bacteria (SRB) strains.

The results of the aforementioned experimental procedure (Table 2) indicate that the four strains were all Gram-negative bacteria. All four strains were negative in the V-P test and were unable to utilize glucose. However, they were able to utilize sucrose and lactose. The strains SRB-1, SRB-6, and SRB-10 exhibited positive results in the gelatin liquefaction experiment, whereas the strain SRB-9 displayed negative results. The four SRB strains exhibit similarities with those reported by other scholars with regard to their physiological and biochemical characteristics.

Table 1. Preliminary screening of SRB strains.

Strain Name	Fermentation pH Value	Blackening Time of Medium
Strain 1	8.71	27
Strain 2	8.60	40
Strain 3	8.03	50
Strain 4	7.89	48
Strain 5	8.52	48
Strain 6	8.88	24
Strain 7	8.60	29
Strain 8	8.21	42
Strain 9	8.74	25
Strain 10	8.82	26

Table 2. Analysis of physiological and biochemical characteristics of SRB strains after preliminary screening.

Strain Name	Gram Stain	Electron Receptor Experiment	Gelatin Liquification Experiments	V-P Test	Sugar Fermentation Experiment			Electron Acceptor Utilization Experiment	
					Glucose	Sucrose	Lactose	Na_2SO_4	Sulfur
SRB-1	-	+	+	-	-	+	+	+	+
SRB-6	-	+	+	-	-	+	+	+	+
SRB-9	-	+	-	-	-	+	+	+	+
SRB-10	-	+	+	-	-	+	+	+	+

+ = Gram-positive or available; - = Gram-negative or unavailable.

2.2. Growth Curve and SO_4^{2-} Reduction Effect of SRB Strain

The results of the experimental investigation into the ability of four strains to remove SO_4^{2-} are presented in Figure 2. The results demonstrate that SRB-1, SRB-6, SRB-9, and SRB-10 were capable of reducing SO_4^{2-}. Of these, SRB-6 exhibited the greatest reducing ability, reducing 88.75% of the SO_4^{2-} in the medium over a 60 h period. The removal rates of SRB-1, SRB-9, and SRB-10 were 33.56%, 32.69%, and 46.05%, respectively, under the same culture time.

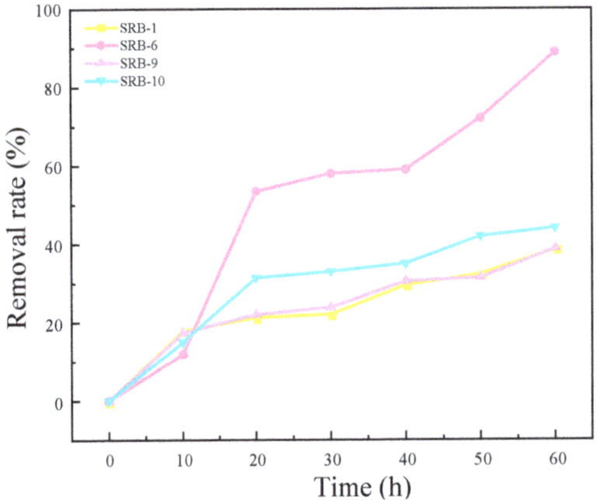

Figure 2. Sulfate removal rates for different SRB strains over time.

As illustrated in Figure 3, the growth curves of the four SRB strains were determined and analyzed following preliminary screening. The removal rate of SO_4^{2-} was also evaluated. It was observed that SRB-6 exhibited the shortest time to enter the logarithmic phase and demonstrated the most effective removal of SO_4^{2-}. Consequently, strain 6 was selected for further investigation.

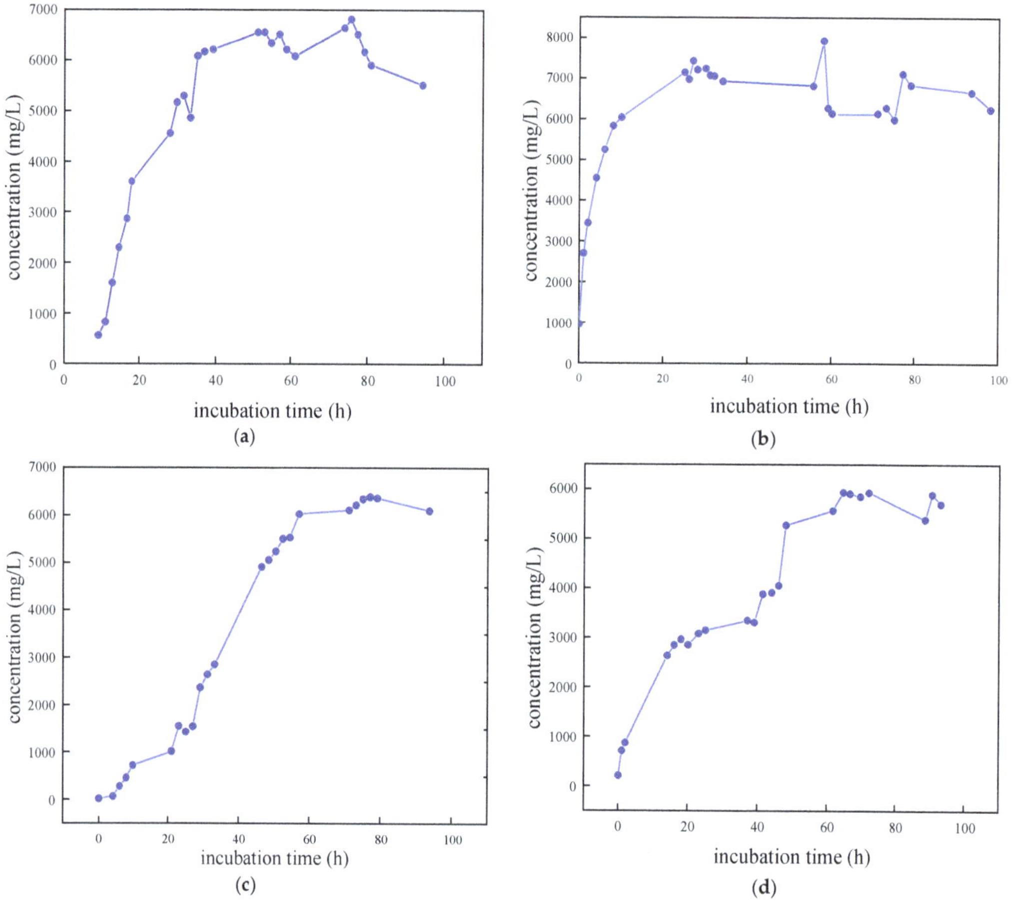

Figure 3. Growth characteristic curves of different SRB strains: (**a**) SRB-1; (**b**) SRB-6; (**c**) SRB-9; and (**d**) SRB-10.

2.3. Analysis of Physical and Chemical Properties of SRB Strains

The cell morphology, cell size, and movement characteristics of SRB-6 were observed using scanning electron microscopy. The results are shown in Figure 4.

Through analysis of the morphology and growth characteristics of the selected SRB strains, it was found that the individuals were rod-shaped, singly arranged non-spores, and the colonies were black.

The rod-shaped morphology and single arrangement of SRB-6 are typical characteristics of many sulfate-reducing bacteria. These morphological features confer an advantage in terms of motility and interaction with the surrounding environment. The absence of spore formation indicates that SRB-6 relies on continuous metabolic activity for survival rather than entering a dormant state, which is a common strategy employed by other bacterial species to withstand unfavorable conditions.

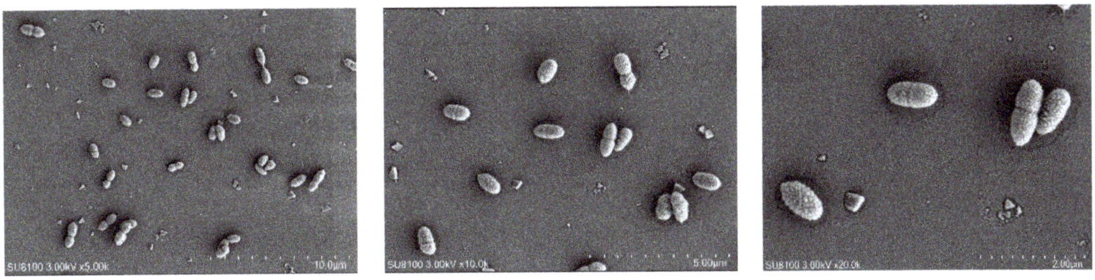

Figure 4. Scanning electron microscopy (SEM) images of SRB strains at different magnifications.

The black coloration of the colonies is a consequence of the production of metal sulfides as a result of sulfate reduction. When SRB-6 reduces sulfate (SO_4^{2-}) to sulfide (S^{2-}), the sulfide can react with metal ions present in the medium, forming black metal sulfide precipitates. This is a common indicator of active sulfate reduction and is frequently employed as a qualitative measure of SRB activity in cultures.

Consequently, the observed morphological and growth characteristics are consistent with the known metabolic functions of SRB-6, particularly its role in sulfate reduction and the subsequent formation of metal sulfide precipitates. These characteristics are fundamental to its ecological function and have the potential to be applied in the fields of bioremediation and bioenergy production.

2.4. The Effect of Different pH Values on the Reduction of SO_4^{2-} by an SRB Strain

The results presented in Figure 5 demonstrate that the removal efficiency of SO_4^{2-} exhibited a pronounced dependence on the pH value. This indicates that the pH value is a crucial factor in determining the activity of an SRB strain. Both excessively low and high pH values had a detrimental impact on the activity of the SRB strains, thereby affecting the reduction of SO_4^{2-}. The results of this experiment demonstrate that when the pH level was between 6 and 7, the removal rate of SO_4^{2-} was closest to 100%. This indicates that the SRB-6 strain, which was the subject of this experiment, is most suitable for growth and activity in a pH range from 6 to 7.

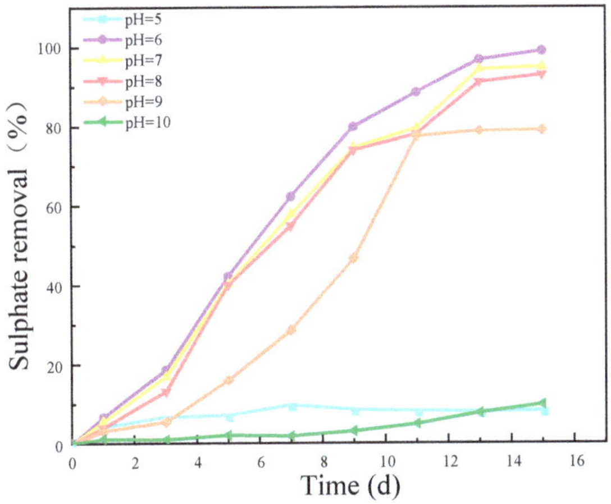

Figure 5. Effect of different pH values on SO_4^{2-} removal rate. (The C/S was 1, the concentration of SO_4^{2-} was 1400 mg/L, and the speed was 50 r/min).

The optimal pH range for the removal of SO_4^{2-} by SRB-6 was between 6 and 7. This is due to the fact that the enzymes involved in the sulfate reduction pathway exhibit optimal catalytic efficiency within this pH range. Furthermore, the stability and integrity of the bacterial cell membrane, which is essential for nutrient uptake and waste expulsion, were maintained best at this pH level. The pH range of 6–7 is conducive to the availability and solubility of essential nutrients and minerals required for bacterial growth and metabolism. Furthermore, extreme pH levels can result in the accumulation of toxic compounds that inhibit bacterial growth. Consequently, the pH range of 6–7 minimizes the toxic effects, thereby creating an optimal environment for SRB activity [24–26].

2.5. The Effect of Different Temperatures on the Reduction of SO_4^{2-} by an SRB Strain

The results of the experimental investigation, presented in Figures 6 and 7, demonstrate that the temperature exerted a significant influence on the growth and metabolic activity of microorganisms. The rate of removal of SO_4^{2-} also varied with the temperature. At 5 °C, the metabolic activity of the SRB strain was inhibited, resulting in a reduction in the growth rate and a significantly diminished removal rate of SO_4^{2-}. In the temperature range of 10–30 °C, the growth rate of the strain increased with rising temperatures, resulting in an increased removal rate of SO_4^{2-}. At 30 °C, the SRB strain exhibited the highest removal efficiency for SO_4^{2-}.

The optimum temperature for SO_4^2 removal by SRB-6 was 30 °C, because the metabolic rate of the SRB strains generally increased with the temperature and reached peak efficiency at this point. At 30 °C, the enzymes and cellular processes involved in sulphate reduction operated at their best. In addition, higher temperatures within this optimum range enhanced essential cellular functions such as nutrient uptake, energy production, and waste removal, which are critical for sustained metabolic activity. This temperature also supported faster bacterial growth and proliferation, resulting in higher biomass amounts capable of reducing more sulphate.

The removal rates of SO_4^{2-} by SRB-6 at 5 °C, 10 °C, 20 °C, and 30 °C were 10.21%, 26.34%, 73.11%, and 88.86%, respectively.

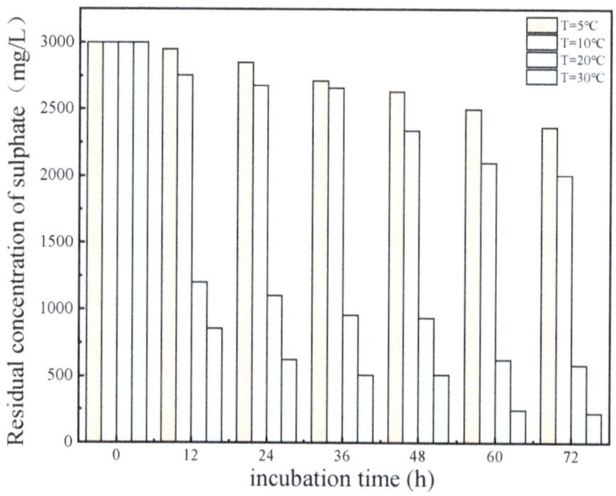

Figure 6. Residual concentration of sulfate (SO_4^{2-}) at different temperatures over incubation time.

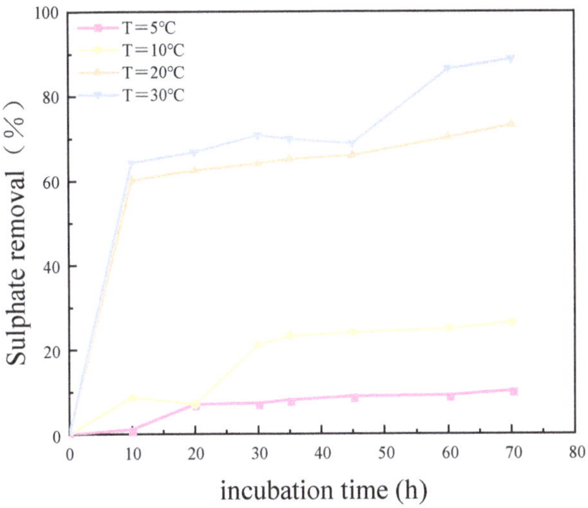

Figure 7. Sulfate removal rate at different temperatures over incubation time.

2.6. Research on SRB Gel Particles

2.6.1. Physical and Chemical Characteristic Analysis

The surface of the SRB gel sphere was smooth, black, and bright, exhibiting a granular structure that is nearly spherical or ellipsoidal (Figure 8). After being stored at 4 °C for 8 h, the SRB gel sphere underwent slight deformation due to minor water loss, causing a slight reduction in volume and a shape that remained close to spherical. Further inspection revealed that the SRB gel spheres had a diameter of approximately 1.5–2.5 mm and surface holes and contained elements such as carbon (C), oxygen (O), nitrogen (N), phosphorus (P), and sulfur (S).

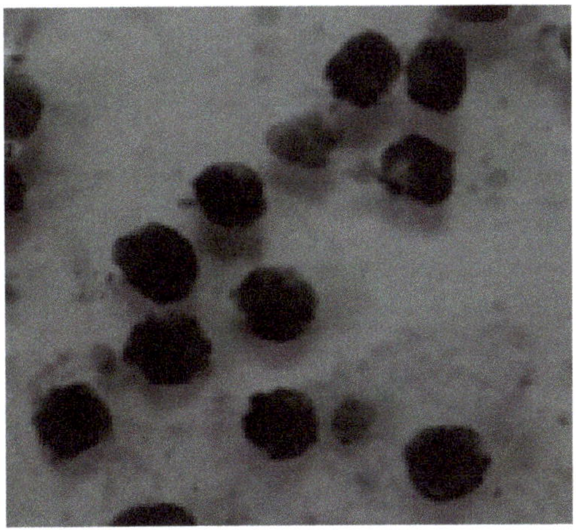

Figure 8. Morphological observation of SRB strains under a microscope.

The authors note that the SRB gel spheres exhibited minor deformation following storage, yet their reusability is a pivotal consideration for practical applications. For repeated use, it was essential to assess the structural integrity and functional performance of the SRB gel spheres through multiple cycles of sulfate reduction and recovery. To ensure the reusability of the SRB gel spheres, several factors had to be evaluated. Firstly, the spheres must maintain their shape and avoid significant degradation or deformation after each use. Secondly, their sulfate reduction capability must remain consistent across multiple cycles. Thirdly, the presence of essential elements (C, O, N, P, and S) must not diminish significantly. Finally, the gel spheres' ability to retain water and prevent excessive drying out during storage must be monitored to maintain their structural and functional integrity.

This paper presents the results of DSC-TGA of sulfate-reducing bacteria gel spheres, as illustrated in Figure 9. The provided DSC-TGA analysis offers valuable insights into the thermal properties and stability of sulfate-reducing bacteria (SRB) gel spheres. The TGA curve indicates that the SRB gel spheres maintained their weight stability up to approximately 100 °C, which is indicative of a low moisture content. Between 100 °C and 200 °C, slight weight loss accompanied by an endothermic peak in the DSC curve suggests the release of adsorbed water and low molecular weight volatiles. The most significant thermal event occurred between 200 °C and 400 °C, where major weight loss was accompanied by a pronounced endothermic peak, indicating the thermal degradation of organic components, including the bacterial biomass and gel matrix. Following decomposition, beyond 400 °C, the weight stabilized, and a minor exothermic peak suggests secondary thermal events such as recrystallization or further chemical transformations of the residual material. This comprehensive thermal analysis is essential for understanding the thermal behavior and stability of SRB gel spheres and demonstrates that this potent SRB gel sphere can slowly dissolve in an aqueous medium, effectively controlling the growth of any microorganisms in an aquatic environment.

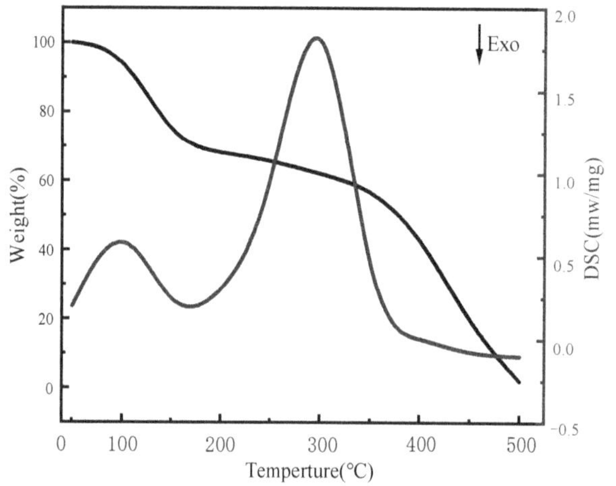

Figure 9. DSC-TGA of SRB gel spheres.

2.6.2. Conductivity

According to the above results (Figure 10), the conductivity of the SRB gel balls first increased and the decreased with the increase in CMC density, and when the CMC density was 4.5%, the conductivity value reached its maximum (2.563×10^{-4} s/cm). When the concentration of CMC was low, the network structure formed by the gel balls was loose and sparse, which is not conducive to the attachment of PAn, thus making it difficult to form a conductive channel. Consequently, the gel balls exhibited low conductivity. When the CMC density was high, a coordinated reaction between the CMC and Fe^{3+} in the gel balls would occur, resulting in blockage of the channel and hindering the entry of PAn, thereby reducing the conductivity.

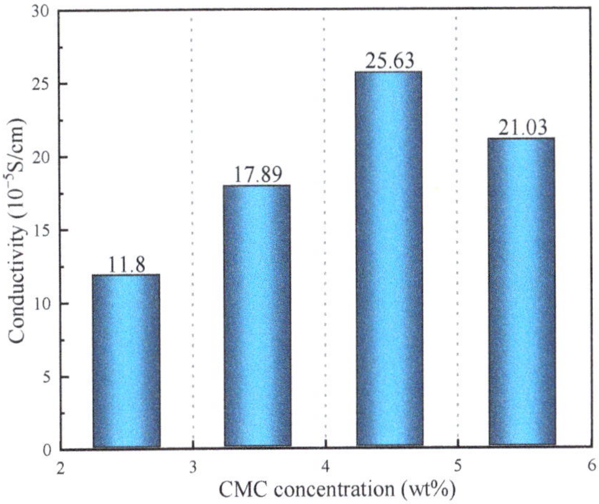

Figure 10. Conductivity of CMC solutions at various concentrations (wt%).

The results presented in Figure 11 demonstrate that the macromolecules in the gel balls intertwined with each other to form a uniform and fine pore structure. As the concentration of BSNa increased, the cross-linked network exhibited a tendency towards greater density. Following the analysis of the rate of sulfate removal by the gel balls at varying densities of BSNa, it was determined that at a BSNa density of 0.4 mol/L, the highest conductivity was 5.62×10^{-3} s/cm. However, an excessively high density of BSNa may result in damage to the structure of PAn, potentially leading to a reduction in conductivity due to the non-conductive nature of BSNa.

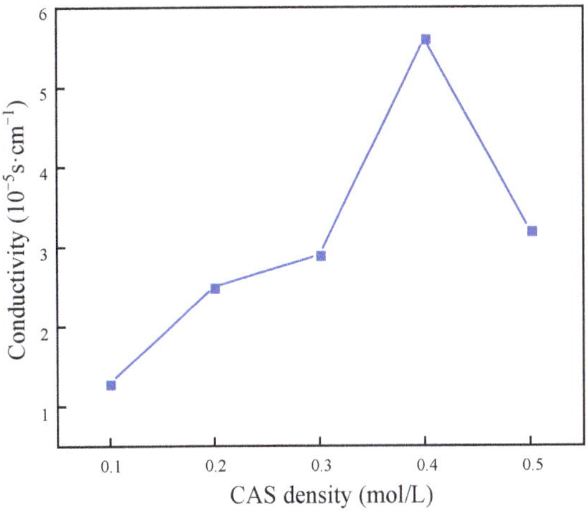

Figure 11. Influence of sodium benzenesulfonate concentration on electrical conductivity of gel spheres.

2.6.3. Water Loss Rate and Cross-Linking Time

The results demonstrate that an increase in CMC density was associated with a reduction in the water loss rate of the SRB gel balls. This occurred primarily within the first hour following the cross-linking process (Figure 12). The experimental results demonstrate that the water loss rate of the CMC density of 5.5% was significantly lower than those of the other three densities of samples. This was primarily due to the low water content of the gel balls themselves, which increased with an increase in the CMC density. Additionally,

CMC is a hydrophilic substance which reduces the free water lost during the water binding process. Consequently, the water loss rate of the gel balls was lowest when the CMC density was 5.5%. Figure 12 illustrates that the degree of cross-linking of the gel balls was essentially saturated at 8 h, which ensured that the gel balls had a certain hardness. Consequently, the cross-linking time to ensure a complete and efficient reaction was determined to be 9 h.

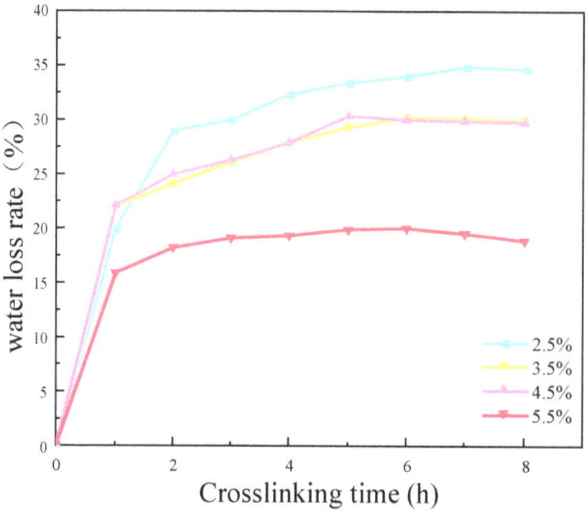

Figure 12. Effect of cross-linking time and sodium benzenesulfonate concentration on weight loss rate of gel spheres.

2.6.4. The Removal Rate of Sulfate Ions in the Treatment of AMD

In light of the aforementioned determination experiments, SRB-6 was selected as the experimental bacterium, and the requisite environment was established. This environment was defined by the following parameters: pH = 6, C/S = 1.5, T = 30 °C, CMC = 4.5%, BSNa = 0.4 mol/L, and a cross-linking time of 9 h. The sulfate-reducing bacteria gel particles prepared under the conditions of CMC = 4.5% and BSNa = 0.4 mol/L were reacted with the AMD. The sulfate radical removal rate of the AMD by the immobilized sulfate-reducing bacteria gel particles was found to reach 91.6%, with a correspondingly positive treatment effect.

2.7. Microbial Abundance Analysis

Figure 13 illustrates the prevalence of several microbial taxa across different experimental strains. The color of the heatmap indicates the level of microbial abundance. The abundance of *Bacteroides_vadinHA17* remained at 25% in all of the experimental strains, indicating that this microbial taxon maintains a high level of stability under different experimental conditions. *Carnobacteriaceae* demonstrated greater abundance (22%) under the SRB-6 experimental conditions, in comparison with a relatively lower abundance (20%) under other conditions. This may be indicative of more favorable conditions for this taxon in SRB-6. *Lentimicrobiaceae* exhibited the lowest abundance (10%) under the SRB-9 experimental conditions, while it exhibited relatively high abundance (15%) under the other conditions. The lowest abundance of *Prolixibacteraceae* was observed in the SRB-9 experimental conditions (8%), with an increase observed in the SRB-6 conditions (12%). This may indicate that different conditions have significantly different effects on the growth of this microorganism. *Anaerolineaceae* showed the greatest abundance (22%) under the SRB-9 experimental conditions and slightly lower abundance (18–20%) under the other conditions. This suggests that the SRB-9 conditions in particular may have facilitated the growth of this taxon. *Desulfobacteraceae* also showed the highest abundance (22%) under the SRB-6 experimental conditions and the lowest abundance (5% and 10%) under the

SRB-1 and SRB-10 experimental conditions, respectively, reflecting the different effects of different conditions on this group of microorganisms. *Desulfobacteraceae* is a key microbial group representing sulfate-reducing bacteria. These microorganisms play an important role in wastewater treatment and ecosystems and are effective in AMD degradation and sulfur cycling. *Desulfobacteraceae* showed the greatest abundance (22%) under the SRB-6 experimental conditions. This suggests that these conditions may have provided a more suitable environment to promote the growth of sulfate-reducing bacteria. Possible facilitating factors include nutrient availability, the environmental pH level, temperature, or other biochemical conditions. This may indicate that the growth of this microorganism is significantly influenced by different conditions. *Anaerolineaceae* exhibited the greatest abundance (22%) under the SRB-9 experimental conditions, with a slightly lower (18–20%) abundance observed under the other conditions. This indicates that the SRB-9 conditions may have been particularly conducive to the growth of this taxon. *Desulfobacteraceae* also exhibited the greatest abundance (22%) under the SRB-6 experimental conditions and the lowest (5% and 10%) abundance under the SRB-1 and SRB-10 experimental conditions, respectively, reflecting the differential effects of the various conditions on this group of microorganisms. *Desulfobacteraceae* represents a key microbial group, comprising sulfate-reducing bacteria. These microorganisms play a pivotal role in wastewater treatment and ecosystems and are effective in AMD degradation and sulfur cycling. *Desulfobacteraceae* exhibited the greatest abundance (22%) under the SRB-6 experimental conditions. This indicates that the condition may have provided a more conducive environment for the proliferation of sulfate-reducing bacteria. It is possible that the facilitating factors include the availability of nutrients, the environmental pH level, temperature, or other biochemical conditions.

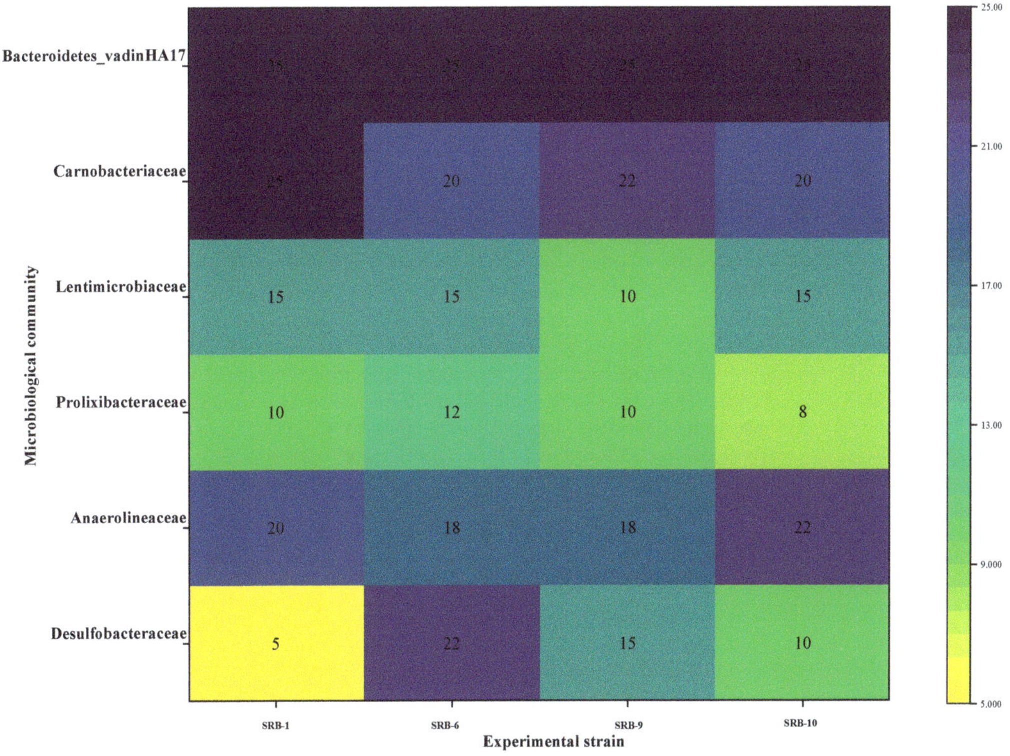

Figure 13. Heat map of microbial community composition across different experimental strains.

3. Materials and Methods

3.1. Materials

Yeast extract, sodium lactate, trisodium citrate, resazurim, ascorbic acid, $MgSO_4 \cdot 7H_2O$, Na_2SO_4, sodium thioglycolate, sodium alginate (SA), carboxymethyl cellulose (CMC), aniline (AN), activated carbon (GAC), graphene oxide (GO), sodium benzene sulfonate (BSNa), and $CaCl_2$ were used in this study. All experimental drugs were analytically pure, and the experimental water was deionized water.

3.2. Preparation of the Materials

It was necessary to collect sludge and contaminated water, which was produced in an anaerobic environment at the water gushing point of the AMD and in nearby rivers and wetlands that had been polluted by AMD, as well as in the pretreatment process of AMD. A total of 20–30 samples were collected from 8–10 sites. The samples were immediately placed into pre-sterilized glassware, sealed to ensure as anaerobic an environment was maintained as possible, and then stored in a refrigerator at a temperature of −20 °C for subsequent analysis.

3.3. Experimental Methods

In this experiment, sludge was anaerobically incubated in a constant temperature shaker at 35 °C and 150 r/min for 4–6 days. This was followed by inoculation into modified Postgate medium, which was repeated five times to select the best-growing SRB suspensions. The suspensions were then incubated by the dilution coating layered sandwich method, with the objective of isolating small, individual black colonies of various morphologies, sizes, and colors. Finally, the suspensions were anaerobically cultured. Three distinct gel particles were prepared at a solid-to-liquid ratio (M/V) of 10% for the treatment of acid mine wastewater. The adsorption of sulphate was then investigated under static conditions, with three sets of parallel experiments set up for each group.

3.4. Analytical Method

Determination of the pH level was carried out using a pH meter (PHS-3C; Leici, Shanghai, China). The mass concentration of sulfate was determined through ion chromatography using an ion chromatograph (ICS-5000; Thermo Scientific, Waltham, MA, USA).

DSC (SeikoDSC6100, Chiba, Japan) was used to determine the thermal transition properties of resistant dextrins. Each sample (3 mg, dry base) was moistened with distilled water (7 μL) in a DSC dish and equilibrated at 4 °C for 24 h prior to analysis. The DSC disc was heated at a rate of 5 °C/min in the range of 0–500 °C, with the empty disc as a reference. A thermal analysis system (EXSTAR-6000; Seiko, Chiba, Japan) was utilized.

This experiment adopted the model of the scanning electron microscope (SEM, HITACHISU8010; Hitachi, Japan) for electron microscopy (SEM) analysis. Scanning electron microscopy (SEM) employs the use of high-energy electron beams focused into extremely fine dots which are then scanned point by point on the surface of a sample. This process allows for the excitation and collection of a diverse range of physical information. The surface structure features of the tested sample are revealed by the process of receiving, magnifying, and displaying the aforementioned information [27–29].

4. Conclusions

This study demonstrates that sulfate-reducing bacteria (SRB) gel particles are an effective type of treatment for acid mine drainage (AMD). The SRB strains were observed to demonstrate high sulfate removal efficiency under optimal conditions. In particular, the SRB gel particles, when prepared with sodium alginate (SA) and carboxymethyl cellulose (CMC), demonstrated sulfate removal rates of up to 91.6% at a pH of six, C/S ratio of 1.5, temperature of 30 °C, CMC concentration of 4.5%, and BSNa concentration of 0.4 mol/L. These findings illustrate the potential of SRB gel particles for practical applications in the bioremediation of AMD.

Nevertheless, it is essential to address several limitations in future research. Firstly, the long-term stability and reusability of SRB gel particles must be thoroughly evaluated under a range of environmental conditions. The slight deformation observed in the SRB gel spheres after storage indicates the potential for structural integrity to be compromised over extended periods and repeated use. Furthermore, the scalability of this treatment method must be evaluated to ascertain its suitability for large-scale industrial applications.

Future research should also investigate the genetic and metabolic pathways of SRB in order to enhance their sulfate reduction efficiency. An investigation into the interactions between SRB and other microbial communities in AMD-affected environments could provide insights into the optimization of treatment conditions. Moreover, the integration of SRB gel particles with other treatment technologies could enhance the overall effectiveness and efficiency of AMD remediation.

In conclusion, while the SRB gel particles showed promise for the treatment of age-related macular degeneration (AMD), it is essential that the limitations and challenges outlined in this study be addressed through future research and development if their full potential for practical applications is to be realized.

Author Contributions: Conceptualization, C.L. and H.W.; methodology, F.L. and H.W.; software, X.H., Y.L. and L.S.; validation, C.Z., S.Y. and F.L.; formal analysis, C.L., S.Y. and H.W.; investigation, L.Y.; resources, C.L. and H.W.; data curation, H.W.; writing—original draft preparation, C.L., X.H. and H.W.; writing—review and editing, F.L. and H.W.; visualization, C.Z.; supervision, F.L.; project administration, C.L. All authors have read and agreed to the published version of the manuscript.

Funding: This work was supported by Tangshan Basic Research Science and Technology Project (23130212E), Hebei Province Doctoral Graduate Student Innovation Ability Training Funding Project (CXZZBS2024131), Hebei Natural Science Foundation Project (E2022209036 and E2023209067), and Key R&D Project in Hebei Province (22323601D).

Institutional Review Board Statement: Not applicable.

Informed Consent Statement: Not applicable.

Data Availability Statement: The experimental data used to support the findings of this study are included in the article.

Conflicts of Interest: The authors declare no conflicts of interest.

References

1. Yang, Y.; Li, B.; Li, T.; Liu, P.; Zhang, B.B.; Che, L.L. A review of treatment technologies for acid mine drainage and sustainability assessment. *J. Water Process Eng.* **2023**, *55*, 104213. [CrossRef]
2. Liu, Y.; Xie, X.M.; Wang, S.; Hu, S.M.; Wei, L.Z.; Wu, Q.H.; Luo, D.G.; Xiao, T.F. Hydrogeochemical evolution of groundwater impacted by acid mine drainage (AMD) from polymetallic mining areas (South China). *J. Contam. Hydrol.* **2023**, *259*, 104254. [CrossRef]
3. Luo, J.Z.; Cai, Y.Y. Mechanism of the pore and molecular structure evolution of coal exposed to acid mine drainage (AMD). *Sci. Total Environ.* **2024**, *906*, 167836. [CrossRef]
4. Ngole-Jeme, V.M.; Ndava, J. The implications of AMD induced acidity, high metal concentrations and ochre precipitation on aquatic organisms. *Pol. J. Environ. Stud.* **2023**, *32*, 2959–2980. [CrossRef]
5. Jiao, Y.N.; Zhang, C.H.; Su, P.D.; Tang, Y.H.; Huang, Z.P.; Ma, T. A review of acid mine drainage: Formation mechanism, treatment technology, typical engineering cases and resource utilization. *Process Saf. Environ.* **2023**, *170*, 1240–1260. [CrossRef]
6. Mosai, A.K.; Ndlovu, G.; Tutu, H. Improving acid mine drainage treatment by combining treatment technologies: A review. *Sci. Total Environ.* **2024**, *919*, 170806. [CrossRef]
7. Shi, W.Z.; Zhao, C.H.; Liang, Y.P.; Han, Z.T.; Xie, H.; Tang, C.L. Genetic mechanism analysis of low Ca/Mg value of acid goaf water in coal mine drainage. *Carsologica Sin.* **2022**, *41*, 511–521.
8. Liang, Y.P.; Shen, H.Y.; Gao, X.B. Review of research progress of karst groundwater in northern China. *Bull. Geol. Sci. Technol.* **2022**, *41*, 199–219.
9. Zhao, C.H.; Liang, Y.P.; Wang, Z.H.; Tang, C.L.; Shen, H.Y. Dynamic characteristics and evolution mechanism of "goaf water" of coal mine in Shandi River Basin, Yangquan, Shanxi Province and its environmental effects on Niangziguan Spring Area. *Geol. China* **2023**, *50*, 1471–1485.
10. Zhang, H.G.; Li, M.; Yang, Z.Q.; Sun, Y.Q.; Yan, J.; Chen, D.Y.; Chen, Y.H. Isolation of a non-traditional sulfate reducing-bacteria *Citrobacter freundii* sp. and bioremoval of thallium and sulphate. *Ecol. Eng.* **2017**, *102*, 397–403. [CrossRef]

11. Pape, P.L.; Battaglia-Brunet, F.; Parmentier, M.; Joulian, C.; Gassaud, C.; Fernandez-Rojo, L.; Guigner, J.M.; Ikogou, M.; Stetten, L. Complete removal of arsenic and zinc from a heavily contaminated acid mine drainage via an indigenous SRB consortium. *J. Hazard. Mater.* **2017**, *321*, 764–772. [CrossRef] [PubMed]
12. Yan, J.; Zhong, K.Q.; Wang, S.J.; Chen, Z.X.; Hu, H.S.; Jian, Z.Y.; Wen, H.J.; Zhang, H.G. Carbon metabolism and sulfate respiration by a non-conventional *Citrobacter freundii* strain SR10 with potential application in removal of metals and metalloids. *Int. Biodeter. Biodegr.* **2018**, *133*, 238–246. [CrossRef]
13. Miao, Y.H.; Qi, S.Y.; Chen, J.; Wang, J.; Tian, B.Y.; Xin, B.P. Application of sulphate reducing bacteria in acid mine drainage treatment: A review. *Appl. Chem. Ind.* **2021**, *50*, 3074–3078+3086.
14. Zhang, Z.; Zhang, C.H.; Yang, Y.; Zhang, Z.W.; Tang, Y.H.; Su, P.D.; Lin, Z.W. A review of sulfate-reducing bacteria: Metabolism, influencing factors and application in wastewater treatment. *J. Clean. Prod.* **2022**, *376*, 134109. [CrossRef]
15. Guo, X.Y.; Hu, Z.Y.; Dong, Y.G.; Fu, S.O.; Li, Y. Study of the preparation of Maifan stone and SRB immobilized particles and their effect on treatment of acid mine drainage. *RSC Adv.* **2022**, *12*, 4595–4604. [CrossRef] [PubMed]
16. Novair, S.B.; Atigh, Z.B.Q.; Lajayer, B.A.; Shu, W.X.; Price, G.W. The role of sulphate-reducing bacteria (SRB) in bioremediation of sulphate-rich wastewater: Focus on the source of electron donors. *Process Saf. Environ.* **2024**, *184*, 190–207. [CrossRef]
17. Hwang, S.K.; Jho, E.H. Heavy metal and sulfate removal from sulfate-rich synthetic mine drainages using sulfate reducing bacteria. *Sci. Total Environ.* **2018**, *635*, 1308–1316. [CrossRef]
18. Magowo, W.E.; Sheridan, C.; Rumbold, K. Global co-occurrence of acid mine drainage and organic rich industrial and domestic effluent: Biological sulfate reduction as a co-treatment-option. *J. Water Process Eng.* **2020**, *38*, 101650. [CrossRef]
19. Chen, X.Y.; Feng, J.; Wang, H.M.; Liu, D.; Dou, Y.F.; Zhang, J.H.; Ma, L.Y. Isolation and identification of sulfate-reducing bacteria in goaf water in Yangquan of Shanxi Province and domestication for the reduction. *Acta Microbiol. Sin.* **2022**, *62*, 2104–2118.
20. Diao, C.Y.; Ye, W.Z.; Yan, J.; Hao, T.W.; Huang, L.; Chen, Y.H.; Long, J.Y.; Xiao, T.F.; Zhang, H.G. Application of microbial sulfate-reduction process for sulfate-laden wastewater treatment: A review. *J. Water Process Eng.* **2023**, *52*, 103537. [CrossRef]
21. Zhang, H.G.; Li, M.; Pang, B.; Wu, Y.J.; Sun, Y.Q.; Chen, D.Y.; Chen, Y.H. Bioremoval of Tl (I) by PVA-immobilized Sulfate-Reducing Bacteria. *Pol. J. Environ. Stud.* **2017**, *26*, 1865–1873. [CrossRef] [PubMed]
22. Zhang, H.G.; Li, M.; Li, H.S.; Li, M.; Luo, D.G.; Chen, Y.H.; Chen, D.Y.; Luo, H.L.; Chen, Z.X.; Li, K.K. Immobilizing metal-resistant sulfate-reducing bacteria for cadmium removal from aqueous solutions. *Pol. J. Environ. Stud.* **2018**, *27*, 2851–2859. [CrossRef] [PubMed]
23. Xin, Z.J.; Wang, X.Y.; Li, L.; Li, Y.; Deng, M.; Yao, Z. Components optimization of immobilized SRB-embeded particles and study on the treatment effect of sulfate-containing wastewater. *Ind. Water Treat.* **2024**, *44*, 112–119.
24. Zhang, L.; Chen, Y.; Zhang, L.; Song, R.K.; Wang, Q.; Liu, C.J.; Mu, Y.D. Preparation of $ZrO_2/\gamma\text{-}Al_2O_3$ albumen type catalyst and its catalytic performance for carbonyl sulfide hydrolysis. *Appl. Phys. A* **2024**, *130*, 170. [CrossRef]
25. Zhang, L.; Wang, Q.; Zhao, H.C.; Song, R.K.; Chen, Y.; Liu, C.J.; Han, Z.K. Synthesis and surface strengthening modification of Silica aerogel from fly ash. *Materials* **2024**, *17*, 1614. [CrossRef] [PubMed]
26. Zhang, L.; Song, R.K.; Jia, Y.; Zou, Z.R.; Chen, Y.; Wang, Q. Purification of quinoline insolubles in heavy coal tar and preparation of meso-carbon microbeads by catalytic polycondensation. *Materials* **2024**, *17*, 143. [CrossRef]
27. Yuan, Y.G.; Leng, C.P.; Zhou, Y.L.; Yuan, Y.; Niu, Y.X.; Xu, R.Y.; Zhong, H.Y.; Li, F.P.; Zhou, H.X.; Wang, H. Impact of separate concentrations of polyethylene microplastics on the ability of pollutants removal during the operation of constructed wetland-microbial fuel cell. *J. Environ. Manag.* **2023**, *341*, 118107. [CrossRef] [PubMed]
28. Xu, R.Y.; Yang, Z.N.; Niu, Y.X.; Xu, D.; Wang, J.; Han, J.L.; Wang, H. Removal of microplastics and attached heavy metals from secondary effluent of wastewater treatment plant using interpenetrating bipolar plate electrocoagulation. *Sep. Purif. Technol.* **2022**, *290*, 120905. [CrossRef]
29. Wang, H.; Zhong, Y.L.; Bo, G.Z. Existing forms and changes of nitrogen inside of horizontal subsurface constructed wetlands. *Environ. Sci. Pollut. Res.* **2018**, *25*, 771–781. [CrossRef]

water

MDPI

Article

Brackish Water Phytodesalination by the Euhalophyte *Sesuvium portulacastrum*

Abdulaziz Alharbi [1,*], Mokded Rabhi [2,3] and Ahmed Alzoheiry [4,5]

1 Department of Environment and Natural Resources, College of Agriculture and Food, Qassim University, Buraydah 52454, Saudi Arabia
2 Department of Plant Production, College of Agriculture and Food, Qassim University, Buraydah 51452, Saudi Arabia; m.rabhi@qu.edu.sa
3 Laboratory of Extremophile Plants, Centre of Biotechnology of Borj-Cedria, P.O. Box 901, Hammam-Lif 2050, Tunisia
4 Department of Agricultural and Biosystems Engineering, College of Agriculture and Food, Qassim University, Buraydah 51452, Saudi Arabia; a.ezoheiry@qu.edu.sa
5 Department of Natural Resources and Agricultural Engineering, Faculty of Agriculture, Damanhour University, Damanhour 5836271, Egypt
* Correspondence: abb.alharbi@qu.edu.sa

Abstract: In the Middle East and North Africa as well as in numerous countries in South America and Southeast Asia, water scarcity is a real concern. Therefore, water desalination has become a key solution and an important source of freshwater production. Solar stills are used for water desalination but they require low depth of sea or brackish water and sufficient solar radiation to evaporate water. In this investigation, a phytodesalinator is presented for the first time. The halophyte used in this work is *Sesuvium portulacastrum* L., a heat-tolerant euhalophyte. The presented phytodesalinator can replace basic solar stills during cold seasons if there is sufficient sunlight to ensure the transpiration process in the plant. The euhalophyte *S. portulacastrum* was tested for its ability to desalinate reject brine as grown for two subsequent phytodesalination cycles. Several factors were found to affect the productivity of the phytodesalinator, in particular, solar radiation, phytodesalination duration, and plant density. Nevertheless, it exhibited an average productivity of 2.44 $kg/m^2/d$ and showed several advantages in comparison with basic solar stills.

Keywords: phytodesalinator; plant; productivity; water; transpiration

1. Introduction

Water scarcity is considered as an imbalance condition that occurs when demands for freshwater become higher than its availability [1]. Several factors can influence water scarcity, in particular, population growth and urbanization, individual consumption, pollution, and climate change [2]. In the Middle East and North Africa as well as in numerous countries in South America and Southeast Asia, water scarcity is already a real concern [3].

During the last seven decades, water desalination has become a key solution and an important source of freshwater production [4], in particular, in the Middle East, considered as the leader of water desalination, which produces 50% of the total desalinated water in the world for only 2.9% of the world population [3]. The two most common desalination technologies used to desalinize seawater, brackish water, and contaminated water are reverse osmosis and thermal-based desalination [5]. However, these two techniques are expensive, energy consuming, and not environmentally friendly [3,5]. For these reasons, the use of solar energy in water desalination is considered as a promising alternative [5].

The systems used for water desalination using solar energy are referred to as solar stills and are grouped into passive and active types [6]. Solar stills are simple and cheap devices that require only simple maintenance [7]. In addition, they can be developed in different ways, and a variety of types have been described in the literature [7]. An emerging desalination technology using algae has recently been receiving more and more attention [8].

In the context of water desalination, the term biodesalination is defined "as utilizing living organisms or biological elements directly or indirectly, mimicking their structures and mechanisms, borrowing concepts or inspiration from their desalination mechanisms for the production of sustainable fresh water" [9]. According to this definition, not only the direct use of organisms is considered as a biodesalination process but also the inspiration from them or their mimetics. In fact, the use of halophilic organisms in desalination has been attracting the attention of researchers for several decades, in particular, the use of halophytes in soil desalination, also called phytodesalination [10].

Sesuvium portulacastrum L. is a euhalophyte or obligate halophyte that requires salt for optimal growth. It is known for its high salt tolerance [11] and its ability to accumulate enormous sodium and potassium quantities within its shoots [12]. This halophyte showed a high phytodesalination capacity by removing high sodium quantities from soil [13,14]. In addition, *S. portulacastrum* tolerates high temperatures. The aim of the present investigation is therefore to check whether this euhalophyte can be used in brackish water phytodesalination by growing it into a solar-still-like phytodesalinator.

2. Materials and Methods

2.1. Description of the Phytodesalinator

The device used in the present work received the Saudi patent number 10822 B1 in 2022. It was designed to function like a phytodesalinator with a base dimension of 100 × 100 cm (Figure 1). The upper part of the device consisted of three straight sides and one side sloping with the slope of the longitude angle of the experiment site. The phytodesalinator was almost a closed system; there were only two small holes of 1 cm diameter in the back side. The phytodesalinator was positioned facing south to ensure the maximum sun light exposure every day throughout the experiment period. The base of the device had a slope of 2% towards its front to allow the collection of desalinated water into a tank. Plants were set into a floating foam board placed on the top of small rectangular containers filled with 6 L of reject brine each. With such a device, plants were grown in a closed system during the experiment.

Figure 1. The phytodesalinator used in the phytodesalination experiment with *S. portulacastrum* plants inside.

2.2. Plant Material

S. portulacastrum cuttings were taken from mother plants grown in the garden of Qassim University. After rooting, obtained seedlings were hydroponically grown for 4 weeks on a fourth-strength Hoagland's nutrient solution [15]. Thereafter, plants were transferred into plastic pots (30 cm × 24 cm × 12 cm) containing 6 L of reject brine each.

2.3. Reject Brine Characterization

The reject brine used in this investigation was obtained from a local desalination plant. It was added with macro- and micronutrients according to fourth-strength Hoagland's nutrient solution regardless of their initial concentrations. Two phytodesalination cycles that lasted 21 and 18 days, respectively, were performed. At the beginning of each cycle, reject brine samples were taken for EC determination and mineral analysis (Table 1).

Table 1. Na^+, K^+, Ca^{++}, and Mg^{++} concentrations (mg/L) in brine water at the beginning of each phytodesalination cycle. LD: low density (1140 plants/m); HD: high density (2900 plants/m). Numbers following LD and HD indicate the cycle number.

	LD1	HD1	LD2	HD2
Na^+	2673 ± 138	2673 ± 138	4300 ± 203	4309 ± 339
K^+	1215 ± 119	1215 ± 119	1735 ± 35	1761 ± 75
Ca^{++}	1006 ± 87	1006 ± 87	1514 ± 77	1479 ± 168
Mg^{++}	701 ± 30	701 ± 30	1037 ± 56	1036 ± 96

2.4. Phytodesalination Experiment

The experiment was conducted on Qassim University Campus (26°35′07″ N, 43°76′75″ W) during the period of 15 January–22 February 2022 under sunlight conditions and did not need any kind of added energy (average direct normal solar radiation and average air temperature during the experiment period are given as supplementary data). Two different plant densities with 3 replicates each were used:

- Low density (LD): 82 plants/container, which corresponds to 1140 plants/m^2;
- High density (HD): 209 plants/container, which corresponds to 2900 plants/m^2.

The volume of obtained desalinated water was measured twice a week, and its EC was measured. At the end of each phytodesalination cycle, the remaining volume of reject brine and its EC were determined. No mineral analysis was performed for the remaining reject brine since, at the end of each desalination cycle, crystals appeared in the bottom of the containers and on the plant roots. In addition, at the beginning of the experiment as well as at the end of each phytodesalination cycle, plants were washed several times with distilled water to remove the salt crystals and root-adsorbed salts. Then, samples were taken for mineral analysis.

2.5. Plant Mineral Analysis

Shoot and root samples were oven-dried at 70 °C until they were a constant weight then ground to fine powder. After that, aliquots of about 20 mg ground plant material were digested with 0.5% HNO_3. Potassium (K^+) and sodium (Na^+) concentrations were determined using a flame photometer apparatus (PFP7, Jenway LTD, Felsted, UK). Calcium and magnesium were analyzed by an atomic absorption flame emission spectrophotometer (AA-6200, Shimadzu, Kyoto, Japan).

2.6. Calculation

The following quantities were calculated using the measured data.

Cumulative volumes of desalinated water were calculated by adding the desalinated water volumes of all collects to each other. The daily productivity was calculated by dividing the volume of desalinated water obtained in each collect by the time of the collect (3 or 4 days), then scaled to a desalinating surface (reject brine surface) area of 1 m^2. Each cycle started with 6 L brine water per container. At the end of the cycle, the remaining brine water volume was determined, then subtracted from the initial 6 L, giving the volume of desalinated water per container. The volume of desalinated water per plant during each phytodesalination cycle was obtained by dividing the volume of desalinated water in a given container by the number of plants in that container.

2.7. Statistical Analysis

When appropriate, data were subjected to One-Way ANOVA using IBM SPSS Statistics 25, and means were compared according to Duncan's multiple-range test at $p \leq 0.05$.

3. Results and Discussion

3.1. The First Time a Phytodesalinator Has Been Presented

As far as we know, this is the first time a phytodesalinator has been described in detail with an evaluation of its productivity. Indeed, the only work that discussed the use of similar plants (mangroves) was a study of the desalination of seawater carried out in a greenhouse [16]. However, the system described by this author needed a heat exchanger, a condenser funnel, and a fan in addition to light sources. A thermal still with *Spirulina* algae was also used to improve seawater evaporation and condensation as compared to a thermal still without algae [17]. The present phytodesalinator beneficiated from the capacity of halophytes to desalinate brackish water through transpiration as in [16] and used a euhalophyte instead of algae as in [17]. The used euhalophyte does not need an added source of carbon as needed by algae. Moreover, the halophyte has the potential to be used for prolonged periods of time if new brackish water is added to replace the transpired water amounts.

3.2. Cumulative Desalinated Water and Daily Productivity

The cumulative quantities of desalinated water increased linearly with time during the experiment (Figure 2). The obtained cumulative quantities were 21.3 and 41.1 kg at the end of the first and the second phytodesalination cycles, respectively.

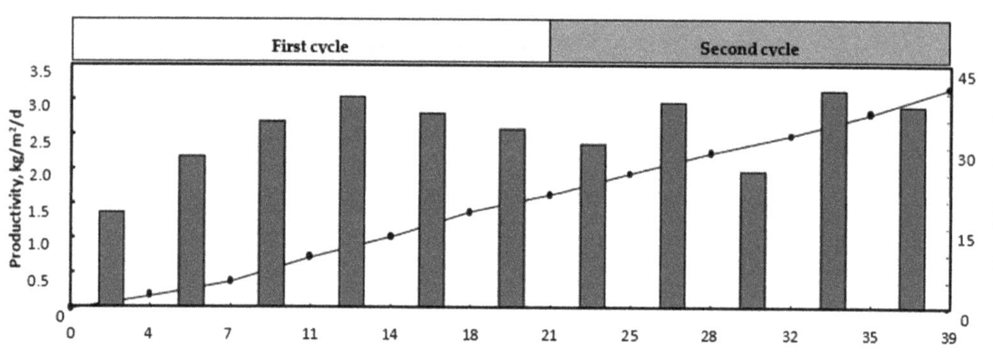

Figure 2. Cumulative volumes of desalinated water (line) and phytodesalination rate (bars) upon two cycles of brine water phytodesalination by the halophyte *S. portulacastrum* plants grown in a glass solar still.

The phytodesalinator daily productivity (daily average quantities of desalinated water per area unit) increased from 1.33 kg/m^2/d at the beginning of the experiment to 2.96 kg/m^2/d on days 11–14 of the first cycle, then decreased to 2.5 kg/m^2/d at the end of the first cycle (Figure 2). The second cycle showed also the same trend for productivity. It started with 2.29 kg/m^2/d and finished with 2.82 kg/m^2/d, the highest productivity (3.05 kg/m^2/d) being recorded on days 32–35 of the experiment (days 11–14 of the second cycle).

During 39 d, the phytodesalinator used in this experiment produced 41.1 kg desalinated water, which corresponds to an average productivity of 2.44 kg/m^2/d. To our knowledge, there is no similar solar still using a halophyte that has been evaluated for productivity. A floating solar still simulating water uptake by plant roots was recently presented [5]. These authors compared the productivity of their "bionic floating solar still" (about 1.5 kg/m^2/d) to those of four other different floating solar stills and found it the highest. The daily productivity of the modified basin solar still was water depth dependent; it increased from 1200 mL/m^2/d at 5 cm water depth to 1800 mL/m^2/d at 2 cm water depth [18]. A circular, finned, single-basin, dual-slope solar still showed a daily productivity of 1491.7 mL/m^2/d at 10 mm water depth [6].

3.3. *Factors That Can Affect Daily Productivity*

3.3.1. Plant Species

The choice of plant species that can be used in a phytodesalinator for phytodesalination of brackish water should be a euhalophyte. In addition, it should be an "includer"; it should not have salt glands such as *Aeluropus littoralis* (Willd) Parl [19] or glandular trichomes like *Atriplex halimus* L. [20]. Indeed, the existence of salt crystals due to such salt-secreting structures on the leaf surface may salinize the distillate. In addition, the chosen plant species should tolerate high temperatures and transpire at a high rate as it will be subjected to both high salinity and temperature. In the present study, the heat-tolerant euhalophyte *S. portulacastrum* was efficient in desalinating reject brine through its high transpiration rate.

3.3.2. Phytodesalination Duration

During each phytodesalination cycle, the daily productivity of the phytodesalinator started with a relatively low value, increased to an optimum, then decreased (Figure 2). The relatively low productivities at the beginning of the first cycle may be explained by two factors: (i) plant acclimation and (ii) low relative humidity inside the phytodesalinator. The productivity with which the second cycle started was 73% higher than that of the first cycle despite the higher reject brine salinity. This means that plants were already acclimated, and the opening of the phytodesalinator was responsible for the decrease in daily productivity because of the decline in relative humidity inside the phytodesalinator. The optimum daily productivity obtained in each phytodesalination cycle was followed by a decline, indicating that conditions, in particular, reject brine salinity, became increasingly unfavorable for the halophyte.

3.3.3. Solar Radiation

Figure 3A shows the average direct normal solar radiation (W/m^2) from 6:00 a.m. to 6:00 p.m. during the second cycle.

During this cycle, there were 3 days (days 29, 30, and 31) of low solar radiation (solar radiation 2) contrary to the remaining days (solar radiation 1). These 3 days coincided with a decrease in productivity from 2.88 to 1.91 kg/m^2/d (Figure 3B).

According to the present study, it is evident that solar radiation affected the phytodesalinator daily productivity through its effect on *S. portulacastrum* transpiration. Upon soilless culture under greenhouse conditions, cucumber showed a 4-fold higher diurnal canopy transpiration rate at high solar radiation (up to 20 MJ/m^2/d) than at low solar radiation (up to 9 MJ/m^2/d) [21]. Nevertheless, it was found that the same solar radiation levels had different transpiration rates in winter and summer, which may be partially due to differences in stomatal and aerodynamic conductance and in vapor pressure deficit [22].

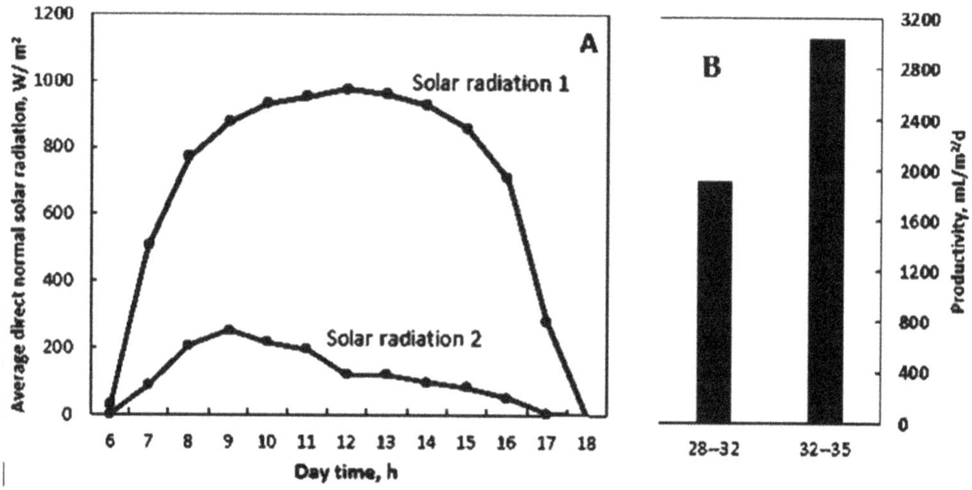

Figure 3. (**A**) Average direct normal solar radiation (W/m^2) during the experiment. Solar radiation 2 stands for values recorded on days 28–32; Solar radiation 1 stands for values recorded on the remaining days of the experiment. (**B**) Phytodesalination rates on days 25–28 and on days 28–32.

3.3.4. Plant Density

The comparison between the cumulative quantities of desalinated water per container obtained in each phytodesalination cycle with low (LD1 and LD2) and high (HD1 and HD2) plant densities showed a constant value for low plant density (3.1 kg/container) in both cycles (Figure 4). As for high plant density, a high value was observed in the first cycle (4.4 kg/container) followed by a lower one (3.8 kg/container) in the second cycle.

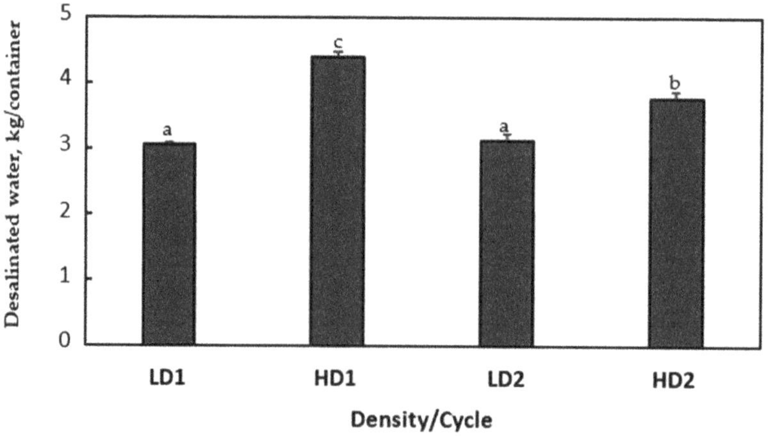

Figure 4. Effect of plant density on the volume of desalinated water per container during two cycles of brine water phytodesalination by the halophyte *S. portulacastrum* plants grown in a glass solar still. Bars are means of 3 replicates ± SE. Bars labeled with at least one similar letter are not significantly different according to Duncan's multiple-range test. LD: low density (1140 plants/m^2); HD: high density (2900 plants/m^2). Numbers following LD and HD indicate the cycle number.

Considering the cumulative volumes of desalinated water per plant obtained in each phytodesalination cycle, it was obvious that, although the total volume of desalinated water was higher for HD plants, plants grown in low density were more able to desalinate brine water (37.3–38.2 g/plant) as compared to plants grown in high density (18.2–21.2 g/plant; Figure 5).

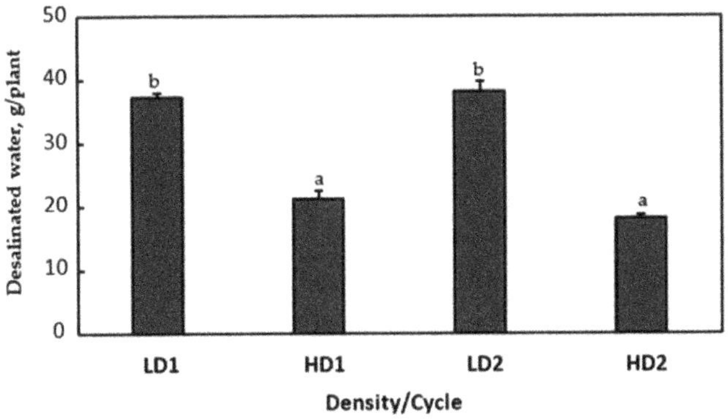

Figure 5. Effect of plant density on the volume of desalinated water per plant during two cycles of brine water phytodesalination by the halophyte *S. portulacastrum* plants grown in a glass solar still. Bars are means of 3 replicates ± SE. Bars labeled with at least one similar letter are not significantly different according to Duncan's multiple-range test. LD: low density (1140 plants/m^2); HD: high density (2900 plants/m^2). Numbers following LD and HD indicate the cycle number.

With a simple estimation, it was found that LD exhibited throughout the two cycles an average productivity of 2.21 kg/m^2/d, while HD displayed an average productivity of 2.91 kg/m^2/d. The difference between the two values was not high considering the huge difference in densities, HD being 2.5-fold higher than LD. Consequently, the individual plant phytodesalination yield was much higher in LD compared to HD (Figure 5). This can be partially explained by the faster increase in reject brine salinity in HD since the phytodesalination process was faster at the level of the whole container.

3.4. Reject Brine Salinity after the Two Cycles of Phytodesalination

At the beginning of the experiment and after the addition of nutrients to the reject brine, EC was about 16 dS/m in both LD and HD containers (Table 2). After the first phytodesalination cycle, it increased by 61 and 123%, respectively. At the beginning of the second cycle, desalinated water quantities were replaced by equal quantities of reject brine, then nutrients were added. The second cycle started with EC values of about 20 dS/m for both LD and HD containers. These values increased by, respectively, 79 and 145% at the end of the experiment.

Table 2. Reject brine EC (dS/m) at the beginning and at the end of each phytodesalination cycle. For each cycle, means (n = 3) followed by at least the same letter are not significantly different according to Duncan's multiple-range test at $p \leq 0.05$. LD: low density (1140 plants/m^2); HD: high density (2900 plants/m^2). Numbers following LD and HD indicate the cycle number.

	Initial EC	Final EC
First cycle		
LD1	16.0 a	25.7 b
HD1	16.1 a	35.9 c
Second cycle		
LD2	19.8 A	35.5 B
HD2	20.3 A	49.8 C

3.5. Plant Mineral Composition

Plant Na^+, K^+, Ca^{++}, and Mg^{++} contents at the beginning of the experiment and at the end of each phytodesalination cycle are presented in Table 3. It seems that this halophyte maintained a relatively constant mineral composition during the experiment, except in the case of its K^+ content, which increased from 30.2 mg/g DW at the beginning of the experiment to 51.6 mg/g DW at its end with no significant difference between the two phytodesalination cycles. Hence, plant tissues were not invaded by salt ions, which indicated its ability to control ion uptake and transport [23].

Table 3. Na^+, K^+, Ca^{++}, and Mg^{++} contents (mg/g DW) in shoots of *S. portulacastrum* plants at the beginning of the experiment (initial harvest) and at the end of each phytodesalination cycle. In each row, means (n = 6) followed by at least the same letter are not significantly different according to Duncan's multiple-range test at $p \leq 0.05$.

	Initial Harvest	Cycle 1 Harvest	Cycle 2 Harvest
Na^+	126.2 a	125.1 a	120.3 a
K^+	30.2 a	44.9 b	51.6 b
Ca^{++}	25.5 ab	22.9 a	31.3 b
Mg^{++}	18.3 a	17.3 a	16.6 a

3.6. Cost Analysis

The device cost/meter squared was USD 125. The expected device life is about 20 years. It was estimated also that one patch of fresh plants per 3 months costs USD 15. The plant nutrient solution was estimated to cost USD 10/month. Hence, the total annual cost could be about USD 0.45/kg of desalinated water. This unit is a prototype, and a lot of its parts were manually manufactured. If commercially produced, the cost/kg of water is expected to decrease.

3.7. Advantages of the Phytodesalinator

This device showed relatively high daily productivity. It needs only sunlight, which is a sustainable energy, but, alternatively, any other source of light can be used. It was conceived to be used during cold periods when basic solar stills cannot work. It is based on the process of plant transpiration and not on the process of evaporation. It is a closed system that does not need any kind of maintenance. It does not require a given water depth. After harvest, plants can be used for several purposes (ornamental, fodder, dune fixation, etc.). It is not an expensive device.

4. Conclusions

The presented phytodesalinator is designed to operate in regions with sunny winters and to replace basic solar stills since it does not require high temperatures for desalination. Unlike basic solar stills that need to heat water to initiate evaporation, this system maintains evaporation through a plant's physiological process (transpiration) as long as the plant is exposed to light. However, the system's performance is limited by the plant's ability to tolerate heat, making it unsuitable for use under high solar radiation intensity. Despite several factors influencing its productivity, the phytodesalinator has notable advantages over basic solar stills. It can function under low-solar-radiation conditions and produces very little high-salinity water at the end of the desalination cycle compared to traditional solar stills, which require running water and produce twice the amount of saline water for each unit of desalinated water. The total annual cost is estimated to be about USD 0.45 per kilogram of desalinated water, with the cost per kilogram expected to decrease with larger units. Further research should explore the use of different types of euhalophytes and various improved designs of this prototype to reduce costs and increase productivity per unit area.

Supplementary Materials: The following supporting information can be downloaded at: https://www.mdpi.com/article/10.3390/w16131798/s1, Table S1: Daily weather data during the experiment collected by Qassim University weather station (N26.351200, E43.766810).

Author Contributions: All authors contributed equally to the work from the design to the final version of the manuscript. All authors have read and agreed to the published version of the manuscript.

Funding: The authors gratefully acknowledge Qassim University, represented by the Deanship of Scientific Research, for the financial support for this research under the number (10199-cavm-2020-1-3-I) during the academic year 1441AH/2020 AD.

Data Availability Statement: Data will be made available on request.

Conflicts of Interest: The authors declare no conflicts of interest.

References

1. Savenije, H.H.G. Water scarcity indicators; the deception of the numbers. *Phys. Chem. Earth—Part B Hydrol. Oceans Atmos.* **2000**, *25*, 199–204. [CrossRef]
2. FAO. *Coping with Water Scarcity—An Action Framework for Agriculture and Food Security*; FAO Water Reports 38; Food and Agriculture Organization of the United Nations: Rome, Italy, 2012.
3. Morillo, J.; Usero, J.; Rosado, D.; El Bakouri, H.; Riaza, A.; Bernaola, F.J. Comparative study of brine management technologies for desalination plants. *Desalination* **2014**, *336*, 32–49. [CrossRef]
4. Greenlee, L.F.; Lawler, D.F.; Freeman, B.D.; Marrot, B.; Moulin, P. Reverse osmosis desalination: Water sources, technology, and today's challenges. *Water Res.* **2009**, *43*, 2317–2348. [CrossRef] [PubMed]
5. Chen, S.; Zhao, P.; Xie, G.; Wei, Y.; Lyu, Y.; Zhang, Y.; Yan, T.; Zhang, T. A floating solar still inspired by continuous root water intake. *Desalination* **2021**, *512*, 115–133. [CrossRef]
6. Jani, H.K.; Modi, K.V. Experimental performance evaluation of single basin dual slope solar still with circular and square cross-sectional hollow fins. *Sol. Energy* **2019**, *179*, 186–194. [CrossRef]
7. Sharshir, S.W.; Ellakany, Y.M.; Algazzar, A.M.; Elsheikh, A.H.; Elkadeem, M.R.; Edreis, E.M.A.; Waly, A.S.; Sathyamurthy, R.; Panchal, H.; Elashry, M.S. A mini review of techniques used to improve the tubular solar still performance for solar water desalination. *Process Saf. Environ. Prot.* **2019**, *124*, 204–212. [CrossRef]
8. Gao, L.; Zhang, X.; Fan, L.; Gray, S.; Li, M. Algae-Based Approach for Desalination: An emerging energy-passive and environmentally friendly desalination technology. *ACS Sustain. Chem. Eng.* **2021**, *9*, 8663–8678. [CrossRef]
9. Taheri, R.; Razmjou, A.; Szekely, G.; Hou, J.; Ghezelbash, R. Biodesalination—On harnessing the potential of Nature's desalination processes. *Bioinspir. Biomim.* **2016**, *11*, 041001. [CrossRef] [PubMed]
10. Rabhi, M.; Atia, A.; Abdelly, C.; Smaoui, A. New parameters for a better evaluation of vegetative bioremediation, leaching, and phytodesalination. *J. Theor. Biol.* **2015**, *383*, 7–11. [CrossRef] [PubMed]
11. Rabhi, M.; Giuntini, D.; Castagna, A.; Remorini, D.; Baldan, B.; Smaoui, A.; Abdelly, C.; Ranieri, A. Sesuvium portulacastrum maintains adequate gas exchange, pigment composition, and thylakoid proteins under moderate and high salinity. *J. Plant Physiol.* **2010**, *167*, 1336–1341. [CrossRef] [PubMed]
12. Rabhi, M.; Alharbi, M.; Alzoheiry, A.; Alharbi, A. Ability of Sesuvium portulacastrum to accumulate sodium and potassium from saline media. *Asian J. Plant Sci.* **2022**, *21*, 99–105. [CrossRef]
13. Rabhi, M.; Ferchichi, S.; Jouini, J.; Hamrouni, M.H.; Koyro, H.W.; Ranieri, A.; Abdelly, C.; Smaoui, A. Phytodesalination of a salt-affected soil with the halophyte *Sesuvium portulacastrum* L. to arrange in advance the requirements for the successful growth of a glycophytic crop. *Bioresour. Technol.* **2010**, *101*, 6822–6828. [CrossRef] [PubMed]
14. Sassi, H.; Farhat, N.; Jendoubi, R.; Elkhouni, A.; Zorrig, W.; Smaoui, A.; Abdelly, C.; Rabhi, M. Optimal soil salinity levels for the highest phytodesalination parameters in the obligate halophyte *Sesuvium portulacastrum* L. *Agrochimica* **2017**, *61*, 329–339. [CrossRef]
15. Hoagland, D.R.; Arnon, D.I. The water-culture method for growing plants without soil. In *California Agricultural Experiment Station*; The College of Agriculture University of California: Berkeley, CA, USA, 1950; Circular 347.
16. Finck, C. A bio-thermic seawater desalination system using halophytes. *Water Supply* **2014**, *14*, 657–663. [CrossRef]
17. Baskaran, V.; Saravanane, R. Experimental investigation on *Spirulina* algae based thermal still for effective bio-desalination. *Environ. Nanotechnol. Monit. Manag.* **2021**, *15*, 100434.
18. El-Sebaey, M.S.; Ellman, A.; Hegazy, A.; Panchal, H. Experimental study and mathematical model development for the effect of water depth on water production of a modified basin solar still. *Case Stud. Therm. Eng.* **2022**, *33*, 101925. [CrossRef]
19. Barhoumi, Z.; Djebali, W.; Smaoui, A.; Chaïbi, W.; Abdelly, C. Contribution of NaCl excretion to salt resistance of *Aeluropus littoralis* (Willd) Parl. *J. Plant Physiol.* **2007**, *164*, 842–850. [CrossRef] [PubMed]
20. Smaoui, A.; Barhoumi, Z.; Rabhi, M.; Abdelly, C. Localization of potential ion transport pathways in vesicular trichome cells of *Atriplex halimus* L. *Protoplasma* **2011**, *248*, 363–372. [CrossRef]
21. Medrano, E.; Lorenzo, P.; Sánchez-Guerrero, M.C.; Montero, J.I. Evaluation and modelling of greenhouse cucumber-crop transpiration under high and low radiation conditions. *Sci. Hort.* **2005**, *105*, 163–175. [CrossRef]

22. Katsoulas, N.; Stanghellini, C. Modelling crop transpiration in greenhouses: Different models for different applications. *Agronomy* **2019**, *9*, 392. [CrossRef]
23. Messeddi, D.; Sleimi, N.; Abdelly, C. Some physiological and biochemical aspects of salt tolerance of *Sesuvium portulacastrum*. In *Cash Crop Halophytes: Recent Studies*; Lieth, H., Ed.; Kluwer Academic Publishers: London, UK, 2003; pp. 71–77.

molecules

MDPI

Article

An Amphiphilic Surface with Improved Thermal Radiation for Water Harvesting

Han Wang [1,2], Shengtao Li [1], Ye Zhang [1], Weihui Wu [2], Khaled Abdeen Mousa Ali [1,3,*] and Changyou Li [1,*]

[1] College of Engineering, South China Agricultural University, Guangzhou 510642, China; wanghan603@scau.edu.cn (H.W.); lishengtao@stu.scau.edu.cn (S.L.); zhangye@scau.edu.cn (Y.Z.)

[2] School of Intelligent Engineering, Shaoguan University, Shaoguan 512158, China

[3] College of Agricultural Engineering, Al-Azhar University, Cairo 11751, Egypt

* Correspondence: khaledabdeen@azhar.edu.eg (K.A.M.A.); lichyx@scau.edu.cn (C.L.)

Abstract: Water scarcity poses a significant challenge for people living in arid areas. Despite the effectiveness of many bioinspired surfaces in promoting vapor condensation, their water-harvesting efficiency is insufficient. This is often exacerbated by overheating, which decreases the performance in terms of the micro-droplet concentration and movement on surfaces. In this study, we used a spotted amphiphilic surface to enhance the surfaces' water-harvesting efficiency while maintaining their heat emissivity. Through hydrophilic particle screening and hydrophobic groove modifying, the coalescence and sliding characteristics of droplets on the amphiphilic surfaces were improved. The incorporation of boron nitride (BN) nanoparticles further enhanced the surfaces' ability to harvest energy from condensation. To evaluate the water-harvesting performance of these amphiphilic surfaces, we utilized a real-time recording water-harvesting platform to identify microscopic weight changes on the surfaces. Our findings indicated that the inclusion of glass particles in hydrophobic grooves, combined with 1.0 wt.% BN nanoparticles, enhanced the water-harvesting efficiency of the amphiphilic surfaces by more than 20%.

Keywords: water harvesting; surface wettability; thermal radiation; amphiphile structure

1. Introduction

Water scarcity is a pressing concern for numerous countries [1–5]. Despite advancements, water harvesting from bioinspired surfaces, such as the back textures of desert beetles, has shown limited efficiency when used as a supplementary water source. This primarily arises from the low harvesting efficiency of droplets on amphiphilic surfaces due to low droplet growth and sliding rates. Improving these rates will require the use of surfaces with controllable textures and continuous access to energy. Furthermore, the properties of droplet harvesting under the interactions between droplet sliding and a sur-face's thermal properties need to be studied. This would require detecting the vapor condensation details under humid circumstances to understand the water-harvesting dynamics on amphiphilic surfaces. These investigations are crucial for advancing our understanding of water-harvesting technologies to address the challenges posed by water scarcity.

A liquid droplet's wettability on solid surfaces is its response to multiple forces, such as surface tension, Laplace pressure, and gravity. The surface polarity and free energy of solids are vital indicators influencing a droplet's nucleation efficiency during water harvesting [6–10]. Wang et al. [11] found that the wettability of different surfaces was due to droplets sliding with different rhythms. Pogorzelski et al. [12,13] suggested that clean metal surfaces have a higher surface tension and hydrophilicity, which can improve the heat exchange efficiency between water vapor and solid surfaces. Song et al. [14] explored the wettability transformation of 304 steel with different surface roughness values under different conditions (temperature, pressure, etc.). They showed that roughness dominated the surface wettability in low-temperature environments (T < 120 °C). However, Zhang et al. [15] proposed that rough heterogeneous surfaces caused contact angle hysteresis. To modify the surface wettability of water-harvesting solids, Zhu et al. [16] proposed a pattern of micro-square holes on a copper surface to amplify the Cassie–Baxter effect and transform the surface from hydrophilic to hydrophobic. Khan et al. [17,18] demonstrated that flash laser scanning changed the wettability of a surface without a chemical process. Romário et al. [19] produced a superhydrophobic material composed of vertically aligned carbon nanotubes, which also improved the water-harvesting efficiency. Yan et al. [20] used dichlorodimethylsilane to modify the surface wettability of SiO_2 nanoparticles.

To explore the influence of diverse surface wettability properties on water-harvesting efficiency, Deng et al. [21,22] determined the relationship between the macroscopic thermal properties of solids and the phase transition efficiency of vapor condensation by analyzing the nucleation, growth, merging, and removal of condensed droplets on surfaces with different wettability values. Liu et al. [23] found that the hydrophilicity or hydrophobicity of inclined surfaces promoted droplet sliding during water harvesting. By constructing nanostructures on the surface of an aluminum metal mesh to control the surface wettability, Kang et al. [10] proved that a hydrophilic surface significantly promoted the droplet sliding efficiency. Especially for surfaces with an active cooling function, a hydrophobic surface promoted their water-harvesting efficiency in a high-humidity environment.

Many studies have prepared bionic surface structures inspired by special surface structures, such as the back of desert beetles [24–27], the surface of lotus leaves [28], and the surface of pitcher plants [29]. These approaches have concentrated condensed droplets on surfaces [30]. The inflation and migration of droplets can be modified to improve the water-harvesting efficiency, which is expected to become an important supplemental water source [31]. Benda et al. [32] prepared an amphiphilic surface called PMMDBAB-PDMS using UV curing. The PDMS content in the pre-polymers affected the contact angle of the solid surfaces. Yan [33] fabricated a prototype fog collector to improve the droplet shedding frequency by capturing mini droplets spontaneously and rapidly passing them through the pores. Because the chemical properties and morphology of a solid surface greatly affect the condensation efficiency of the droplets, Huang et al. [34] simulated the condensation process of droplets on an amphibious surface by using the Lattice–Boltzmann method. They proposed a micro-nanostructure with a higher condensation efficiency. Sun et al. [35] proposed a coating and immersion method to prepare a hydrophilic/superhydrophobic bionic water-harvesting surface and then proved that adjusting the size and distance of the hydrophilic spots promoted the water-harvesting efficiency. Inspired by the gas trapping of underwater spiders, Zhang et al. [36] proposed an amphiphilic biological hydrogel with adjustable wettability. They demonstrated that the amphiphilic mesh had more water-harvesting advantages than a traditional mesh. Inspired by the water-collecting structure of cactus spines, Zhang et al. [37] proposed an amphiphilic surface with an asymmetrical structure. They found that the smooth water-collecting channels promoted the sliding of coalesced droplets. However, water harvesting is the result of continuous energy exchange among water vapor, a solid medium, and the environment. Simply optimizing the surface texture to facilitate droplet movement has a limited impact on the water-harvesting efficiency overall.

Zhang et al. [38] designed a 3D sandwich structure composed of a hydrophilic hydroxy-rich cellulose layer, a water-conducting layer, and a hydrophobic layer to obtain a water-harvesting structure with mycelium filaments that collected water from the air. Nguyen et al. [39] synthesized a super-amphiphilic microgroove surface with adjustable wettability using fused deposition modeling (FDM). The super-amphiphilic surface improved the mobility of water droplets. Overall, although amphibious surfaces are beneficial for water harvesting, the water-harvesting efficiency, related to the condensation and movement of droplets, which are influenced by the amphibious surface texture and wettability, requires further study.

This paper presents a novel approach to enhancing the water-harvesting efficiency through the development of a new process to construct a spotted amphiphilic surface with superior thermal radiation properties. By accelerating the droplet sliding and augmenting the surface emission power, the water-harvesting ability of an amphiphilic surface was significantly improved. A new experimental platform, equipped with a self-designed C++ program for automated recognition of minute weight fluctuations caused by droplet movements, was established to investigate the imperceptible process of vapor condensation and its correlation with the solid surface's characteristics. Our findings indicate that improving the surface energy emission of the hydrophobic region, particularly through the addition of nanoparticles, play a crucial role in enhancing the water-harvesting efficiency in a natural environment. However, changes to the thermal radiation properties of the hydrophobic regions may lead to a surface wettability fluctuation. Specifically, the incorporation of boron nitride (BN) nanoparticles (1.0 wt.%) with Polydimethylsiloxane (PDMS) increased the droplet mobility on the hydrophilic spots and enhanced the emission power of the surfaces, thereby promoting water harvesting on spotted amphiphilic surfaces.

2. Theoretical Basis

2.1. Design Rationale 1: Thermal Radiation and Emission Features on Amphiphilic Surfaces

Although amphiphilic surfaces accelerate the vapor concentration and sliding processes according to texture optimization, it is difficult to develop an amphiphilic surface that maintains a constantly high water-harvesting efficiency because the surface temperature dramatically increases after receiving energy from condensed vapor. Thus, amphiphilic surfaces must deliver energy by thermal radiation. Therefore, the thermal output efficiency of the surfaces was vital for their water harvesting, which was determined by the emissivity of the amphiphilic surfaces according to the Stefan–Boltzmann law [40]:

$$E_{amphiphilic} = \varepsilon_{amphiphilc}\sigma T^4$$

where $\varepsilon_{amphiphilc}$ is the emissivity of an amphiphilic surface.

Figure 1 shows that the surfaces received different amounts of energy from condensation and atmospheric radiation under dry and wet conditions. Figure 1a displays that the thermal radiation input from the hydrophilic region (gray areas) and the hydrophobic region (blue area) was dominated by the thermal properties and textures of the dry surfaces. Then, the hydrophilic and hydrophobic parts, displaying high reflectivity, prevented energy from entering via the ambient atmosphere. The emissivity of the amphiphilic surfaces followed the equation $\varepsilon_{amphiphilie} = \varepsilon_{hydrophilie} f_{hydrophilie} + \varepsilon_{hydrophobe} f_{hydrophobe}$, where $\varepsilon_{hydrophilie}$ and $\varepsilon_{hydrophobe}$ are the emissivity of the hydrophilic and hydrophobic surfaces, and $f_{hydrophilie}$ and $f_{hydrophobe}$ are the area fractions of different substances ($f_{hydrophilie} + f_{hydrophobe} = 1$) on the surface, respectively.

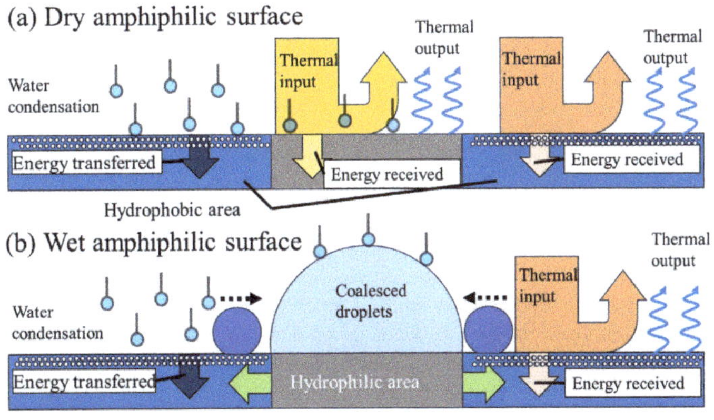

Figure 1. Energy interaction between moisture.

Figure 1b indicates that thermal radiation's absorption by the hydrophilic region (gray areas) was prevented by concentrated droplets, and the emissivity of the amphiphilic surfaces was mainly determined by the hydrophobic region on the wet regions according to the equation $\varepsilon_{amphiphilie} = \varepsilon_{hydrophobe} f_{hydrophobe}$. Meanwhile, the energy received from condensed vapor on the hydrophilic area was transferred to the hydrophobic area because the emissivity of the hydrophilic area was diminished by the droplets on its surface.

2.2. Design Rationale 1: Surface Mechanicss

The island-like amphiphilic surfaces in Figure 2a require careful design of both their surface wettability and droplet mobility to obtain a higher water-harvesting efficiency. When the surface tension of solids is less than the attracted force among the water molecules, the coalesced droplets tend to form a spherical shape, as shown in Figure 2b. This surface is considered non-wettable. When the surface tension of the solids is larger than the molecular forces, the droplets on the solid surface will show a tendency to expand outwards, as shown in Figure 2c. This surface is considered wettable. Here, we refer to a surface as "hydrophobic" when $\theta > 90°$ or "hydrophilic" when $\theta < 90°$ [41].

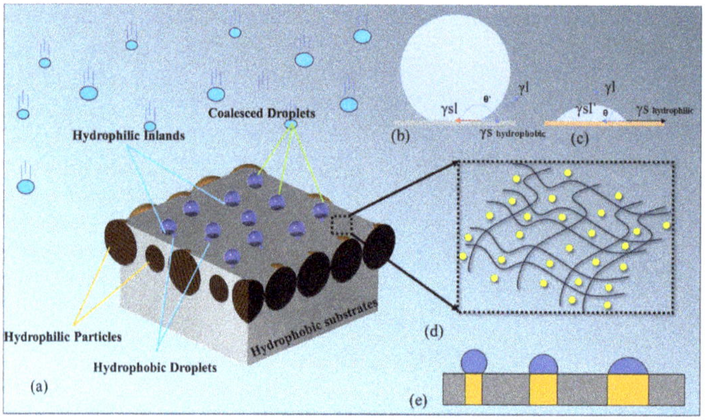

Figure 2. (**a**) Diagram of coalesced droplets on amphiphilic surfaces. (**b**) Diagram of droplets adhered to hydrophobic surface. (**c**) Diagram of droplets adhered to hydrophilic surface. (**d**) Hydrophobic groove mixed with nanoparticles. (**e**) Diagram of droplets' status on different hydrophilic islands.

The contact angle of droplets on a solid surface was used to evaluate the surface wettability and droplet mobility [42]. Figure 2e illustrates the relationship between the surface texture and droplet mobility. The droplets coalesced on a smaller hydrophilic island can slide off earlier, as the surrounding hydrophobic grooves prevent the droplets from spreading [43]. Figure 2d shows a hydrophobic area with mixed nanoparticles to improve its energy emission abilities [44,45].

3. Results and Discussion

3.1. SEM Results

SEM was used to observe the morphological characteristics of the prepared amphiphilic surfaces with 0.5 wt.% and 1.0 wt.% of the nano-admixtures to enhance the mobility of the droplets in the hydrophobic grooves. The results in Figure 3a,b show that islands were closely surrounded by PDMS solids to form amphiphilic surfaces with different textures. The EDS results displayed the Ca and Na elements distributed on the amphiphilic surfaces to indicate the locations of the emerged glass islands. However, the diameters of the hydrophilic islands on the surface mixed with 0.5 wt.% of the BN nano-admixture (Figure 3c) were significantly larger than those of the BN 1.0 wt.% mixture (Figure 3d). This will have affected the water-harvesting efficiency. We measured the parameters of the hydrophilic islands and the hydrophobic grooves on the different surfaces using the Nano Measurer program and obtained their normal distributions.

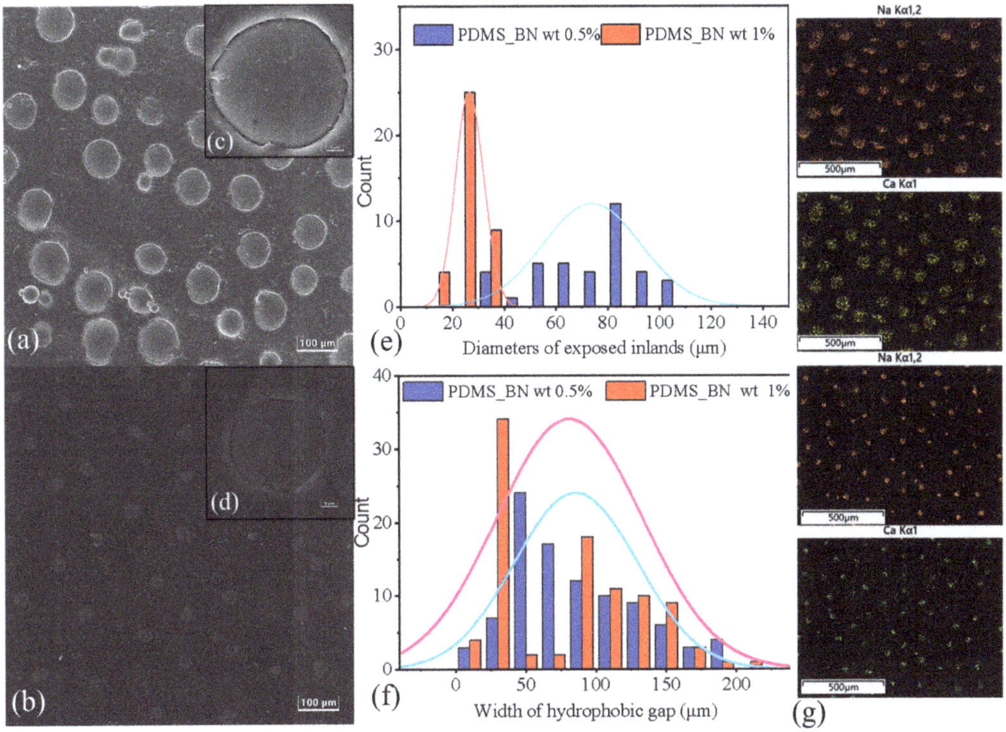

Figure 3. (**a**) SEM image of an amphiphilic surface with an admixture of 0.5 wt.% BN, (**b**) SEM image of an amphiphilic surface with an admixture of 1.0 wt.% BN, (**c**) SEM image of hydrophilic islands on an admixture of 0.5 wt.% BN on an amphiphilic surface, (**d**) SEM image of hydrophilic islands on an admixture of 1.0 wt.% BN within amphiphilic surface, (**e**) normal distribution of hydrophilic island diameters, (**f**) normal distribution of hydrophobic grooves, (**g**) EDS results of Na and Ca in amphiphilic surfaces.

The statistical results (Figure 3e,f) showed that the hydrophilic island size was much smaller than the hydrophilic particle diameter (100 μm) in both cases. Therefore, the hydrophilic particles could become tightly stuck in the cured PDMS. The size distribution of the hydrophilic islands of the 0.5 wt.% BN sample was mainly concentrated in the range of 30–100 μm, while the size distribution of the hydrophilic islands of the 1.0 wt.% BN sample was uniformly distributed in the range of 20–40 μm. The former was significantly smaller than the latter. Although the hydrophobic groove widths of the 0.5 wt.% BN sample and the 1.0 wt.% BN sample were in the same range, the groove widths of the former were closer to a normal distribution, while the latter was mainly concentrated at 30 μm and 50 μm. Therefore, the latter provided more precise control of the width of the inlet hydrophobic grooves. We predicted that the structural features on the amphiphilic surface of the 0.5 wt.% BN sample would negatively impact the droplet sliding and thermal emissivity when wet.

3.2. AFM Result

The SEM images of the amphiphilic surfaces showed that the morphological features of the hydrophobic region in the amphiphilic surface were impacted by the cured PVP adhesive on the pre-templated surface (Figure 4a-1). The roughness of the hydrophobic surface was significantly reduced after adding 0.5 wt.% BN (Figure 4a-2). The nanoscale morphological features of the hydrophobic region on the amphiphilic surface were observed using AFM, which showed that the addition of different proportions of BN (0.5 wt.%, 1.0 wt.%, and 2.0 wt.%) to the PDMS pre-polymers produced different structures on the amphiphilic surfaces. The surface structure of the cured pure PDMS pre-polymer (Figure 4b-1) demonstrated a surface roughness of 10.4 nm and a maximum height difference of 104 nm (Figure 4c). The hydrophobic surface of PDMS_BN 0.5 wt.% (Figure 4b-2) was the smoothest, with a surface roughness of only 0.689 nm and the lowest maximum height difference (5.9 nm). In contrast, the 1.0 wt.% BN surface (Figure 4b-3) was moderately rougher (6.39 nm), with a maximum height difference of 46.4 nm. The hydrophobic surface of the PDMS_BN 2.0 wt.% sample (Figure 4b-4) was rougher than all the other surfaces (18.2 nm), with a maximum height difference of 160.5 nm. We compared the characteristics of continuous height variation for different surfaces by extracting the section profiles in the center of the sample height map (Figure 4d) and found that PDMS_BN 2.0 wt.% significantly reinforced the surface topography of the pre-polymerized template. The peak and valley values of its grooves were significantly higher than those of the original rough PDMS sample. This was because the pre-polymers of the PDMS_BN 1.0 wt.% and PDMS_BN 0.5 wt.% samples filled the surface nanogrooves of the pre-polymerized templates during polymerization, which reduced the surface roughness of the PDMS. The PDMS_BN 0.5 wt.% sample had the most obvious smoothing effect for hydrophobic grooves. Mixing PDMS with BN increased the smoothness of the cured PDMS surface and intensified the thermal radiation efficiency of the hydrophobic area, which was more conducive to droplet sliding and cooling of the surface.

3.3. Static Contact Angle Test and XRD Results

During the preparation of amphiphilic surfaces, any additives (glass particles and BN nanoparticles) theoretically reduce the degree of polymerization of PDMS, thus affecting the contact angle. However, when glass beads emerge on the surface, their inherent hydrophilicity influences the contact angle test results. To test the effect of hydrophilic particles and BN nanoparticles on the hydrophobic regions (PDMS), an equal amount of hydrophilic glass beads was mixed into the PDMS. Subsequently, XRD tests and contact angle tests were performed on the surface. The results showed that although XRD only detected the influence of BN on the characteristic peaks of PDMS at the PDMS surface, the internally mixed glass beads also affected the wettability of the hydrophobic regions.

The contact angle test was used to reveal the surface wettability and mobility of droplets. We randomly chose three points on each sample to test the wettability of the smooth PDMS surface and the surfaces with different ratios of nanoparticles. This was performed to evaluate the influence on the surface wettability caused by inserting the nanoparticles and glass particles. We characterized the static CA of these amphiphilic surfaces with different volumes of BN nanoparticles using an automated goniometer at ambient conditions with a 5 mL water droplet. The results in Figure 5a show that the green columns represent the results for the PDMS composite surfaces prepared by mixing PDMS with BN admixtures (different ratios), and the blue columns represent the PDMS composite surfaces prepared by mixing PDMS with BN admixtures (different ratios) and glass particles (same mass). The wettability test results indicated that nearly all contact angle of the samples with additives were lower than those of the pure PDMS surface. However, for the samples without added glass particles, the contact angle tended to decrease and then increase as the percentage of BN increased from 0.5% to 2%. The contact angle of the sample with 2.0% BN was very close to the hydrophobicity of the original PDMS (0.4°). The contact angle trend for the samples with added glass particles (blue column) was it being significantly higher than that of the original sample (green column). The wettability of the PDMS mixed surface was already 1.3° higher than that of the original

PDMS at a BN content of 1.0 wt.%. However, a significant decrease in the contact angle (100.6°) was observed when the BN content was 2.0 wt.%.

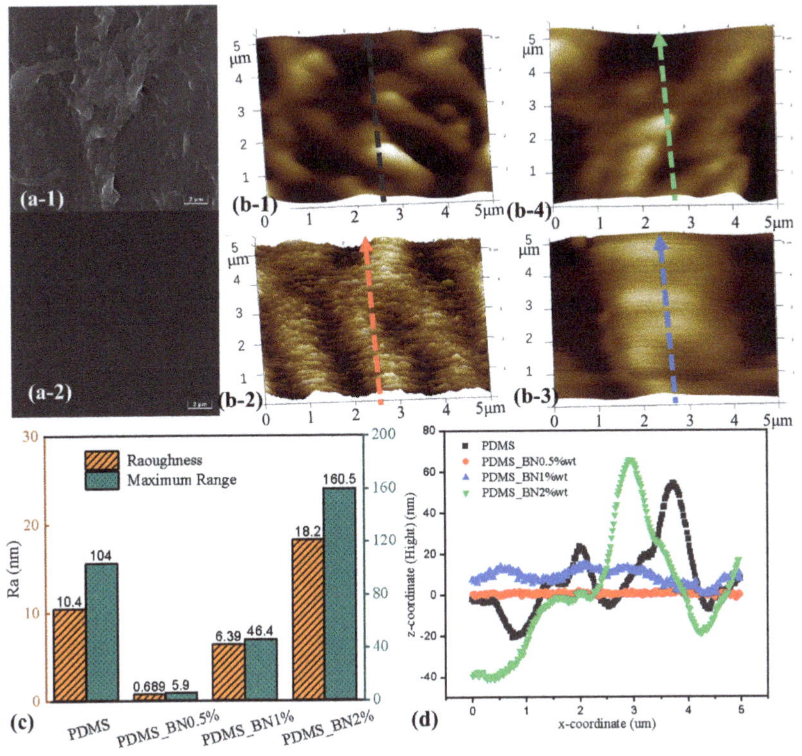

Figure 4. (**a-1**) SEM images of hydrophobic grooves on an amphiphilic surface without the nano-admixture and (**a-2**) with an admixture of 0.5 wt.% BN. (**b-1**) AFM images of hydrophobic grooves on an amphiphilic surface without the nano-admixture and (**b-2**) with an admixture of 0.5 wt.% BN. (**b-3**) SEM images of hydrophobic grooves on an amphiphilic surface with an admixture of 1.0 wt.% BN and (**b-4**) with an admixture of 2.0 wt.% BN. (**c**) Statistics of roughness and maximum altitude difference on amphiphilic surfaces. (**d**) The surface fluctuation details of 4 lines on amphiphilic surfaces.

To further investigate the influence of additives on the PDMS surface, we investigated the crystalline components of the PDMS_BN composite surface using XRD, and the results are shown in Figure 5b. The sharp peak on the right side represents crystalline BN, and the height of the crystalline peak increased significantly upon increasing the addition ratio. The broad peak on the left side is the amorphous peak of PDMS. The amorphous peak of PDMS_BN 0.5% was significantly weaker than that of the pure PDMS samples, while the amorphous peak of PDMS_BN 1% was the strongest among all four samples. The amorphous peak of PDMS_BN 2% had the lowest height among all four samples. The heights of the crystalline peaks increased and then decreased upon increasing the BN addition. This was consistent with the trend observed from the contact angle tests.

The results showed that the additives interfered with the crosslinking reaction of the pre-polymer and increased the hydrophobicity and thermal emissivity of the amphiphilic surfaces. The inserted hydrophilic particles also affected the surface tension of the PDMS area. The hydrophobicity of the PDMS on the PDMS_Glass_BN 1.0 wt.% sample was higher than that of pure PDMS, which increased the mobility of the droplets on the amphiphilic surfaces. However, adding excess nanoparticles may expose BN, which may degrade the hydrophobicity of the PDMS area.

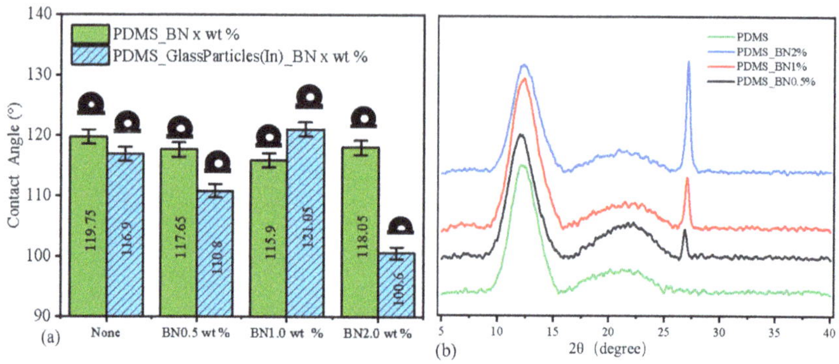

Figure 5. (**a**) The static contact angle results for hydrophobic grooves. (**b**) XRD results of hydrophobic grooves.

3.4. Droplet Coalescence Test Results

The experimental samples were placed on a copper plate with a lower temperature to observe the vapor concentrating and droplet coalescence processes on the amphiphilic surfaces. Figure 6a shows that the cold panel is equipped with metal tubes to allow cold water to flow, maintaining plate cooling and thereby accelerating the water vapor condensation process on the amphiphilic surface. The water vapor condensation process was observed and recorded using an optical microscope, which provided important evidence for evaluating the water vapor concentrating and merging properties on the amphiphilic surfaces.

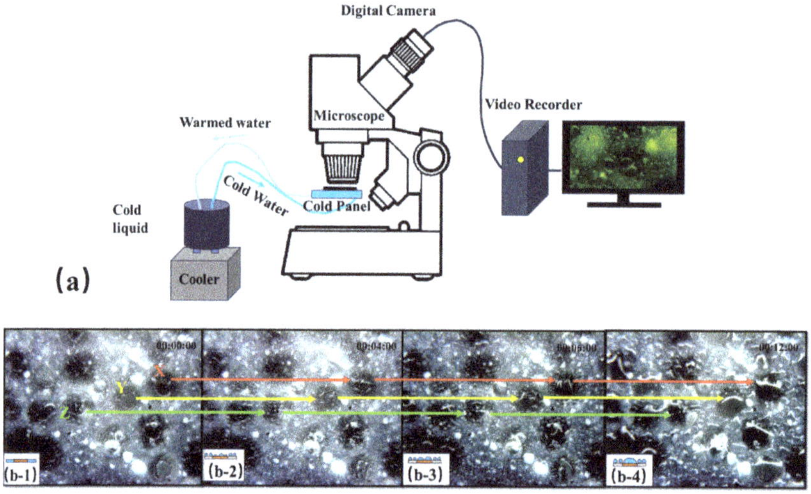

Figure 6. (**a**) Diagram of droplet coalescence test platform. (**b-1–b-4**) Droplet inflation and coalescence processes of the amphiphilic surface.

The vapor concentration efficiency and droplet sliding behavior were observed over a period of 1 h. Initially, before the vapor concentration (Figure 6b-1), there was a distinct boundary between the hydrophilic and hydrophobic regions of the amphiphilic surface. Subsequently, as water vapor continuously landed on the surface, no significant difference in the droplet sizes between the hydrophilic islands and hydrophobic grooves was observed at around 4 min into the experiment (Figure 6b-2). However, at approximately 6 min (Figure 6b-3), it became evident that droplet coalescence on the hydrophilic islands occurred more frequently than in the hydrophobic region, resulting in larger droplets in the former. By the 12th minute of the experiment (Figure 6b-4), the droplets on all the hydrophilic islands had merged into large droplets, with overflow observed on larger islands (X and Y) and successful coalescence on smaller islands (Z). At this stage, the boundary between the water islands and the hydrophobic gullies became indistinct, and the main driving force for droplet movement in this phase was attributed to the hydrophobic region.

3.5. *Water-Harvesting Test Results*

Figure 7a,c show the maximum weight of the droplets on the samples (bar graph) before sliding events. The line graph represents the static water contact angles of different PDMS surfaces and hydrophilic particulates previously reported [13,44,45]. Figure 7a shows that bigger droplets holding onto the samples was significantly positively correlated with the wettability of the sample surface. Meanwhile, the droplet mass carried by the surface of sample PDMS(GlassP)_BNwt 1.0% was the largest (0.56293 g), which also matched the lowest (121°) wettability of the hydrophobic samples. Conversely, the PDMS(GlassP)_BNwt 0.5% surface carried the smallest droplet (0.38769 g), and the contact angle of PDMS(GlassP)_BNwt 2% was the smallest of all the samples (100.6°). Figure 7c reveals that the higher the frequency of the droplet sliding, the better the wettability of the hydrophilic islands on the amphiphilic surfaces. The highest droplet slippage efficiency (52 times/h) was found on the sample whose hydrophilic island material was glass, and this sample also displayed the most hydrophilic islands. Apart from hydrophobic surfaces (42 times/h), the lowest droplet fall-off rating was observed for the amphiphilic surface with iron particles (48 times/h), which also was the hydrophilic island with the lowest wettability.

Figure 7b,d show the mean vapor concentration speed (bar graph) for each droplet-refreshing cycle on different surfaces. The line graph represents the mass of water vapor collected by different surfaces during the test. Figure 7b shows that the vapor concentration efficiency on the sample surfaces was vital for determining the water-harvesting efficiency of the amphiphilic surfaces. Moreover, the water-harvesting quantity and vapor condensation rate on the amphiphilic surfaces were overall higher than those of the purely hydrophobic samples (2.68 g and 7.73×10^{-4} g/s). The surface with the highest water vapor condensation efficiency was sample PDMS(GlassP)_BNwt 1.0% (12.99×10^{-4} g/s), which also collected the most water (4.2262 g) during the test (1 h). Except for the hydrophobic samples, PDMS(GlassP)_BNwt 2% collected the smallest amount of water (3.5764 g), and its water-harvesting rate was the same as that of the sample PDMS(GlassP) without the BN admixture (10.97×10^{-4} g/s and 11.08×10^{-4} g/s). Figure 7d shows a strong relationship between the condensation vapor concentration and water harvesting on the amphiphilic surfaces. The water-harvesting mass and vapor landing efficiency of the amphiphilic surfaces with different hydrophilic island compositions were equally higher than those of the purely hydrophobic samples (2.68 g and 7.73×10^{-4} g/s). When the hydrophilic/homophobic ratio of the amphiphilic surface was constant, increasing the wettability dissimilation of the hydrophilic island and hydrophobic grooves improved the water-harvesting efficiency. Although the composite surface with added tungsten particles showed the highest water-harvesting efficiency (11.417×10^{-4} g/s), the amphiphilic surface with the highest amount of water harvesting was the one containing added glass particles (3.5915 g). Apart from the hydrophobic samples, the amphiphilic surface with Fe particles harvested the smallest amount of water (3.01 g), and it also concentrated vapor at a lower efficiency than the other two amphiphilic surfaces.

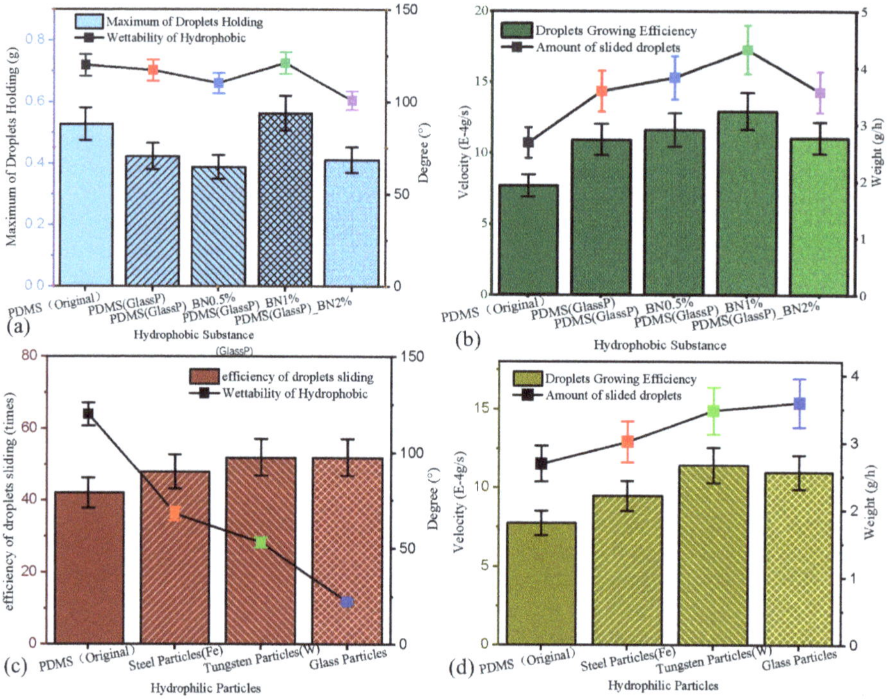

Figure 7. (**a**) The results for surface wettability and droplet dragging ability on amphiphilic surfaces with different volumes of BN nanoparticles. (**b**) The result of droplets sliding and growing on amphiphilic surfaces with different volumes of BN nanoparticles. (**c**) The result of surface wettability and droplet sliding ability on amphiphilic surfaces with different hydrophilic particles inserted. (**d**) The result of droplets sliding and growing on amphiphilic surfaces with different hydrophilic particles inserted.

4. Materials and Methods

Given these physical insights, we investigated a wide range of nanoparticles and hydrophilic islands. To design an amphiphilic surface with emerged islands, three engineering criteria must be satisfied. First, the candidate particles must be wettable but not soluble in water. Second, the hydrophilic particles must be closely connected to an adhesive until the end of the polymer curing process to produce a surface with emerged islands. Third, the nano-admixture should be inserted and evenly dispersed in the pre-polymer liquid and must also maintain its structural stability and high thermal reflectivity and emissivity. The design principle of such a surface includes the careful selection of a base admixture and particles, as well as structural design. We investigated several particles with different surface tension values, including glass [46], iron [13], tungsten [47], and a different volume of admixtures (BN nanoparticles). These particles can be closely bundled by selecting an appropriate assembly process. All the candidate particles were divided into different groups based on their different diameters.

4.1. Design Rationale 3: Amphiphilc Surface Assmbly Method

An island-type amphiphilic water-harvesting surface was prepared using a templating method proposed to synthesize a water-harvesting surface that imitated a beetle's back. The assembly process involved three steps (Figure 8):

(1) Hydrophilic particle pinning: First, PVP semi-solid glue was uniformly applied on the surface of a glass plate. Then, size-screened hydrophilic particles were sprin-

kled onto the glass glazing plate, which was put into an oven at 60 °C for 30 min to obtain prefabricated templates with fixed hydrophilic islands. The diameter range of the hydrophilic particles was limited to 100–125 μm to eliminate the influence of the amphiphilic surface's morphology due to the different diameters of the hydrophilic particles.

(2) The preparation and solidification of the hydrophobic prepolymer: The PDMS prepolymer, crosslinker, diluent, and nanoparticles were mixed at different ratios and mechanically stirred for 10 min at a low temperature until they formed a homogeneous dispersion, which was poured onto the prefabricated templates. The samples were placed into a vacuum-drying oven to perform a two-step solidification process. Firstly, a vacuum was drawn for 30 min at room temperature. After all of the air bubbles were removed, the pre-polymer was cured at 85 °C for 30 min. Finally, solid PDMS composites with an internal amphiphilic structure were obtained.

(3) The purification of the amphiphilic surface: The glass plate applied with solid PDMS was immersed in alcoholic liquid for two hours until the PVP colloid melted and separated from the glass plate. Then, the fallen PDMS solid was placed in an ultrasonic cleaner to clean off the PVP colloid adhered to the PDMS surface to obtain an island-type amphiphilic surface.

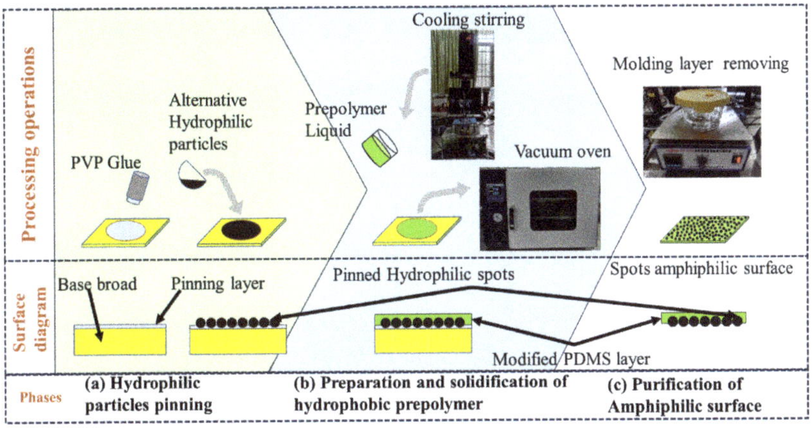

Figure 8. The process flow of amphiphilic surface assembling.

4.2. Water-Harvesting Performance Test

Amphiphilic surface samples 50 mm × 50 mm in size were suspended in the water collection test rig at an inclination angle of 30° to investigate the influence of the wettability of the hydrophobic grooves and the wettability of the hydrophilic islands for water harvesting. The results on the vapor concentration efficiency and droplet sliding behavior were tested for 1 h. Because the water concentration process is slow and affected by the environment, we designed a water-harvesting performance test platform (Figure 9c) to determine the vapor concentration and droplet sliding on the amphiphilic surfaces. A temperature control function (Figure 9b) was added based on its original structure to reduce the influence of the environmental variables. The stability of the testing environment was improved by controlling the temperature of the liquid in the nebulizer and the temperature of the inlet air.

The total mass of the droplets ($G_{harvesting}$) collected from the different amphiphilic surface structures during the test was used to evaluate the growth and movement of the droplets on amphiphilic surfaces, $G_{Harvesting} = \sum_{i=1}^{n} G_i$, where $G_{Harvesting}$ represents the total weight of the droplets dropped per unit of time, and G_i represents the mass of the droplets dropped each time. Water harvesting is a discontinuous process from water vapor condensation to growth and then to sliding from the surface. The sample weight continues to grow during vapor's concentration on a surface unless a droplet falls off, in which case the sample weight will drop sharply. These data are detected using an electronic balance and saved in a computer. The weight of the test sample will periodically increase until the next droplets fall. We designed a C++17 program (Figure 9a) to distinguish the weight chain into subsets regarding the weight increasing or decreasing due to the triggers. Then, we recorded the difference between the first and last data points in the subsets as the mass of the droplets dropped (G_i). Finally, the results of the recorded data were output as a report following the chronological order of the droplets dropping. The vapor concentration efficiency during the droplet dropping intervals and the maximum droplets that could be carried on the surfaces were calculated through data post-processing. This provided experimental support to provide a comparison of the water vapor concentrations and sliding efficiency on the amphiphilic surfaces.

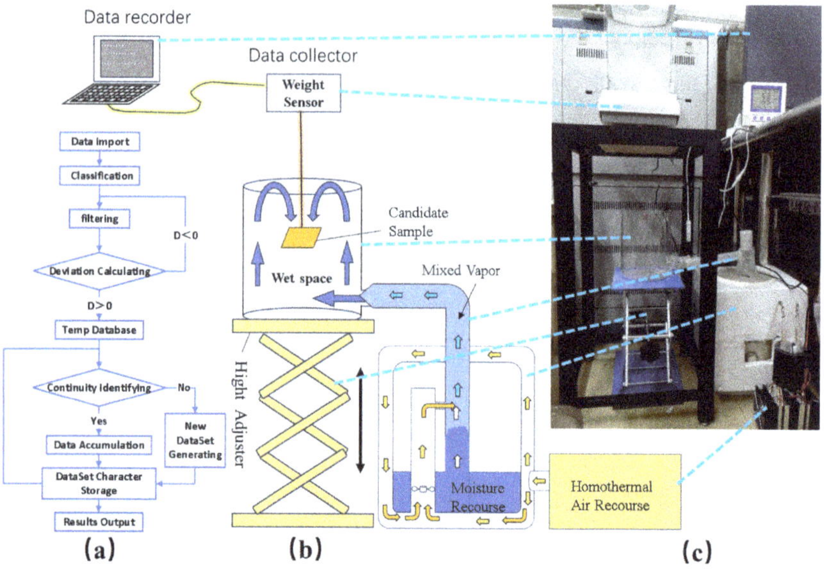

Figure 9. (**a**) Schedule of C++ program design logic; (**b**) diagram of water-harvesting platform; (**c**) set plan of water-harvesting platform.

Detailed data on the amphiphilic surfaces, such as their nanostructure and composition, were obtained using SEM, XRD, and AFM to identify the key factors of the amphiphilic surfaces that affected their water-harvesting performance.

5. Conclusions

In this study, we systematically screened raw materials and nanoparticles mixed into PDMS to improve the droplet movements and thermal radiation properties. A comparative analysis of various amphiphilic surfaces was carried out to evaluate the droplet coalescence and sliding process to facilitate their water-harvesting efficiency. Our findings highlight the pivotal role of surfaces' textures and energy emissive ability in promoting complete water harvesting on amphiphilic surfaces. We also observed that the mobility of droplets in misty environments varied according to the surfaces' wettability. Our research also underscores the significance of the surface tension of hydrophobic grooves, which we modified by incorporating BN nanoparticles and inserting hydrophilic particles during hydrophobic groove curing. This adjustment allowed us to increase the contact angle of PDMS, thereby enhancing the sliding frequency of droplets. Moreover, our investigation revealed that the quantity of added nanoparticles had a crucial effect on the thermal emission power of the amphiphilic surfaces during water-harvesting tests.

Author Contributions: Conceptualization, H.W. and C.L.; methodology, H.W.; software, S.L.; validation, Y.Z., W.W. and H.W.; formal analysis, S.L.; investigation, K.A.M.A.; resources, H.W.; data curation, H.W.; writing—original draft preparation, H.W.; writing—review and editing, K.A.M.A.; visualization, S.L.; supervision, C.L.; project administration, C.L.; funding acquisition, C.L. All authors have read and agreed to the published version of the manuscript.

Funding: This research was funded by the National Natural Science Foundation of China, grant number 32171906.

Institutional Review Board Statement: Not applicable.

Informed Consent Statement: Not applicable.

Data Availability Statement: The data presented in this study are available in article.

Conflicts of Interest: The authors declare no conflicts of interest.

References

1. Overpeck, J.T. The challenge of hot drought. *Nature* **2013**, *503*, 350–351. [CrossRef]
2. McGrath, G.S.; Sadler, R.; Fleming, K.; Tregoning, P.; Hinz, C.; Veneklaas, E.J. Tropical cyclones and the ecohydrology of Australia's recent continental-scale drought. *Geophys. Res. Lett.* **2012**, *39*. [CrossRef]
3. Dai, A. Increasing drought under global warming in observations and models. *Nat. Clim. Chang.* **2013**, *3*, 52–58. [CrossRef]
4. Trenberth, K.E.; Dai, A.; van der Schrier, G.; Jones, P.D.; Barichivich, J.; Briffa, K.R.; Sheffield, J. Global warming and changes in drought. *Nat. Clim. Chang.* **2014**, *4*, 17–22. [CrossRef]
5. Grubert, E.A.; Stillwell, A.S.; Webber, M.E. Where does solar-aided seawater desalination make sense? A method for identifying sustainable sites. *Desalination* **2014**, *339*, 10–17. [CrossRef]
6. Song, Y.-Y.; Liu, Y.; Jiang, H.-B.; Li, S.-Y.; Kaya, C.; Stegmaier, T.; Han, Z.-W.; Ren, L.-Q. Temperature-tunable wettability on a bioinspired structured graphene surface for fog collection and unidirectional transport. *Nanoscale* **2018**, *10*, 3813–3822. [CrossRef]
7. Hou, B.; Wu, C.; Li, X.; Huang, J.; Chen, M. Contact line-based model for the Cassie-Wenzel transition of a sessile droplet on the hydrophobic micropillar-structured surfaces. *Appl. Surf. Sci.* **2021**, *542*, 148611. [CrossRef]
8. Qin, S.; Jin, Y.; Yin, F.; Wang, Z.; Bai, G. Can solid surface energy be a predictor of ice nucleation ability? *Appl. Surf. Sci.* **2022**, *602*, 154193. [CrossRef]
9. Bao, X.; Zhang, M.; Li, P.; Lu, J.; Yuan, G.; Zuo, Y. Investigating the surface wettability and surface free energy of sodium silicate-impregnated poplar wood. *Wood Mater. Sci. Eng.* **2021**, *18*, 141–150. [CrossRef]
10. Kang, J.H.; Lee, J.-W.; Kim, J.Y.; Moon, J.W.; Jang, H.S.; Jung, S.Y. Effect of Mesh Wettability Modification on Atmospheric and Industrial Fog Harvesting. *Front. Phys.* **2021**, *9*, 1–6. [CrossRef]
11. Wang, S.-W.; Peng, L.; Chen, J.-W.; Li, L. A comparative study of the self-propelled jumping capabilities of coalesced droplets on RTV surfaces and superhydrophobic surfaces. *Chin. Phys. B* **2021**, *30*, 046501. [CrossRef]
12. Wu, S.-C.; Lin, Z.-H.; Lo, C.-K. Study of super-hydrophilic nanoscale bilayer assembly surface modification and its application to enhance evaporation. *Therm. Sci. Eng. Prog.* **2022**, *27*, 101133. [CrossRef]
13. Pogorzelski, S.; Boniewicz-Szmyt, K.; Grzegorczyk, M.; Rochowski, P. Wettability of Metal Surfaces Affected by Paint Layer Covering. *Materials* **2022**, *15*, 1830. [CrossRef] [PubMed]
14. Song, J.-W.; Fan, L.-W. Understanding the effects of surface roughness on the temperature and pressure relevancy of water contact angles. *Colloids Surf. A Physicochem. Eng. Asp.* **2023**, *656*, 130391. [CrossRef]
15. Zhang, X.; Qin, Y. Contact angle hysteresis of a water droplet on a hydrophobic fuel cell surface. *J. Colloid Interface Sci.* **2019**, *545*, 231–241. [CrossRef] [PubMed]
16. Zhu, D.; Shi, Z.; Tan, X.; Zhang, J.; Zhang, S.; Zhang, X. Accelerated wetting transition from hydrophilic to hydrophobic of sputtered Cu films with micro-scale patterns. *Appl. Surf. Sci.* **2020**, *527*, 146741. [CrossRef]
17. Khan, S.A.; Boltaev, G.S.; Iqbal, M.; Kim, V.; Ganeev, R.A.; Alnaser, A.S. Ultrafast fiber laser-induced fabrication of superhydrophobic and self-cleaning metal surfaces. *Appl. Surf. Sci.* **2021**, *542*, 148560. [CrossRef]
18. Liu, M.; Li, M.-T.; Xu, S.; Yang, H.; Sun, H.-B. Bioinspired Superhydrophobic Surfaces via Laser-Structuring. *Front. Chem.* **2020**, *8*, 835. [CrossRef] [PubMed]
19. Pinheiro, R.A.; Rosa, F.M.; Volú, R.M.; de Vasconcelos, G.; Trava-Airoldi, V.J.; Corat, E.J. Vertically aligned carbon nanotubes (VACNT) surfaces coated with polyethylene for enhanced dew harvesting. *Diam. Relat. Mater.* **2020**, *107*, 107837. [CrossRef]
20. Yan, Y.-L.; Cai, Y.-X.; Liu, X.-C.; Ma, G.-W.; Lv, W.; Wang, M.-X. Hydrophobic Modification on the Surface of SiO_2 Nanoparticle: Wettability Control. *Langmuir* **2020**, *36*, 14924–14932. [CrossRef]
21. Deng, Z.; Gao, S.; Wang, H.; Liu, X.; Zhang, C. Visualization study on the condensation heat transfer on vertical surfaces with a wettability gradient. *Int. J. Heat Mass Transf.* **2022**, *184*, 122331. [CrossRef]

22. Kim, D.E.; Oh, J.S. Local phase and thermal behaviors in pool boiling on different wettability surfaces. *Exp. Therm. Fluid Sci.* **2022**, *139*, 110728. [CrossRef]
23. Liu, X.; Zhao, X.; Lu, L.; Li, J. Liquid bridges between particles and the hydrophobic or hydrophilic surfaces of solar photovoltaic glass. *Sci. Total Environ.* **2022**, *822*, 153552. [CrossRef] [PubMed]
24. Parker, A.R.; Lawrence, C.R. Water capture by a desert beetle. *Nature* **2001**, *414*, 33–34. [CrossRef]
25. Yu, Z.; Yun, F.F.; Wang, Y.; Yao, L.; Dou, S.; Liu, K.; Jiang, L.; Wang, X. Desert Beetle-Inspired Superwettable Patterned Surfaces for Water Harvesting. *Small* **2017**, *13*, 1701403. [CrossRef] [PubMed]
26. Chen, Z.; Zhang, Z. Recent progress in beetle-inspired superhydrophilic-superhydrophobic micropatterned water-collection materials. *Water Sci. Technol.* **2020**, *82*, 207–226. [CrossRef]
27. Zhu, H.; Huang, Y.; Lou, X.; Xia, F. Beetle-inspired wettable materials: From fabrications to applications. *Mater. Today Nano* **2019**, *6*, 100034. [CrossRef]
28. Dai, X.; Sun, N.; Steven, O.N.; Birgitt, B.S.; Wang, J.; Yang, S.; Wong, T.-S. Hydrophilic directional slippery rough surfaces for water harvesting. *Sci. Adv.* **2018**, *4*, eaaq0919. [CrossRef] [PubMed]
29. Yue, H.; Zeng, Q.; Huang, J.; Guo, Z.; Liu, W. Fog collection behavior of bionic surface and large fog collector: A review. *Adv. Colloid Interface Sci.* **2022**, *300*, 102583. [CrossRef]
30. Jarimi, H.; Powell, R.; Riffat, S. Review of sustainable methods for atmospheric water harvesting. *Int. J. Low-Carbon Technol.* **2020**, *15*, 253–276. [CrossRef]
31. Wang, B.; Zhou, X.; Guo, Z.; Liu, W. Recent advances in atmosphere water harvesting: Design principle, materials, devices, and applications. *Nano Today* **2021**, *40*, 101283. [CrossRef]
32. Benda, J.; Narikiyo, H.; Stafslien, S.J.; VanderWal, L.J.; Finlay, J.A.; Aldred, N.; Clare, A.S.; Webster, D.C. Studying the Effect of Pre-Polymer Composition and Incorporation of Surface-Modifying Amphiphilic Additives on the Fouling-Release Performance of Amphiphilic Siloxane-Polyurethane Coatings. *ACS Appl. Mater. Interfaces* **2022**, *14*, 37229–37247. [CrossRef]
33. Yan, D.; Chen, Y.; Liu, J.; Song, J. Super-Fast Fog Collector Based on Self-Driven Jet of Mini Fog Droplets. *Small* **2023**, *19*, e2301745. [CrossRef]
34. Huang, B.; Zhang, X.; Yao, Z. Condensation on solid surfaces with amphiphilic micro-nanostructures by lattice Boltzmann simulation. *Chem. Phys.* **2018**, *513*, 258–265. [CrossRef]
35. Sun, R.; Zhao, J.; Liu, C.; Yu, N.; Mo, J.; Pan, Y.; Luo, D. Design and optimization of hybrid superhydrophobic–hydrophilic pattern surfaces for improving fog harvesting efficiency. *Prog. Org. Coat.* **2022**, *171*, 107016. [CrossRef]
36. Zhang, M.; Zhao, T.; Yu, C.; Liu, Q.; Wang, G.; Yang, H.; Yang, M.; Jiang, L.; Liu, M. Amphiphilic Pd@micro-organohydrogels with controlled wettability for enhancing gas-liquid-solid triphasic catalytic performance. *Nano Res.* **2022**, *15*, 557–563. [CrossRef]
37. Zhang, S.; Chi, M.; Mo, J.; Liu, T.; Liu, Y.; Fu, Q.; Wang, J.; Luo, B.; Qin, Y.; Wang, S.; et al. Bioinspired asymmetric amphiphilic surface for triboelectric enhanced efficient water harvesting. *Nat. Commun.* **2022**, *13*, 4168. [CrossRef]
38. Zhang, Y.; Zhu, C.; Shi, J.; Yamanaka, S.; Morikawa, H. Bioinspired Composite Materials used for Efficient Fog Harvesting with Structures that Consist of Fungi-Mycelia Networks. *ACS Sustain. Chem. Eng.* **2022**, *10*, 12529–12539. [CrossRef]
39. Nguyen, V.-T.; Park, E.; Nguyen, N.-A.; Omelianovych, O.; Larina, L.L.; Hussain, S.S.; Choi, H.-S. 3D-printed plasma-treated super-amphiphilic microgroove surface for outperformance of liquid vertical transportation. *Appl. Surf. Sci.* **2023**, *615*, 156418. [CrossRef]
40. Kozbial, A.; Trouba, C.; Liu, H.; Li, L. Characterization of the Intrinsic Water Wettability of Graphite Using Contact Angle Measurements: Effect of Defects on Static and Dynamic Contact Angles. *Langmuir* **2017**, *33*, 959–967. [CrossRef]
41. Chen, Y.; Yan, D.; Liu, R.; Lu, Y.; Zhao, D.; Deng, X.; Song, J. Green self-propelling swimmer driven by rain droplets. *Nano Energy* **2022**, *101*, 107543. [CrossRef]
42. Suaste, E.; Castillo, V.; González, R. Determination of the phase transition in $Pb_{0.88}Ln_{0.08}Ti_{0.98}Mn_{0.02}O_3$ (Ln=La, Sm, Eu) piezoceramics based on the Stefan–Boltzmann law. *Mater. Charact.* **2004**, *52*, 279–282. [CrossRef]
43. Wang, H.; Wang, D.T.; Zhang, X.Y.; Zhang, Z.Z. Modified PDMS with inserted hydrophilic particles for water harvesting. *Compos. Sci. Technol.* **2021**, *213*, 108954. [CrossRef]
44. Cai, Q.; Scullion, D.; Gan, W.; Falin, A.; Zhang, S.; Watanabe, K.; Taniguchi, T.; Chen, Y.; Santos, E.J.G.; Li, L.H. High thermal conductivity of high-quality monolayer boron nitride and its thermal expansion. *Sci. Adv.* **2019**, *5*, eaav0129. [CrossRef] [PubMed]
45. Brodu, E.; Balat-Pichelin, M. Emissivity of Boron Nitride and Metals for the Solar Probe Plus Mission. *J. Spacecr. Rocket.* **2016**, *53*, 1119–1127. [CrossRef]
46. Vukovic, T.; Røstad, J.; Farooq, U.; Torsæter, O.; van der Net, A. Systematic Study of Wettability Alteration of Glass Surfaces by Dichlorooctamethyltetrasiloxane Silanization—A Guide for Contact Angle Modification. *ACS Omega* **2023**, *8*, 36662–36676. [CrossRef]
47. He, H.; Qu, N.; Zeng, Y. Lotus-leaf-like microstructures on tungsten surface induced by one-step nanosecond laser irradiation. *Surf. Coat. Technol.* **2016**, *307*, 898–907. [CrossRef]

molecules

MDPI

Article

Magnesium Ion Gated Ion Rejection through Carboxylated Graphene Oxide Nanopore: A Theoretical Study

Jianjun Jiang [1,2], Yusong Tu [1,3,*] and Zonglin Gu [3,*]

[1] College of Chemistry and Chemical Engineering, Yangzhou University, Yangzhou 225002, China; jianjunjiang@126.com
[2] Department of Physics, Sanjiang College, Nanjing 210012, China
[3] College of Physical Science and Technology, Yangzhou University, Yangzhou 225009, China
* Correspondence: ystu@yzu.edu.cn (Y.T.); guzonglin@yzu.edu.cn (Z.G.)

Abstract: While nanoporous graphene oxide (GO) is recognized as one of the most promising reverse osmosis desalination membranes, limited attention has been paid to controlling desalination performance through the large GO pores, primarily due to significant ion leakage resulting in the suboptimal performance of these pores. In this study, we employed a molecular dynamics simulation approach to demonstrate that Mg^{2+} ions, adhered to carboxylated GO nanopores, can function as gates, regulating the transport of ions (Na^+ and Cl^-) through the porous GO membrane. Specifically, the presence of divalent cations near a nanopore reduces the concentration of salt ions in the vicinity of the pore and prolongs their permeation time across the pore. This subsequently leads to a notable enhancement in salt rejection rates. Additionally, the ion rejection rate increases with more adsorbed Mg^{2+} ions. However, the presence of the adsorbed Mg^{2+} ions compromises water transport. Here, we also elucidate the impact of graphene oxidation degree on desalination. Furthermore, we design an optimal combination of adsorbed Mg^{2+} ion quantity and oxidation degree to achieve high water flux and salt rejection rates. This work provides valuable insights for developing new nanoporous graphene oxide membranes for controlled water desalination.

Keywords: graphene oxide membrane; molecular dynamics simulation; water desalination

1. Introduction

Freshwater scarcity has become a widespread issue due to rapid urbanization and industrialization [1–3]. The ever-increasing demand for freshwater necessitates the cost-effective purification of seawater, which constitutes 97.5% of the Earth's water [4,5]. Reverse osmosis (RO) has been proven to be an efficient method for seawater desalination, characterized by lower energy consumption and capital costs [1,6–8]. Currently, RO accounts for 70% of the global seawater desalination capacity [9]. Membrane thickness is a critical factor influencing the quality of the RO water desalination process [4,10]. Traditional RO membranes have limited water flux, making freshwater production energy-intensive. This is mainly because commercial RO membranes require adequate crosslinking to achieve high salt rejection rates, often resulting in relatively thick membranes [11–14]. In the actual desalination process, using membranes with excellent selectivity and effective pre-treatment and post-treatment processes is crucial for improving RO water quality [15,16]. Nanoporous graphene membranes have garnered significant interest in the development of nanostructured materials [10,17–30]. It has been demonstrated that nanoporous graphene can function as a synthetic water channel, holding great promise for water desalination due to its remarkable properties. These properties include its high aspect ratio and meager transport resistance, attributed to its monoatomic thickness [31].

Prior research has shown that the pore diameters of nanoporous graphene membranes can influence the transport of substances during the water desalination process [10,32]. It has been observed that a critical pore diameter exists below which the permeation of salt ions is effectively blocked [10,32]. While increasing pore size enhances water flux, it simultaneously reduces salt rejection. In practical applications, most synthesized porous graphene membranes possess pore sizes larger than this critical threshold [27,33]. Consequently, researchers often modify the functionalization of the pore periphery or introduce charges to the graphene surfaces to enhance purification efficiency and ion selectivity [4,24,27,30,34–38]. For example, ionizing the functionalization groups on the nanopores' perimeters can increase water flux at the expense of reduced salt rejection [35]. Utilizing fluorine functionalization has proven effective in preventing the passage of Cl^- ions through the membrane, leading to higher salt rejection rates [38]. Due to various interactions such as electrostatic and cation–π interactions [39], divalent cations can be absorbed onto the surfaces of graphene membranes or graphene oxide (GO) membranes, which are synthesized by oxidizing pure graphene membranes. In reference [40], Chen demonstrated precise control of the interlayer spacing of GO membranes through cationic manipulation. It has been established that GO membranes with surface cation adsorption can be effectively employed for water desalination. In this study, we designed a GO membrane with divalent cations (Mg^{2+}) adsorbed on the rim of a large pore. Without adsorbed Mg^{2+} ions, the GO membrane with a large pore exhibited high water permeability but low salt rejection, rendering it unsuitable for water desalination. Our results indicate that the nanoporous graphene membrane with adsorbed Mg^{2+} ions acts as a gate, effectively controlling the passage of salt ions, including Na^+ ions and Cl^- ions, across the pore. As a result, such a membrane demonstrates significantly improved salt rejection and meets the requirements for desalination. Moreover, the water permeability of the large pore with adsorbed Mg^{2+} ions remains much higher than that of many other nanopores.

Pore diameter, temperature, applied pressure, and the oxidation degree of the GO membrane are three significant factors influencing water desalination performance. While extensive research has been conducted on the roles of pore diameter, temperature, and applied pressure in water desalination [4,26,33,38], there has been a relative lack of detailed investigation into the impact of the oxidation degree. In this study, we focused on elucidating the influence of the oxidation degree on the transport of water molecules and salt ions through the pore. Our results emphasize the importance of maintaining an appropriate oxidation degree to achieve high water flux and salt rejection rates. Furthermore, we compared the water desalination performance between pores with and without an Mg^{2+} cation, revealing that its presence makes the membrane more robust against changes in oxidation degree.

Recently, molecular dynamics (MD) simulations have proven to be effective in exploring the microscopic desalination mechanism of nanoporous GO membranes, providing insights and details that are not easily attainable through experimental means. In this study, we employed MD simulations to investigate the effect of Mg^{2+} cations adsorbed at the pore's edge on the desalination behavior of the pore.

2. Results and Discussion

2.1. Salt Rejection Rate

In this section, we begin by using a GO membrane with an oxidation degree of 7% as a representative case to elucidate how the pore with adsorbed Mg^{2+} ions regulates the passage of salt ions. One of the critical factors in water desalination is the membrane's capability to reject ions as the primary objective of water desalination is the separation of ions from saline water. We quantify the salt rejection rate as follows to assess the membrane's performance in allowing ions to transport through the gated pore [10].

$$\begin{aligned} SR_{Na(Cl)} &= (1 - n_{Na(Cl)}/N_{Na(Cl)}) \\ SR_{NaCl} &= (1 - n_{NaCl}/N_{NaCl}) \end{aligned} \quad (1)$$

In the above equation, $N_{Na(Cl)}$ and N_{NaCl} represent the initial numbers of Na^+ (Cl^-) ions and total salt ions (the sum of Na^+ ions and Cl^- ions) in the saltwater side at $t = 0$ while $n_{Na(Cl)}$ and n_{NaCl} denote the numbers of Na^+ (Cl^-) ions and total salt ions that have passed through the pore at $t = t_{1/2}$ (defined as the time when half of the water molecules in the saltwater side have transported through the pore [10]). Figure 1 illustrates the salt rejection rates of the GO membrane with pores adsorbed by zero, one, and two Mg^{2+} ions. Notably, in the absence of the Mg^{2+} ion, the salt rejection rate is statistically equal to the Cl^- ion salt rejection rate. However, in the presence of Mg^{2+} cation, the Na^+ rejection rate exceeds the Cl^- rejection rate. This difference arises primarily from the repulsion of Na^+ ions from the divalent cations. The discrepancy between the two salt rejection rates becomes more pronounced as the number of adsorbed Mg^{2+} ions increases. Both the Cl^- ion salt rejection rate and the Na^+ ion salt rejection rate increase when more Mg^{2+} ions are adsorbed by carboxyl groups. Consequently, the overall salt rejection rate follows the same trend.

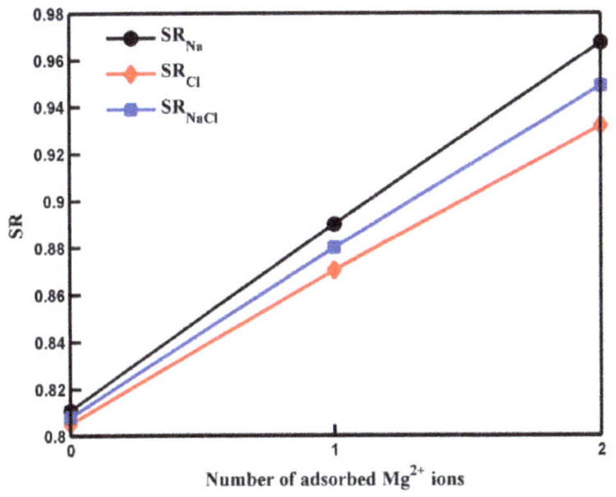

Figure 1. The salt rejection rates of pores with different numbers of adsorbed Mg^{2+} ions. The solid lines are a guide to the eye.

To illustrate the observed behavior more clearly, we conducted non-equilibrium MD simulations to analyze the distribution and dynamics of salt ions around the center of the pore. Figure 2 presents the results, showing the number of salt ions accumulating in a cylindrical region (more illustration below) as a function of the cylinder height (h). As depicted in Figure 2, an increase in the number of adsorbed Mg^{2+} ions decreases the quantity of the Na^+ ions around the pore. This decrease can be attributed to the electrostatic repulsion force between the adsorbed Mg^{2+} ions and Na^+ ions. While the adsorbed Mg^{2+} ions exert an attractive force on Cl^- ions, the quantity of Cl^- ions near the pore displays a slightly nonmonotonic variation with an increased number of adsorbed Mg^{2+} ions. This phenomenon is partly because the attractive force acting on Cl^- ions through the Na^+ ions gathering near the pore decreases as the number of adsorbed Mg^{2+} ion ions rises. For a more in-depth investigation into the mechanisms underlying the distribution of salt ions around the pore and their passage through it at the molecular level, we calculated the potential of mean force (PMF) of the ions along the *z*-axis of the simulation box. These calculations were performed under an applied pressure of 100 MPa, and the results are depicted in Figure 3. The PMF profiles indicate that the permeation of Na^+ and Cl^- ions across the pore is energetically unfavorable, making the spontaneous passage of Na^+ and Cl^- ions impossible. When Mg^{2+} ions are adsorbed on the rim of the pore, the free energy barrier for Na^+ ions to be transported through the pore increases significantly. This accounts for the observed decrease in the number of Na^+ ions around the pore in the presence of adsorbed Mg^{2+} cations. Notably, the free energy barrier for Cl^- ions remains nearly unchanged in the presence of Mg^{2+}. The behavior of the PMF profile for Cl^- ions clarifies that the adsorption of Mg^{2+} ions on the pore does not result in the aggregation of Cl^- ions inside the pore.

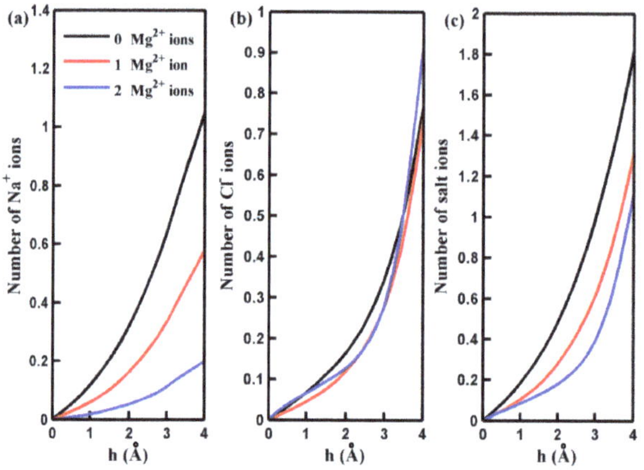

Figure 2. The number of the Na^+ ions (**a**), Cl^- ions (**b**), and salt ions (**c**) accumulating in a cylindrical region below and around the centers of the pores with different numbers of adsorbed Mg^{2+} ions as a function of the height of the cylinder h.

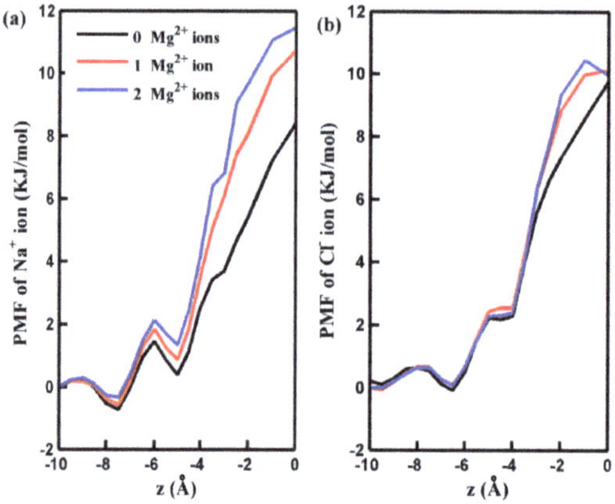

Figure 3. The PMF of Na^+ ion (**a**) and Cl^- ion (**b**) along the *z*-axes of GO membranes with different numbers of adsorbed Mg^{2+} ions. The pore is located at $z = 0$.

In summary, the analysis of salt ion distribution near the pore and the PMF profiles strongly suggest that the Mg^{2+} cation adsorbed on the edge of the pore impedes the passage of Na^+ ions. Consequently, the Na^+ ion salt rejection rate increases. However, the slight change in Cl^- ion concentration near the pore is not the primary cause of the increase in the Cl^- ion salt rejection rate. To further investigate the influence of the dynamics of salt ions on their transport through the pore, we conducted non-equilibrium MD simulations. The dynamics of salt ions were examined by calculating the permeation time autocorrelation function $C(t)$, defined as follows.

$$\begin{gathered} C_{Na(Cl)}(t) = \langle H_{Na(Cl)}(t) H_{Na(Cl)}(0) \rangle / \langle H_{Na(Cl)}(0)^2 \rangle \\ C_{NaCl}(t) = \langle H_{NaCl}(t) H_{NaCl}(0) \rangle / \langle H_{NaCl}(0)^2 \rangle \end{gathered} \quad (2)$$

Here, $C_{Na(Cl)}(t)$ and $C_{NaCl}(t)$ represent the permeation time autocorrelation function of Na^+ (Cl^-) ions and total salt ions, and $H_{Na(Cl)}(t)$ ($H_{NaCl}(t)$) is a binary function that equals 1 if one ion is initially located inside the pore at $t = 0$ and continuously exists above the center of the pore at any time greater than t and equals 0 otherwise. A water molecule or ion within a cylindrical region with a diameter of 1.6 nm and a height of 0.2 nm below and around the center of the pore is considered to be within the pore [30], also demonstrated below. This permeation time autocorrelation function (Equation (2)) is a suitable measure of the permeation probability of ions or water molecules passing through the pore. The results of the permeation time autocorrelation function $C(t)$ are presented in Figure 4. The figure shows that the rate of increase in $C(t)$ for ions decreases over simulation time. This phenomenon arises because, as shown in Table 1, ions have finite residence times inside the pore. Furthermore, the residence time of ions increases with an increasing number of adsorbed Mg^{2+} cations. Therefore, as the number of adsorbed cations increases, the decrease in the rate of $C(t)$ with time becomes less pronounced. The evolution of $C(t)$ with time suggests that estimating the permeation probability of ions using $C(t)$ with small time intervals (t) is more appropriate. Figure 4b demonstrates that the adsorption of Mg^{2+} ions on the pore significantly hinders the mobility of Cl^- ions near the pore center and prolongs their permeation time through the pore. This dynamic behavior of Cl^- ions helps explain why an increase in the number of adsorbed Mg^{2+} cations contributes to the enhancement of the Cl^- ion salt rejection rate. It should also be noted that the permeation probability of Na^+ ions, similar to Cl^- ions, decreases with increasing adsorbed Mg^{2+} cation quantity. Extending the permeation time for Na^+ ions improves the Na^+ ion salt rejection rate. The results presented in this section demonstrate that the system can function as a gated system for controlling and regulating the passage of salt ions through the pore.

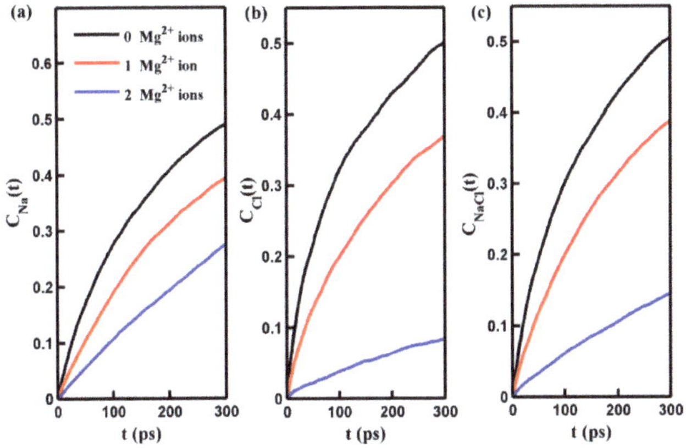

Figure 4. The permeation time autocorrelation function of Na^+ ion ($C_{Na}(t)$) (**a**), Cl^- ion ($C_{Cl}(t)$) (**b**), and salt ion ($C_{NaCl}(t)$) (**c**) in pores with different numbers of adsorbed Mg^{2+} ions as a function of time t.

Table 1. The residence times of ions and water molecules located in the pores with an oxidization degree R = 0.07.

Number of Adsorbed Mg^{2+} Ions	Na^+ (ps)	Cl^- (ps)	Water Molecule (ps)
0 Mg^{2+} ions	10.7316	8.1324	3.6078
1 Mg^{2+} ion	12.2970	8.9998	8.2718
2 Mg^{2+} ions	21.5172	10.9348	12.4690

2.2. Water Flux

Like the salt rejection rate, water flux is another crucial factor in water desalination using membrane materials. The water flux was measured by counting the net number of water molecules passing through the pore during the course of the simulation time [41]. Figure 5 illustrates the results for water flux as a function of the number of adsorbed Mg^{2+} cations. It is observed that the water flux decreases as the number of adsorbed Mg^{2+}

cations increases. Nevertheless, as indicated in Table 2, even in the case of a pore with two adsorbed Mg^{2+} cations, its water permeability still reaches an impressive value of 58.7 L/cm^2/day/MPa. This level of water permeability significantly surpasses that of many other nanopores, including MoS_2 nanopores, hydrogenated graphene nanopores, phosphorene, and g-C_3N_4 nanopores. Furthermore, the salt rejection of the pore with two adsorbed Mg^{2+} cations meets the requirements for desalination purposes. As explained in Supplementary Materials Section S2, the variation in water flux is attributed to the fact that the adsorption of Mg^{2+} ions on the pore's rim can influence the permeation probability of water molecules passing through the pore.

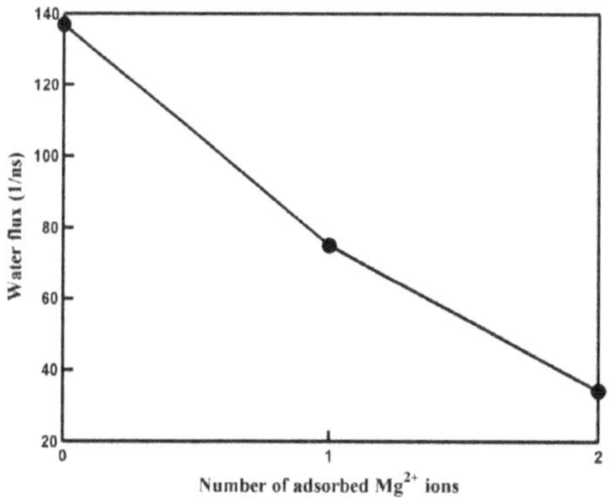

Figure 5. The water flux of pores with different numbers of adsorbed Mg^{2+} ions. The solid line is a guide to the eye.

Table 2. The comparison of the desalination performances of the pore and other RO membranes.

RO Membrane	Water Permeability (L/cm^2/Day/MPa)	Salt Rejection (%)
MoS_2 nanopore [11]	16	100
Hydrogenated graphene nanopore [10]	39	100
Phosphoerne nanopore [42]	11	100
g-C_3N_4 nanopore [43]	15	100
Hydroxylated graphene nanopore [10]	66	100
Pore without adsorbed Mg^{2+} ions (current job)	234	81
Pore with one adsorbed Mg^{2+} ion (current job)	128	88
Pore with two adsorbed Mg^{2+} ions (current job)	59	95

2.3. *Effects of the Oxidization Degree on Desalination*

Now, let us discuss the influence of the oxidization degree on the transport of water molecules and salt ions through the pore. Figure 6 presents the salt rejection rate and water flux of the GO membrane with various numbers of Mg^{2+} ions as a function of the oxidization degree. As depicted in Figure 6, neither the salt rejection rate nor water flux exhibits a monotonic variation with oxidization degree. Furthermore, the variation in the salt rejection rate or water flux becomes smaller as the number of adsorbed Mg^{2+} cations increases. These results reveal that when the oxidization degree changes, the pore with adsorbed Mg^{2+} cations exhibits a more robust water desalination performance than those without adsorbed Mg^{2+} cations. To understand the mechanism behind the effect of oxidization degree on water desalination performance, we provide the results for all systems regarding the number of salt ions (water molecules) located inside the pore and the ion (water molecule) permeation time autocorrelation function $C(t = 2\ \text{ps})$ in Figure 7.

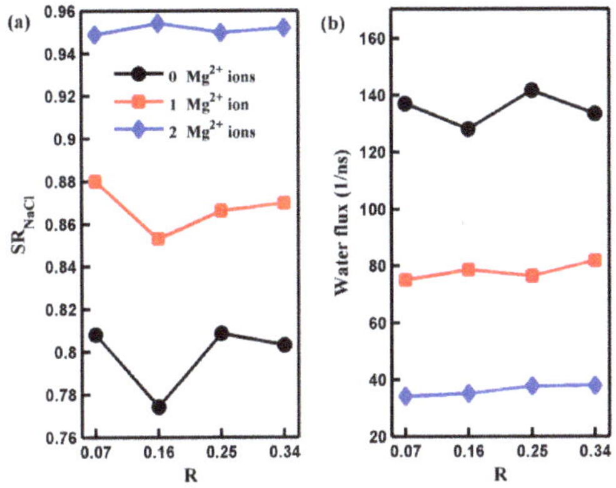

Figure 6. The salt rejection rate (**a**) and the water flux (**b**) of the nanoporous GO membranes with different numbers of adsorbed Mg^{2+} ions and oxidization degree R. The solid lines are a guide to the eye.

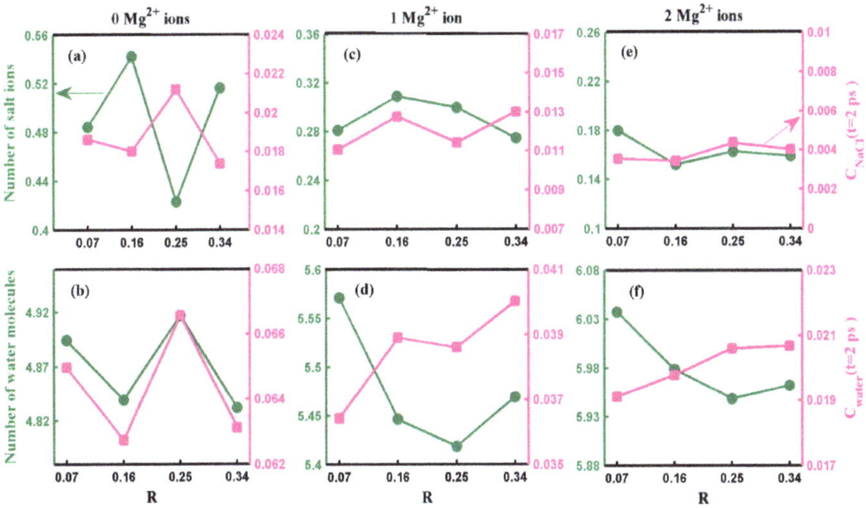

Figure 7. (**a**–**f**) show the number of the salt ions (water molecules) located inside the pore and the salt ion (water molecule) permeation time autocorrelation function at time t = 2 ps for each oxidization degree R. The figures along the parallel direction represent the pores with 0, 1, and 2 adsorbed Mg^{2+} ions. The solid lines are a guide to the eye.

In the case of the pore without an Mg^{2+} cation, the salt rejection rate and water flux exhibited peaks and troughs at R = 0.25 and R = 0.16, respectively, as shown in Figure 6. The result displayed in Figure 7a reveals that the number of salt ions inside the pore without an Mg^{2+} cation changes significantly but not monotonically with the oxidization degree. In the case of R = 0.25 (R = 0.16), although the permeation probability of salt ions is the highest (the second smallest) among the four cases, the number of ions inside the pore is the smallest (largest). Thus, the substantial difference in the number of ions inside the pore between R = 0.25 and R = 0.16 leads to an extensive range of variation in the salt rejection rate without adsorbed Mg^{2+} cations. As demonstrated in Figure 7b, the distribution of water inside the pore is nearly the smallest (the largest) among all the systems when R = 0.16 (R = 0.25). These results suggest the presence of competitive penetration between salt ions and water molecules [32]. Although the number of water molecules in the pore is not greatly affected by changes in the oxidization degree, the water flux of the pore without adsorbed Mg^{2+} cations exhibits a significant change with the oxidization degree. This is mainly because the variation trend of the permeation probability of water with respect to the oxidization degree is similar to the variation trend of the number of water molecules inside the pore as the oxidization degree changes. Therefore, both the permeation time of water molecules and the number of water molecules within the pore play crucial roles in water desalination. However, the situation is somewhat different when considering the salt rejection rate of the pore with adsorbed Mg^{2+} cations. The salt rejection rate of the pore with adsorbed cations is primarily influenced by two significant factors: the number of salt ions inside the pore and the permeation time of the salt ions. However, since the oxidization degree of the GO membrane has minimal impact on the number of salt ions inside the pore and the permeation probability of ions, the salt rejection rate of the pore with adsorbed Mg^{2+} cations fluctuates within a narrow range. On the other hand, the water flux of the pore with adsorbed Mg^{2+} cations is co-determined by the permeation time of water molecules across the pore and the number of water molecules inside the pore, similar to the case of the pore without adsorbed Mg^{2+} cations. However, due to the pronounced difference between the change in the number of water molecules inside the pore and the permeation probability of water molecules with respect to the oxidization degree, the small variations in either the number of water molecules inside the pore or the permeation probability of water molecules alone cannot result in a significant change in the water flux.

2.4. Water Desalination Performance of Another Porous GO Membrane

The pore diameter is a crucial factor in determining the separation performance of the GO membrane. To investigate the role of the adsorbed divalent cation on the water desalination performance of the GO membrane with a large pore diameter of 2.15 nm (as shown in Figure S6), we conducted additional non-equilibrium MD simulations. The results indicate that the influence of the adsorbed Mg^{2+} cation on the water desalination performance of the pore with a diameter of 2.15 nm is similar to that of the adsorbed Mg^{2+} cation observed for the pore with a diameter of 1.29 nm. Further details can be found in Supplementary Materials Section S3 of the paper.

3. Materials and Methods

Figure 8a illustrates the gated system used in our study. This system consists of two slabs: one containing a saltwater (NaCl solution) region and the other containing pure water, separated by a nanoporous GO membrane at $z = 0$. The two graphene plates are free to move in the z-direction to control the system's pressure. The simulation model's dimensions are 3.30 × 3.52 × 10.50 nm in the x, y, and z directions, respectively, with periodic boundary conditions applied in all directions. Without Mg^{2+} cations, the saltwater portion contains 46 Na^{+} ions, 46 Cl^{-} ions, and 1126 water molecules, resulting in a salt concentration of 132.8 g/L. This elevated salinity level is advantageous for obtaining precise statistical results regarding ion permeation through the pore [10,30]. The pure water section comprises 1120 water molecules.

Given that GO layers predominantly consist of hydroxyl groups [30,44], the GO membrane utilized in our study features hydroxyl groups on both sides of the graphene basal plane. Figure 8b shows the placement of a single nanopore at the center of the GO membrane. The pore diameter, calculated using the formula $d = 2\sqrt{A/\pi}$, measures 1.29 nm [10]. This pore size is sufficiently large to facilitate the transport of salt ions from saltwater to pure water and can be readily created experimentally [27]. Three carboxyl groups are also attached to the carbon atoms on the pore's edge. As oxidized regions tend to form a continuous network across the GO membrane and consistently coexist with nanopores in the same area, the hydroxyl groups are distributed in a circular region surrounding the pore [32]. The oxidation degree of the GO membrane is characterized by the R = n_{oh}/n_c ratio, which ranges from 7% to 34%, where n_{oh} represents the number of hydroxyl groups and n_c denotes the number of carbon atoms. Figure 8b provides top views of the GO membranes. The figure illustrates that the Mg^{2+} ions were initially positioned near the carboxylic oxygen (double-bonded) sites. Due to the strong electrostatic interaction between Mg^{2+} ions and carboxylic oxygen, the Mg^{2+} ions consistently remain near the carboxylic oxygen throughout all MD simulations (for further details, please refer to Supplementary Materials Section S1). Figure 8b shows that the number of Mg^{2+} ions adsorbed to the carboxyl groups varies between 1 and 2. Except the counter ions of Na^+, there are another 2 or 4 Cl^- ions that should be added to maintain the electroneutrality of the system with Mg^{2+} ions.

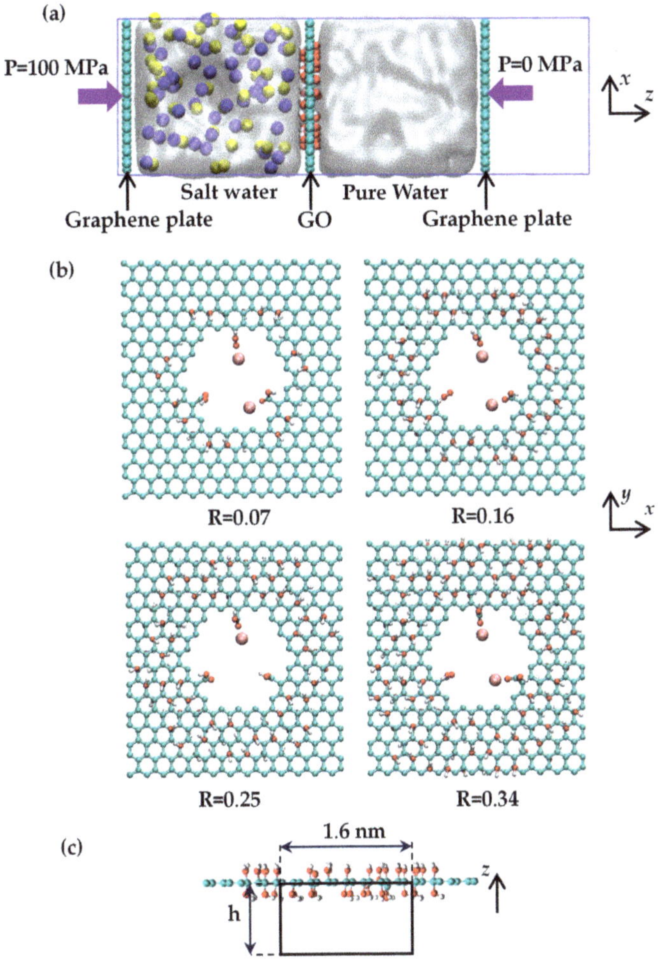

Figure 8. (**a**) Side view of the gated system for investigating desalination through the GO membrane. The yellow and blue balls denote the sodium and chloride ions. (**b**) Atomistic structure of the nanoporous GO membrane with a variation in oxygen degree. One or two Mg^{2+} ions are placed near the carboxylic oxygen. The red, cyan, white, and pink spheres represent the oxygen, carbon, hydrogen, and magnesium atoms, respectively. (**c**) The black box in the figure denotes a cutout view of a cylindrical region with a diameter of 1.6 nm and a height of h below and around the center of the pore in the GO membrane.

All MD simulations were conducted using the Gromacs simulation package in the canonical (NVT) ensemble at a temperature of 300 K, employing a velocity-rescale thermostat for temperature control [45]. The GROMOS53a6 force field was utilized to describe both the GO membrane and the ions [46,47] while water molecules were represented using the standard SPC model [48]. The LINCS algorithm was employed to constrain all bonds [49], and the classical equations of motion were integrated using the leapfrog algorithm with a time step of 1 fs [50]. The particle mesh Ewald method and Lennard–Jones potential were implemented to calculate electrostatic interactions and non-bonded interactions between different particles [51,52]. The cutoff for Lennard–Jones or electrostatic interactions was set at 1.2 nm. All atoms in the GO membrane, except those belonging to oxidation functional groups, were fixed throughout all simulations to mitigate the impact of mechanical deformation phenomena on saltwater permeation behavior [10].

Initially, we equilibrated the systems at 300 K while keeping water molecules near the center of the GO membrane fixed to separate the saltwater and pure water regions. The pressures on the two graphene plates were maintained at zero during this process. Following equilibration for 1 ns, as illustrated in Figure 8a, we applied pressures of 100 MPa and 0 MPa to the left and right graphene plates, respectively to conduct non-equilibrium MD simulations. Such a significant pressure difference ensures that accurate data for water flux and salt rejection are acquired within a finite simulation time accessible through MD simulations [10,26]. The pressure applied to the graphene plate can be calculated using the moving acceleration rate of the plates as follows [32]:

$$P = Nma/S \tag{3}$$

Here, N is the total number of carbon atoms in the plate, m is the mass of a carbon atom, S is the cross-sectional area, and P is the applied pressure. Throughout the non-equilibrium simulation, coordinates were saved every 2 ps. To obtain converged quantities for each system, such as salt rejection and water flux, 10–20 non-equilibrium runs of 5–20 ns were performed with independent initial configurations.

The PMF was used to provide a semi-quantitative understanding of the energetics for ion permeation through the pore [30,38]. The PMF was calculated through 10–20 non-equilibrium simulations. The adaptive biasing forcing algorithm was employed to determine the PMF as usual [53–55]. The reaction coordinate for the GO membrane was set in the z-direction, ranging from -10 Å to 0 Å (center of the membrane pore), preceding the pore.

4. Conclusions

In conclusion, our molecular dynamics simulations demonstrate that the nanoporous GO membrane adsorbing the Mg^{2+} cation onto the rim of the pore can effectively act as a gate to control the passage of salt ions across the large pore. This significantly improves the desalination performance of the nanoporous GO membrane by adsorbing Mg^{2+} cations. The simulation results reveal that the transport of salt ions (and water molecules) through the pore is closely correlated with both the distribution of salt ions (and water molecules) near the center of the pore as well as the salt ion (water molecule) permeation time across the pore. The Mg^{2+} cations absorbed on the pore reduce the concentration of salt ions around the center of the pore due to strong repulsive forces between the Na^+ ions and the Mg^{2+} cations. Moreover, the increase in the number of adsorbed Mg^{2+} cations increases the time it takes for salt ions to permeate through the pore, thereby improving the salt rejection rate. However, the Mg^{2+}-adsorbed pore compromises the passage of water molecules through the pore. As the divalent cation can be adsorbed on the surface of the GO membrane very stably owing to their interactions (e.g., the density functional theory (DFT) calculation results in Supplementary Materials Section S1 disclose that there is strong interaction between the Mg^{2+} cation and the nanoporous GO membrane), one possible way to obtain the nanoporous GO membrane with the adsorption of Mg^{2+} cations is to soak the freestanding nanoporous GO membrane in magnesium-ion water solution for some time during the membrane fabrication process, similar to previous research [40].

The role of the oxidation degree on the desalination performance of the nanoporous GO membrane was also investigated. The mechanisms underlying the variation in the salt rejection rate and water flux with the oxidation degree were elucidated. The number of water molecules (ions) near the center of the pore and the permeation time of water molecules (ions) across the pore are the two key factors influencing water flux (and salt rejection rate) as the oxidation degree of GO varies. The simulations reveal the presence of competitive penetration between salt ions and water molecules in the system, which can be beneficial for the nanoporous GO membrane to achieve high water flux and a perfect salt rejection rate at certain oxidation degrees. Our work suggests that adsorbed Mg^{2+} cations on the nanoporous GO membrane offer a promising avenue for optimizing GO membranes for improved water desalination performance.

Supplementary Materials: The following supporting information can be downloaded at: https://www.mdpi.com/article/10.3390/molecules29040827/s1, Figure S1: The distance between the Mg^{2+} ion and the carboxylic oxygen of the system; Figure S2: The configuration of the graphene sheet with one pore and one adsorbed magnesium ion [56–58]; Figure S3: The number of water molecules accumulating in a cylindrical region; Figure S4: The PMF of a water molecule along the z-axis of GO membranes with different numbers of adsorbed Mg^{2+} ions; Figure S5: The permeation time autocorrelation function of water molecules in pores with different numbers of adsorbed Mg^{2+} ions; Figure S6: Atomistic structure of the GO membrane; Figure S7: The salt rejection rate and the water flux of the GO membranes with a large pore diameter of 2.15 nm vs. the oxidization degree R.

Author Contributions: Conceptualization, Y.T. and Z.G.; methodology, J.J.; formal analysis, J.J.; investigation, J.J.; writing—original draft preparation, J.J.; writing—review and editing, Y.T. and Z.G.; supervision, Y.T. and Z.G.; funding acquisition, Y.T. and Z.G.; All authors have read and agreed to the published version of the manuscript.

Funding: This research was funded by National Natural Science Foundation of China (NSFC) grant number 12075201 and 12104394, Natural Science Foundation of Jiangsu Province grant number BK20201428, and Natural Science Research of the Jiangsu Higher Education Institutions of China grant number 21KJB140024.

Data Availability Statement: The raw data supporting the conclusions of this article will be made available by the authors on request.

Conflicts of Interest: The authors declare no conflict of interest.

References

1. Shannon, M.A.; Bohn, P.W.; Elimelech, M.; Georgiadis, J.G.; Mariñas, B.J.; Mayes, A.M. Science and technology for water purification in the coming decades. *Nature* **2008**, *452*, 301–310. [CrossRef] [PubMed]
2. Elimelech, M. The global challenge for adequate and safe water. *Aqua* **2006**, *55*, 3–10. [CrossRef]
3. Jang, J.H.; Woo, J.Y.; Lee, J.; Han, C.S. Ambivalent effect of thermal reduction in mass rejection through graphene oxide membrane. *Environ. Sci. Technol.* **2016**, *50*, 10024–10030. [CrossRef] [PubMed]
4. Jafarzadeh, R.; Azamat, J.; Erfan-Niya, H. Water desalination across functionalized silicon carbide nanosheet membranes: Insights from molecular simulations. *Struct. Chem.* **2020**, *31*, 293–303. [CrossRef]
5. Yang, Y.Z.; Mu, L.H.; Chen, L.; Shi, G.S.; Fang, H.P. Precise control of the interlayer spacing between graphene sheets by hydrated cations. *Phys. Chem. Chem. Phys.* **2019**, *21*, 7623–7629. [CrossRef]
6. Greenlee, L.F.; Lawler, D.F.; Freeman, B.D.; Marrot, B.; Moulin, P. Reverse osmosis desalination: Water sources, technology, and today's challenges. *Water Res.* **2009**, *43*, 2317–2348. [CrossRef] [PubMed]
7. Werber, J.R.; Osuji, C.O.; Elimelech, M. Materials for next-generation desalination and water purification membranes. *Nat. Rev. Mater.* **2016**, *1*, 16018. [CrossRef]
8. Khawaji, A.D.; Kutubkhanah, I.K.; Wie, J.-M. Advances in seawater desalination technologies. *Desalination* **2008**, *221*, 47–69. [CrossRef]
9. Jones, E.; Qadir, M.; van Vliet, M.T.H.; Smakhtin, V.; Kang, S.-mu. The state of desalination and brine production: A global outlook. *Sci. Total Environ.* **2019**, *657*, 1343–1356. [CrossRef]
10. Cohen-Tanugi, D.; Grossman, J.C. Water desalination across nanoporous graphene. *Nano Lett.* **2012**, *12*, 3602–3608. [CrossRef]
11. Heiranian, M.; Farimani, A.B.; Aluru, N.R. Water desalination with a single-layer MoS_2 nanopore. *Nat. Commun.* **2015**, *6*, 8616. [CrossRef] [PubMed]

12. Xue, M.M.; Qiu, H.; Guo, W.L. Exceptionally fast water desalination at complete salt rejection by pristine graphyne monolayers. *Nanotechnology* **2013**, *24*, 505720. [CrossRef] [PubMed]
13. Lim, Y.J.; Goh, K.; Kurihara, M.; Wang, R. Seawater desalination by reverse osmosis: Current development and future challenges in membrane fabrication—A review. *J. Membr. Sci.* **2021**, *629*, 119292. [CrossRef]
14. Lim, Y.J.; Ma, Y.; Chew, J.W.; Wang, R. Assessing the potential of highly permeable reverse osmosis membranes for desalination: Specific energy and footprint analysis. *Desalination* **2022**, *533*, 115771. [CrossRef]
15. Prihasto, N.; Liu, Q.F.; Kim, S.H. Pre-treatment strategies for seawater desalination by reverse osmosis system. *Desalination* **2009**, *249*, 308–316. [CrossRef]
16. Shatat, M.; Riffat, S.B. Water desalination technologies utilizing conventional and renewable energy sources. *Int. J. Low-Carbon Technol.* **2014**, *9*, 1–19. [CrossRef]
17. Surwade, S.P.; Smirnov, S.N.; Vlassiouk, I.V.; Unocic, R.R.; Veith, G.M.; Dai, S.; Mahurin, S.M. Water desalination using nanoporous single-layer graphene. *Nat. Nanotechnol.* **2015**, *10*, 459–464. [CrossRef]
18. Celebi, K.; Buchheim, J.; Wyss, R.M.; Droudian, A.; Gasser, P.; Shorubalko, I.; Kye, J.-I.; Lee, C.; Park, H.G. Ultimate permeation across atomically thin porous graphene. *Science* **2014**, *344*, 289–292. [CrossRef]
19. Han, Y.; Xu, Z.; Gao, C. Ultrathin graphene nanofiltration membrane for water purification. *Adv. Funct. Mater.* **2013**, *23*, 3693–3700. [CrossRef]
20. Wang, Y.H.; He, Z.J.; Gupta, K.M.; Shi, Q.; Lu, R.F. Molecular dynamics study on water desalination through functionalized nanoporous graphene. *Carbon* **2017**, *116*, 120–127. [CrossRef]
21. Dreyer, D.R.; Park, S.; Bielawski, C.W.; Ruoff, R.S. The chemistry of graphene oxide. *Chem. Soc. Rev.* **2010**, *39*, 228–240. [CrossRef]
22. Huang, H.B.; Ying, Y.L.; Peng, X.S. Graphene oxide nanosheet: An emerging star material for novel separation membranes. *J. Mater. Chem. A.* **2014**, *2*, 13772–13782. [CrossRef]
23. Golchoobi, A.; Tasharrofi, S.; Taghdisian, H. Functionalized nanoporous graphene membrane for water desalination; Effect of feed salinity on permeability and salt rejection, a molecular dynamics study. *Comput. Mater. Sci.* **2020**, *172*, 109399. [CrossRef]
24. Sun, C.Z.; Bai, B.F. Improved CO_2/CH_4 separation performance in negatively charged nanoporous graphene membranes. *J. Phys. Chem. C* **2018**, *122*, 6178–6185. [CrossRef]
25. Sun, C.Z.; Boutilier, M.S.H.; Au, H.; Poesio, P.; Bai, B.F.; Karnik, R.; Hadjiconstantinou, N.G. Mechanisms of molecular permeation through nanoporous graphene membranes. *Langmuir* **2014**, *30*, 675–682. [CrossRef] [PubMed]
26. Cohen-Tanugi, D.; Lin, L.C.; Grossman, J.C. Multilayer nanoporous graphene membranes for water desalination. *Nano Lett.* **2016**, *16*, 1027–1033. [CrossRef] [PubMed]
27. Tian, Z.Q.; Mahurin, S.M.; Dai, S.; Jiang, D.E. Ion-gated gas separation through porous graphene. *Nano Lett.* **2017**, *17*, 1802–1807. [CrossRef] [PubMed]
28. Liu, Q.; Chen, M.G.; Sun, L.L.; Liu, G.P.; Xu, R. Pore density effect on separations of water/ethanol and methanol/ethanol through graphene oxide membranes: A theoretical study. *Sep. Purif. Technol.* **2023**, *308*, 122975. [CrossRef]
29. Zhang, M.X.; Sun, B.; Shah, K.W.; Zhang, X.S. Molecular dynamics study on electric field-facilitated separation of H_2O/O_2 through nanoporous graphene oxide membrane. *J. Mol. Liq.* **2022**, *351*, 118634. [CrossRef]
30. Fang, C.; Yu, Z.; Qiao, R. Impact of surface ionization on water transport and salt leakage through graphene oxide membranes. *J. Phys. Chem. C* **2017**, *121*, 13412–13420. [CrossRef]
31. Lim, Y.J.; Goh, K.; Wang, R. The coming of age of water channels for separation membranes: From biological to biomimetic to synthetic. *Chem. Soc. Rev.* **2022**, *51*, 4537–4582. [CrossRef] [PubMed]
32. Liu, Q.; Wu, Y.Y.; Wang, X.; Liu, G.P.; Zhu, Y.D.; Tu, Y.S.; Lu, X.H.; Jin, W.Q. Molecular dynamics simulation of water-ethanol separation through monolayer graphene oxide membranes: Significant role of O/C ratio and pore size. *Sep. Purif. Technol.* **2019**, *224*, 219–226. [CrossRef]
33. Köhler, M.H.; Bordin, J.R.; Barbosa, M.C. Ion flocculation in water: From bulk to nanoporous membrane desalination. *J. Mol. Liq.* **2019**, *277*, 516–521. [CrossRef]
34. Sint, K.; Wang, B.Y.; Král, P. Selective ion passage through functionalized graphene nanopores. *J. Am. Chem. Soc.* **2008**, *130*, 16448–16449. [CrossRef] [PubMed]
35. Sun, C.Z.; Zhu, S.H.; Liu, M.C.; Shen, S.H.; Bai, B.F. Selective molecular sieving through a large graphene nanopore with surface charges. *J. Phys. Chem. Lett.* **2019**, *10*, 7188–7194. [CrossRef] [PubMed]
36. Fang, C.; Wu, H.Y.; Lee, S.Y.; Mahajan, R.L.; Qiao, R. The ionized graphene oxide membranes for water-ethanol separation. *Carbon* **2018**, *136*, 262–269. [CrossRef]
37. Azamat, J.; Khataee, A.; Joo, S.W. Molecular dynamics simulation of trihalomethanes separation from water by functionalized nanoporous graphene under induced pressure. *Chem. Eng. Sci.* **2015**, *127*, 285–292. [CrossRef]
38. Hosseini, M.; Azamat, J.; Erfan-Niya, H. Water desalination through fluorine-functionalized nanoporous graphene oxide membranes. *Mater. Chem. Phys.* **2019**, *223*, 277–286. [CrossRef]
39. Mahadevi, A.S.; Sastry, G.N. Cation-π interaction: Its role and relevance in chemistry, biology, and material science. *Chem. Rev.* **2013**, *113*, 2100–2138. [CrossRef]
40. Chen, L.; Shi, G.S.; Shen, J.; Peng, B.Q.; Zhang, B.W.; Wang, Y.Z.; Bian, F.G.; Wang, J.J.; Li, D.Y.; Qian, Z.; et al. Ion sieving in graphene oxide membranes via cationic control of interlayer spacing. *Nature* **2017**, *550*, 380–383. [CrossRef]

41. Shahbabaei, M.; Kim, D. Transport of water molecules through noncylindrical pores in multilayer nanoporous graphene. *Phys. Chem. Chem. Phys.* **2017**, *19*, 20749–20759. [CrossRef] [PubMed]
42. Pathania, Y.; Gaganpreet. Self-passivated nanoporous phosphorene as a membrane for water desalination. *Desalination* **2021**, *497*, 114777. [CrossRef]
43. Liu, Y.C.; Cheng, Z.W.; Song, M.R.; Jiang, L.Z.; Fu, G.; Liu, L.; Li, J.Y. Molecular dynamics simulation-directed rational design of nanoporous graphitic carbon nitride membranes for water desalination. *J. Membr. Sci.* **2021**, *620*, 118869. [CrossRef]
44. Wei, N.; Peng, X.S.; Xu, Z.P. Understanding water permeation in graphene oxide membranes. *ACS Appl. Mater. Interfaces* **2014**, *6*, 5877–5883. [CrossRef] [PubMed]
45. Bussi, G.; Donadio, D.; Parrinello, M. Canonical sampling through velocity rescaling. *J. Chem. Phys.* **2007**, *126*, 014101. [CrossRef] [PubMed]
46. Oostenbrink, C.; Villa, A.; Mark, A.E.; van Gunsteren, W.F. A biomolecular force field based on the free enthalpy of hydration and solvation: The GROMOS force-field parameter sets 53A5 and 53A6. *J. Comput. Chem.* **2004**, *25*, 1656–1676. [CrossRef] [PubMed]
47. Tu, Y.S.; Lv, M.; Xiu, P.; Huynh, T.; Zhang, M.; Castelli, M.; Liu, Z.R.; Huang, Q.; Fan, C.H.; Fang, H.P.; et al. Destructive extraction of phospholipids from Escherichia coli membranes by graphene nanosheets. *Nat. Nanotechnol.* **2013**, *8*, 594–601. [CrossRef]
48. Jorgensen, W.L.; Chandrasekhar, J.; Madura, J.D.; Impey, R.W.; Klein, M.L. Comparison of simple potential functions for simulating liquid water. *J. Chem. Phys.* **1983**, *79*, 926–935. [CrossRef]
49. Hess, B.; Bekker, H.; Berendsen, H.J.C.; Fraaije, J.G.E.M. LINCS: A linear constraint solver for molecular simulations. *J. Comput. Chem.* **1997**, *18*, 1463–1472. [CrossRef]
50. Hockney, R.W. The potential calculation and some applications. *Methods Comput. Phys.* **1970**, *9*, 136–211.
51. Darden, T.; York, D.; Pedersen, L. Particle mesh Ewald: An N·log (N) method for Ewald sums in large systems. *J. Chem. Phys.* **1993**, *98*, 10089–10092. [CrossRef]
52. Essmann, U.; Perera, L.; Berkowitz, M.L.; Darden, T.; Lee, H.; Pedersen, L.G. A smooth particle mesh Ewald method. *J. Chem. Phys.* **1995**, *103*, 8577–8593. [CrossRef]
53. Comer, J.; Gumbart, J.C.; Hénin, J.; Lelièvre, T.; Pohorille, A.; Chipot, C. The adaptive biasing force method: Everything you always wanted to know but were afraid to ask. *J. Phys. Chem. B* **2015**, *119*, 1129–1151. [CrossRef] [PubMed]
54. Prasad, V.; Kannam, S.K.; Hartkamp, R.; Sathian, S.P. Water desalination using graphene nanopores: Influence of the water models used in simulations. *Phys. Chem. Chem. Phys.* **2018**, *20*, 16005–16011. [CrossRef]
55. Prasad, V.; Sathian, S.P. The effect of temperature on water desalination through two-dimensional nanopores. *J. Chem. Phys.* **2020**, *152*, 164701. [CrossRef]
56. Frisch, M.J.; Trucks, G.W.; Schlegel, H.B.; Scuseria, G.E.; Robb, M.A.; Cheeseman, J.R.; Scalmani, G.; Barone, V.; Mennucci, B.; Petersson, G.A.; et al. Uranyl Extraction by N, N-Dialkylamide Ligands Studied by Static and Dynamic DFT Simulations. In *Gaussian 09*; Gaussian Inc: Wallingford, CT, USA, 2009; Volume 19, p. 227.
57. Tu, Y.S.; Li, P.; Sun, J.J.; Jiang, J.; Dai, F.F.; Li, C.Z.; Wu, Y.Y.; Chen, L.; Shi, G.S.; Tan, Y.W.; et al. Remarkable antibacterial activity of reduced graphene oxide functionalized by copper ions. *Adv. Funct. Mater.* **2021**, *31*, 2008018. [CrossRef]
58. Wendler, K.; Thar, J.; Zahn, S.; Kirchner, B. Estimating the Hydrogen Bond Energy. *J. Phys. Chem. A* **2010**, *114*, 9529–9536. [CrossRef] [PubMed]

molecules

MDPI

Article

Evaluation of Cr(VI) Removal from Tanning Effluents Using Magnetic Nanoparticles of Fe_3O_4 Synthesized with *Olea europaea* Bone Extract

Maria Bejarano-Meza *, Fabricio Eduardo Deza-Carrasco, Sofia Salinas-Herrera, Susan Flores-Calla and Hugo Guillermo Jimenez-Pacheco

Universidad Católica de Santa María, Urbanization San Jose s/n, Umacollo, Arequipa 04013, Peru; 72192130@ucsm.edu.pe (F.E.D.-C.); 75363738@ucsm.edu.pe (S.S.-H.)

* Correspondence: mbejarano@ucsm.edu.pe

Abstract: The tanning industry generates effluents with high chromium content, which require treatment prior to discharge into the sewage system. This article explores the use of magnetic magnetite nanoparticles (MNPs) to remove Cr(VI) from aqueous solutions, such as tanning effluents. The MNPs were synthesized by coprecipitation reaction using the *Olea europaea* extract as a reducing agent. Subsequently, they were characterized by dynamic light scattering spectroscopy (DLS), scanning electron microscopy (SEM) and energy dispersive X-ray spectroscopy (EDS). MNPs with irregular morphology and diameters ranging from 73.28 to 162.90 nm were obtained. Cr(VI) removal was performed using jar test methodology, and its efficiency was evaluated in the laboratory for different initial Cr(VI) (mg/L) concentration and nanoparticle (g/L) concentration. A kinetic study was developed and indicated that the equilibrium adsorption mechanism corresponds to a pseudo-second-order model. Furthermore, the isotherm analysis revealed that chromium adsorption best fits the Langmuir isotherm. Finally, Cr(VI) removal rates from 85% to 100% were achieved in tanning and retanning effluents.

Keywords: magnetic nanoparticles; magnetite; olive extract; chromium removal; tanning effluent

1. Introduction

The toxic waste generated makes the tanning industry one of the most polluting industries [1]. During leather production, different chemical inputs and water volumes between 50 and 100 L per kg of salted leather are used [2], which generates wastewater with high concentrations of pollutants, including heavy metals such as chromium, which, in its hexavalent form (VI), becomes a contaminant due to its high mobility and solubility in aqueous media, tends to bioaccumulate and has a carcinogenic effect [2,3].

In some cities, such as Arequipa in the Rio Seco Industrial Park (PIRS), tanning effluents are discharged into a sewage system or river, causing contamination of the environment and flowing into bodies of water.

These effluents have high concentrations of toxic heavy metals, especially hexavalent chromium (Cr(VI)). It also stands out that Cr(VI), due to its high toxicity associated with high mobility and solubility in aqueous media, tends to bioaccumulate, which can lead to serious effects on human health, such as carcinogenic effects [3]. Nowadays, various environmental problems can be addressed by nanotechnology. Metal and metal oxide nanoparticles, due to their more efficient adsorption and reduction capabilities than their macro or micro counterparts, have shown outstanding properties for adsorption/reduction of a large number of heavy metal ions, such as Ni^{2+}, Cu^{2+}, Cd^{2+}, Zn^{2+} and Cr^{6+}, and catalytic degradation of some organic pollutants [4,5].

The synthesis of metal and metal oxide nanoparticles has been carried out using conventional methods and green synthesis. Generally, the conventional methods use toxic materials, such as solvents and surfactants, which can accumulate in the environment and pollute it. This has motivated the research and development of new, more reliable and sustainable strategies for the synthesis of nanoparticles known as green synthesis [6]. A feature common to the methods of green synthesis is to reduce the use of toxic inputs in the production of nanoparticles by replacing them with clean precursors, such as plants, bacteria, actinomycetes, fungi, yeasts, viruses, among others [7]. Among these precursors, Moringa oleifera leaves have been used for the removal of lead from an aqueous source, Catharanthus roseus leaves for the removal of cadmium and chromium, date palm bone for the degradation of methylene blue and Ipomoea aquatica roots for the adsorption of methyl violet, among others. The bone of *Olea europaea* stands out due to its capacity to act as a biosorbent of heavy metals, as well as the presence of various polyphenols in its composition, which make it feasible to use it as a reducing agent in the production of magnetic nanoparticles. The advantages of green synthesis of metal nanoparticles not only reduce the environmental impact but also exhibit high economic efficiency compared to some conventional methods. Other advantages include reduction in energy consumption, reduction in chemical inputs and ease of implementation on a large scale [8].

Among the metal oxide nanoparticles, iron oxides are one of the most applied because they exist in various forms in nature, such as magnetite (Fe_3O_4), maghemite (γ-Fe_2O_3) and hematite (α-Fe_2O_3). In recent years, different authors have reported the efficiency of iron-based nanoparticles in the removal of heavy metals, such as Cr(VI), Pb(II), Cd(II) and As(III), from aqueous solutions [9]. Magnetite has been shown to be more useful in different applications [10]. Magnetite structures at the nanoscale, called nanoparticles, have attracted much interest due to their unique magnetic property known as superparamagnetism; this property is present in particles with a size of order of magnitude of nanometers, having a behavior similar to ferromagnetic and ferrimagnetic materials, [10] being the main characteristic, which allows the MNP to be very efficient in the adsorption of heavy metals, such as chromium [11]. Aftabtalab et al. [12] discussed Cr(VI) removal from wastewaters using high adsorption iron oxide (Fe_3O_4) nanoparticles synthesized by the sol–gel method. In a more recent article, Hao et al. [13] showed that a green tea extract was successfully used to prepare iron nanoparticles (nFe) to remove Cr(VI) from aqueous solutions.

Accordingly, this article reports the green synthesis of magnetite (Fe_3O_4) nanoparticles using olive pit extract. In addition, the efficiency of these nanoparticles in Cr(VI) removal is evaluated by varying the initial Cr(VI) (mg/L) concentration and nanoparticle (g/L) concentration. The material characteristics of the nanoparticles are also presented, as well as the adsorption behavior, kinetic and isotherm models.

In general, the present research provides a sustainable alternative to contribute to solving the problem of the lack of treatment of Cr(VI) loaded effluents from tanneries before discharge into the environment. The results obtained show that the development and application of nanoparticles obtained by green synthesis using *Olea europaea* bone extract as a reducing agent are an efficient treatment for the adsorption of Cr(VI) from tanning effluents.

2. Results and Discussion

2.1. Characteristics of the Tanning and Retanning Effluents

The tanning and retanning effluent samples were analyzed in the laboratory. Table 1 shows the physico-chemical characteristics of the samples.

The results indicate that for the retanning effluent, the parameters of pH, total chromium, hexavalent chromium and total suspended solids exceed the maximum permissible limits (MPL). In the case of the tanning effluent, the parameters of total chromium, hexavalent chromium and chemical oxygen demand (COD) exceed the maximum permissible limits (MPL), with total chromium and hexavalent chromium being the most relevant parameters in this study. These parameters exceed the MPL by almost 42% in the case of total chromium, 1% for hexavalent chromium in the retanning stage and 258% in the tanning stage.

Table 1. Characteristics of tanning and retanning effluents.

Stage	Type	Parameter	Value (mg/L)	MPL Value (mg/L)
Retanning	In situ	pH	3.567	6.5–9.5
		Dissolved oxygen	5.81	-
		Temperature	18.7	
	Ex situ	Total chromium	1599.14	5
		Hexavalent chromium	4.913	0.5
		Oils and fats	64.9	
		Total suspended solids	53951	1000
Tanning	Ex situ	Total chromium	1610.65	5
		Hexavalent chromium	1287.72	0.5
		Oils and fats	15.5	
		Total suspended solids	241	1000
		BODS	199	1000
		COD	4155	

2.2. Favorable Operating Conditions and Validation of the Method for Cr(VI) Quantification

2.2.1. Linearity

To determine the linearity of the method, a calibration plot of concentration vs. absorbance is presented in Figure 1.

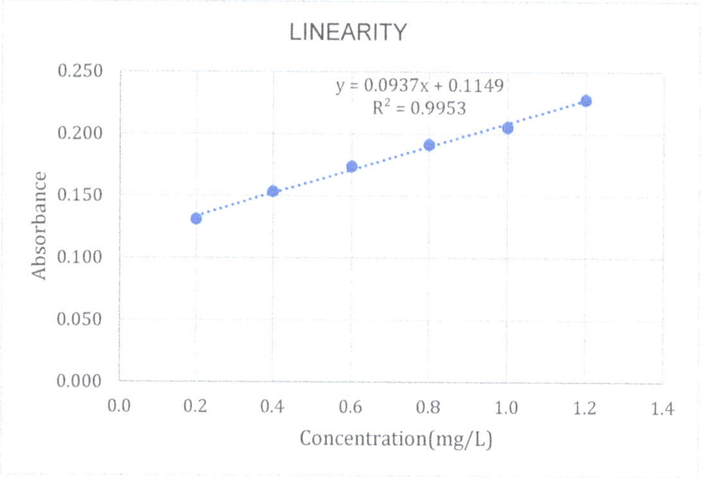

Figure 1. Calibration curve.

The R^2 value obtained from the equation of the line is 0.9953, which is very close to the value obtained by Pacheco-Portugal [14], whose R^2 was 0.9964. Similarly, an R^2 value of 0.9996 was reported by Cañazaca and Ccama [15] in the validation of Cr(VI) in wastewater.

According to international standards (AOAC), values greater than 0.99 are considered acceptable; therefore, we conclude that our method is linear.

2.2.2. Precision and Repeatability

To determine the precision and repeatability of the method, the measurement of the analyte Cr(VI) was performed in triplicate under the same working conditions and with the same instrument. The results are shown in Table 2.

Table 2. Precision test results.

Concentration (mg/L)	Absorbance	Cf (mg/L)
0.6	0.18	0.6947705
0.6	0.181	0.7054429
0.6	0.177	0.6627535
0.6	0.179	0.6840982
0.6	0.181	0.7054429
0.6	0.179	0.6840982
Average		0.689
Standard deviation		0.0162
RSD %		0.023

According to the ICH validation method, the acceptance criterion for repeatability and precision is that the %RSD is less than 2%; therefore, comparing this criterion with our results, it is concluded that the method is precise and accurate. The RSD value obtained is 0.023%, which is below 2%.

2.2.3. Detection Limit (LOD) and Limit of Quantification (LOQ)

The detection limit allows finding the lowest concentration, which can be detected with statistical significance using the analytical procedure studied.

The results obtained with the calculated values for both the limit of detection and the limit of quantification are shown in Table 3.

Table 3. Results of target readings.

Targets	X	Y
1	0.2	0.131
2	0.4	0.154
3	0.6	0.174
4	0.8	0.192
5	1	0.206
6	1.2	0.228
Average	0.7	0.181
Standard deviation	0.374	0.035
Typical error xy		0.0027
LOD (mg/L)		0.0528
LOQ (mg/L)		0.1760

Based on the results obtained, shown in the table above, it is established that the method can be used for samples whose Cr(VI) concentration is above 0.0528 mg/L and quantifies samples whose concentration is equal to or greater than 0.1760 mg/L.

2.2.4. Accuracy

The accuracy was determined by calculating the percentage of recovery (% R); the results obtained are shown in Table 4.

Table 4. Results of the readings for the recovery percentage.

Concentration (mg/L)	Absorbance	Average
0.6	0.174 0.174 0.174	0.174
0.6	0.173 0.173 0.173	0.173
0.6	0.171 0.171 0.172	0.171
Average absorbance		0.173

According to the ICH validation method, the acceptance criterion for accuracy is 95% < Percent Recovery < 103%; therefore, it is considered that the validated analytical method is accurate, since the recovery percentage of 102.949 is within the range of the acceptance criterion.

2.3. *Characteristics of the Magnetite Nanoparticles (MNPs)*

Dynamic light scattering (DLS) is a hydrodynamic technique, which provides insight into the particle size distribution in a liquid system. The measurement is based on the detection of variations in the intensity of light scattered by particle scattering. The DLS results revealed that most of the MNPs had a size of 161.8 nm, corresponding to 98.5% of the total population, while a smaller population, corresponding to 1.5%, presented a size of 5362 nm.

The characterization of nanoparticles using SEM (Figures 2 and 3) showed that the nanoparticles exhibit sizes within the range of 73.28–162.9 nm and an irregular shape presenting agglomerations (Figure 3). This is due to the synthesis method of the nanoparticles using plant extract, which is a mixture of reducing and stabilizing agents with complex chemistry, producing heterogeneous MNPs and not homogeneous, as is the case when using chemical agents [16]. Such heterogeneity is an advantage, given that it facilitates the adsorption process by providing active sites for physical exchanges and chemical reactions, as discussed by Es'haghi et al. [17], who synthesized indefinitely shaped and highly agglomerated magnetic iron nanoparticles coated with olive oil.

Our results show that, although the particle diameter found with DLS is within the range measured by SEM, the difference in the measurements can be attributed to the formation of a hydrodynamic radius characteristic of dilute solutions, such as those used for DLS characterization, while for SEM, the sample used is completely dry, which allows for a much more accurate measurement.

Figure 4 shows the EDS spectrogram of the nanoparticles. The presence of elemental iron with an atomic percentage of 9.4% can be observed, corresponding to the intense peak between 6 keV and 7 keV. The presence of oxygen is also detected with an atomic percentage of 43.72%, corresponding to the most intense peak between 0 keV and 1 keV, suggesting the oxidation of nanoparticles due to their exposure to water or air. Peaks of trace element contamination are also observed, attributed to the sample holder.

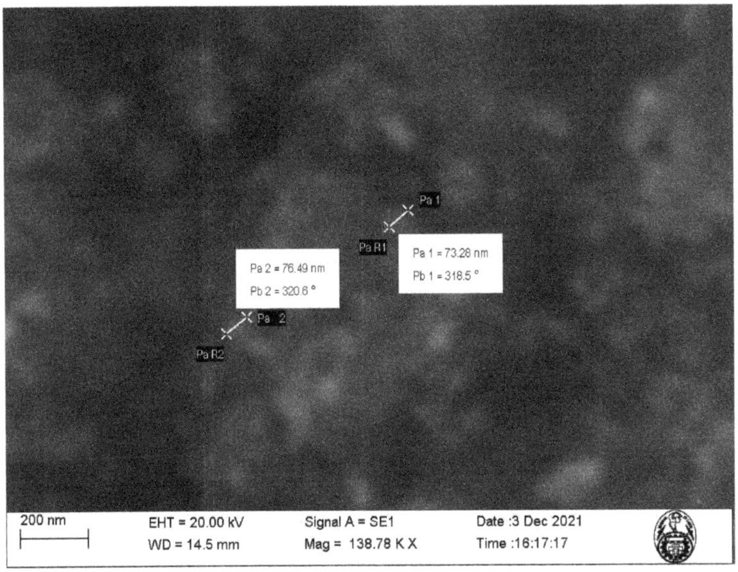

Figure 2. SEM image of the magnetite nanoparticles (MNPs) at 138.78 K magnification.

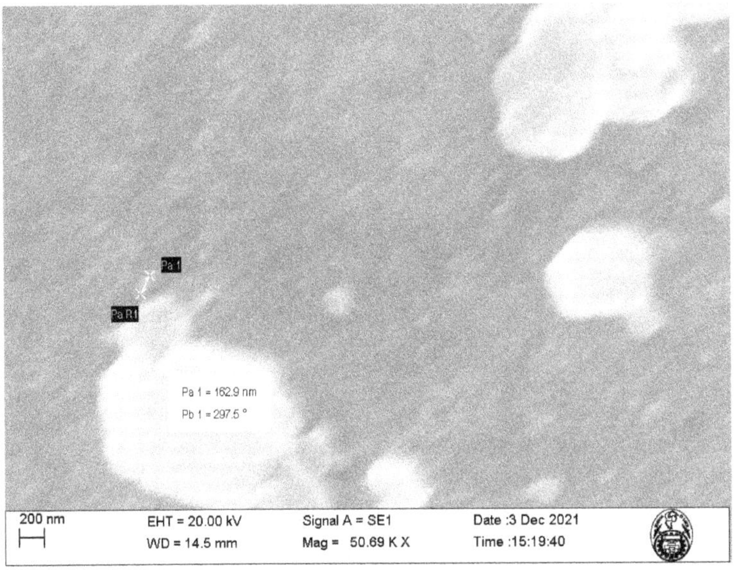

Figure 3. SEM image of the magnetite nanoparticles (MNPs) at 50.69 K magnification.

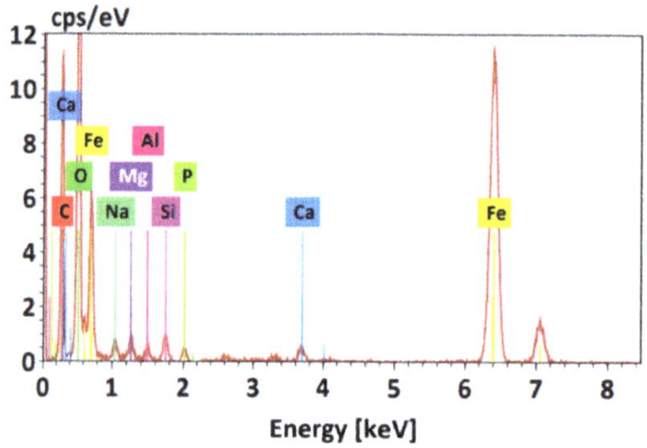

Figure 4. EDS spectrum of magnetite nanoparticles (MNPs).

2.4. Removal Efficiency of Cr(VI)

After the experiments, it was observed that the optimal operating pH was pH = 2, which achieved a removal percentage of 100%, while with a basic pH, only 30% removal was achieved. These findings corroborate those obtained by Fazlzadeh et al. [18] and Assi et al. [19] and can be attributed to the fact that pH determines the adsorption of metal ions, since it modifies the surface charge of the sorbent; therefore, it modifies the stability of hexavalent chromium, causing its reduction and turning it to trivalent chromium.

The equilibrium time at which the adsorption process became constant was 30 min, where the removal percentage was 100%, remaining constant and registering a stagnation in the adsorption process.

The optimum agitation speed was 140 rpm, and a removal of 90% was obtained. This was the maximum value reached with the jar equipment used, indicating that the higher the agitation speed, the higher the percentage of Cr(VI) removal.

With the adsorption process using jar tests, removal percentages of 100% for the initial concentration of Cr(VI) of 10 mg/L and 20 mg/L were reached. On the other hand, for 50 mg/L, percentages of only up to 75% were reached due to the saturation of the MNPs, which reached their highest performance with concentrations of up to 20 mg/L. The most efficient combinations were obtained with initial concentrations of Cr(VI) of 10 mg/L and 20 mg/L and a nanoparticle concentration of 1 g/L. These values are in accordance with the values obtained by Hao, Li, Zhang and Jiao [13], who achieved a removal efficiency of 91.6%.

The effect of a factor can be defined as the observed change in the response variable (% removal) due to the change in the level of that factor. It should be noted that the main effects are the changes in the mean value of the response variable, which is due to the individual action of each factor. Based on the Pareto diagram obtained using the OriginLab program, it can be evidenced that the initial concentration of Cr(VI) is a limiting factor for the nanoparticles, since these will saturate faster, thus limiting their adsorption capacity when the initial concentration of the pollutant increases. However, the statistical analysis also allowed us to determine that the performance of MNP in contaminant removal has a relationship directly proportional to MNP concentration.

In relation to the percentage of Cr(VI) removal from tanning and retanning effluents, filtration was initially carried out before the removal process using magnetite nanoparticles to eliminate suspended solids in the tanning effluent sample. The results obtained from the process showed a removal of 85% and 100% for the tanning and retanning effluents, respectively, since the tanning sample showed a higher initial Cr(VI) concentration. Following the adsorption process, the nanoparticles were recovered with a magnet to be characterized by EDS and to determine their final composition, obtaining mainly iron with a presence of 76.35%.

2.5. Isothermal Models

2.5.1. Langmuir Isotherm

The Langmuir and Freundlich adsorption isotherm models were used to evaluate the experimental data of the adsorption equilibrium of dichromate ion adsorption using MNPs. Figure 4 shows the graph of the experimental values of the specific adsorption (C_eq/q_e) as a function of the concentration of the metal ion in equilibrium (C_eq) and the linear form of the Langmuir equation. These results show an R^2 of 0.9194, representing the dispersion around the regression line, indicating that the phenomenon has monolayer adsorption, as shown in Figure 5.

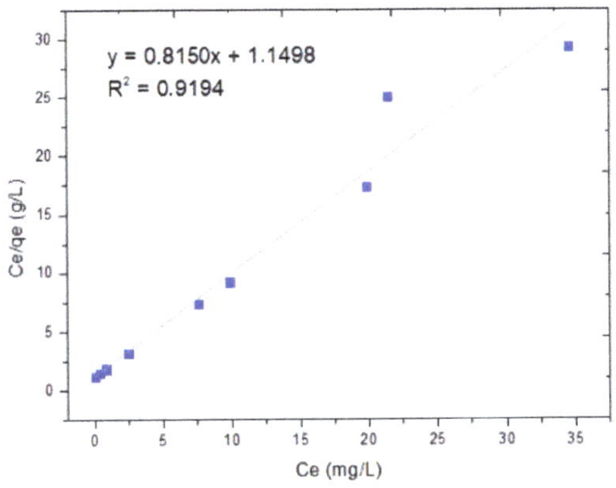

Figure 5. Langmuir isotherm.

The linearization of the isotherm using the Langmuir equation comes from the clearance of the equation; it is based on relating the values of C_e and C_e/q_e in a linear regression of these points in order to obtain the slope and the intercept, which will allow us to find the values of q_{max} and b.

$$\frac{C_e}{q_e} = \frac{1}{bq_{max}} + \frac{1}{q_{max}}C_e, \tag{1}$$

where C_e: adsorbate Cr(VI) equilibrium concentration in solution (mg/L); b: constant referring to the adsorption affinity between the adsorbent (nanoparticle) and adsorbate Cr(VI); q_{max}: the maximum amount of sorbate Cr(VI) per unit mass of adsorbent (nanoparticulate) to complete the monolayer (mg/g).

In Table 5, the value of the equilibrium parameter (R_L) indicates the type of isotherm.

Table 5. Values of the equilibrium parameter (R_L) derived from the Langmuir model.

R_L Value	Type of Isotherm
$R_L > 1$	Unfavorable
$R_L = 1$	Linear
$0 < R_L < 1$	Favorable
$R_L = 0$	Irreversible

To calculate the value of q max, which is the inverse of the slope, the following result is obtained: 1.226 y b = 0.709. To calculate R_L, each C_0 concentration value of 0, 10, 20, 50 and 80 is replaced:

$$R_L = \frac{1}{1 + bC_0}, \tag{2}$$

According to Table 6, the R_L values are less than zero, except for C_0 of 0 mg/L, with an average R_L of 0.2468, which indicates that the absorption is favorable compared to the R_L values in Table 5, where the types of isotherms are found.

Table 6. Results of R_L to determine the type of isotherm.

C_0 mg/L	b	R_L	Average R_L
0		1	
10		0.124	
20	0.709	0.066	0.2468
50		0.027	
80		0.017	

2.5.2. Freundlich Isotherm

Figure 6 shows the plot of $\log(q_e)$ as a function of $\log(C_e)$ and the linear form of the Freundlich equation. These results show an R^2 of 0.5621, which represents or gives indications of multi-layer adsorption.

$$q_e = K_f * C_e^{1/n}, \quad (3)$$

where C_e: the Cr(VI) concentration at equilibrium in (mg/L); K_f: the Freundlich counter, which is related to the incorporation of all factors affecting adsorption capacity (it is a parameter known to be related to bond strength and adsorption, as well as adsorbent capacity); n: an indication of favorability of metal adsorption by the nanoparticles.

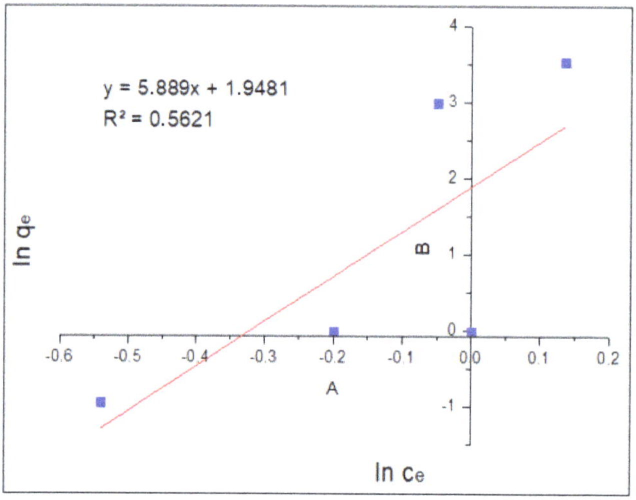

Figure 6. Freundlich isotherm.

Linearization using Freundlich's method focuses on relating the values of $\log(C_e)$ and $\log(q_e)$ in a linear regression of these points in order to obtain the slope and the intercept, which will allow us to obtain the values of K_f and n.

$$\log(q_e) = \log(K_f) + \frac{1}{n} * \log(C_e), \quad (4)$$

The values of K_f and 1/n are obtained from the intercept and the slope, resulting from plotting $\log(q_e)$ vs. $\log(C_e)$. The constant "n" is an indication of metal adsorption favorability by the nanoparticles.

The value 1/n is the slope related to surface heterogeneity and bond distribution. It is dimensionless and is related to the intensity of adsorption. In general, in systems following the Freundlich isotherm, adsorption occurs with the formation of multi-layers instead of only a monolayer, which is the case in the Langmuir isotherm.

To calculate the values of 1/n and K_f, the Freundlich equation is cleared with the values of the linear regression found. To calculate the value of K_f, the antilogarithm of the intercept value is used.

$$\text{Log}(K_f) = 0.9018$$
$$K_f = 7.967$$
$$n = 0.1698$$

The Langmuir isotherm model was obtained by graphing C_e/q_e vs. C_e, resulting in an equation Y = 0.815X + 1.1498 with a correlation coefficient R^2 = 0.9194, while the Freundlich model was obtained by graphing log(q_e) vs. log(C_e), as shown in Figure 5, obtaining an equation Y = 5.889X + 1.9481 with a correlation coefficient R^2 = 0.5621. When comparing the correlation coefficients (R^2 = 0.0.9194 vs. R^2 = 0.5621), it is observed that the Langmuir model shows a better fit with the data, since it has the highest value. This means that our process is limited to monolayer adsorption, and there is no interaction between the adsorbed particles.

The q_{max} value was 1.226 mg/g. This determined the number of moles of solute required to form a monolayer on the surface of the sorbent. While the value of b was 0.709—which indicates that the adsorption of Chromium (VI) is favorable, since increasing the value of variable b causes the adsorption strength to increase proportionally—this result was similar to that obtained by Cañazaca and Ccama [15], who obtained in their adsorption process of Cr(VI), using zerovalent nanoparticles, a fit of their data with the Langmuir isotherm with a value of b of 1.061.

Table 7 shows the parameters of the Langmuir and Freundlich models for the Cr(VI) adsorption equilibrium using the MNPs.

Table 7. Parameters of the isothermal models.

Langmuir	q_{max} (mg/g)	b	R^2
	1.226	0.709	0.9194
Freundlich	K_f (mg1 − n·Ln/g)	n	R^2
	7.0153	0.1698	0.5621

The results obtained show that the Langmuir model is the one, which best fits our data, presenting a coefficient of determination R^2 of 0.9194, higher than that obtained for the Freundlich isotherm, whose R^2 is 0.5621. The value of q_{max} was 1.226 mg/g, determining the number of moles of solute required to form a monolayer on the surface of the absorbent. We also observed that the value of b was 0.709, which indicates that the adsorption of chromium (VI) is favorable, since an increase in the value of variable b allows for the adsorption strength to increase proportionally. To evaluate adsorption according to the shape of the isotherm, the equilibrium parameter or Langmuir parameter (R_L) is used, which must be calculated in the amplitude of the initial concentrations (C_0) of the process.

$$R_L = \frac{1}{1 + bC_0}, \tag{5}$$

C_0 is considered as the reference concentration for equilibrium, such that $C_e < C_0$ in the range of C_e concentrations. In addition, C_e and C_0 must satisfy the Langmuir equation. The values taken by R_L are less than zero, except for C_0 of 0 mg/L, with an average R_L value = 0.2468, which indicates that the adsorption is favorable compared to the values of R_L.

2.6. Kinetics

The application of the kinetic models to the experimental data of the adsorption of the metal ion as a function of time allowed us to evaluate which of the proposed models best describes the kinetics and the limiting step in the adsorption of the metal ion on the adsorbent. These are shown in Tables 8 and 9.

Table 8. Absorbance vs. time results.

Time (min)	Absorbance (nm)			Concentration (mg/L)		
	R1	**R2**	**R3**	**R1**	**R2**	**R3**
0	1.722	1.722	1.504	17.1515	17.1515	14.8250
5	0.596	0.956	0.956	5.1345	8.9765	8.9765
10	0.360	0.566	0.360	2.6158	4.8143	2.6158
20	0.054	0.262	0.262	0.0000	1.5699	1.5699
30	0.061	0.083	0.061	0.0000	0.0000	0.0000
60	0.041	0.043	0.043	0.0000	0.0000	0.0000
90	0.034	0.034	0.036	0.0000	0.0000	0.0000
120	0.033	0.035	0.032	0.0000	0.0000	0.0000
150	0.036	0.034	0.032	0.0000	0.0000	0.0000
180	0.033	0.034	0.031	0.0000	0.0000	0.0000
210	0.031	0.032	0.032	0.0000	0.0000	0.0000

Table 9. Adsorption capacity (QE) (mg/g) depending on residence time.

Time (min)	Cr(VI) Concentration (mg/L)				Adsorption Capacity (QE) (mg/g)
	R1	**R2**	**R3**	**Average**	
0	17.1515	17.1515	14.8250	16.3760	0
5	5.1345	8.9765	8.9765	7.6958	0.2604
10	2.6158	4.8143	2.6158	3.3486	0.3908
20	−0.6499	1.5699	1.5699	0.8300	0.4664
30	0.0000	0.0000	0.0000	0.0000	0.4913
60	0.0000	0.0000	0.0000	0.0000	0.4913
90	0.0000	0.0000	0.0000	0.0000	0.4913
120	0.0000	0.0000	0.0000	0.0000	0.4913
150	0.0000	0.0000	0.0000	0.0000	0.4913
180	0.0000	0.0000	0.0000	0.0000	0.4913
210	0.0000	0.0000	0.0000	0.0000	0.4913

Figure 7 shows the results of the change in Cr(VI) concentration as the contact time with the magnetic nanoparticles increases. It can be seen that the equilibrium time was reached after 30 min, where the chromium concentration was 0 mg/L and remained constant from that moment on.

Figures 8 and 9 show the application of the pseudo-first-order models and the application of the pseudo-second-order model to the experimental data of Cr(VI) adsorption kinetics using MNP, respectively.

The results show that, for the pseudo-first-order model, the R^2 value is 0.9792, while for the pseudo-second-order model, the value of the coefficient of determination R^2 is 0.9907. This indicates that the experimental data obtained fit the pseudo-second-order kinetic model better; therefore, it is preferable to have a chemisorption process, where the rate limiter of the adsorption process is the same reaction and adsorption and not a mass transfer of the metal ion from within the solution to the surface of the nanoparticle.

Table 10 shows the parameters of the pseudo-first-order and pseudo-second-order models. As can be noted, the adsorption velocity for the pseudo-second-order mechanism (K_2) is higher than that obtained in the pseudo-first-order model (K_1).

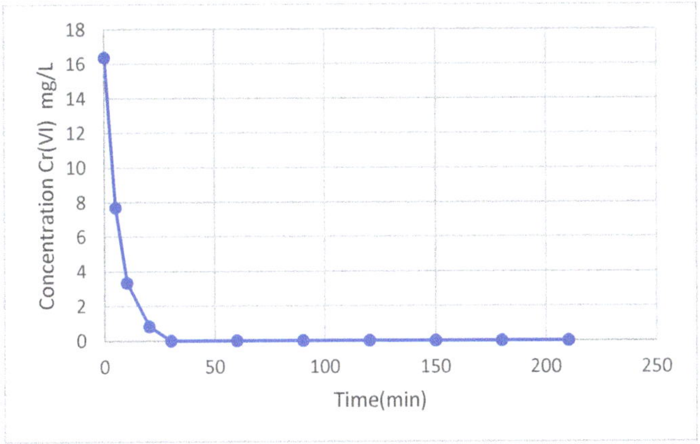

Figure 7. Cr(VI) concentration (mg/L) vs. contact time.

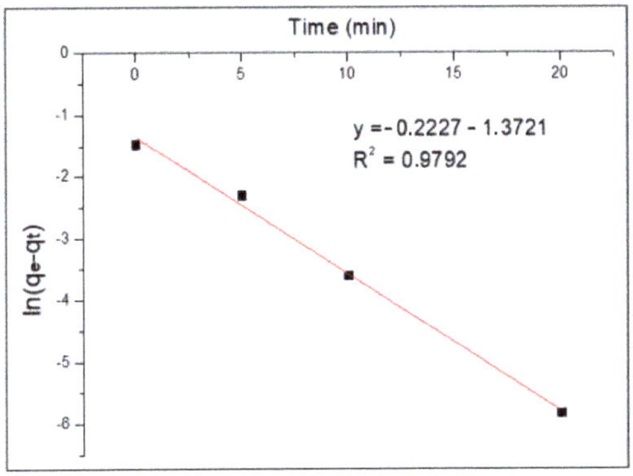

Figure 8. Pseudo-first-order kinetic model.

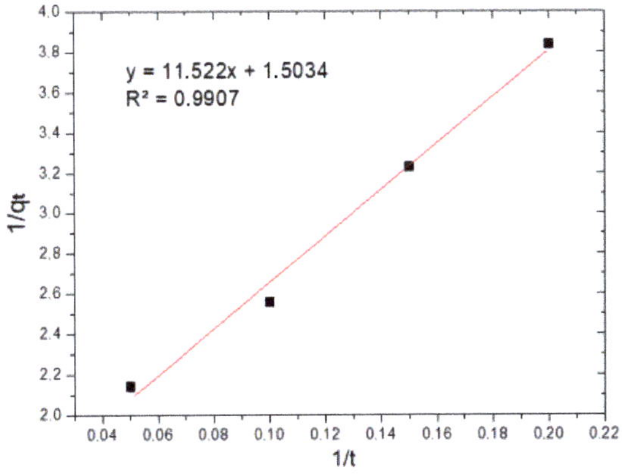

Figure 9. Pseudo-second-order kinetic model.

Table 10. Parameters of the kinetic model.

Kinetic Model					
Pseudo First Order			Pseudo Second Order		
q_e, mg/g	K_1, m^{-1}	R^2	q_e, mg/g	K_2, $mg^{-1}min^{-1}$	R^2
0.2536	0.2227	0.9792	0.6652	0.1962	0.9907

3. Experimental Procedure

The collection and characterization of the tanning and retanning effluents prior to their discharge into the sewage system were carried out in situ and ex situ (see Section 2.1).

3.1. Effluent Characterization

The collection of effluent samples from the tanning stage was carried out at the Isidro Viza tannery located in the Río Seco Industrial Park, following the *Manual for Inspectors: Industrial Effluent Control* [20]. Samples were collected before effluent discharge to the sewer and placed in previously sterilized PVC bottles for transport in a thermal container (cooler) (away from sunlight) at a temperature of 4 °C using ice (or ice pack).

In situ measurements of pH, dissolved oxygen, conductivity and temperature were performed using a portable multi-parameter WTW Model 3620.

Quantification of ex situ parameters was performed using the international standard methodology for waters and effluents APHA-AWWA-WEF 2017. The parameters of total suspended solids, hexavalent chromium, total chromium, oils and fats, chemical oxygen demand (COD) and biological oxygen demand (BOD) were measured using the methods included in the *Standard Methods for the Examination of Water and Wastewater.23rd Ed. 2017* [21].

3.2. Preparation of the Olive Pit Extract

For the preparation and determination of the total phenolic content of *Olea europaea* bone extract, 500 g of olive pits was washed and dried and then crushed. Then, 10 g of the crushed biological material was added to a 1:1 solution of alcohol and water and left to stand for 24 h at room temperature. The resulting solution was filtered using slow filter paper, and the alcohol was eliminated by phase separation using a rotary evaporator. The final extract obtained was used for the determination of phenolic concentration using standard methods of the American Public Health Association.

3.3. Synthesis of Magnetite (Fe_3O_4) Nanoparticles

In a beaker, 0.07 mL of 40% ferric chloride was mixed with 50 mL of ultrapure water and 0.07 g of ferrous sulfate hepta-hydrate. The mixture was stirred at 200 rpm using a shaker (model MS-H280-PRO) at room temperature for 20 min. Subsequently, this solution was brought to pH = 12 by slowly adding 2M sodium hydroxide (NaOH); then, the solution was brought to a temperature of 80 °C, raising the temperature of the shaker. The temperature was controlled with the help of a thermometer. Forty milliliters of phenolic extract was added to the solution and left in constant agitation at 200 rpm for 2 h. Then, the solution was subjected to an ultrasound field at a temperature of 70 °C for 20 min. Finally, the nanoparticles were washed 2 times in a centrifuge at a revolution of 40 rpm for 10 min using ethyl alcohol and dried at 70 °C for 8 h in an oven. The dried nanoparticles were crushed with a mortar and pestle and stored for batch characterization and adsorption experiments.

3.4. Characterization of the Produced Nanoparticles

The synthesized nanoparticles were characterized using different techniques. The size of the nanoparticles in solution was evaluated by dynamic light scattering spectroscopy (DLS), using the DLS-Zetasizer Malvern ZS 90. To carry out the analysis, the dry nanoparticles were ground using a mortar until obtaining a very fine powder; of this powder, 0.002 g was taken in a test tube with 10 mL of deionized water, and then, it was dispersed by sonicating the mixture for 3 min. To record the measurement, a drop of the dispersion was taken and placed in a cuvette, adding 1.5 mL of deionized water; the ready cuvette was placed in the sample holder of the equipment, and a reading was taken.

The morphology and size distribution of the nanoparticles were determined by scanning electron microscopy (SEM), using the ZEISS EVO LS10 scanning electron microscope. For the analysis, the nanoparticles were dried and crushed into a fine powder for reading. The nanoparticles were placed on the sample holder using carbon tape. The elemental composition of the nanoparticles was determined by energy dispersive spectroscopy (EDS), using Hitachi equipment (model SU-8230).

3.5. Favorable Operating Conditions and Validation of the Method for Cr(VI) Quantification

For the quantification of hexavalent chromium, the colorimetric method was used, using a UV-VIS spectrophotometer.

3.5.1. Validation of Diphenyl Carbazide

The following solutions were prepared for validation. Chromium stock solution: Potassium dichromate ($K_2Cr_2O_7$) was dried for 1 h in a desiccator to constant weight. Then, 0.071 g of potassium dichromate was weighed, placed in a vial and filled up to the mark with 250 mL of distilled water. This solution was brought to pH = 2 by adding 37.5 mL of 0.5% sulfuric acid for preservation. One milliliter of the solution contained one-tenth milligram of Cr(VI). 1.5-Diphenylcarbazide solution 0.5% (*m*/*v*): 125 mg of 1.5-Diphenylcarbazide was weighed and then dissolved in 25 mL of acetone. Sulfuric acid solution 1:1 or 50% for the method: 25 mL of concentrated sulfuric acid was taken and filled up to the mark in a 50 mL flask with distilled water. N sulfuric acid solution for pH adjustment: 0.275 mL was taken—taking into account the concentration and purity of the acid—to finally attain a final volume of 200 mL.

Subsequently, the 100 ppm Cr(VI) standard solution was prepared using 1.5-Diphenylcarbazide (known concentration of Cr(VI)) by placing 100 μL of H_2SO_4 50% and 200 μL of diphenyl carbazide in a 10 mL flask, which was filled up to the mark with distilled water. Finally, spectrophotometric scanning was performed at 540 nm of the Cr(VI) solution at a concentration of 10 mg/L.

3.5.2. Linearity

An absorbance vs. concentration calibration curve was constructed using 7 chromium concentrations: 0.0, 0.2, 0.4, 0.6, 0.8, 1.0 and 1.2 mg/L. Each point was evaluated in triplicate; the data were statistically processed with the linear regression test. Linear correlation coefficient (R), the coefficient of determination (R^2), intercept (a) and slope (b) were determined with 95% confidence.

3.5.3. Accuracy and Repeatability

Six measurements of 0.6 mg/L chromium concentration were recorded under the same conditions (operator, apparatus, laboratory and short time interval). In a 25 mL vial, 174 mL of chromium stock solution was used. Afterward, 2500 μL of the previous solution, 250 μL of 0.1 N sulfuric acid and 500 μL of the 1.5-Diphenyl carbazide solution were mixed in a 25 mL vial. See Table 11.

3.5.4. Limit of Detection (LOD) and Limit of Quantification (LQO)

Eight blank readings (containing all reagents, except the chromium (VI) standard) were performed on the same day. The average and standard deviation of the results were then calculated. In addition, the slope found in the calibration curve used for the linearity of the method was used.

$$\mathrm{LOD} = \frac{\mathrm{Ybl} + 3\mathrm{DSbl}}{\mathrm{m}} \times \frac{1}{\sqrt{\mathrm{n}}}, \tag{6}$$

$$LOQ = \frac{Ybl + 10DSbl}{m} \times \frac{1}{\sqrt{n}}, \quad (7)$$

where DSbl: standard deviation of targets; Ybl: average of blank readings; m: slope of the calibration curve; n: data number.

Table 11. Chromium concentration.

Chromium Concentration (mL)	Cr(VI) Stock Solution (mL)	H_2O (mL)
0.0	0.000	10.00
0.2	0.023	9.977
0.4	0.046	9.954
0.6	0.070	9.930
0.8	0.093	9.907
1.0	0.116	9.884
1.2	0.139	9.861

3.5.5. Accuracy

The percentage recovery of the average absolute analyte was analyzed; three measurements were performed in triplicate at a concentration of 0.6 mg/L of chromium (VI).

3.6. *Adsorption Experiments*

3.6.1. Synthetic Solutions

The adsorption efficiency of the produced magnetite nanoparticles was evaluated in jar tests using synthetic solutions and tanning and retanning effluents. The operating conditions, such as pH, equilibrium time and agitation speed, were optimized. For the adsorption process using synthetic solutions, different initial Cr(VI) concentrations (10, 20 and 50 mg/L) and different nanoparticle concentrations (0.5–1 and 2 g/L) were used. In 50 mL beakers, volumes of 30 mL of the synthetic solutions were placed, and then, different concentrations of nanoparticles were added. The samples were taken to the jar equipment under the optimized operating conditions: pH = 2, equilibrium time = 30 min, speed = 140 rpm. After the adsorption process, 0.1 mL aliquots were taken and filtered to remove the nanoparticles from the solution. Finally, 4 aliquots were read using the diphenyl carbazide colorimetric method to determine the final Cr(VI) concentration.

3.6.2. Tanning and Retanning Effluents

Initially, for the adsorption process with tanning and retanning effluents, the samples were filtered under vacuum in order to eliminate particles or other suspended materials. Subsequently, three aliquots of the filtered samples of 30 mL each were taken and placed in beakers to carry out the adsorption process in triplicate using optimized nanoparticle concentrations and operating conditions. Aliquots of 0.1 mL were taken from the final solutions and filtered using a nanopore (1–100 nm) to finally quantify the Cr(VI) concentration by spectrophotometry using the diphenyl carbazide method. Based on the final Cr(VI) concentrations found in the tanning and retanning effluents, the Cr(VI) removal percentage was calculated for each of the treated effluents.

3.6.3. Thermodynamic Study

In order to obtain the isotherms, synthetic chromium solutions (0, 10, 20, 20, 50 and 80 mg/L) were put in contact with 0.03 g of MNPs. Each one of these solutions, together with the concentration of MNPs, was placed in beakers and left in agitation for 30 min at pH = 2 until reaching equilibrium; experiments were carried out in triplicate in order to quantify the error. The results obtained from the readings of the five concentrations of Cr(VI) placed in contact with the concentration of MNPs, in triplicate, were used to perform the linearization of the Langmuir and Freundlich isotherm models.

The Langmuir isotherm was modeled according to the following equation:

$$q_e = q_{max}\frac{C_e * b}{(1 + b * C_e)}, \tag{8}$$

where q_e represents the amount of adsorbate on the surface at equilibrium (mg/g); C_e represents the equilibrium concentration of adsorbate (Cr(VI)) in solution (mg/L); b represents the constant, referring to the adsorption affinity between the nanoparticle and Cr(VI); and q_{max} represents the maximum amount of Cr(VI) per unit mass of nanoparticle to complete the monolayer (mg/g). The numerical values of q_{max} and b are obtained from the slope and intercept, respectively, of the linear form of the isotherm.

The Freundlich isotherm was modeled with the following equation:

$$\log(q_e) = \log(K_f) + \frac{1}{n}\log(C_e) \tag{9}$$

where q_e represents the adsorption capacity of magnetite nanoparticles at equilibrium (mg/g); C_e represents the Cr(VI) concentration at equilibrium (in mg/L); and K_f is the Freundlich counter, which is related to the incorporation of all factors affecting adsorption capacity.

3.7. Kinetic Study

The adsorption kinetics of Cr (VI) metal ion can be influenced by the adsorption reaction and mass transfer. In order to identify the controlling mechanism in the adsorption rate of Cr (VI), the experimental data obtained from the equilibrium time were linearized to pseudo-first-order and pseudo-second-order models. The coefficient of determination R^2 is the parameter, which indicates the best model, and it is obtained from the equation of the line. The highest R^2 determines the model, which best fits the adsorption process.

4. Conclusions

The employment of green synthesis achieved magnetic nanoparticles with a diameter between 73.28 and 162.9 nm. It was also observed that the adsorption process was monolayer, which corresponded to the Langmuir isotherm. The optimum removal results were obtained at pH = 2 and agitation speed = 140 rpm for a contact time = 30 min. Finally, the removal percentage achieved in the tanning effluent was 79.13%, while in the retanning effluent, a value close to 100% removal was reached. Overall, these results demonstrate the efficiency of the proposed method, positioning it as a viable alternative for the industrial treatment of effluents with high concentrations of Cr(VI) from tanneries.

Author Contributions: Conceptualization, M.B.-M., F.E.D.-C. and S.S.-H.; methodology, M.B.-M., F.E.D.-C. and S.S.-H.; software, S.F.-C. and H.G.J.-P.; validation, M.B.-M., F.E.D.-C. and S.S.-H.; formal analysis, M.B.-M., F.E.D.-C. and S.S.-H.; investigation, M.B.-M., F.E.D.-C. and S.S.-H.; resources, M.B.-M., F.E.D.-C., S.S.-H., S.F.-C. and H.G.J.-P.; data curation, M.B.-M., F.E.D.-C. and S.S.-H.; writing—original draft preparation, M.B.-M., F.E.D.-C. and S.S.-H.; writing—review and editing, M.B.-M., F.E.D.-C. and S.S.-H.; visualization, M.B.-M., F.E.D.-C. and S.S.-H.; supervision, M.B.-M., F.E.D.-C. and S.S.-H.; project administration, M.B.-M.; funding acquisition, M.B.-M. All authors have read and agreed to the published version of the manuscript.

Funding: This research was funded by Universidad Católica de Santa María through the project "Concurso Fondo para la Investigación 2020", approved under resolution 27574-R-2020.

Institutional Review Board Statement: Not applicable.

Informed Consent Statement: Not applicable.

Data Availability Statement: The findings of this research are supported by data from the corresponding author, M.B., which are available upon reasonable request.

Conflicts of Interest: The authors declare no conflict of interest.

References

1. Bharagava, R.; Saxena, G.; Mulla, S.; Patel, D. Characterization and identification of recalcitrant organic pollutants (rops) in tannery wastewater and its phytotoxicity evaluation for environmental safety. *Arch. Environ. Contam. Toxicol.* **2018**, *75*, 259–272. [CrossRef] [PubMed]
2. Esparza, E.; Gamboa, N. Contaminación Debida a la Industria Curtiumbre. *Rev. Química* **2001**, *15*, 41–63.
3. Sharma, P.; Singh, S.; Parakh, S.; Tong, Y. Health hazards of hexavalent chromium (Cr(VI)) and its microbial reduction. *Bioengineered* **2022**, *13*, 4923–4938. [CrossRef] [PubMed]
4. Ge, F.; Li, M.-M.; Ye, H.; Zhao, B.-X. Effective removal of heavy metal ions Cd^{2+}, Zn^{2+}, Pb^{2+}, Cu^{2+} from aqueous solution by polymer-modified magnetic nanoparticles. *J. Hazard. Mater.* **2012**, *211–212*, 366–372. [CrossRef] [PubMed]
5. Wang, B.; Li, Q.; Lv, Y.; Fu, H.; Liu, D.; Feng, Y.; Xie, H.; Qu, H. Insights into the mechanism of peroxydisulfate activated by magnetic spinel CuFe2O4/SBC as a heterogeneous catalyst for bisphenol S degradation. *Chem. Eng. J.* **2021**, *416*, 129162. [CrossRef]
6. Saif, S.; Tahir, A.; Chen, Y. Green Synthesis of Iron Nanoparticles and Their Environmental Applications and Implications. *Nanomaterials* **2016**, *6*, 209. [CrossRef] [PubMed]
7. Alsaiari, N.; Alzahrani, F.; Amari, A.; Osman, H.; Harharah, H.; Elboughdiri, N.; Tahoon, M. Plant and Microbial Approaches as Green Methods for the Synthesis of Nanomaterials: Synthesis, Applications, and Future Perspectives. *Molecules* **2023**, *28*, 463. [CrossRef] [PubMed]
8. Gautam, P.; Shivalkar, S.; Banerjee, S. Synthesis of *M. oleifera* leaf extract capped magnetic nanoparticles for effective lead [Pb (II)] removal from solution: Kinetics, isotherm and reusability study. *J. Mol. Liq.* **2020**, *305*, 112811. [CrossRef]
9. Aragaw, T.; Bogale, F.; Aragaw, B. Iron-based nanoparticles in wastewater treatment: A review on synthesis methods, applications, and removal mechanisms. *J. Saudi Chem. Soc.* **2021**, *25*, 101280. [CrossRef]
10. Yew, Y.; Shameli, K.; Miyake, M.; Khairudin, N.; Mohamad, S.; Naiki, T.; Lee, K. Green biosynthesis of superparamagnetic magnetite Fe_3O_4 nanoparticles and biomedical applications in targeted anticancer drug delivery system: A review. *Arab. J. Chem.* **2020**, *13*, 2287–2308. [CrossRef]
11. Niculescu, A.-G.; Chircov, C.; Grumezescu, A. Magnetite nanoparticles: Synthesis methods—A comparative review. *Methods* **2022**, *199*, 16–27. [CrossRef] [PubMed]
12. Aftabtalab, A.; Chakra, C.; Sadabadi, H.; Rao, V. Magnetite nanoparticles (Fe_3O_4) synthesis for removal of Chromium (VI) from waste water. *Int. J. Sci. Eng. Res.* **2014**, *5*, 1419–1423.
13. Hao, R.; Li, D.; Zhang, J.; Jiao, T. Green Synthesis of Iron Nanoparticles Using Green Tea and Its Removal of Hexavalent Chromium. *Nanomaterials* **2021**, *11*, 650. [CrossRef] [PubMed]
14. Pacheco-Portugal, J.-D. Evaluación Del Proceso De Biosorción De Cr(VI) Usando Residuos Agroindustriales De La Región Arequipa (Cascarilla De Arroz Y Chala De Maiz). Bachelor's Thesis, Universidad Católica de Santa María, Repositorio Institucional—Universidad Católica de Santa María, Yanahuara, Peru, 2019.
15. Cañazaca, C.; Ccama, W. Biosíntesis De Nanopartículas De Hierro Cero Valente (Nzvi) Usando Hojas De Eucalipto (*Eucalyptus* Sp.) Para La Remoción De Cromo Hexavalente. Bachelor's Thesis, Repositorio Institucional UNSA, Arequipa, Peru, 2017.
16. Shafey, A.E. Green synthesis of metal and metal oxide nanoparticles from plant leaf extracts and their applications: A review. *Green Process. Synth.* **2020**, *9*, 304–339. [CrossRef]
17. Es'haghi, Z.; Vafaeinezhad, F.; Hooshmand, S. Green synthesis of magnetic iron nanoparticles coated by olive oil and verifying its efficiency in extraction of nickel from environmental samples via UV–vis spectrophotometry. *Process Saf. Environ. Prot.* **2016**, *102*, 403–409. [CrossRef]
18. Fazlzadeh, M.; Rahmani, K.; Zarei, A.; Abdoallahzadeh, H.; Nasiri, F.; Khosravi, R. A novel green synthesis of zero valent iron nanoparticles (NZVI) using three plant extracts and their efficient application for removal of Cr(VI) from aqueous solutions. *Adv. Powder Technol.* **2017**, *28*, 122–130. [CrossRef]
19. Assi, N.; Aberoomand-Azar, P.; Saber-Tehrani, M.; Husain, S.; Darwish, M.; Pourmand, S. Selective solid-phase extraction using 1,5-diphenylcarbazide-modified magnetic nanoparticles for speciation of Cr(VI) and Cr(III) in aqueous solutions. *Int. J. Environ. Sci. Technol.* **2018**, *16*, 4739–4748. [CrossRef]
20. Galán, P. *Manual para Inspectores Control de Efluentes Industriales*, 1st ed.; Ministerio de Desarrollo Social y Medio Ambiente, Ed.; Ministerio de Desarrollo Social y Medio Ambiente, Secretaría de Desarrollo Sustentable y Políticas Ambiental: Bogotá, Colombia, 1999; Volume 2, p. 220.
21. *International Standard Methods for the Examination of Water and Wastewater*, 23rd ed.; American Public Health Association, American Water Works Association, Water Environment Federation: Denver, CO, USA, 2023.

water

MDPI

Article

Effect of Cone-Plate Radius on the Separation Performance of Cyclone Clarifier

Yulong Zhang, Hongxu Bai *, Linjing Xiao, Jingqi Huang, Yu Wang, Zhengwei Yang, Kuoyue Zhang and Weiping Du

College of Mechanical & Electronic Engineering, Shandong University of Science and Technology, Tsingtao 266590, China; yulong_89@126.com (Y.Z.); xiaolj2008@126.com (L.X.); halohalo1208@163.com (J.H.); wangyu@sdust.edu.cn (Y.W.); yzw9824@163.com (Z.Y.); 17852755962@163.com (K.Z.); handsomeqiqi@foxmail.com (W.D.)

* Correspondence: baihongxu98@163.com; Tel.: +86-156-1032-1832

Abstract: In order to further reduce the number of particles in the overflow port of a cyclone clarifier, a cone-plate structure with an equiproportionally varying cone-plate radius is proposed. This includes two structures, namely, an equal–proportional gradually shrinking cone-plate radius and an equal–proportional gradually expanding cone-plate radius. In this paper, numerical simulation is used to comparatively study the flow field characteristics and particle separation inside the traditional equal radius, the gradually shrinking radius, and the gradually expanding radius cone-plate cyclone clarifier. The simulation results show that compared with the traditional equal radius cone-plate structure, the gradual shrinking of the cone-plate structure, due to the bottom of the cone-plate radius being small, can better give full play to the cone-plate settling capacity. The gradually expanding cone-plate cyclone clarifier, due to the bottom of the cone-plate radius being large, results in more fine particles entering the overflow pipe and being discharged from the overflow port. Compared with the traditional cone-plate cyclone clarifier, the gradually shrinking cone-plate cyclone clarifier has a significant increase in the removal efficiency of particles of different sizes than the traditional cyclone clarifier overflow port. The removal efficiency of all particles at the overflow of the gradually shrinking cyclone clarifier was increased by 10.32% compared to the conventional cone-plate cyclone clarifier.

Keywords: cyclone clarifier; coal water treatment; cyclone separation; cone-plate settlement

1. Introduction

Coal mines generate a large amount of mine water during development, construction, and production [1]. The suspended particles contained in mine water are small in particle size and light in specific gravity, which makes them difficult to settle under natural sedimentation methods. Direct discharge of mine water without treatment will cause serious damage to the mining area and the surrounding ecological environment [2,3]. Commonly used methods for removing suspended particles from water include inclined plate sedimentation and centrifugal sedimentation.

Wang Keyuan et al. [4] conducted an in-depth study on the sedimentation process of an inclined tube gravity sedimentation tank with a high fine particle removal rate in practical applications. It was found that the suitable height of the inclined tube was 70~90 cm, and the removal rate of fine sediment in the inclined tube gravity sedimentation tank was 64.7~69.7%, which was higher than that in the gravity sedimentation tank without an inclined tube by 20.7~32%.

Kirpa Hirom et al. [5] investigated the effect of inlet bias, outlet bias, and center bias of inclined plates on settling efficiency using numerical simulation. It was found that the location of outlet bias was most effective in improving the settling efficiency of the settling tank. It was also observed that the settling efficiency increased with the number of inclined plates within a certain range, after which the settling efficiency decreased rapidly. The highest total efficiency was achieved when only 39 inclined plates with outlet bias positions were installed.

In a study by Peng Yaoli et al. [6], in order to reduce the ash content of the concentrate, inclined plates were also incorporated in the froth zone of the cyclone microbubble flotation column. The inclination angle and spacing of the inclined plate were discussed. The ash content of the concentrate was lowest when the inclination angle of the inclined plate was 70° and the spacing was 15 mm. It was found that the inclined plates and the form in which they were present had a significant effect on both concentrate yield and ash content.

Coal mine underground space is limited, but the settling pond covers a large area, so it is difficult to build in the coal mine underground space, the mine water needs to be lifted from the underground to the surface for treatment, and the process increases the cost of pipeline laying and mine water lifting; the untreated mine water being lifted directly will cause serious wear and tear on the pipeline, reducing the service life of the equipment. Therefore, there is a need for equipment with a small footprint to treat mine water.

Cyclone clarifier is a kind of equipment that can effectively separate the insoluble particles in wastewater under the action of the centrifugal sedimentation principle [7–9]. It is widely used due to its small size, simple structure, and absence of moving parts [10–12]. However, the cyclone clarifier cyclone separation speed is fast and the settling area of the particles is small. Fine particles will be discharged from the overflow port under the entrainment of the inner vortex flow, resulting in poor water quality.

Cone-plate cyclone clarifier is a new type of solid–liquid separation equipment that synthesizes the principle of inclined plate settling and centrifugal settling. Figure 1 is the working principle diagram of the cone-plate cyclone clarifier. Extend the overflow pipe of the cyclone downward, and add a series of cone-plate structures on the outside of the overflow pipe to obtain the traditional cone-plate cyclone clarifier [13]. The role of the cone-plate is to increase the settling area of the particles. Cone-plate cyclone clarifiers not only have a small footprint but also have a large settling area for particles. Therefore, they are suitable for installation in coal mine underground.

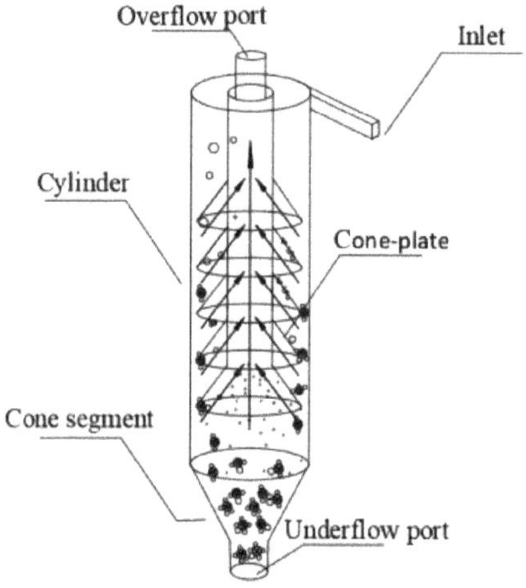

Figure 1. Cone-plate cyclone clarifier working principle diagram.

When the mine water flows tangentially into the cyclone clarifier cylinder from the inlet, the pressure energy is converted into kinetic energy, forming a rotating flow field. Coarse particles and fine particles have different particle sizes and are subjected to different forces such as gravity, centrifugal force, fluid drag force, etc., in the cyclone [14–16]. Fine particles are subject to less centrifugal force. A part of the mine water flows through the cone-plate into the overflow pipe and is discharged from the overflow port. Most of the fine particles will be entrained by the mine water and settle on the cone-plate. The centrifugal force acting on the coarse particles is large. The coarse particles are separated by centrifugal force into the outside vortex flow of the cyclone clarifier. Among them, the coarser particles are discharged from the bottom flow outlet under the effect of gravity, and the smaller particles flow into the overflow pipe with the inner vortex flow and are discharged from the overflow port [17,18]. Due to the complex flow field environment inside the cyclone clarifier, it is easy to cause excessive fine particles in the inner vortex flow to overflow [19]. Conventional cone-plate cyclone clarifiers have cone-plate radii of the same size from top to bottom, which does not give full play to the role of the cone-plate. Therefore, equiproportionally varying cone-plate structures are proposed to investigate the effect of cone-plate structure on the effluent quality of the cyclone clarifier.

In this paper, a cyclone clarifier with a gradually shrinking and gradually expanding cone-plate is proposed. The COMSOL turbulence module is used to compare the distribution of streamlines and velocity fields of mine water at different locations in conventional, gradually shrinking, and gradually expanding cone-plate cyclone clarifiers, and theoretically analyze the motion behavior of mine water inside the cyclone clarifier. By coupling the turbulence module with the fluid flow particle tracking module, the motion and separation rows of the particles in three kinds of cyclone clarifiers are visually described. Thus, by further reducing the particle content in the cyclone clarifier overflow and improving the quality of the effluent at the overflow, this paper also provides a new method for improving the cone disc structure of the cyclone clarifier.

2. Numerical Methods

2.1. Geometric Structure

The three-dimensional model of the cone-plate cyclone clarifier is established with the help of SOLIDWORKS 2020 software. Take the center of the bottom flow opening of the cone-plate cyclone clarifier as the coordinate origin. The X-axis points to the direction of the inlet. The Z-axis points to the direction of the overflow pipe.

The gradually shrinking cone-plate cyclone clarifier is an improvement over the cone-plate structure of the traditional cone-plate cyclone clarifier. The cone-plate group adopts an upper large and lower small structure in order to ensure a sufficiently large area of settlement on the basis of a certain proportion of the cone-plate radius being gradually reduced. The purpose is to reduce the number of fine particles inside the cyclone into the overflow pipe, forcing more fine particles for secondary separation.

The gradually expanding cone-plate cyclone clarifier is an inversion of the top and bottom of the cone-plate structure of the gradually shrinking cone-plate cyclone clarifier, and the cone-plate group adopts the top small and bottom large cone-plate structure. The purpose is to verify the effect of the radius of the bottom cone-plate on the separation performance of the cyclone clarifier under the condition that the volume of the cone-plate remains unchanged.

Figure 2 shows the three-dimensional diagrams of three types of cone-plate. Type A is a traditional cone-plate cyclone clarifier. Type B is a gradually expanding cone-plate cyclone clarifier, and Type C is a gradually shrinking cone-plate cyclone clarifier. The dimensional parameters of the different cone-plate configurations are shown schematically in Figure 3. The dimensional parameters are shown in Table 1.

2.2. Model Description

The cone-plate cyclone clarifier has a simple structure, but the internal flow field contains an unpredictable inner and outside vortex flow, etc. The extremely complex flow field poses a serious challenge in predicting particle separation [20–22].

With the development of computer technology, computational fluid dynamics is widely used in the study of cyclones [23]. Firstly, the k-ε model is used to simulate the liquid phase, and the flow field of the liquid phase is obtained, which is combined with the theoretical knowledge to predict the cyclone separation. Then, the particle tracking module of the fluid flow is used to obtain the results of the separation of the particles in the liquid phase. For space reasons, only the key equations involved in the model are listed.

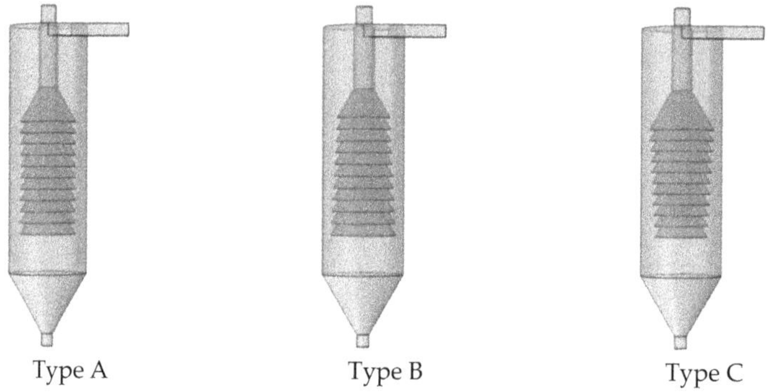

Figure 2. Three-dimensional drawings of three types of cone-plate rotary clarifiers.

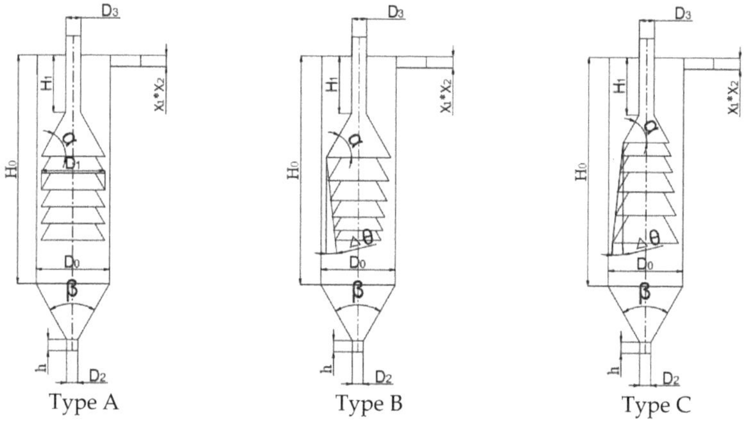

Figure 3. Schematic dimensional parameters of different cone-plate configurations.

Table 1. Overall structural parameters of the cyclone clarifiers.

Structural Parameter	Value
Cylinder diameter D_0/mm	200
Cylinder height H_0/mm	600
Cone-plate diameter D_1/mm	145
Cone-plate insertion depth H_1/mm	150
Underflow port diameter D_2/mm	30
Overflow port diameter D_3/mm	40
Underflow height h/mm	30
Inlet size $X_1 \times X_2$/mm	20×30
Cone-plate dip angle $\alpha/^\circ$	60
Cone segment angle $\beta/^\circ$	60
Angle of change in cone–plate set $\theta/^\circ$	3.3

The k-epsilon model is a two-equation model under RANS (Reynolds-Averaged Navier–Stokes) that solves two independent transport equations to solve the COMSOl "Turbulence, k-ε" interface which is used to simulate high Reynolds number single-phase flows for both incompressible and low Mach number (typically, <0.3 k-ε) compressible flows. The interface solves the momentum-conserving Navier–Stokes equations and the mass-conserving continuity equations, and turbulence effects are modeled by a standard k-ε two-equation model with realizability constraints. Near-wall flow is modeled using wall functions. It has the advantages of model stability, a fast convergence rate, low memory requirement, and an easy solution, and is widely used in industrial generation.

The turbulent kinetic energy k transport equation is shown below.

$$\rho\frac{\partial k}{\partial t} + pu\cdot\nabla k = \nabla\cdot\left(\left(\mu + \frac{\mu_t}{\sigma_k}\right)\nabla k\right) + P_k - \rho\varepsilon \quad (1)$$

The formula for the turbulent viscosity μ_t is as follows:

$$\mu_t = \rho C_\mu \frac{k^2}{\varepsilon} \quad (2)$$

The formula for the generating term P_k is as follows:

$$P_k = \mu_t\left(\nabla \boldsymbol{u} : \left(\nabla \boldsymbol{u} + (\nabla \boldsymbol{u})^T\right) - \frac{2}{3}(\nabla\cdot \boldsymbol{u})^2\right) - \frac{2}{3}\rho k \nabla\cdot \boldsymbol{u} \quad (3)$$

The turbulent kinetic energy dissipation rate ε transport equation is as follows:

$$\rho\frac{\partial k}{\partial t} + pu\cdot\nabla\varepsilon = \nabla\cdot\left(\left(\mu + \frac{\mu_t}{\sigma_\varepsilon}\right)\nabla\varepsilon\right) + C_{\varepsilon 1}\frac{\varepsilon}{k}P_k - C_{\varepsilon 2}\rho\frac{\varepsilon^2}{k} \quad (4)$$

Here, σ_k and σ_ε are the turbulent Prandtl numbers for the turbulent kinetic energy and turbulent kinetic energy dissipation rate, respectively. The constants in the transport equations as well as in the turbulent viscosity formulation are semi-empirical, and the empirical values derived from the experiments have a good accuracy.C_μ = 0.09, $C_{\varepsilon 1}$ = 1.44, $C_{\varepsilon 2}$ = 1.92, σ_k = 1.0, and σ_ε = 1.3.

2.3. Simulation Conditions

An unstructured grid was used to delineate the cone-plate cyclone clarifier. Grid generation is a prerequisite for simulation to ensure the accuracy of the simulation results while making full use of computational resources, so it is necessary to verify the grid independence [24]. The tangential velocity is selected to verify the grid independence, and the results are shown in Figure 4 The tangential velocity of the mine water increases with the increase in the number of grids, and when the number of grids reaches 4.6×10^5, the tangential velocity does not change with the increase in grids, which proves that continuing to increase the number of grids will not have an effect on the results of the numerical simulation. Therefore, in this study, 4.6×10^5 grids are used as the cells of the computational domain. It contains 273,171 tetrahedral meshes, 16,260 pyramidal meshes, 162,296 prismatic meshes, 46,964 triangular meshes, 160 quadrilateral meshes, 4089 edge cells, and 191 vertex cells.

Set boundary conditions in COMSOL for three types of cone-plate cyclone clarifiers. In the steady state study, the inlet condition of the cyclone clarifier is velocity inlet, which flows along the normal direction, where inlet velocity is 2 m/s. The fluid is incompressible, the material is water, and the density and viscosity of the fluid properties come from the material. The overflow port and the underflow port are pressure outlets; the outlet pressure is 0 MPa, and the wall condition is no slip.

Coal dust in mine water has a small particle size and low content, so only the effect of fluids on solid particles is considered. In the transient simulation, the "fluid flow particle tracking" is coupled to the "k-ε" model, the particle density is set to 1500 kg/m^3, the particle sizes are set to 5 μm, 10 μm, 15 μm, and 20 μm, and the particle type is solid. The release starts from moment 0, and the release time is 10 s. Particles with particle sizes from large to small release 5, 10, 15, and 20 particles per second, respectively, and the inlet velocity is based on the velocity field. The overflow and underflow port wall conditions are defined as freezing, and the other wall conditions are defined as rebounding. The drag force is chosen to obey Stokes' law, the kinetic viscosity and density are according to the material, the velocity is based on the velocity field, the gravity is defined as the Z-axis vertically downward, and the gravitational acceleration is g.

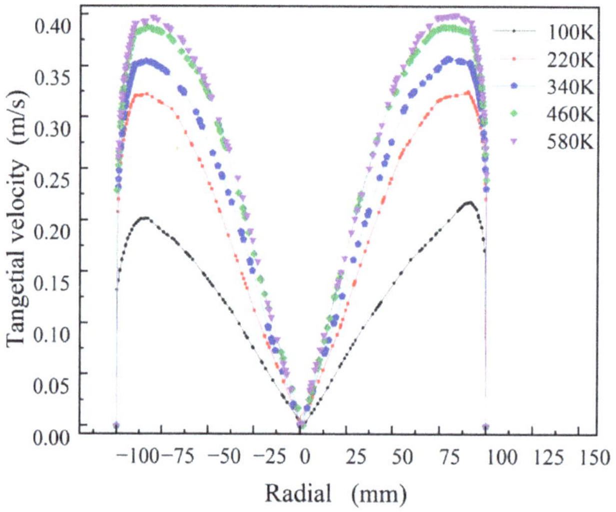

Figure 4. Grid independence verification.

3. Results and Discussion

The three-dimensional model in this paper is based on the original traditional cone-plate cyclone clarifier with an improved cone-plate structure. Due to the unavailability of all post-processing information of the original model, the comparison and discussion of the results will be based on the existing three cone-plate structure cyclone clarifiers.

3.1. *Flow Field*

The flow field is the carrier of particle movement; changes in the structure of the cone-plate of the cyclone clarifier will inevitably cause changes in the behavior of the flow field, thus affecting cyclone separation of the particles [25,26]. Therefore, the effect of structural changes in the cone-plate of the cyclone clarifier on the flow field is first analyzed. This study will be based on the variation in the flow field in a cyclonic clarifier to predict the trajectory of particles.

3.1.1. Streamline Analysis

The streamline is a curve tangent to the fluid vector field at an arbitrary position at a certain moment, which applies to the analysis of the results of the fluid flow and can be used with the help of streamlining tools to describe the trajectory of the fluid in the cyclone clarifier. Take the inlet as the positioning boundary of the streamlines and the number of streamlines to be set at 15. The streamline diagrams of fluid in the conventional cone-plate cyclone clarifier, the gradually shrinking cone-plate cyclone clarifier, and the gradually expanding cone-plate cyclone clarifier are shown in Figure 5.

The fluid enters the cyclone clarifier and rotates downwards. A portion of the fluid flows through the cone-plate and is discharged through the overflow port. A portion of the fluid flows downwards after forming an outside vortex flow. A portion of the outside vortex flow is discharged from the underflow port. The rest of the outside vortex flow is restricted by the smaller space in the cone segment of the cyclone clarifier and rotates upwards to form an inner vortex flow, which is discharged from the overflow port. The flow line of the type A cyclone clarifier becomes gradually denser from the top to the bottom. The flow lines are densest at the bottom of the cone-plate group (V zone), which results in most of the fluid in the inner vortex flow flowing directly into the overflow pipe and discharging from the overflow port. Type B cyclone clarifiers have a small upper and large lower cone-plate structure, which intensifies the denseness of the flow lines. There will be more inner vortex flow directly into the overflow pipe and discharging from the overflow port. The type C cyclone clarifier cone-plate adopts the upper large and lower small cone-plate structure so as to make the distribution of the flow line of the fluid in the cyclone clarifier more uniform. In particular, the flow line under the type C cyclone clarifier cone-plate group (V zone) is relatively sparse, indicating that more fluid from the inner vortex flow is able to flow between the cone-plates. Increasing the effective settling area of fine particles in the fluid promotes the secondary settling of fine particles on the cone-plate, reduces the possibility of fine particles entering the overflow pipe, and improves the quality of the effluent.

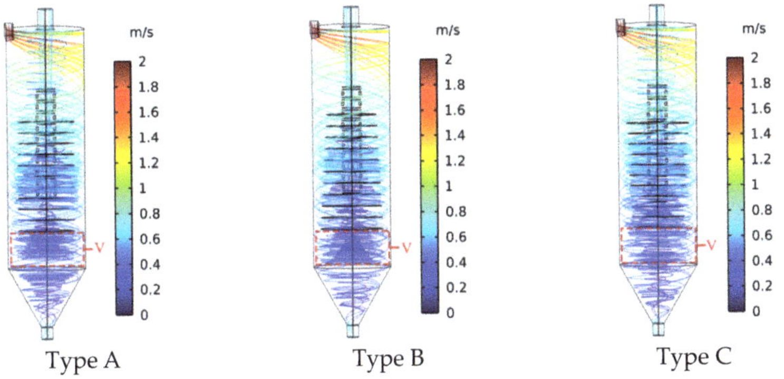

Figure 5. Flow line diagram of fluid in cyclone clarifiers with different cone-plate structures.

3.1.2. Analysis of Tangential Velocity

Fluid enters the inlet along the normal direction and then flows tangentially into the cylinder of the cyclone clarifier. The pressure energy possessed by the fluid is transformed into the kinetic energy to support the rotation of the fluid, so the fluid has a certain tangential velocity. The tangential velocity is the largest one among the velocity components of the fluid [27,28]. The centrifugal force acting on the particles is the main driving force for the radial migration of the particles, and the centrifugal force F_c can be expressed as follows:

$$F_c = \frac{\pi d^3}{6}\rho_s\frac{{u_t}^2}{r} \tag{5}$$

where d is the particle size of the particles, ρ_s is the density of the particles, and u_t is the tangential velocity of the particles. It can be seen that the centrifugal force on the particles is proportional to the square of the tangential velocity, and the larger the size of the particles and the tangential velocity are, the easier the particles migrate to the outside wall of the cyclone clarifier [29,30].

Figure 6 shows the principle of shallow pool sedimentation in a cyclone clarifier. V_a, V_g, and V_t denote the velocities generated by trailing force, gravity, and centrifugal force on the particles, respectively. V_0 is the combined velocity of the particles. β is the angle between the combined velocity of the particles and the horizontal direction. α is the angle between the cone-plate and the horizontal direction. When β is less than α, the particles can settle on the cone-plate. All other conditions being equal, the smaller the tangential velocity of the fluid between the cone-plates, the smaller the centrifugal force on the particles, and the more favorable it is for the particles to settle.

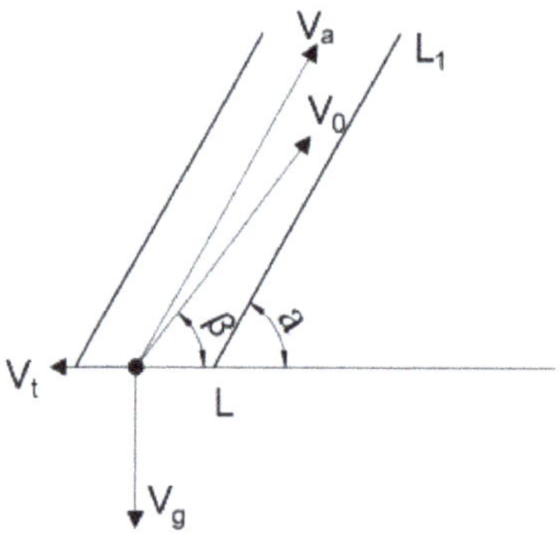

Figure 6. Principles of shallow pool sedimentation in cyclone clarifiers.

Figure 7 shows the tangential velocity contour of three kinds of cone-plate structure cyclone clarifiers. It can be seen that the tangential velocity contour of the three kinds of cone-plate structure cyclone clarifiers is roughly about the axis in an axisymmetric distribution, and the tangential velocity near the inlet of the column section is the largest. The tangential velocity decreases gradually from the top to the bottom. The larger tangential velocity above the column section near the outer wall of the cyclone clarifier favors particle separation in the radial direction into the outside vortex flow. Smaller tangential velocities between the cone-plates favor particles settling on the cone-plates.

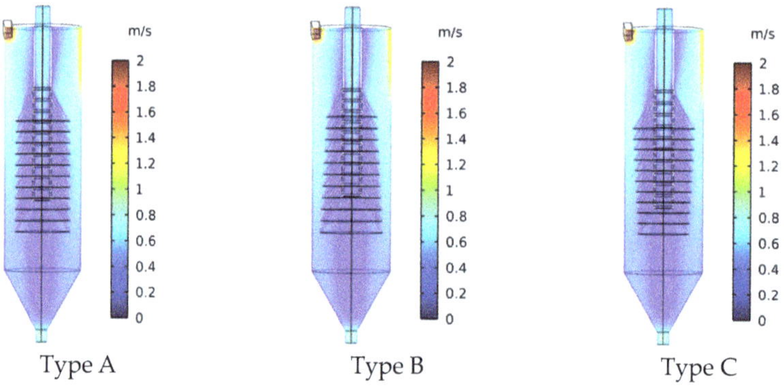

Figure 7. Tangential velocity contours of fluid in cyclone clarifiers with three cone-plate configurations.

In order to describe in more detail the trend of the velocity change in fluid inside the cyclone clarifiers with different cone-plate structures, the cross-section Z = 500 mm and the cross-section Z = 200 mm were selected to analyze the velocity of the fluid. The location of the cross-sections is shown in Figure 8.

The tangential velocities at the cross-sections are shown in Figure 9. The tangential velocity of the fluid in the cyclone clarifiers with three kinds of cone-plate structures increases rapidly with the decrease in the radius from the wall surface and then decreases gradually. Section Z = 500 mm, due to the existence of the cone-plate, will make the tangential velocity fluctuate within a certain range, but the tangential velocity generally shows a trend of gradually becoming smaller. The tangential velocity in the overflow pipe also increases, then decreases, and reaches the minimum at the center of the axis. The tangential velocities between the plates of the three cone-plate configurations are smallest for the type C cyclone clarifiers and largest for the type B cyclone clarifiers at the cross-section of Z = 500 mm. The small tangential velocity between the cone-plates of the type C cyclone clarifier facilitates the settling of particles on the cone-plates. At the cross-section of Z = 200 mm, the tangential velocity of the fluid in the type C cyclone clarifier is slightly greater than that in the type A cyclone clarifier, and the tangential velocity is the smallest in the type B cyclone clarifier. The larger tangential velocity outside the cone-plate favors the radial separation of particles. Taken together, the cyclone clarifier type C has the strongest cyclone separation ability.

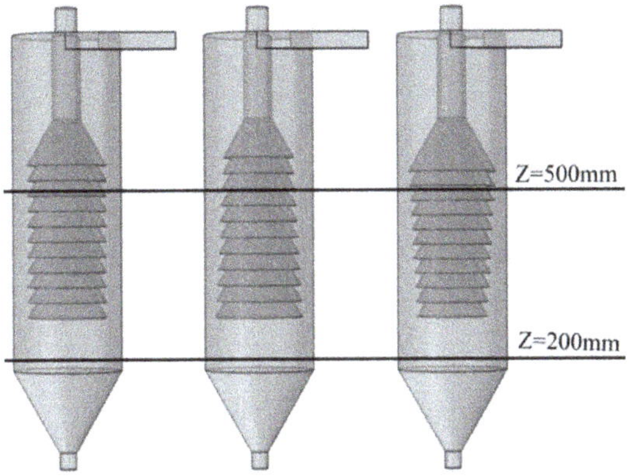

Figure 8. Cross-sections of cyclone clarifiers at different positions.

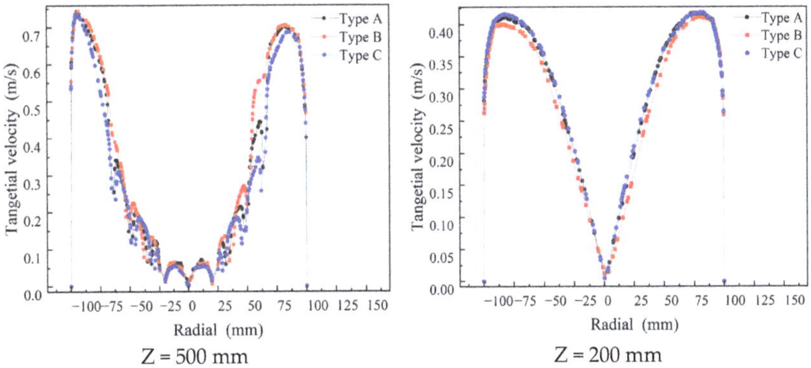

Figure 9. Comparison of tangential velocity in cyclone clarifiers with three cone-plate configurations of Z = 500 mm and Z = 200 mm sections.

3.1.3. Analysis of Axial Velocity

Axial velocity directly affects the movement direction of particles and the separation time within the cyclone clarifier. The size of the axial velocity determines the settling performance of the particles; smaller axial velocities can prolong the time that the fluid stays in the cyclone clarifier and promote the separation of particles in the fluid [31,32].

Figure 10 shows the axial flow velocity plots for three cyclone clarifiers of cone-plate construction. The axial velocity plots are also roughly axisymmetrically distributed around the axis, with maximum axial velocities near the overflow and underflow ports. Near the outer wall of the cyclone clarifier, the axial flow velocity is negative and the fluid moves downwards. It is outside of the vortex flow and discharges from the underflow port. The axial velocity is positive near the center of the shaft, and the fluid moves upwards. It is inside of the vortex flow and discharges from the overflow port. By the law of fluid continuity, the mass of fluid flowing through a horizontal cross-section remains the same while the inlet flow remains the same, so the larger the cross-section area, the lower the axial velocity. Type C cyclone clarifier in the cone-plate group in the lower position of the cone-plate radius is the smallest, resulting in the maximum cross-sectional area between the cone-plate and the outer wall of the cyclone clarifier; so, the type C cyclone clarifier in the cone-plate group in the lower position of the axial velocity of the fluid is the smallest. This prolongs the separation time of the particles in the clarifier and promotes the full separation of the particles.

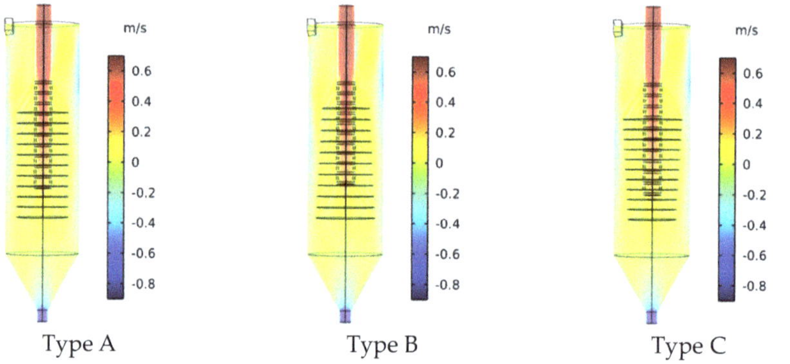

Figure 10. Axial velocity contours of fluid in cyclone clarifiers with three cone-plate configurations.

Two cross-sections Z = 500 mm and Z = 200 mm were selected to analyze the axial velocity of fluid in three types of cyclone clarifiers with cone-plate structures, as shown in Figure 11. Near the outer wall of the cyclone clarifier with the cross-section Z = 500 mm and cross-section Z = 200 mm, the axial velocity of the fluid in the type C cyclone clarifier is the lowest. This can prolong the separation time of the particles and reduce the number of particles in the overflow port. At the axial position of the cross-section Z = 200 mm, the area where the axial velocity of the fluid in the type C cyclone clarifier is close to 0 is wider. Lower axial velocities allow for adequate separation of particles and improved effluent quality at the overflow port.

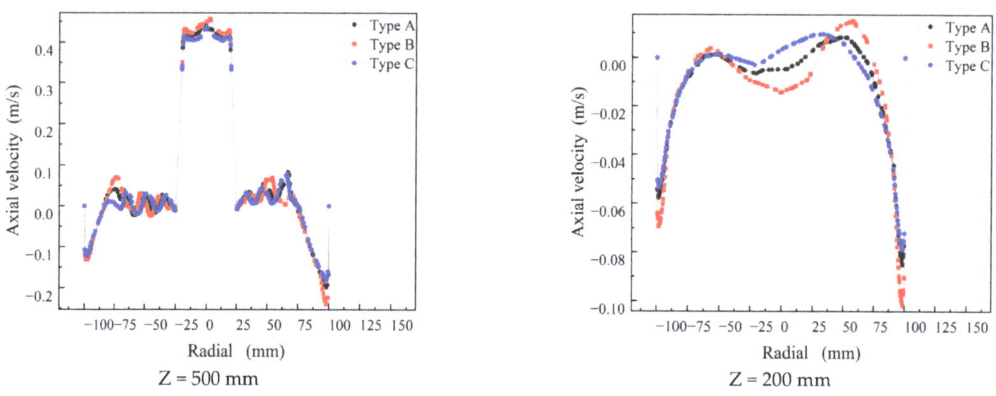

Figure 11. Comparison of axial velocity of fluid in cyclone clarifiers with three cone-plate configurations of Z = 500 mm and Z = 200 mm sections.

The surface formed by connecting points where the axial velocity is zero is known as the zero-speed envelope surface [33]. Figure 12 shows the zero-speed envelope surfaces of three different cone-plate structures for cyclone clarifiers. Inside the zero-speed envelope surface is the inner vortex flow, and outside the zero-speed envelope surface to the outer wall of the cyclone clarifier is the outside vortex flow.

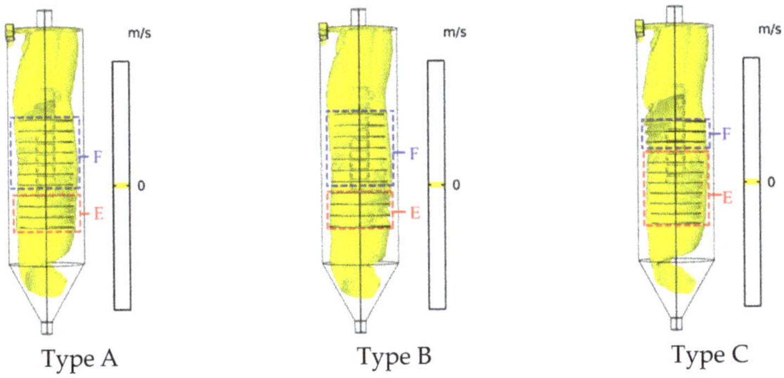

Figure 12. Clouds of zero-velocity envelope surfaces of fluid in cyclone clarifiers with three cone-plate configurations.

The spacing between the zero-velocity envelope surface of the fluid and the cone-plate in the type A cyclone clarifier is relatively uniform. Fluid enters between the cone-plate and flows upwards, while the fluid near the cone-plate flows downwards. At this time, the direction of axial velocity changes from downwards to upwards, where the formation of zero-velocity envelope surfaces.

The zero-velocity envelope surface of the fluid in the type B cyclone clarifier is more widely spaced from the upper portion of the cone-plate (F zone). This results in larger particles that have not been adequately cyclonically separated flowing into the cone-plate and out of the overflow port. The zero-velocity envelope surface of the fluid in the B cyclone clarifier is less spaced from the lower part of the cone-plate (E zone). The particles in the fluid are sufficiently separated in the vicinity of the E zone, and less fluid flows into the cone-plate (E zone) and is discharged through the overflow port. At the same time, the vast majority of fluids in the inner vortex flow directly to the overflow port without secondary separation, so the quality of the water out of the overflow port is poor.

The spacing between the zero-velocity envelope surface of the fluid and the upper portion of the cone-plate (zone F) is smaller in the type C cyclone clarifier, reducing the likelihood that particles not adequately separated by the cyclone clarifier will enter the cone-plate from the overflow. The spacing between the zero-velocity envelope surface of the fluid and the lower portion of the cone-plate (zone E) is larger in the type C cyclone clarifier. More fluid flows into the cone-plate (Zone E). Here, the particles in the fluid are fully separated. As a result, more fluid enters the overflow pipe, but the number of particles decreases. At the same time, the inner vortex flow portion of the fluid can enter the inside of the cone-plate along the outer edge of the cone-plate, allowing for further settling of the particles in the fluid. The quality of the water discharged from the overflow is greatly improved.

3.1.4. Analysis of Velocity Vector

In order to analyze the reasons for the variation in tangential and axial velocities in cyclonic clarifiers with different cone-plate constructions, velocity vectors are introduced, as shown in Figure 13.

After the fluid enters the column section of the cone-plate cyclone clarifier, the fluid will form a circulating flow between the first cone-plate and the top of the cyclone clarifier (H area) due to the cone-plate's restriction of the fluid's movement. The fluid rotates downwards under the action of gravity and centrifugal force. As can be seen in the figure, a small amount of fluid flows upwards near the cone-plate at the bottom (G area) of the type A cyclone clarifier and flows into between the two cone-plates. The fluid flowing between the two cone-plates of the downside position of the cone-plate group is mainly from the inner vortex flow, while the fluid flowing between the two cone-plates of the upper portion of the cone-plate group is mainly from the outside vortex flow. The cone-plate at the bottom of the type B cyclone clarifier has a large radius, and the inner vortex flow basically flows to the overflow pipe. The fluid flowing between the cone-plates is mainly from the outside vortex flow. The type C cone-plate cyclone clarifier has the smallest radius of the cone-plate at the bottom. More inner vortex fluid is able to flow between the middle and lower portions of the cone-plate set for particle settling. Fluid flowing into the cone-plates of the upper portion of the cone-plate group is mainly from the outside vortex flow. The larger tangential velocity in the type C cyclone clarifier at the following positions of the cone-plate group also results in a larger centrifugal force on the particles in the inner vortex flow. This causes more particles to flow between the two cone-plates (G zone) and promotes the secondary settling of fine particles. The greater inner vortex flow on both sides of the cone-plate of the type C cyclone clarifier generates a higher energy loss by interacting with the outside vortex flow. This reduces the tangential velocity of the type C cyclone clarifier in the vicinity of the cone-plate. It also reduces the axial velocity near the outer wall of the cyclone clarifier. This is consistent with the previous analysis.

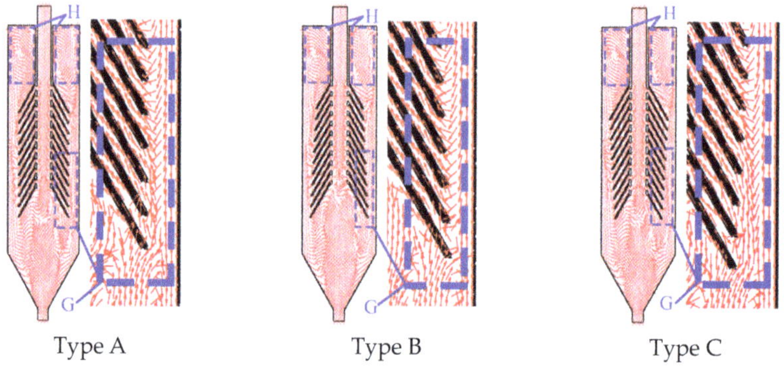

Figure 13. Fluid velocity vectors in cyclone clarifiers with three cone-plate configurations.

3.2. Particle Tracking

Simulation of the flow field only qualitatively analyzes the movement of particles in the cyclone clarifier. In order to visualize the number, position, and velocity of the particles in the cyclone clarifiers with different cone-plate structures at each moment, transient simulations of fluid were carried out with the aid of the COMSOL particle tracking module.

In the post-processing work, the optimization of the structural parameters of the cyclone clarifier was greatly assisted by the addition of particle counters to count the number of particles flowing through the bottom of the cone-plate and the bottom of overflow tube to the overflow port. Figure 14 shows the velocity distributions of particles of 10 μm size in the three cone-plate cyclone clarifiers in the transient simulation.

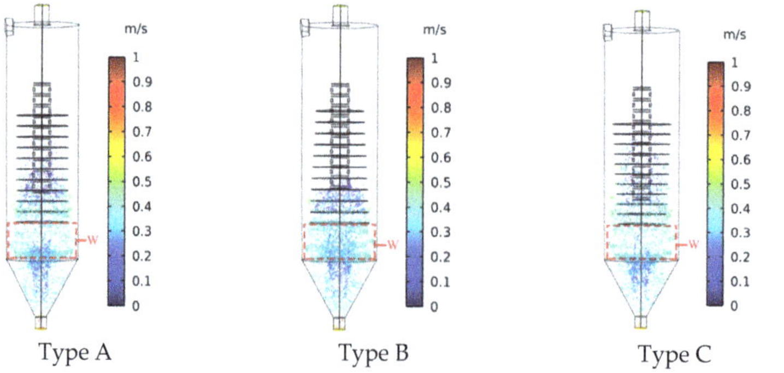

Figure 14. Particle velocity contours in cyclone clarifiers with three cone-plate structures.

The particles in the type A cyclone clarifier are mainly settled between the two cone-plates of the lower half of the cone-plate group, and the upper half of the cone-plate group is not fully utilized, thus not giving full play to the settling performance of all of the cone-plates. At the same time, the fine particles that fail to settle enter the inner vortex flow in the W zone with a higher number of particles, resulting in a higher number of fine particles flowing out through the overflow pipe. The particles in the type B cyclone clarifier are mainly settled between the two cone-plates of the lowest three cone-plates; these cone-plates cannot be fully utilized, further aggravating the phenomenon. The W area of the inner vortex flow into the overflow pipe and through the overflow port discharge of the number of particles is further increased. The number of particles settling between the cone-plates in the type C cyclone clarifier are more uniform, and the settling performance of each cone-plate can be fully utilized. The number of particles entering the overflow pipe in W area of the cyclone are reduced, which improves the quality of the water discharged from the overflow.

The number of particles in the overflow are a direct reflection of the capacity of the cyclone clarifier to treat mine water. Figure 15 is a graph of the number of particles in the overflow port as a function of time. The number of particles in the overflow port of the type A cyclone clarifier basically remained unchanged after about 31 s. The number of particles in the overflow port of the type B cyclone clarifier remained unchanged until 34 s. The number of particles in the overflow of the type C cyclone clarifier remained unchanged after about 25 s. The number of particles in the overflow of the type C cyclone clarifier are much smaller than those of the type B cyclone clarifier, and the number of particles in the overflow of the type A cyclone clarifier are in between type B and type C.

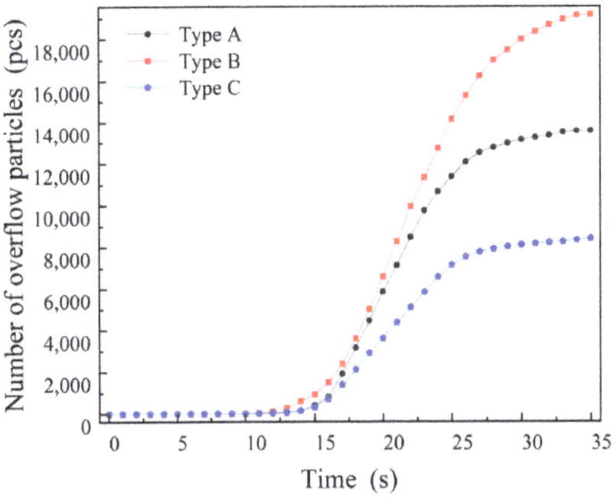

Figure 15. Plots of particles at the overflow of cyclone clarifiers with three cone-plate structures as a function of time.

The fluid contains particles with four particle sizes of 5 μm, 10 μm, 15 μm, and 20 μm. Figure 16 shows the number of particles of different particle sizes at the overflow port in three cone-plate configurations of cyclone clarifiers. In the three cone-plate structure cyclone clarifiers, the number of small particles at the overflow are the highest, and the number of particles decrease as the particle size increases. This indicates that large particles are more easily separated and settled by centrifugal force and gravity. The type B cyclone clarifier has the highest number of 5 μm particles in the overflow port, the type A cyclone clarifier has the middle number of 5 μm particles in the overflow port, and the type C cyclone clarifier has the lowest number of 5 μm particles in the overflow port. The removal efficiency of 5 μm particles at the overflow of the type C cyclone clarifier is 11.51% higher than that of the type A cyclone clarifier. This is due to the small tangential velocity of the fluid between the cone-plates in the type C cone-plate cyclone clarifier, which is conducive to the settling of particles on the cone-plate. In the cyclone clarifier outer wall and near the axis, the type C cone-plate cyclone clarifier in the fluid axial velocity is small and is conducive to the full separation of particles. At the same time, the type C cyclone clarifier adopts the cone-plate structure which is gradually becoming smaller from the top to the bottom, forcing the inner vortex flow particles to flow into the cone-plate for secondary settlement. The number of 20 μm particles at the overflow port of the cyclone clarifiers with the three cone-plate configurations were similar. This is due to the fact that large-sized particles receive high centrifugal force and flow directly out of the underflow port with the outside vortex flow. The cone-plate structure has little effect on the large-sized particles. Type C has increased the removal efficiency of 5 μm, 10 μm, 15 μm, and 20 μm particles by 11.51%, 9.99%, 9.83%, and 7.62%, respectively, compared with that of the overflow of the Type A cyclone clarifier. The removal efficiency of all particles at the overflow of the type C cyclone clarifier was increased by 10.32% compared to the type A cyclone clarifier.

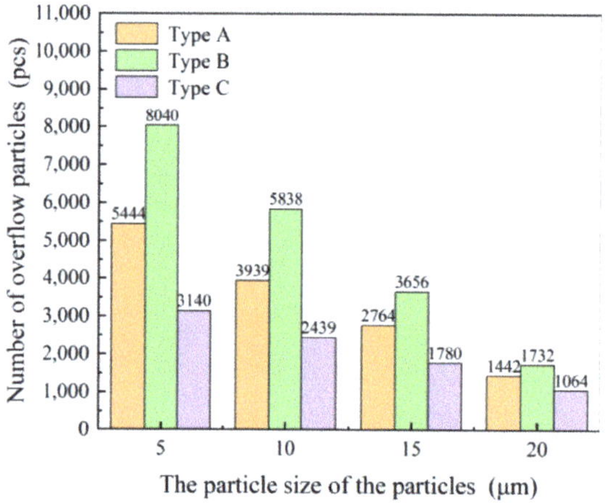

Figure 16. Number of particles of different sizes in the overflow of cyclone clarifiers with three cone-plate configurations.

4. Conclusions

In this paper, CFD numerical simulation has been used to study the cone-plate cyclone clarifier. The effect of cone-plate radius on the flow field and particle separation performance was analyzed. The following conclusions are drawn:

(1) The change in cone-plate structure has a significant effect on the flow field of the cyclone clarifier. There is poor effluent quality in gradually expanding cone-plate cyclone clarifiers compared to traditional cone-plate cyclone clarifiers. The effluent quality is improved in the gradually shrinking cone-plate cyclone clarifier where particles settle in large numbers on the cone-plate.

(2) The gradually shrinking cone-plate cyclone has a small tangential velocity be-tween the cone-plates, which is favorable for particles to settle on the cone-plates. The gradually shrinking cone-plate cyclone clarifier has a small axial velocity near the outer wall and at the bottom of the cylinder section, which can prolong the time that the mine water stays in the cyclone clarifier and promote the full separation of particles.

(3) The zero-velocity envelope surface and velocity vectors of the three types of cone-plate cyclone clarifiers were analyzed in comparison. Particles in the gradually shrinking cone-plate cyclone clarifier were able to enter between the cone plates for secondary settling.

(4) The gradually shrinking cone-plate cyclone clarifier has the highest removal efficiency for particles. The removal efficiency of all particles at the overflow of the gradually shrinking cone-plate cyclone clarifier was increased by 10.32% compared to the traditional cone–plate cyclone clarifier. The gradually shrinking cone-plate cyclone clarifier, rather than

the traditional cone-plate cyclone clarifier, overflow port increased removal efficiency of the 5 μm, 10 μm, 15 μm, and 20 μm particles by 11.51%, 9.99%, 9.83%, and 7.62%, respectively.

Author Contributions: Conceptualization, Y.Z.; data curation, L.X. and J.H.; formal analysis, Y.W.; investigation, Z.Y.; methodology, H.B.; resources, Y.Z.; software, H.B.; validation, K.Z.; visualization, W.D.; writing—original draft, H.B.; writing—review and editing, Y.Z., L.X. and H.B. All authors have read and agreed to the published version of the manuscript.

Funding: This work was supported by the Shandong Natural Science Foundation, China (ZR2023QE121).

Data Availability Statement: Due to privacy concerns, we do not upload our data to the web.

Conflicts of Interest: The authors have declared no conflicts of interest.

References

1. Liu, Y.; Liu, P.; Bo, L.; Li, Q.; Quan, G.; Zhuo, Y.; Han, Y.; Wang, Y. Research on Mine Water Dispatching Mode Based on Maximization of Reuse Rate. *Sustainability* **2022**, *14*, 9289. [CrossRef]
2. Wang, X.; Gao, Y.; Jiang, X.; Zhang, Q.; Liu, W. Analysis on the Characteristics of Water Pollution Caused by Underground Mining and Research Progress of Treatment Technology. *Adv. Civ. Eng.* **2021**, *2021*, 9984147. [CrossRef]
3. Guminska, J.; Plewa, F.; Grodzicka, A.; Guminski, A.; Rozmus, M.; Michalak, D. Economic Analysis of the Application of the Technological System for Removing Suspended Solids from Mine Drainage Waters. *Energies* **2021**, *14*, 232. [CrossRef]
4. Wang, K.; Li, Y.; Ren, S.; Yang, P. A Case Study on Settling Process in Inclined-Tube Gravity Sedimentation Tank for Drip Irrigation with the Yellow River Water. *Water* **2020**, *12*, 1685. [CrossRef]
5. Hirom, K.; Devi, T.T. Determining the Optimum Position and Size of Lamella Packet in an Industrial Wastewater Sedimentation Tank: A Computational Fluid Dynamics Study. *Water Air Soil Pollut.* **2022**, *233*, 261. [CrossRef]
6. Peng, Y.; Ni, C.; Tan, J.; Sha, J.; Xie, G. Separation performance of flotation column with inclined plates in the froth zone. *Int. J. Miner. Process.* **2016**, *148*, 124–127. [CrossRef]
7. Garcia, V.A.; Lobato, F.S.; Martins Vieira, L.G. Design of high performance thickener hydrocyclones using robust optimization. *J. Pet. Sci. Eng.* **2020**, *191*, 107144. [CrossRef]
8. Salvador, F.F.; Ascendino, G.G.; de Faria, E.V.; Barrozo, M.A.d.S.; Vieira, L.G.M. Geometric optimization of fi ltering cylindrical hydrocyclones. *Powder Technol.* **2021**, *381*, 611–619. [CrossRef]
9. He, L.; Ji, L.; Sun, X.; Chen, S.; Kuang, S. Investigation of mini-hydrocyclone performance in removing small-size microplastics. *Particuology* **2022**, *71*, 1–10. [CrossRef]
10. Jokovic, V.; Morrison, R.; Alexander, D. Can the performance of semi-inverted hydrocyclones be similar to fine screening? *Miner. Eng.* **2020**, *146*, 106147. [CrossRef]
11. Ji, L.; Paul, P.; Shanbhag, B.K.; Dixon, I.; Kuang, S.; He, L. Emerging application of hydrocyclone in biotechnology and food processing. *Sep. Purif. Technol.* **2023**, *309*, 122992. [CrossRef]
12. Gorobets, A.V.; Tarabara, V.V. Separation performance of desanding and deoiling hydrocyclones treating three-phase feeds: Effect of oil-particle aggregates. *Sep. Purif. Technol.* **2020**, *237*, 116466. [CrossRef]
13. Yang, X.; Liu, P.; Zhang, Y.; Jiang, L. Numerical simulation and experimental study on a cone-plate clarifier. *Adv. Mech. Eng.* **2019**, *11*, 1687814019826788. [CrossRef]
14. Zhao, Q.; Hou, D.; Cui, B.; Wei, D.; Song, T.; Feng, Y. Development of an integrated multichannel inlet for improved particle classification in hydrocyclones. *Adv. Powder Technol.* **2021**, *32*, 4546–4561. [CrossRef]
15. Xiong, Z.; Xu, J.; Liu, C. Interaction effects of inlet velocity and apex diameter on the separation performance of two-stage cone hydrocyclones. *Powder Technol.* **2023**, *422*, 118446. [CrossRef]
16. Liu, L.; Zhao, L.; Wang, Y.; Zhang, S.; Song, M.; Huang, X.; Lu, Z. Research on the Enhancement of the Separation Efficiency for Discrete Phases Based on Mini Hydrocyclone. *J. Mar. Sci. Eng.* **2022**, *10*, 1606. [CrossRef]
17. Goncalves, S.M.; Ullmann, G.; Morimoto, M.G.; de Souza Barrozo, M.A.; Martins Vieira, L.G. Effect of rheology and solids concentration on hydrocyclones performance: A study involving the design variables of an optimized hydrocyclone. *J. Pet. Sci. Eng.* **2022**, *210*, 110019. [CrossRef]
18. Hou, D.; Zhao, Q.; Cui, B.; Wei, D.; Song, Z.; Feng, Y. Geometrical configuration of hydrocyclone for improving the separation performance. *Adv. Powder Technol.* **2022**, *33*, 103419. [CrossRef]
19. Padhi, M.; Mangadoddy, N.; Mainza, A.N.; Anand, M. Study on the particle interaction in a hydrocyclone classifier with multi-component feed blend at a high solids content. *Powder Technol.* **2021**, *393*, 380–396. [CrossRef]
20. de Faria, E.V.; Salvador, F.F.; Ascendino, G.G.; Barrozo, M.A.d.S.; Vieira, L.G.M. Geometric optimization of filtering conical hydrocyclones for thickening purposes with low energy consumption. *Chem. Eng. Res. Des.* **2022**, *178*, 168–178. [CrossRef]
21. Dianyu, E.; Fan, H.; Su, Z.; Xu, G.; Zou, R.; Yu, A.; Kuang, S. Numerical study of the multiphase flows and separation performance of hydrocyclone with tapered cross-section inlet. *Powder Technol.* **2023**, *416*, 118208. [CrossRef]
22. Hou, D.; Cui, B.; Zhao, Q.; Wei, D.; Song, Z.; Feng, Y. Research on the structure of the cylindrical hydrocyclone spigot to mitigate the misplacement of particles. *Powder Technol.* **2021**, *387*, 61–71. [CrossRef]

23. Jiang, J.; Ying, R.; Feng, J.; Wang, W. Computational and Experimental Study of the Effect of Operating Parameters on Classification Performance of Compound Hydrocyclone. *Math. Probl. Eng.* **2018**, *2018*, 7596490. [CrossRef]
24. Jiang, L.; Liu, P.; Zhang, Y.; Yang, X.; Wang, H. The Effect of Inlet Velocity on the Separation Performance of a Two-Stage Hydrocyclone. *Minerals* **2019**, *9*, 209. [CrossRef]
25. Fang, X.; Wang, G.; Zhong, L.; Wang, D.; Qiu, S.; Li, X. Analysis of weakly cemented gas hydrate bearing sediments particles movement and de-cementation behavior in hydrocyclone separator. *Powder Technol.* **2023**, *424*, 118174. [CrossRef]
26. Hou, D.; Zhao, Q.; Liu, P.; Jiang, L.; Cui, B.; Wei, D. Effects of bottom profile on the circulation and classification of particles in cylindrical hydrocyclones. *Adv. Powder Technol.* **2023**, *34*, 104050. [CrossRef]
27. Davailles, A.; Climent, E.; Bourgeois, F.; Majumder, A.K. Analysis of swirling flow in hydrocyclones operating under dense regime. *Miner. Eng.* **2012**, *31*, 32–41. [CrossRef]
28. Wang, Z.B.; Yi, M.; Jin, Y.H. Simulation and experiment of flow field in axial-flow hydrocyclone. *Chem. Eng. Res. Des.* **2011**, *89*, 603–610. [CrossRef]
29. Jiang, L.; Liu, P.; Yang, X.; Zhang, Y.; Wang, H.; Xu, C. Numerical analysis of flow field and separation characteristics in hydrocyclones with adjustable apex. *Powder Technol.* **2019**, *356*, 941–956. [CrossRef]
30. Jing, J.; Zhang, S.; Qin, M.; Luo, J.; Shan, Y.; Cheng, Y.; Tan, J. Numerical simulation study of offshore heavy oil desanding by hydrocyclones. *Sep. Purif. Technol.* **2021**, *258*, 118051. [CrossRef]
31. Kyriakidis, Y.N.; Silva, D.O.; Souza Barrozo, M.A.; Martins Vieira, L.G. Effect of variables related to the separation performance of a hydrocyclone with unprecedented geometric relationships. *Powder Technol.* **2018**, *338*, 645–653. [CrossRef]
32. Liow, J.-L.; Oakman, O.A. Performance of mini-axial hydrocyclones. *Miner. Eng.* **2018**, *122*, 67–78. [CrossRef]
33. Jiang, L.; Liu, P.; Zhang, Y.; Yang, X.; Wang, H.; Gui, X. Design boundary layer structure for improving the particle separation performance of a hydrocyclone. *Powder Technol.* **2019**, *350*, 1–14. [CrossRef]

molecules

MDPI

Article

Fabrication of Superhydrophobic Porous Brass by Chemical Dealloying for Efficient Emulsion Separation

Yanbiao Zhou, Qingqing Ye, Yongjun Han *, Guoxu He and Changdong Chen *

School of Chemistry and Environmental Engineering, Pingdingshan University, Pingdingshan 467000, China; 2738@pdsu.edu.cn (G.H.)
* Correspondence: hyj@pdsu.edu.cn (Y.H.); 2749@pdsu.edu.cn (C.C.)

Abstract: By taking advantage of typical dealloying and subsequent aging methods, a novel homogeneous porous brass with a micro/nano hierarchical structure was prepared without any chemical modification. The treatment of commercial brass with hot concentrated HCl solution caused preferential etching of Zn from $Cu_{62}Zn_{38}$ alloy foil, leaving a microporous skeleton with an average tortuous channel size of 1.6 μm for liquid transfer. After storage in the atmosphere for 7 days, the wettability of the dealloyed brass changed from superhydrophilic to superhydrophobic with a contact angle > 156° and sliding angle < 7°. The aging treatment enhanced the hydrophobicity of the brass by the formation of Cu_2O on the surface. By virtue of the opposite wettability to water and oil, the aged brass separated surfactant-stabilized water-in-oil emulsions with separation efficiency of over 99.4% and permeate flux of about 851 $L \cdot m^{-2} \cdot h^{-1}$ even after recycling for 60 times. After 10 times of tape peeling or sandpaper abrasion, the aged brass maintained its superhydrophobicity, indicating its excellent mechanical stability. Moreover, the aged brass still retained its superhydrophobicity after exposure to high temperatures or corrosive solutions, displaying high resistance to extreme environments. The reason may be that the bicontinuous porous structure throughout the whole foil endows stable mechanical properties to tolerate extreme environments. This method should have a promising future in expanding the applications of alloys.

Keywords: dealloying; porous brass; Cu_2O; superhydrophobic; stability

1. Introduction

Confronted with the increasing oily wastewater produced from daily life and industrial processes, the removal and recovery of oil stains from water have become a serious global problem [1,2]. Traditional oil/water separation techniques typically suffer from the disadvantages of low separation efficiency, high energy consumption, and sophisticated operation processes. With the development of interface theory, superwetting technology is considered an effective and inexpensive approach for the separation of oily wastewater. The proper porous materials with opposite superwettability toward water and oil, such as superhydrophobic/superoleophilic, superoleophilic/underoil superhydrophobic, and superhydrophilic/underwater superoleophobic properties, have been manufactured and utilized for effective oil/water separation [3]. However, constructing highly uniform and compact porous structures with opposite wettability to oil and water is still an enormous challenge.

The design and development of novel porous metal materials with superwettability have drawn intensive attention to the separation of oily sewage. "Coating" and "etching" are two effective strategies to fabricate porous membranes with opposite wettability. By "coating", a hierarchically rough film covers or grows out of porous hosts to obtain superwettable membranes [4]. This strategy mainly includes physical coating [5], surface redox [6], and chemical modification [7]. These coated membranes can efficiently purify oily wastewater. However, physical surface coating by polymer is unstable for long-term operation, and surface grafting is too complex for practical application, especially for separating water-in-oil emulsions [8]. The coating of inorganics usually has no robust adhesion to the host and can peel off when subjected to heat or external forces, which is mainly due to the relatively weak interfacial bonding between the inorganics and the host [9]. Therefore, more efforts are required to develop facile methods for preparing membranes with robust structural stability.

Conversely, the "etching" strategy has the potential to avoid the above problems. By "etching", etching specific components and areas can yield stable micro/nanostructures to some extent [4]. For example, the laser irradiation technique is utilized to etch specific areas for fabricating a microscale vertical pore structure [10]. Chemical etching [11], electrochemical etching [12], and plasma etching [13] methods usually employ corrosive media or plasma to unselectively etch the surface of materials, resulting in the formation of hierarchical rough structures on the surface. As far as we know, only ununiform throughout microporous aluminum sheets with intrinsic superwettability have been fabricated via monolithic chemical etching and electrochemical etching [14,15].

Dealloying, as one of the etching methods, is the process of selective dissolution of more active components from homogeneous alloy, which has become one of the robust and generic approaches for constructing homogeneous porous structures with high surface area [16]. The dealloying method has been successfully applied to produce various porous materials in alloys, such as CuZn [17–20], CuAl [21], CuFe [22], AuSi [23], LiSn [24], etc. These uniform bicontinuous nanopore structures exhibit unusual properties such as tunable porous size, high permeability, and high structural stability, which render them separation materials with great potential for purifying oily wastewater. For example, Jie et al. demonstrated a combining dealloying and heating approach to achieve a superhydrophobic surface on brass [25]. Li et al. prepared a superhydrophobic copper substrate through a dealloying and subsequent coating method [26]. Although the dealloying technique without surface coating has been utilized to acquire the superhydrophobicity on alloy surfaces [26,27], the development of the superhydrophobic membrane only by dealloying for oil/water separation has not been reported previously.

Herein, a facile chemical dealloying and subsequent aging process was first developed to produce a highly uniform and compact microporous brass for water-in-oil emulsion separation (Scheme 1). The dissolution of zinc from a commercial bimetallic $Cu_{62}Zn_{38}$ alloy brass foil by HCl solution resulted in a bicontinuous structure of well-interconnected pores with a diameter of about 1.6 μm. The bicontinuous Cu framework can act as a channel for liquid transport. After aging for 7 days, the brass changed from superhydrophilic to superhydrophobic. Such a superhydrophobic brass could separate surfactant-stabilized water-in-oil emulsions with a high efficiency of over 99.4% and flux of about 851 $L{\cdot}m^{-2}{\cdot}h^{-1}$. Even after 60 separation cycles, both separation efficiency and flux remained stable, demonstrating the excellent reusability of the porous brass. Moreover, the as-prepared porous brass exhibited exceptional mechanical, thermal, and chemical stability. This work may provide a new way to construct controllable structures for oil/water separation materials.

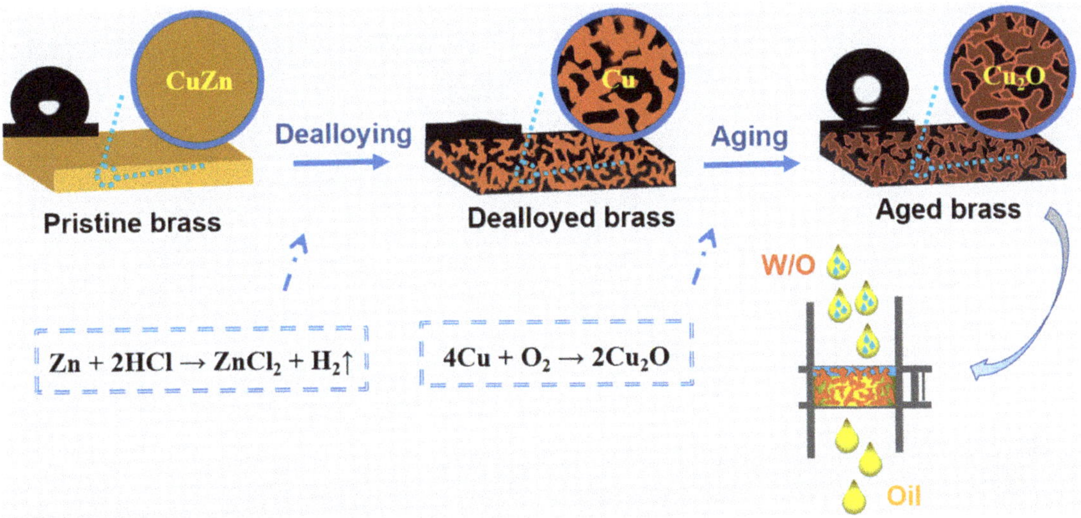

Scheme 1. Schematic diagram of dealloying and aging of CuZn alloy foil to prepare a superhydrophobic porous brass for water-in-oil emulsion separation.

2. Results and Discussion

2.1. Construction of Bicontinuous Porous Structure by Dealloying

The open bicontinuous porous structure was developed by selectively monolithic etching of brass foil ($Cu_{62}Zn_{38}$). Typically, chemical dealloying of the commercially available CuZn alloy foil was conducted in a hot concentrated HCl solution. During the dealloying process, the surface color of the CuZn alloy gradually changed from golden–yellow (Figure 1a) to reddish–brown, in accordance with Cu color (Figure 1d). The high concentration and temperature of HCl solution help to form deep and large pores [28]. In the macroscopic view, the Zn phase had been removed from the CuZn alloy, and the indissolvable Cu phase was preserved.

$$Zn + 2HCl \rightarrow ZnCl_2 + H_2\uparrow \quad (1)$$

Upon taking out from the hydrochloric acid solution, the dealloyed brass gradually turned dark brown as the aging time increased (Figure 1g). This clearly shows that the dealloying and aging processes can cause a dramatic change in the color of the brass.

Microstructural changes in samples in each processing stage were observed by scanning electron microscope (SEM). The pristine brass foil presented a relatively flat and solid surface (Figure 1b,c). During etching, hydrochloric acid etching proceeded from both surfaces of the brass foil to dissolve Zn. Meanwhile, the copper elements diffused, congregated, and formed interconnected skeletons. At 80 °C, corrosion perforation of the brass sheet occurred after 5 h. Because only a small percentage of pores perforated throughout the brass, the flux was too small to effectively transport liquids. After 8 h, the brass completely transformed to bicontinuous porous copper with irregular micro/nanopores. The diameter of these pores was smaller than 4 μm observed in the structure (Figure 1e,f). Such small micropores are difficult to obtain based on frequently used commercial brass or copper mesh [29,30]. After aging, no obvious difference was detected in the SEM images between the aged and dealloyed brass (Figure 1h,i). The widespread pores in size of the aged brass resulted in the formation of micro/nanoscale hierarchical structures, which played an essential role in the generation of superwettability.

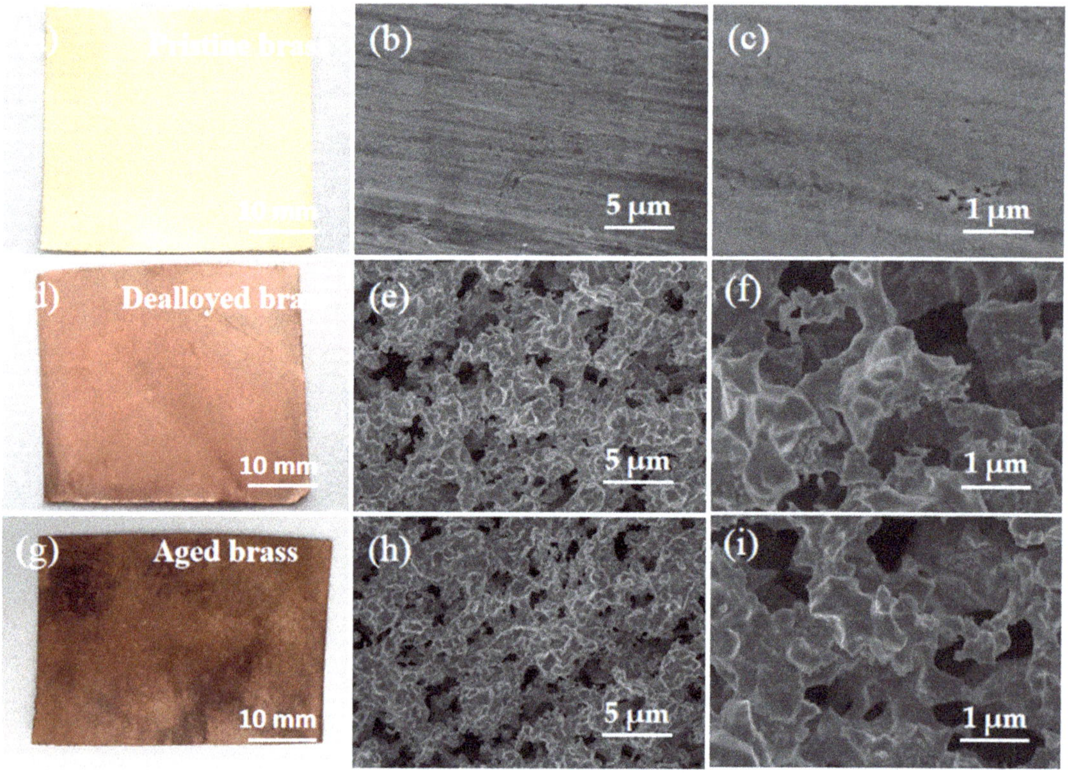

Figure 1. Top-view color photos and scanning electron microscope (SEM) images of samples in different processing stages: (**a**–**c**) pristine brass; (**d**–**f**) dealloyed brass; and (**g**–**i**) aged brass.

Dealloying treatment began on two sides of the CuZn sheet to remove Zn and progressed until corrosion perforation of the brass sheet occurred after 8 h. The cross-section morphology of the dealloyed and aged foil observed by SEM exhibited a highly interconnected porous structure and hierarchical skeleton throughout the whole foil (Figure 2e,f). These micro/nanopores can provide tortuous channels to transport liquids. It is worth noting that the total thickness of the foil after dealloying slightly decreased by about 7% from 100 to 93 μm due to a small amount of dissolution of copper (Figure 2a,d), whereas the thickness and morphology of the foil after aging did not change (Figure 2h,i). To probe the porosity of the dealloyed and aged samples [31], relative bulk density was measured by weighing the mass of the material with a predefined volume to provide insight into the porosity. Compared with that of pristine brass density (8.40 g/cm^3), the relative density values of the dealloyed and aged brass were around 53.23% and 52.28%, respectively. This means that the porosity of aged brass is slightly lower than that of dealloyed brass, which may be attributed to the partial oxidation of surface Cu.

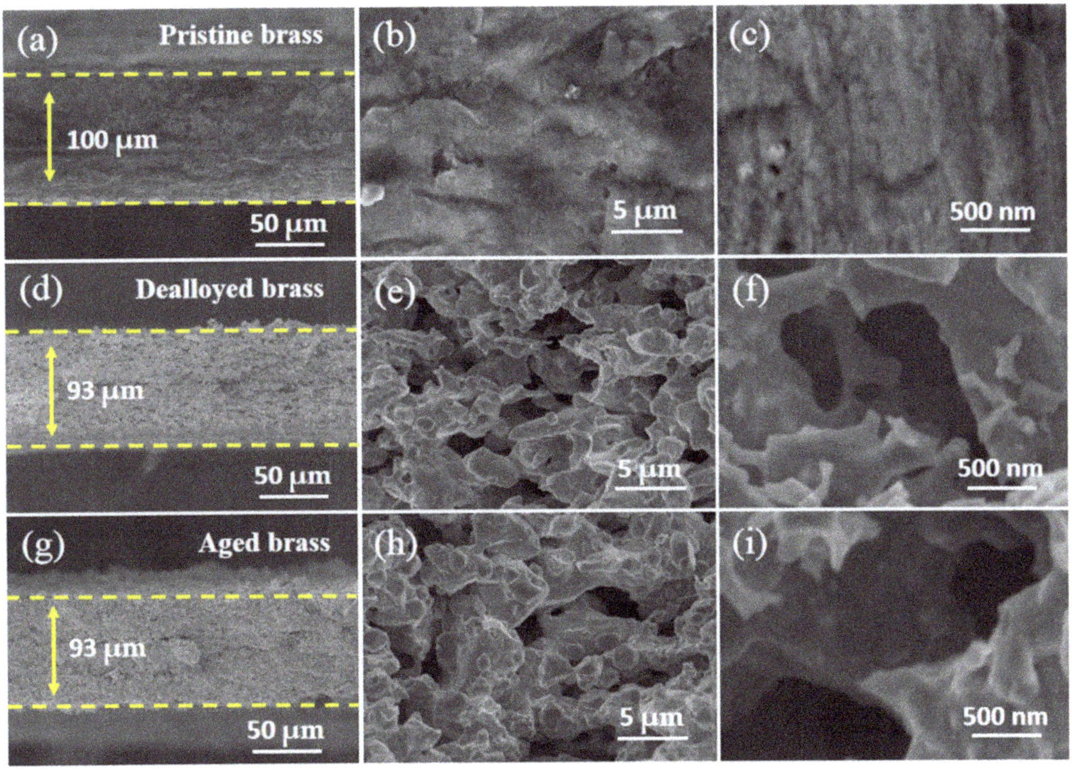

Figure 2. Side-view FESEM images of samples in different processing stages: (**a–c**) pristine brass; (**d–f**) dealloyed brass; and (**g–i**) aged brass. The distance between yellow dotted lines represents sample thicknesses.

To further investigate the macroporous distribution of the aged brass, the interconnected pore size distribution was measured by a mercury injection test. The result showed that the porosity of the aged brass was 59.6%, and the pore size ranged from 1.0 to 4.0 μm, with an average of 1.6 μm (Figure 3a), verifying that the aged brass had a compact structure and narrow pore size distribution. The porosity of the aged brass is larger than that obtained by the above gravimetric method, which could be because a small number of pores in the brass might collapse under the pressure required to inject mercury into the micropores [32]. To evaluate the mesoporous and microporous structures, N_2 adsorption/desorption behaviors were analyzed for the aged brass (Figure 3b). The aged brass presented a surface area of 2.1 m^2/g and a Brunauer–Emmett–Teller pore size distribution of 2–40 nm (inset in Figure 3b). According to the above SEM observations and pore size distribution, it can be concluded that the aged brass holds a hierarchical architecture with micro/nanoscale pores. All these pores were interconnected to result in unimpeded channels for liquid transport.

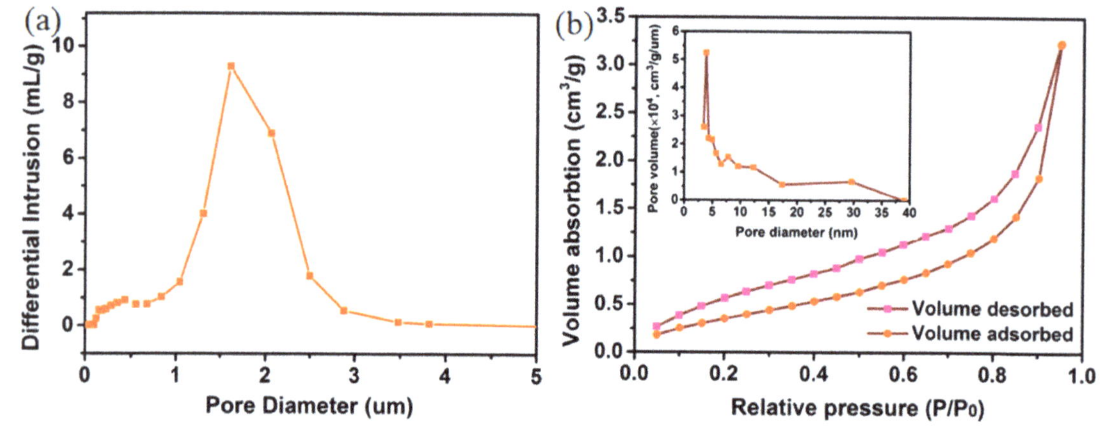

Figure 3. Pore size distribution of the aged brass: (**a**) macroscopic pore size distribution obtained from mercury-intrusion porosimetry measurement; (**b**) N_2 adsorption/desorption isotherm and corresponding nanopore size distribution.

2.2. Effect of Aging on Chemical Composition

The energy-dispersive X-ray spectrometer (EDS) analysis showed that the surface of the pristine brass contained nearly 28 wt% Zn and 59 wt% Cu (Figure 4a). The Zn and Cu elements were evenly distributed in the alloy (Figure S1a), which was very important to constructing a uniform and compact porous structure. After dealloying for 8 h, the zinc content dropped to 2%, while the relative content of Cu increased to 89% (Figure S1b). The element mapping images presented that the Zn in the pristine alloy had been almost completely removed, and porous Cu was retained. The oxygen content of the aged brass was much higher than that of the dealloyed brass but lower than that of the pristine brass, which was consistent with the increase in the relative density for the aged brass, indicating the possible formation of Cu_2O on the remained skeleton (Figure 4a). However, the Cu/O atomic ratio of the superhydrophobic surface was lower than that of Cu_2O. The reason is that the EDS measurement penetrates into a depth of a few micrometers into the subsurface and, thus, the detected % of Cu covers both the oxide layer and the copper metal.

The chemical constituents and chemical bonds of the pristine, dealloyed, and aged brass surface were examined by attenuated total reflection Fourier-transform infrared spectra (ATR-FTIR) (Figure 4b). The two series of adsorption peaks at about 3633 and 1602 cm^{-1} are attributed to the presence of adsorbed water [33,34]. This result indicates the existence of a hydration layer on the pristine and dealloyed surface. Upon dealloying, the new adsorption peak for Cu_2O at 628 cm^{-1} emerged, which implied that the dealloyed brass was coated with Cu_2O. After the aging treatment, the intensity of the peak at 628 cm^{-1} for Cu_2O was obviously enhanced, while the absorption bands at 3600 and 1602 cm^{-1} for absorbed water declined, suggesting that the hydrophobic Cu_2O reduced water absorption.

The crystalline phase of pristine, dealloyed, and aged brass was determined with powder X-ray diffractometer (XRD) patterns (Figure 4c). As expected, the peaks of the pristine brass at 42.3°, 49.2°, and 72.2° matched well with the diffraction peaks of $Cu_{62}Zn_{38}$ (JCPDS, 50-1333), which consisted of a solid solution and intermetallic compound composed of Zn and Cu elements, respectively. After dealloying, the characteristic peaks for $Cu_{62}Zn_{38}$ disappeared while the three diffraction peaks at around 43.4°, 50.5°, and 74.1° for the indices (111), (200), and (220) crystal planes of pure Cu phase (JCPDS, 04-0836) evidently intensified, suggesting the almost complete dissolution of Zn element from the pristine CuZn alloy and the remaining of almost pure porous Cu [35,36]. After aging for 7 days, diffraction peaks of the Cu phase remained sharp and strong, and two new peaks at 36.4° and 61.4° assigned to (111) and (220) planes of Cu_2O (JCPDS, 65-3288) appeared. This indicated that a thin passive Cu_2O layer formed on the porous Cu ligaments [37,38]. Raman spectra were measured to further determine the composition change in the samples at different process stages (Figure 4d). After dealloying, the characteristic peaks of the Cu_2O phase at 148, 219, and 632 cm^{-1} demonstrated the formation of Cu_2O on the dealloyed brass. After aging, the intensity of peak respecting Cu_2O increased significantly, demonstrating that more copper was oxidized over time to cuprous oxide.

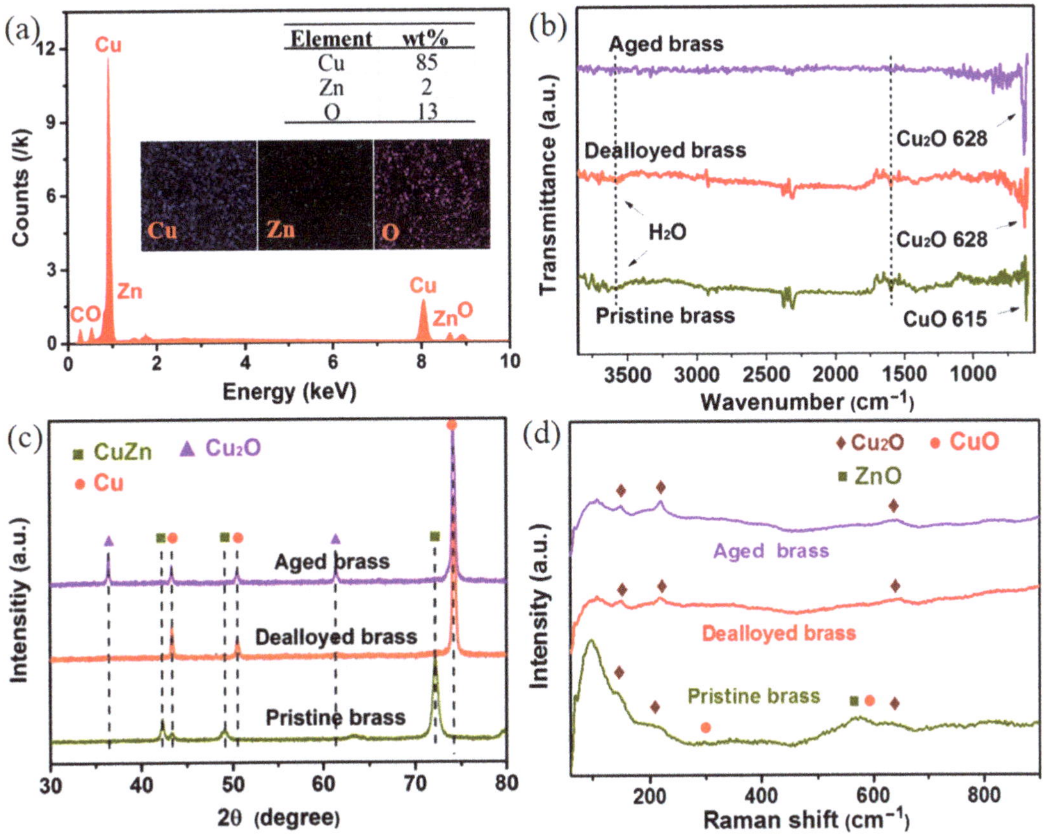

Figure 4. Surface chemistry compositions of samples in different processing stages: (**a**) energy-dispersive X-ray spectrometer (EDS) spectrum analysis and mapping images of the aged brass surface; (**b**) attenuated total reflection Fourier-transform infrared spectra (ATR-FTIR); (**c**) X-ray diffractometer (XRD) patterns; and (**d**) Raman spectra of the pristine, dealloyed and aged brass surface.

The X-ray photoelectron spectroscopy (XPS) analysis was utilized to further determine the chemical states of elements on the sample surface (Figure 5). All the binding energies in the XPS analysis were calibrated with C1s at 284.8 eV as the reference. Organic matter might have been adsorbed on the surface from the ambient air [39]. After the dealloying process, the peak intensities of Zn2p3 at 1021.5 eV and Zn2p1 at 1044.5 eV were dramatically diminished (Figure 5a). This confirmed that almost all Zn was selectively leached out from the alloy, and only a very low percentage of Zn was retained inside the matrix. Cu2p spectra were used to determine the valence state of the Cu element. Two sharp Cu2p peaks of the dealloyed brass at 932.4 and 952.4 eV corresponded to Cu2p3/2 and Cu2p1/2, binding the energy of Cu_2O or Cu (Figure 5b) [40,41]. The LMM Anger spectra could distinguish them with 568 and 570 eV for Cu and Cu_2O, respectively [42]. On the dealloyed and aged brass surfaces, the Cu LMM peak at 570 eV verified the presence of Cu_2O. Aging led to an increase in the intensity of the peak for Cu_2O with time of air exposure, indicating that more copper on the surface of the dealloyed brass gradually oxidized to Cu_2O (Figure S2). The binding energy of the Cu_2O phase generally has a slightly lower binding energy than the signal of metallic Cu [43,44]. The high-resolution spectra of Cu2p3/2 of the dealloyed and aged brass could be fitted into two peaks at 932.2, assigned to Cu_2O, and 932.6 eV, assigned to Cu [45,46]. After aging, the intensity of the Cu peak weakened, while the intensity of the Cu_2O peak strengthened, indicating that the exterior of the structure gradually oxidized over time (Figure 5c,d) [43]. In general, the combination of EDS, ATR-FTIR, XRD, Raman, and XPS analysis results has proved that more Cu on the dealloyed surface oxidized to Cu_2O over time.

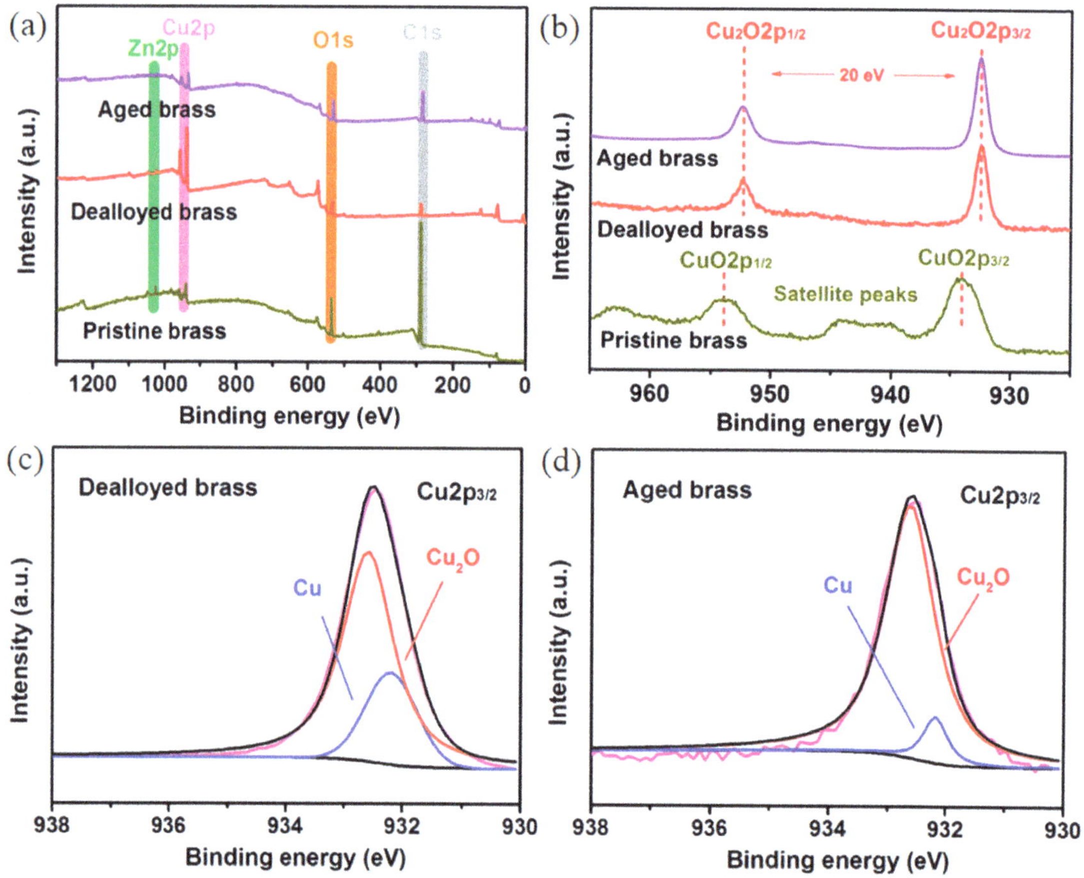

Figure 5. X-ray photoelectron spectroscopy (XPS) spectra of samples in different processing stages: (**a**) wide-spectrum survey; (**b**) high-resolution spectra of Cu2p region; (**c**) deconvolution of Cu2p3/2 of the dealloyed brass; and (**d**) deconvolution of Cu2p3/2 of the aged brass surface.

2.3. *Effect of Aging Time on Wettability Transition*

The freshly dealloyed brass surface exhibited a superhydrophilic behavior in the air with water CA close to 0°. However, when the dealloyed brass was exposed to ambient air, the water CA exhibited a sharp increase during the first 5 days, which then slowed to a gradual growth before finally reaching a steady state (Figure 6a). Prolonging the aging time to 7 days, the water CA increased to 156° and SA decreased to 7°, respectively, indicating that the aged surface possessed good superhydrophobicity. On the contrary, oil droplets could swiftly wet and penetrate through the aged brass (Figure S3). Furthermore, when the brass was continually stored for more than 6 months under ambient conditions, the surface could maintain opposite superwettability toward oil and water (Figure 6a). Compared with some previous reports, the aged brass showed better long-term storage stability in ambient atmosphere [47–49]. Unlike in the air, the freshly dealloyed brass exhibited superhydrophobicity in oil (kerosene) and did not change with increasing exposure time (Figure 6b). Such long-term storage stability is crucial for the practicality of superhydrophobic surfaces. Moreover, the dynamic repellency toward water was evaluated by squeezing, contacting, pressing, and lifting water to the aged brass surface. No obvious adhesion and droplet deformation were detected as the water droplets lifted from the aged brass surface, indicating the low adhesion of the aged surface (Figure 6c). The excellent water repellency had a synergistic effect on the hierarchical roughness induced by dealloying and the low surface energy introduced by the formation of the Cu_2O surface. This non-sticking characteristic of the superhydrophobic surface corresponds to the Cassie–Baxter state.

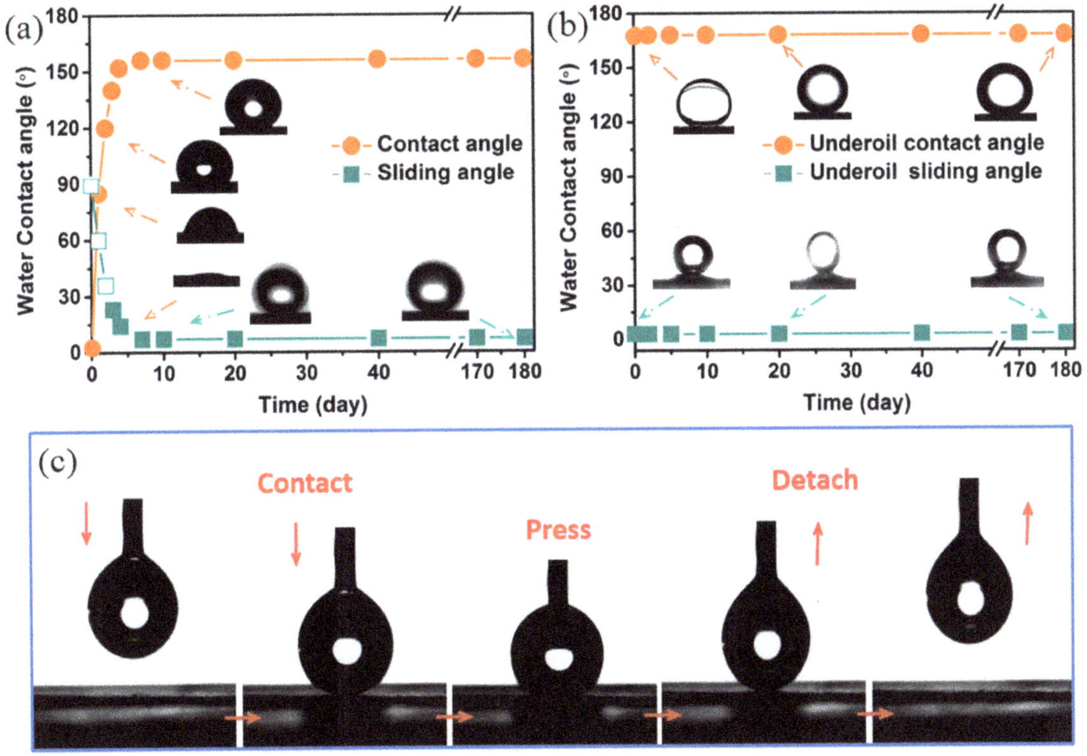

Figure 6. Effect of aging time on surface wettability transition: variation of water CA and SA (**a**) under air. The hollow square represents that an accurate water SA value for water cannot be obtained. (**b**) under oil with aging time; and (**c**) dynamic process of antiwater adhesion in air.

The wettability of the aged brass surface showed underoil superhydrophobicity after being immersed in different oils (Figure 7a). When immersing the aged brass in oil, oil could be trapped into the hierarchical structure to form oil shielding, thereby reducing the contact area between the solid and water. This was due to the higher adhesion work of oil than that of water. The aged brass presented excellent and stable underoil superhydrophobic properties with a water contact angle (CA) greater than 167° and sliding angle (SA) smaller than 3° in various oils, including chloroform, xylene, kerosene, and diesel, indicating the ultralow adhesion between the water droplet and the brass surface. Furthermore, the dynamic process of a stream of droplets bouncing on the aged brass surface was recorded by a digital camera (Figure 7b) and contact angle meter (Figure 7c), respectively. When a stream of water droplets in kerosene was squeezed from a syringe toward the aged brass, water droplets spontaneously bounced off over the horizontal surface with no water droplets left, indicating that the aged brass surface had an outstanding repellent property for water. After that, when the water was lifted from the aged brass surface, no droplet deformation was detected, further suggesting the low adhesion of the aged surface.

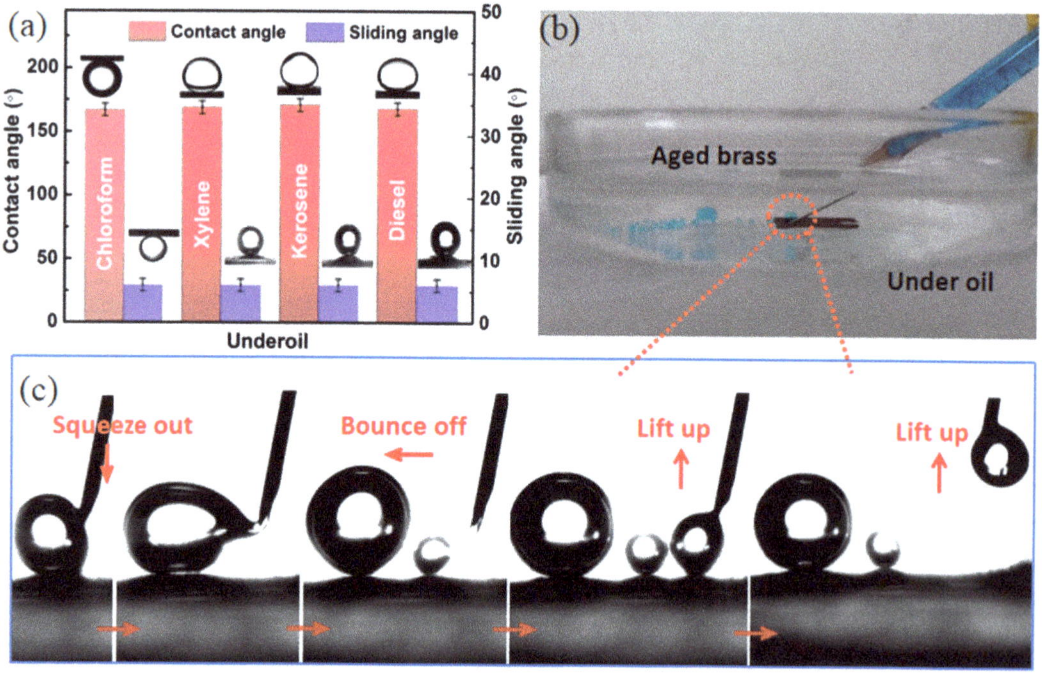

Figure 7. Wetting behavior of the aged brass in oils: (**a**) water CAs and SAs in different types of oils; (**b**,**c**) process of continuous water jet flow under oil.

2.4. Mechanism of Wetting Transition

Surface chemistry and surface topography are two critical factors affecting surface-wetting properties. HCl etching resulted in the dissolution of Zn, and more internal pure Cu atoms were exposed to the ambient air. Cu is a hydrophilic material in nature [50], and the hydrophilicity increases with roughness. According to Wenzel's theory, the introduction of hierarchical roughness can amplify the wettability from hydrophilicity to more hydrophilicity. So, after dealloying, the sample exhibited superhydrophilic behavior. However, when exposed to ambient conditions, the dealloyed porous brass immediately reacted with O_2 from the air to form a solid Cu_2O surface. The Cu_2O was generated through the interfacial reaction between Cu and O_2 by the proposed reaction [51]:

$$4Cu + O_2 \rightarrow 2Cu_2O \quad (2)$$

Aging led to the gradual conversion of the upmost layer of hydrophilic Cu to hydrophobic Cu_2O. Similarly, the hierarchical roughness amplified the hydrophobicity of Cu_2O into superhydrophobicity according to Wenzel's theory. Along with the reaction, the hydrophobic surface became more hydrophobic. After being exposed to ambient air for 7 days, the superhydrophilic brass surface reached a superhydrophobic state. This observation was only attributed to the change in surface chemical composition with time, as there was no change in surface morphology during aging (Figure 2).

Three mechanisms have been proposed by others, such as the decomposition of carbon dioxide into carbon with active magnetite [52], adsorption of hydrophobic organic matter from the atmosphere [53–55], and creation of hydrophobic groups [56] to explain the wettability transition from superhydrophilicity to hydrophobicity. In our case, there is no energy accountable for the creation of an active magnetite to catalyze the decomposition of carbon dioxide into carbon. Meanwhile, XPS analyses were performed on the dealloyed brass during the aging in the air to evaluate organic adsorption on the surface. When organic matter is adsorbed onto the surface, carbon atoms will be detected. The C1s spectra of the organic matters can be resolved into three peaks with binding energies of about 284.7 eV, 286.4 eV, and 288.6 eV, which are ascribed to the C-C (H), C-O-C, and O-C=O species, respectively [57]. These three species may come respectively from the alkyl, ether, and carboxyl/ester matter spontaneously adsorbed from the air. Among the C-C (H), C-O-C, and O-C=O species, C-C (H) is nonpolar and serves as a hydrophobic region, while the C-O-C and O-C=O species are polar and serve as a hydrophilic region on the surface [54]. It can be seen that the percentage content of the hydrophobic C-C (H) decreased from 76.4% for the dealloyed brass to 39.5% for the aged brass in the total C1s peak area, while the percentage content of the hydrophilic C-O-C and O-C=O species significantly increased from 23.4% to 60.5% (Figure S4). The result indicates that the percentage content of the hydrophilic species increases with aging time. This cannot explain why the aged surface exhibits higher hydrophobicity. Above all, the formation of a very thin passivation layer of hydrophobic Cu_2O is the reason for the transition from superhyrophilic to superhydrophobic.

To further support the mechanism behind wettability switching, dried brass and undried brass, after dealloying, were exposed to four types of the atmosphere, including ambient air, petroleum ether, dichloromethane, and n-hexane. For this purpose, eight freshly dealloyed samples were first rinsed thoroughly in distilled water, and then half of them were dried with cold air from a hair dryer. After that, these dried and undried samples were immediately stored in ambient air or vapors produced by three organic compounds in closed glass bottles for 3 days. Every 24 h, the samples were taken out from each container to measure their water CAs. Following the measurement, these samples were immediately returned back to their containers. The results showed that the water CAs increased as the storage time increased. After three days, the water CAs of both dried and undried samples exceeded 131° (Figure 8). Therein, the water CAs of the samples exposed to three organic compound vapors were no more than that of the corresponding sample in the air. The reason may be that the oxygen concentration in three organic vapors was lower than that in the air, resulting in the aged brass having a lower degree of oxidation in the vapors than that in the air. This result supports the transition mechanism that the variation in surface chemical composition substantially induces an increase in hydrophobicity.

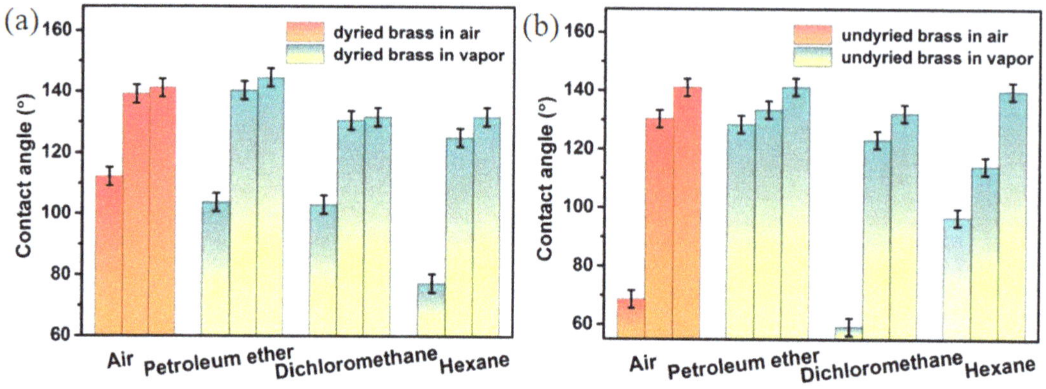

Figure 8. Water CAs of (**a**) dried freshly dealloyed brass and (**b**) undried freshly dealloyed brass after different days of exposure to air.

2.5. Water-in-Oil Emulsion Separation with High Efficiency

To achieve oil/water phase separation, the aged porous brass was mounted onto a separation device with clamps and sealed with silicone gaskets (Figure 9a). Taking the Sudan III-dyed water-in-chloroform emulsion as a typical example, the feed emulsion was milky (inset in Figure 9b), and numerous dispersed droplets of size below 5.0 μm were distributed in the feed (Figure 9b). The filtrate became totally transparent (inset in Figure 9c), and there were no water droplets observed in the microscopic image (Figure 9c), which meant that the dispersed droplets were successfully removed. The corresponding dynamic light scattering (DLS) curve revealed that the water droplet diameters of the feed emulsion were in a wide range of 0.2–3.0 μm (Figure 9d), while only several strong peaks of less than 6.0 nm were observed for the filtrate (Figure 9e), further illustrating the effective separation for water-in-chloroform emulsion.

When the water-in-chloroform emulsion was introduced to touch the aged brass surface, oil was captured quickly while water was repelled completely. An oil film bridging process would rapidly occur among porous skeletons so that pores in the brass were readily filled by oil [58]. So, the oil film could block the large water droplets released by the demulsification from directly passing through the channels in the aged brass (Video S1). Meanwhile, the released small water droplets entered the tortuous pore channels under gravity and coalesced with each other by physical collision, forming large droplets that were easy to prevent [59–61]. Consequently, tortuous channels increased the permeation resistance of the water droplets and permitted more water droplets to coalesce in the porous brass, thus greatly enhancing the probability of demulsification.

The service life of the aged brass was studied for practical applications. The aged brass exhibited excellent separation performance during filtration with a separation efficiency greater than 99.4% and a permeation flux of about 851 $L \cdot m^{-2} \cdot h^{-1}$ for water-in-chloroform emulsion with a height of 10 cm (Figure 10). Obviously, the separation efficiency of the aged brass was superior to that of some reported metal membranes [62,63]. The high efficiency can be attributed to its narrow range of tortuous channel size distribution (Figure 2h), which increases the efficiency of the separation process [64]. After each separation, the aged brass was reused by drying at room temperature. After recycling for 60 cycles, no change was found in the separation efficiency and flux for water-in-chloroform emulsion. Moreover, similar effective separations were also achieved for other surfactant-stabilized emulsions, such as water-in-kerosene and water-in-dichloroethane emulsions. This result clearly demonstrated that the aged brass possessed excellent separation performance for water-in-oil emulsions.

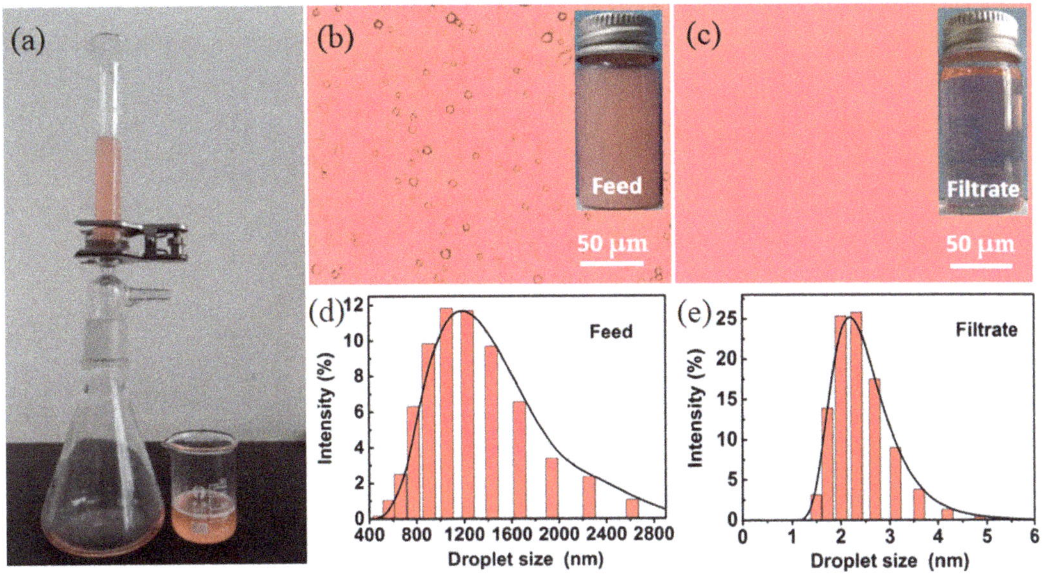

Figure 9. Separation performance of surfactant-stabilized water-in-chloroform emulsion: (**a**) photograph of the separation of water-in-chloroform emulsion; (**b**) microscope photo and optical image of feed; (**c**) the corresponding filtrate; (**d**) the measured droplet size distribution of the feed; and (**e**) the corresponding filtrate.

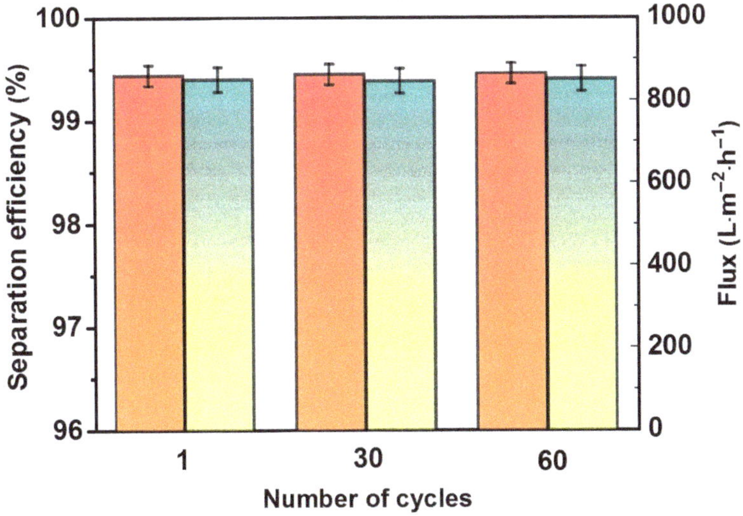

Figure 10. Changes in separation efficiency and permeation flux of water-in-chloroform emulsion over 60 cycles.

2.6. Mechanical, Thermal, and Chemical Stability

Superhydrophobic surfaces are required to be stable in a harsh environment, such as mechanical friction, high temperature, or corrosive media. The tape-peeling test was performed on the aged brass surface using Scotch tape (No. 30403, Deli Group, Ningbo, China). In this test, the tape was pressed onto the aged brass, following which a roller of 200 g weight was rolled forward and then backward once. Only a very weak attachment was observed on the Scotch® tape surface after 10 cycles (Figure 11a). However, the aged brass still had a water CA of 154°, and its underoil superhydrophobicity remained fairly stable (Figure 11b). Moreover, the morphological structure of the surface showed no obvious changes (inset in Figure 11b), indicating its good mechanical stability.

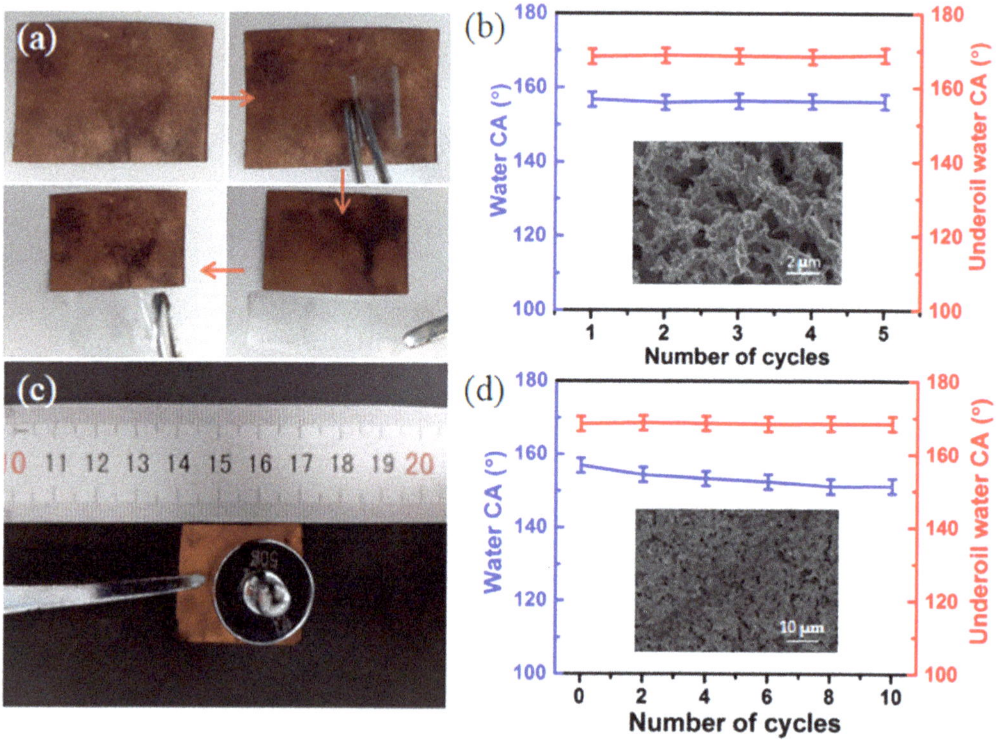

Figure 11. The mechanical stability of the aged brass in different working environments: (**a**) tape-peeling test; (**b**) variation of water CAs over tape-peeling cycles with the corresponding SEM image after 10 cycles in the inset; (**c**) sandpaper abrasion test; and (**d**) influence of abrasion cycles on the water CAs with the corresponding SEM image after 10 cycles in the inset.

Then, an abrasion test was carried out by placing the aged sample face-down on a 1000-grit SiC sandpaper under a 50 g weight and dragging for a distance of 10 cm on the sandpaper, which was defined as one abrasion cycle. After eight abrasion cycles, the water CA slightly reduced from 156° to 151°, which could be attributed to the fact that rough sandpaper wore out the hierarchical structure on the aged surface. However, the exposed structure on the surface had the same phase as the substrate, and the exposed skeleton was gradually oxidized to Cu_2O by O_2 from the air (inset in Figure 11d). Further increase in grinding cycles did not show an obvious downward trend for static water CA. About 7 days after friction, the water CA of the sample again reached its maximum value and then remained constant. Meanwhile, the abraded brass could maintain underoil superhydrophobicity over 167° after 10 abrasion cycles.

Furthermore, the thermal stability was evaluated by heating the aged brass in the air at 100 °C and 150 °C for 12 h. After heating at 100 °C, the aged brass still maintained the water CA above 156° (inset in Figure 12a), and the original structure of the aged brass remained unchanged (Figure 12a). However, when the temperature increased to 150 °C, the water CA decreased to about 0 °C. This result indicates that the aged brass retains thermal stability at least up to 100 °C. Finally, the chemical stability of the aged brass was evaluated by corrosive media. As shown in Figure 12b, the water drops of 3% NaCl, 1 mM HCl, 1 mM NaOH, and 90 °C solution retained a spherical shape with CAs greater than 154°. Correspondingly, when the aged brass was immersed in kerosene, its CAs of salt, acid, base, and hot water still remained above 167°, demonstrating moderate chemical stability and corrosion resistance. The chemical durability of the aged surface was further tested by immersing the aged brass in weak acid (1 mM HCl solution) and alkali (1 mM NaOH solution) solutions for 10 h, respectively. When the aged brass reached the acid/base solution surface, a dense air protection layer could be formed on the surface of the aged brass [65]. The corrosion medium was difficult to approach and diffuse into the superhydrophobic material, resulting in the sample's floating on the liquid surface steadily (Figure S5). So, the aged brass was forced into the corrosive solutions for 10 h, recording the water CAs after drying at room temperature. The water CA remained at about 153°, although there was a slight decrease observed after 10 h in both HCl and NaOH solutions (inset in Figure 12c,d). Compared with the aged brass, there is no apparent damage observed on the sample after the test (Figure 12c,d). This is attributed to the inherent chemical inertness of the Cu_2O protective layer, delivering moderate chemical stability for the aged brass. However, the HCl solution would have a reaction with hydrophobic Cu_2O as follows:

$$2HCl + Cu_2O \rightarrow CuCl_2 + H_2O \quad (3)$$

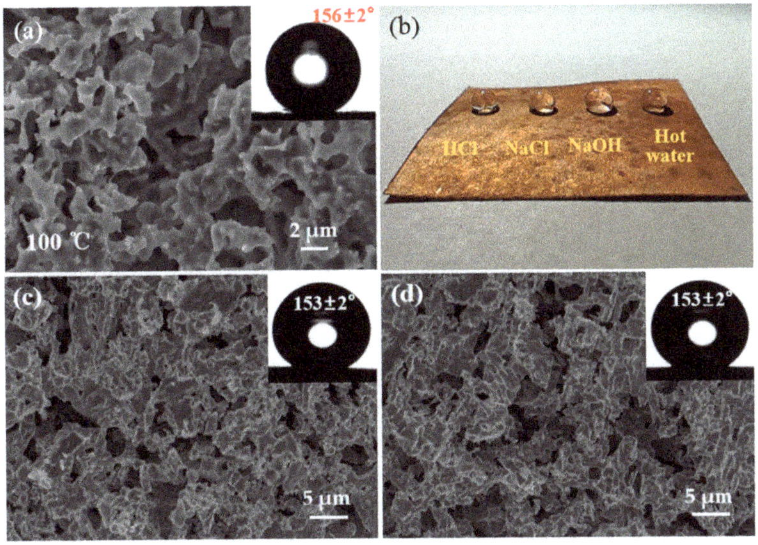

Figure 12. The thermal and chemical stability in different working environments: (**a**) the SEM images of the aged brass after heated at 100 °C with a corresponding water CA image in the inset; (**b**) the chemical stability of the aged brass against various corrosive drops; the SEM images of the aged brass after immersing into (**c**) 1 mM HCl solution; and (**d**) 1 mM NaOH solution for 10 h with corresponding water CA images in the inset.

After the Cu_2O film was dissolved in an acidic solution, the fresh copper atoms were exposed and immediately reacted with oxygen dissolved in water, forming Cu_2O on the aged brass surface. So, another reason why the surface of the aged brass could maintain superhydrophobicity might be that the dissolution rate of Cu_2O was slower than the formation rate due to the low HCl concentration. This observation is basically consistent with the results reported in the previous literature [66,67]. Above all, the aged brass shows the potential practical application for oily sewage separation under harsh conditions.

3. Materials and Methods

3.1. Materials and Chemicals

Commercial brass foil ($Cu_{62}Zn_{38}$) with a thickness of 100 μm was purchased from Jona Hardware Company (Taizhou, China). Hydrochloric acid (HCl, 37%), petroleum ether, dichloromethane, n-hexane, and kerosene were bought from Sinopharm Chemical Reagent Co., Ltd. (Shanghai, China). Diesel was purchased from a local fueling station. Ethanol, xylene, and chloroform were obtained from Zhengzhou Junyi Chemical Reagent Co., Ltd. Sudan III (Zhengzhou, China) as dyed reagent was supplied by Shanghai Zhongqin Chemical Reagent Co., Ltd. (Shanghai, China). Span 80 was purchased from Aladdin Chemistry Co., Ltd. (Shanghai, China). All of the above agents were of analytical grade and used without further purification.

3.2. Superhydrophobic Porous Brass Fabricated by Dealloying and Aging

Briefly, porous brass was fabricated by dealloying and subsequent aging of pristine brass foil. Firstly, a commercially purchased brass foil (Cu62Zn38) was cut into slices with desired dimensions (denoted as pristine brass). The obtained brass was ultrasonically cleaned, in turn, by ethanol and distilled water to remove surface contamination. Then, dealloying was conducted by immersing the cleaned brass foil into a 37% HCl solution for 8 h at 80 °C to remove the Zn element. After dealloying, the reddish–brown sample was washed with water and ethanol (denoted as dealloyed brass). After being exposed to ambient air for 7 days, the aged brass could be used directly as separating material (denoted as aged brass).

3.3. Aging in Ambient Air and Organic Vapors

To study the mechanism behind the wettability transition, dried and undried dealloyed samples were exposed to four types of atmosphere, including ambient air, petroleum ether, dichloromethane, and n-hexane. For this purpose, eight freshly dealloyed samples were first rinsed thoroughly in distilled water, and then half of them were dried with cold air from a hair dryer. After that, these dried and undried samples were immediately stored in ambient air or vapors produced by 60 mL of organic compounds in closed glass bottles for 3 days. Every 24 h, these samples were taken out from each container to measure their water CAs. Following the measurement, these samples were immediately returned back to their containers.

3.4. Preparation and Separation of Water-in-Oil Emulsions

Different surfactant-stabilized water-in-oil emulsions were prepared by emulsifying 1.0 g of water and 0.4 g of Span 80 in 100 mL of oils, including chloroform, kerosene, and 1,2-dichloroethane under intensive stirring for 8 h. For emulsion separation, the aged brass was tightly fixed in the separation device with clamps and sealed with silicone gaskets. The as-prepared emulsion was poured into the device to carry out separation only driven by gravity. The oil phase would easily permeate through the brass, whereas water droplets could be effectively blocked. The water rejection efficiency of the aged brass was quantitatively calculated according to the following equation:

$$R(\%) = (1 - \frac{C_f}{Cp}) \times 100\% \tag{4}$$

where C_f and C_p are the contents of water in the feed and filtrate, respectively. The permeate flux is determined by dividing the permeate volume by the product of the effective area and the penetration time under gravity with the height of feed emulsions of 10 cm according to the following equation:

$$F = \frac{V}{St} \quad (5)$$

where V represents the permeate volume (L); S represents the effective area of the brass (m^2), and t represents the penetration time (h), respectively.

3.5. *Instruments and Characterization*

All optical photographs were taken by a digital camera (Nikon D5000, Tokyo, Japan). The surface microstructure and element content of brass specimens were observed by a field emission scanning electron microscope (FESEM, Hitachi SU8000, Tokyo, Japan) equipped with an energy-dispersive X-ray spectrometer (EDS, AMETEK, Berwyn, PA, USA) attachment. The porosity and pore size distribution of the as-prepared brass were measured and analyzed by an automatic mercury porosimeter (Autopore V 9620, Micromeritics, Norcross, GA, USA). The specific surface area and average pore diameter distribution were analyzed by nitrogen adsorption/desorption isotherms using an automated adsorption apparatus (Autosorb-iQ-MP-C, Quantachrome, Boynton Beach, FL, USA). Attenuated total reflection Fourier-transform infrared spectra (ATR-FTIR) of the samples were recorded on a Brucker Tensor-37 spectrometer (Brucker Tensor, Billerica, MA, USA) equipped with a MIRacle ATR accessory (Pike Technologies, Fitchburg, WI, USA). The crystallinity and phase of the specimens were determined by X-ray diffractometer (XRD, Rikagu SmartLab system, Tokyo, Japan) using a Cu Kα radiation source. The chemical composition of the surface was identified by using a Laser Microscopic Confocal Raman Spectrometer (Senterra, Bruker, Billerica, MA, USA) at a laser wavelength of 532 nm. The surface chemical composition and oxidation state of the as-prepared samples were analyzed by X-ray photoelectron spectroscopy (XPS, Thermo Scientific ESCALAB 250Xi, Waltham, MA, USA), equipped with a monochromatic X-ray source (Al Ka, 1486.6 eV). The XPS spectra were fitted to a Shirley–Linear background with XPS peak 4.1 software. The average contact angle (CA) and sliding angle (SA) were calculated by measuring three different positions of the same sample with a 5 uL droplet by an Attension Theta system (PD-200, Biolin Scientific, Västra Frölunda, Sweden). The dispersion of the feed emulsions and the corresponding filtrates was obtained by taking optical images using an optical microscope (Leica DM4000, Wetzlar, Germany). The droplet diameter distribution in the emulsions and filtrates was calculated by dynamic light scattering (DLS) with Zetasizer Nano ZS (Malvern 3600, Malvern, UK). The water content in the filtrate was accurately determined by Karl Fischer Titrator (MKS-500).

4. Conclusions

A uniform microporous structure with an average pore size of 1.6 μm was produced by a simple chemical dealloying process of CuZn alloy foil. After dealloying, the sample exhibited superhydrophilic behavior but gradually became superhydrophobic over time, with the maximum water CA being achieved after 7 days. This is because the surface of the dealloyed brass was oxidized to Cu_2O by O_2 from the air to induce wettability conversion. Such a superhydrophobic surface showed excellent long-term storage durability for storing the sample in an ambient atmosphere. By combining superhydrophobicity with small pore channels, the novel porous brass could be employed to separate water-in-oil emulsions with high efficiency and repeatability. Thanks to the hierarchical structures on the surface having the same phase as the substrate, the superhydrophobic porous brass demonstrated good mechanical, thermal, and chemical stability. This simple fabrication technique may provide a new avenue for the design and fabrication of porous materials used in oil/water separation.

Supplementary Materials: The following supporting information can be downloaded at https://www.mdpi.com/article/10.3390/molecules28186509/s1, Figure S1: EDS spectra analysis and mapping images: (a) pristine brass and (b) dealloyed brass; Figure S2: CuLMM Auger transitions of pristine brass, dealloyed brass, and aged brass; Figure S3: The kerosene CA of the aged brass surface in air; Figure S4: High-resolution C1s spectra of (a) the dealloyed brass and (b) the aged brass; Figure S5: The aged brass floated on the liquid surface of 1 mM HCl solution: Video S1: Separation process of surfactant-stabilized water-in-chloroform emulsion.

Author Contributions: Conceptualization, G.H. and Y.H.; methodology, Y.Z. and C.C.; data curation, C.C.; writing—original draft, Y.Z.; writing—review and editing, Q.Y.; supervision, G.H. and Y.H. All authors have read and agreed to the published version of the manuscript.

Funding: This work was supported by the Science and Technology Development Program of Henan province (No. 222102320057, 232102321043).

Institutional Review Board Statement: This study does not involve humans and animals, and we have chosen to exclude this statement.

Informed Consent Statement: This study does not involve humans, and we have chosen to exclude this statement.

Data Availability Statement: Our research data will be available on demand.

Conflicts of Interest: The authors declare no conflict of interest.

Sample Availability: Not available.

References

1. Wang, H.; Hu, X.; Ke, Z.; Du, C.Z.; Zheng, L.; Wang, C.; Yuan, Z. Review: Porous metal filters and membranes for oil-water separation. *Nanoscale Res. Lett.* **2018**, *13*, 284. [CrossRef] [PubMed]
2. Lee, H.J.; Choi, W.S. 2D and 3D bulk materials for environmental remediation: Air filtration and oil/water separation. *Materials* **2020**, *13*, 5714. [CrossRef] [PubMed]
3. Lian, Z.; Xu, J.; Wang, Z.; Yu, Z.; Weng, Z.; Yu, H. Nanosecond laser-induced underwater superoleophobic and underoil superhydrophobic mesh for oil/water separation. *Langmuir* **2018**, *34*, 2981–2988. [CrossRef] [PubMed]
4. Zhu, S.; Deng, W.; Su, Y. Recent advances in preparation of metallic superhydrophobic surface by chemical etching and its applications. *Chin. J. Chem. Eng.* **2023**, *61*, 221–236. [CrossRef]
5. Yin, X.; Wu, J.; Zhao, H.; Zhou, L.; He, T.; Fan, Y.; He, Y. A microgel-structured cellulose nanofibril coating with robust antifouling performance for highly efficient oil/water and immiscible organic solvent separation. *Colloids Surf. A* **2022**, *647*, 128875. [CrossRef]
6. Xue, B.; Lin, H.; Chai, G.; Wang, C.; Yang, H.; Lu, H. Micro-arc oxidation enhances the wear resistance and corrosion resistance of oil-water separating mesh. *J. Mater. Sci.* **2022**, *57*, 18370–18384. [CrossRef]
7. Said, A.; Al Abdulgader, H.; Alsaeed, D.; Drmosh, Q.A.; Baroud, T.N.; Saleh, T.A. Hydrophobic tungsten oxide-based mesh modified with hexadecanoic branches for efficient oil/water separation. *J. Water Process Eng.* **2022**, *49*, 102931. [CrossRef]
8. Zhang, X.; Liu, C.; Yang, J.; Huang, X.J.; Xu, Z.K. Wettability switchable membranes for separating both oil-in-water and water-in-oil emulsions. *J. Membr. Sci.* **2021**, *624*, 118976. [CrossRef]
9. Luo, Z.Y.; Chen, K.X.; Wang, J.H.; Mo, D.C.; Lyu, S.S. Hierarchical nanoparticle-induced superhydrophilic and under-water superoleophobic Cu foam with ultrahigh water permeability for effective oil/water separation. *J. Mater. Chem. A* **2016**, *4*, 10566–10574. [CrossRef]
10. Bakhtiari, N.; Azizian, S.; Mohazzab, B.F.; Jaleh, B. One-step fabrication of brass filter with reversible wettability by nanosecond fiber laser ablation for highly efficient oil/water separation. *Sep. Purif. Technol.* **2021**, *259*, 118139. [CrossRef]
11. Tong, Q.; Fan, Z.; Wang, B.; Liu, Q.; Bo, Y.; Qian, L. Preparation and application of superhydrophobic copper mesh by chemical etching and in-situ growth. *Front. Chem.* **2021**, *9*, 737550. [CrossRef] [PubMed]
12. Chen, M.; Huang, L.; Liu, Z.; Liu, J.; Xing, Y.; Liu, X.; Wang, X. Development of superhydrophilic Al foil with micropore arrays via mask electrochemical machining and chemical immersion for efficient oil/water separation. *J. Disper. Sci. Technol.* **2020**, *41*, 1335–1345. [CrossRef]
13. Chen, F.; Song, J.; Liu, Z.; Liu, J.; Zheng, H.; Huang, S.; Liu, X. Atmospheric pressure plasma functionalized polymer mesh: An environmentally friendly and efficient tool for oil/water separation. *ACS Sustain. Chem. Eng.* **2016**, *4*, 6828–6837. [CrossRef]
14. Liu, J.; Aday, X.; Wang, X.; Li, Z.; Liu, J. On demand oil/water separation enabled by microporous ultra-thin aluminum foil with asymmetric wettability. *Colloids Surf. A* **2022**, *648*, 129334. [CrossRef]
15. Álvarez-Gil, L.; Ramirez, J.; Fernandez-Morales, P. Ultra-hydrophobic aluminum foam development for potential application in continuous water-oil separation processes. *Surf. Interfaces* **2021**, *26*, 101362. [CrossRef]
16. An, Y.; Tian, Y.; Wei, C.; Tao, Y.; Xi, B.; Xiong, S.; Qian, Y. Dealloying: An effective method for scalable fabrication of 0D, 1D, 2D, 3D materials and its application in energy storage. *Nano Today* **2021**, *37*, 101094. [CrossRef]

17. Ibrahim, S.; Dworzak, A.; Crespo, D.; Renner, F.U.; Dosche, C.; Oezaslan, M. Nanoporous copper ribbons prepared by chemical dealloying of a melt-spun ZnCu alloy. *J. Phys. Chem. C* **2021**, *126*, 212–226. [CrossRef]
18. Wu, X.; He, G.; Ding, Y. Dealloyed nanoporous materials for rechargeable lithium batteries. *Electrochem. Energy R.* **2020**, *3*, 541–580. [CrossRef]
19. Sajitha, M.; Abraham, B.; Nelliyil, R.B.; Yoosaf, K. Chemically etched nanoporous copper and galvanically displaced silver nanoflowers for SERS sensing. *ACS Appl. Nano Mater.* **2021**, *4*, 10038–10046. [CrossRef]
20. Zhao, H.; Lei, D.; He, Y.B.; Yuan, Y.; Yun, Q.; Ni, B.; Lu, J. Compact 3D copper with uniform porous structure derived by electrochemical dealloying as dendrite-free lithium metal anode current collector. *Adv. Energy Mater.* **2018**, *8*, 1800266. [CrossRef]
21. Liu, D.; Yang, Z.; Wang, P.; Li, F.; Wang, D.; He, D. Preparation of 3D nanoporous copper-supported cuprous oxide for high-performance lithium ion battery anodes. *Nanoscale* **2013**, *5*, 1917–1921. [CrossRef]
22. Zou, L.; Chen, F.; Wang, H.; Shen, Q.; Lavernia, E.J.; Zhang, L. Influence of porosity on mechanical behavior of porous Cu fabricated via de-alloying of Cu-Fe alloy. *Met. Mater. Int.* **2019**, *25*, 83–93. [CrossRef]
23. Reinhardt, H.; Bücker, K.; Yang, F.; Nürnberger, P.; Hampp, N.A. Highly dynamic alloying and dealloying in the model system gold-silicon (AuSi). *J. Phys. Chem. C* **2015**, *119*, 5462–5466. [CrossRef]
24. Han, S.Y.; Lewis, J.A.; Shetty, P.P.; Tippens, J.; Yeh, D.; Marchese, T.S.; McDowell, M.T. Porous metals from chemical dealloying for solid-state battery anodes. *Chem. Mater.* **2020**, *32*, 2461–2469. [CrossRef]
25. Jie, H.; Xu, Q.; Wei, L.; Min, Y. Etching and heating treatment combined approach for superhydrophobic surface on brass substrates and the consequent corrosion resistance. *Corros. Sci.* **2016**, *102*, 251–258. [CrossRef]
26. Li, H.; Sun, Y.; Wang, Z.; Wang, S. Constructing superhydrophobic surface on copper substrate with dealloying-forming and solution-immersion method. *Materials* **2022**, *15*, 4816. [CrossRef] [PubMed]
27. Smith, J.L.; Tran, N.; Song, T.; Liang, D.; Qian, M. Robust bulk micro-nano hierarchical copper structures possessing exceptional bactericidal efficacy. *Biomaterials* **2022**, *280*, 121271. [CrossRef] [PubMed]
28. Lin, B.; Kong, L.; Hodgson, P.D.; Dumée, L.F. Impact of the de-alloying kinetics and alloy microstructure on the final morphology of de-alloyed meso-porous metal films. *Nanomaterials* **2014**, *4*, 856–878. [CrossRef] [PubMed]
29. Luan, C.C.; Zhang, Y.P.; Cui, C.X.; Chen, D.L.; Chen, Y.; Chen, M.J. Controllable water penetration through a smart brass mesh modified by mercaptobenzoic acid and naphthalenethiol. *Coatings* **2022**, *12*, 1729. [CrossRef]
30. Kong, L.; Guo, Z. Fabrication of a durable brass mesh capable of rapid transformation between two modes of liquid transportation. *Appl. Surf. Sci.* **2023**, *621*, 156880. [CrossRef]
31. Yang, S.; Si, Y.; Fu, Q.; Hong, F.; Yu, J.; Al-Deyab, S.S.; Ding, B. Superwetting hierarchical porous silica nanofibrous membranes for oil/water microemulsion separation. *Nanoscale* **2014**, *6*, 12445–12449. [CrossRef] [PubMed]
32. Shi, Y.; Hu, Y.; Shen, J.; Guo, S. Optimized microporous structure of ePTFE membranes by controlling the particle size of PTFE fine powders for achieving high oil-water separation performances. *J. Membr. Sci.* **2021**, *629*, 119294. [CrossRef]
33. Chen, D.; Kang, Z.; Li, W.; Su, F. Simultaneously spray-assisted assembling reversible superwetting coatings for oil-water separation. *Chin. J. Chem. Eng.* **2021**, *32*, 144–150. [CrossRef]
34. Chattopadhyay, P.K.; Singha, N.R. MOF and derived materials as aerogels: Structure, property, and performance relations. *Coord. Chem. Rev.* **2021**, *446*, 214125. [CrossRef]
35. Lu, W.; Sun, Y.; Dai, H.; Ni, P.; Jiang, S.; Wang, Y.; Li, Z. Direct growth of pod-like Cu_2O nanowire arrays on copper foam: Highly sensitive and efficient nonenzymatic glucose and H_2O_2 biosensor. *Sens. Actuators B Chem.* **2016**, *231*, 860–866. [CrossRef]
36. Hamam, A.; Dehchar, C.; Maiza, M.; Chikouche, I.; Merabti, H. Facile synthesis and electrochemical study of CuO thin films for hydrogen peroxide reduction in alkaline medium. *Int. J. Electrochem. Sci.* **2020**, *15*, 3534–3542. [CrossRef]
37. Babouri, L.; Belmokre, K.; Kabir, A.; Abdelouas, A.; Khettabi, R.; El Mendili, Y. Microstructure and crystallographic properties of $Cu_{77}Zn_{21}$ alloy under the effect of heat treatment. *Mater. High Temp.* **2019**, *36*, 165–172. [CrossRef]
38. Azadi, H.; Aghdam, H.D.; Malekfar, R.; Bellah, S.M. Effects of energy and hydrogen peroxide concentration on structural and optical properties of CuO nanosheets prepared by pulsed laser ablation. *Results Phys.* **2019**, *15*, 102610. [CrossRef]
39. Yang, Z.; Liu, X.; Tian, Y. Insights into the wettability transition of nanosecond laser ablated surface under ambient air exposure. *J. Colloid Interf. Sci.* **2019**, *533*, 268–277. [CrossRef] [PubMed]
40. Fantauzzi, M.; Elsener, B.; Cocco, F.; Passiu, C.; Rossi, A. Model protective films on Cu-Zn alloys simulating the inner surfaces of historical brass wind instruments by EIS and XPS. *Front. Chem.* **2020**, *8*, 272. [CrossRef]
41. Li, L.; Zhong, K.; Dang, Y.; Li, J.; Ruan, M.; Fang, Z. Chemical dealloying pore structure control of porous copper current collector for dendrite-free lithium anode. *J. Porous Mat.* **2021**, *28*, 1813–1822. [CrossRef]
42. Pan, L.; Zou, J.J.; Zhang, T.; Wang, S.; Li, Z.; Wang, L.; Zhang, X. Cu_2O film via hydrothermal redox approach: Morphology and photocatalytic performance. *J. Phys. Chem. C* **2014**, *118*, 16335–16343. [CrossRef]
43. Kung, C.H.; Zahiri, B.; Sow, P.K.; Mérida, W. On-demand oil-water separation via low-voltage wettability switching of core-shell structures on copper substrates. *Appl. Surf. Sci.* **2018**, *444*, 15–27. [CrossRef]
44. Zahiri, B.; Sow, P.K.; Kung, C.H.; Mérida, W. Active control over the wettability from superhydrophobic to superhydrophilic by electrochemically altering the oxidation state in a low voltage range. *Adv. Mater. Interface* **2017**, *4*, 1700121. [CrossRef]
45. Biesinger, M.C. Advanced analysis of copper X-ray photoelectron spectra. *Surf. Interface Anal.* **2017**, *49*, 1325–1334. [CrossRef]
46. Sasmal, A.K.; Dutta, S.; Pal, T. A ternary Cu_2O-Cu-CuO nanocomposite: A catalyst with intriguing activity. *Dalton Trans.* **2016**, *45*, 3139–3150. [CrossRef]

47. Wang, Y.; Zhong, Q.; Wu, L.; Liu, P.; Guan, H.; Hu, Y.; Wang, W. Design and fabrication of superhydrophobic layered double hydroxide and oxides composite coating on brass mesh with excellent anticorrosion, delay icing and oil-water separation ability. *J. Coat. Technol. Res.* **2023**, *20*, 275–289. [CrossRef]
48. Niu, L.; Kang, Z. A facile approach for the fabrication of 3D flower-like Cu_2S nanostructures on brass mesh with temperature-induced wetting transition for efficient oil-water separation. *Appl. Surf. Sci.* **2017**, *422*, 456–468. [CrossRef]
49. Pi, P.; Hou, K.; Zhou, C.; Li, G.; Wen, X.; Xu, S.; Wang, S. Superhydrophobic $Cu_2S@Cu_2O$ film on copper surface fabricated by a facile chemical bath deposition method and its application in oil-water separation. *Appl. Surf. Sci.* **2017**, *396*, 566–573. [CrossRef]
50. Wakerley, D.; Lamaison, S.; Ozanam, F.; Menguy, N.; Mercier, D.; Marcus, P.; Mougel, V. Bio-inspired hydrophobicity promotes CO_2 reduction on a Cu surface. *Nat. Mater.* **2019**, *18*, 1222–1227. [CrossRef]
51. Ta, D.V.; Dunn, A.; Wasley, T.J.; Kay, R.W.; Stringer, J.; Smith, P.J.; Shephard, J.D. Nanosecond laser textured superhydrophobic metallic surfaces and their chemical sensing applications. *Appl. Surf. Sci.* **2015**, *357*, 248–254. [CrossRef]
52. Kietzig, A.M.; Hatzikiriakos, S.G.; Englezos, P. Patterned superhydrophobic metallic surfaces. *Langmuir* **2009**, *25*, 4821–4827. [CrossRef] [PubMed]
53. Wang, Q.; Liu, C.; Wang, H.; Yin, K.; Yu, Z.; Wang, T.; Liu, X. Laser-heat surface treatment of superwetting copper foam for efficient oil-water separation. *Nanomaterials* **2023**, *13*, 736. [CrossRef] [PubMed]
54. Li, X.; Jiang, Y.; Jiang, Z.; Li, Y.; Wen, C.; Lian, J. Reversible wettability transition between superhydrophilicity and superhydrophobicity through alternate heating-reheating cycle on laser-ablated brass surface. *Appl. Surf. Sci.* **2019**, *492*, 349–361. [CrossRef]
55. Long, J.; Zhong, M.; Fan, P.; Gong, D.; Zhang, H. Wettability conversion of ultrafast laser structured copper surface. *J. Laser Appl.* **2015**, *27*. [CrossRef]
56. Müller, D.W.; Holtsch, A.; Lößlein, S.; Pauly, C.; Spengler, C.; Grandthyll, S.; Müller, F. In-depth investigation of copper surface chemistry modification by ultrashort pulsed direct laser interference patterning. *Langmuir* **2020**, *36*, 13415–13425. [CrossRef] [PubMed]
57. Khorsand, S.; Raeissi, K.; Ashrafizadeh, F.; Arenas, M.A. Super-hydrophobic nickel–cobalt alloy coating with micro-nano flower-like structure. *Chem. Eng. J.* **2015**, *273*, 638–646. [CrossRef]
58. Cao, C.; Cheng, J. Fabrication of robust surfaces with special wettability on porous copper substrates for various oil/water separations. *Chem. Eng. J.* **2018**, *347*, 585–594. [CrossRef]
59. Yang, J.; Wang, H.; Tao, Z.; Liu, X.; Wang, Z.; Yue, R.; Cui, Z. 3D superhydrophobic sponge with a novel compression strategy for effective water-in-oil emulsion separation and its separation mechanism. *Chem. Eng. J.* **2019**, *359*, 149–158. [CrossRef]
60. Wang, R.; Zhu, X.; Zhu, L.; Li, H.; Xue, J.; Yu, S.; Xue, Q. Multifunctional superwetting positively charged foams for continuous oil/water emulsion separation and removal of hazardous pollutants from water. *Sep. Purif. Technol.* **2022**, *289*, 120683. [CrossRef]
61. Yang, Y.; Li, Y.; Cao, L.; Wang, Y.; Li, L.; Li, W. Electrospun PVDF-SiO_2 nanofibrous membranes with enhanced surface roughness for oil-water coalescence separation. *Sep. Purif. Technol.* **2021**, *269*, 118726. [CrossRef]
62. Singh, H.; Saxena, P.; Puri, Y.M. The manufacturing and applications of the porous metal membranes: A critical review. *CIRP J. Manuf. Sci. Technol.* **2021**, *33*, 339–368. [CrossRef]
63. Ali, N.; Bilal, M.; Khan, A.; Ali, F.; Khan, H.; Khan, H.A.; Iqbal, H.M. Understanding the hierarchical assemblies and oil/water separation applications of metal-organic frameworks. *J. Mol. Liq.* **2020**, *318*, 114273. [CrossRef]
64. Zhang, Y.; Chen, Y.; Hou, L.; Guo, F.; Liu, J.; Qiu, S.; Zhao, Y. Pine-branch-like TiO_2 nanofibrous membrane for high efficiency strong corrosive emulsion separation. *J. Mater. Chem. A* **2017**, *5*, 16134–16138. [CrossRef]
65. Yu, X.; Shi, X.; Xue, F.; Bai, W.; Li, Y.; Liu, Y.; Feng, L. SiO_2 nanoparticle-containing superhydrophobic materials with enhanced durability via facile and scalable spray method. *Colloids Surf. A* **2021**, *626*, 127014. [CrossRef]
66. Singh, V.; Sheng, Y.J.; Tsao, H.K. Facile fabrication of superhydrophobic copper mesh for oil/water separation and theoretical principle for separation design. *J. Taiwan Inst. Chem. E* **2018**, *87*, 150–157. [CrossRef]
67. Xiao, Y.J.; Zhang, Y.J.; Liang, Z.J.; Yue, T.M.; Guo, Z.N.; Liu, J.W.; Chen, X.L. Fabrication of a superhydrophobic mesh via magnetically aided electrode electric discharge machining. *Colloids Surf. A* **2021**, *612*, 125963. [CrossRef]

molecules

MDPI

Article

Earth Abundant Oxidation Catalysts for Removal of Contaminants of Emerging Concern from Wastewater: Homogeneous Catalytic Screening of Monomeric Complexes

Leslie Garcia [1], Makynna R. Koper [1], Somrita Mondal [1], Joshua T. Priddle [1], William A. Truong [1], Elisabeth M. A. Allbritton [1], Ashtyn G. McAdoo [1], Desiray J. Cannon-Smith [1], Neil L. Funwie [1], Tuyet Hoang [1], Inseo Kim [1], David J. Hubin [1], Jeanette A. Krause [2], Allen G. Oliver [3], Timothy J. Prior [4] and Timothy J. Hubin [1,*]

1 Department of Chemistry and Physics, Southwestern Oklahoma State University, Weatherford, OK 73096, USA
2 Department of Chemistry, University of Cincinnati, Cincinnati, OH 45220, USA
3 Department of Chemistry and Biochemistry, University of Notre Dame, Notre Dame, IN 46556, USA
4 Department of Chemistry, School of Natural Sciences, University of Hull, Kingston Upon Hull HU6 7RX, UK
* Correspondence: tim.hubin@swosu.edu

Abstract: Twenty novel Mn, Fe, and Cu complexes of ethylene cross-bridged tetraazamacrocycles with potentially copolymerizable allyl and benzyl pendant arms were synthesized and characterized. Multiple X-ray crystal structures demonstrate the cis-folded pseudo-octahedral geometry forced by the rigidifying ethylene cross-bridge and show that two cis coordination cites are available for interaction with substrate and oxidant. The Cu complexes were used to determine kinetic stability under harsh acidic and high-temperature conditions, which revealed that the cyclam-based ligands provide superior stabilization with half-lives of many minutes or even hours in 5 M HCl at 50–90 °C. Cyclic voltammetry studies of the Fe and Mn complexes reveal reversible redox processes indicating stabilization of Fe^{2+}/Fe^{3+} and $Mn^{2+}/Mn^{3+}/Mn^{4+}$ oxidation states, indicating the likelihood of catalytic oxidation for these complexes. Finally, dye-bleaching experiments with methylene blue, methyl orange, and rhodamine B demonstrate efficient catalytic decolorization and allow selection of the most successful monomeric catalysts for copolymerization to produce future heterogeneous water purification materials.

Keywords: dye bleaching; oxidation catalyst; cross-bridged tetraazamacrocycle

1. Introduction

Organic dyes have extensive applications in several industries such as food, pharmaceuticals, textiles, and cosmetics and thus are often the subject of academic scientific research focused on these same areas [1–4]. However, these dyes are potentially toxic in nature. Consequently, dyes in wastewater can cause severe damage to the aquatic environment as well as living organisms including humans. In order to overcome this environmental harm, dye degradation studies have been a topic of scientific interest over the past several decades [5–7]. Different methods have been implemented for dye degradation, e.g., oxidation of organic dyes using transition metal ions via the "Fenton" mechanism [8] and photocatalysis using semiconductor metal oxides and sulfides [9,10]. However, these conventional methods may suffer from several diverse drawbacks, depending on each specific system. Examples of issues hampering dye degradation systems include stability over a narrow pH range, low surface area, high cost, and limited applicability in visible light [8,11].

Transition metal complexes with cross-bridged macrocyclic ligands have been broadly explored for both research purposes and commercial purposes. Cross-bridged tetraazamacrocyclic transition metal complexes have been a prevalent candidate for substrate- and mechanism-flexible oxidation catalysis over the past two decades [12–22]. The ethylene cross-bridged tetraazamacrocycle transition metal complexes have been well-explored in biomedical imaging [23–31] and for anti-cancer therapy [32,33], owing to their extreme kinetic stability. However, studies pertaining to their dye-bleaching application have been limited.

The Busch catalyst, a Mn complex of the di-methylated ethylene cross-bridged cyclam **Me$_2$Bcyclam** (Scheme 1) [14], is a patented catalyst used for activation of peroxide or oxygen in water. In our previous study, we reported the functionalization of the original Busch catalyst with methyl and -OAc groups at the non-cross-bridged nitrogen atoms (Scheme 1) and their application in the degradation of three model dyes, namely, methyl orange, Rhodamine B, and methylene blue [34].

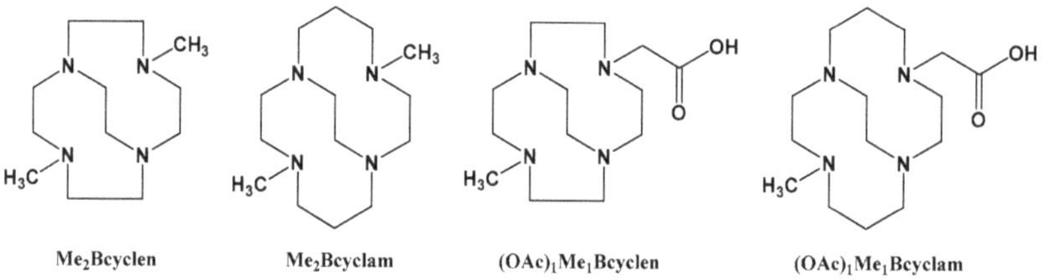

Scheme 1. Previously studied ethylene cross-bridged tetraazamacrocyclic ligands.

Building on our previous studies, our current study reports on the functionalization of the tetraazamacrocyclic ligand with different combinations of allyl, benzyl, and methyl groups at the non-cross-bridged nitrogen atoms (Scheme 2) and their respective Fe, Mn, and Cu complexes. The synthetic routes are facile and cost effective with significant synthetic yield. The ligands and their corresponding complexes were characterized using ^{1}H and ^{13}C NMR (Figures S1–S12), LCMS, elemental analysis, X-ray crystallography, and cyclic voltammetry. Further, we have performed acid decomplexation studies on the copper complexes in order to determine their kinetic stability (Figure S13).

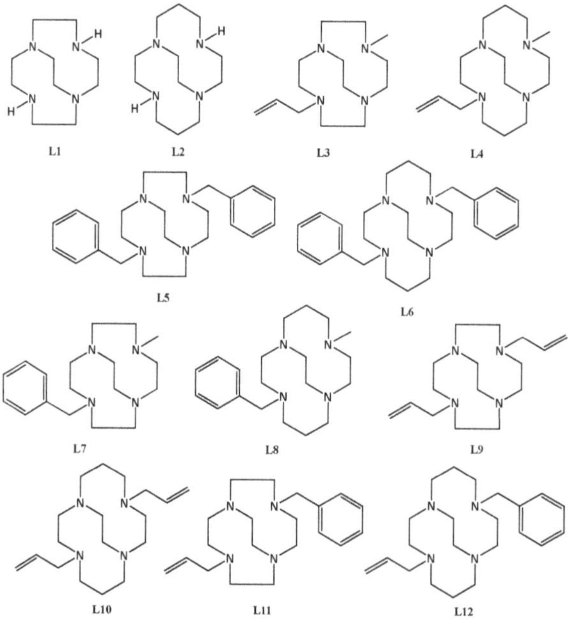

Scheme 2. L1–L12 structures used in this work.

In addition, we have expanded our efforts in exploring the ability of the metal complexes (typically Fe and Mn) to degrade the model dyes identified above. Most of the catalysts displayed significant activity in the degradation of all three dyes at 1 mM dye concentration. Thus, these metal complex catalysts can be practically applied for cheap and convenient bleaching of organic dyes, which can overcome the disadvantage of existing traditional photocatalysts.

Heterogeneous catalysis includes some liquid- or gas-phase reaction on a solid surface. Currently, heterogeneous catalysis is widely explored both in academic research and industry owing to its advantage of recyclability and easy separation of product [35]. In such cases, catalytic activity depends on many factors, such as particle size, surface area, stoichiometry, and topology of the designed catalysts. Different methods, e.g., non-covalent immobilization of catalysts within porous zeolites, clays, or embedding into silica or carbon; using a metal-organic framework (MOF) structure; and copolymerization through formation of covalent bonds, have been used to make catalysts heterogeneous [36–39]. The method of grafting or immobilization has drawbacks such as poor selectivity and catalytic activity. Therefore, direct heterogenization through covalent bond formation is a potential method to synthesize robust heterogeneous catalysts [36]. Rigidity of the building blocks associated with strong covalent bonds is highly desirable for stability of heterogeneous polymeric frameworks [40].

Attaching allyl and benzyl functional groups furthers the goal to explore the polymerization of our catalysts through covalent bond formation. These ligands were chosen as potential co-monomers that could be used to transform the homogeneous catalysts discussed in this paper into polymeric, heterogeneous catalysts. Thus, this study may lead to the successful design of a series of potential heterogenous catalysts for degradation of organic pollutants leading to wastewater purification in a simple, convenient, and cost-effective way. The polymerization, characterization, and efficacy of these heterogeneous materials are the subject of ongoing work and will be disclosed in later publications. Below, we comment on each planned polymerization reaction type to explain why benzyl and allyl pendant arms have been utilized for this future work.

The Friedel–Crafts reaction is a historically important method of activating and substituting aromatic rings through the action of a metal catalyst [41]. When a di-functional alkyl halide, such as dichloromethane, is used, a Friedel–Crafts polymerization can be used to link monomeric aromatics, such as macrocyclic ligands that contain a benzyl group [42]. Specifically, the benzyl pendant arms may be polymerized by Friedel–Crafts polymerization following the work of Yoon, where salphen coordination complexes were polymerized through their aromatic rings [36]. Application of this method of polymerization to our benzyl ligands is ongoing in our lab.

Alkene functional groups are the prototypical polymerizable entity, with the ability to polymerize as catalyzed by heat, acid, base, and other prompts [43]. Allyl pendant arms have a terminal alkene functional group for this purpose within our experimental design. The allyl-pendant-armed ligands and complexes can copolymerize following the work of Zuo for ethylene glycol di-methacrylate (EGDMA) polymerization [44]. This copolymerization has been quite successful in our hands and will be the subject of future manuscripts.

2. Results and Discussion

2.1. Synthesis

Ligands **L1** [45], **L2** [45], **L5** [45], **L6** [45], **L7** [33], and **L8** **[32]** were synthesized by literature procedures. Metal complexes Mn**L1**Cl_2, Fe**L1**Cl_2, Mn**L2**Cl_2, Fe**L2**Cl_2, Mn**L5**Cl_2, Fe**L5**Cl_2, Mn**L6**Cl_2, Fe**L6**Cl_2 [13], and Fe**L8**Cl_2 [46] were prepared as previously published.

The homogeneous ligands and catalysts were produced according to Scheme 3, following synthetic methods that are now common for synthesizing cross-bridged tetraazamacrocycles. Generally, the tetracyclic glyoxal-tetraazamacrocycle condensate (Scheme 3, structures **5** or **6**) can be mono- or di-alkylated by haloalkyl electrophiles. Mono-alkylation is often controlled by low numbers of equivalents of the alkylating agent and non-polar solvent (THF), which causes the mono-quaternary salt to precipitate and avoid further reaction. The mono-methylated compounds **1** and **2** were prepared according to the literature [47,48], as were mono-benzylated compounds **9** and **10** [48]. Mono-alkylated compounds were then reacted with excess allyl bromide in more polar solvents (CH_3CN), at longer reaction times, and in some cases at higher temperatures, to produce di-quaternary mono-allylated products **3**, **4**, **11**, and **12**. Reductive ring cleavage with $NaBH_4$ produced the cross-bridged ligands **L3**, **L4**, **L11**, and **L12**. Di-allyl ligands **L9** and **L10** were made from **5** and **6** using excess allyl bromide to produce **7** and **8**, which were also reduced to the corresponding cross-bridged tetraazamacrocycle ligand.

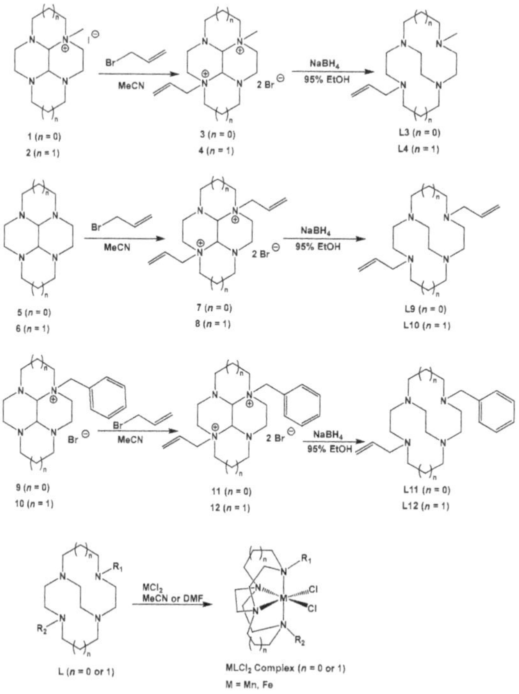

Scheme 3. Combined synthetic scheme for the synthesis of new ligands and all metal complexes.

Complexation of all ligands was carried out using anhydrous $CuCl_2$, $FeCl_2$, or $MnCl_2$ in an inert atmosphere glovebox in CH_3CN or DMF. Cross-bridged tetraazamacrocycles are proton sponges, so anhydrous conditions and aprotic solvents are generally required to successfully make their Fe and Mn complexes [49]. Yields were often in the 50% range, as the products in most cases precipitated directly from the reaction. Identity and purities were established by electrospray mass spectroscopy and elemental analysis. Copper complexes were synthesized in order to probe the kinetic stability imparted by these ligands. This was accomplished by the published method using HCl at high concentrations (1 M or 5 M) and high temperatures up to 90 °C to decomplex the ligand. Comparisons to other known ethylene cross-bridged tetraazamacrocycle copper complexes can then place a new ligand in the hierarchy of kinetic stabilization ability.

2.2. X-ray Crystal Structures

Crystal data for all structures are found in Table 1, while selected bond lengths and angles are compiled in Table 2.

Table 1. Crystal data for X-ray crystal structures.

Identification code	[Mn**L1**Cl_2]	[Mn**L3**Cl_2]	[Fe**L3**Cl_2]	**L8**-precursor	[Mn**L9**Cl_2]	[Mn**L11**Cl_2]	[Fe**L11**Cl_2]. 2MeCN
Empirical formula	$C_{10}H_{22}Cl_2MnN_4$	$C_{14}H_{28}Cl_2MnN_4$	$C_{14}H_{28}Cl_2FeN_4$	$C_{20}H_{34}I_2N_4O_2$	$C_{16}H_{30}Cl_2MnN_4$	$C_{20}H_{32}Cl_2MnN_4$	$C_{22}H_{35}Cl_2FeN_5$
Formula weight	324.15	378.24	379.15	616.31	404.28	454.33	496.30
Temperature/K	120	120	120	150	120	120	120
Crystal system	monoclinic	orthorhombic	orthorhombic	monoclinic	triclinic	monoclinic	orthorhombic
Space group	*Cc*	$P2_12_12_1$	$P2_12_12_1$	$P2_1/n$	*P*-1	$P2_1/c$	*Pnma*
a/Å	12.3081(13)	7.6390(4)	7.6198(5)	9.5389(10)	8.5540(11)	8.4238(4)	14.4316(7)
b/Å	10.1119(13)	12.0399(6)	11.9637(8)	16.9155(17)	13.2482(17)	13.5803(6)	8.9132(5)
c/Å	11.2029(12)	18.6843(10)	18.6306(14)	14.7768(14)	17.920(2)	18.6065(8)	19.5405(10)
α/°	90	90	90	90	107.492(2)	90	90
β/°	96.578(2)	90	90	91.479(4)	95.600(2)	98.507(2)	90
γ/°	90	90	90	90	97.994(2)	90	90
Volume/Å^3	1385.1(3)	1718.45(15)	1698.4(2)	2383.5(4)	1897.0(4)	2105.12(16)	2513.5(2)
Z	4	4	4	4	4	4	4
ρ_{calc}/gcm^{-3}	1.554	1.462	1.483	1.717	1.416	1.434	1.311
μ/mm^{-1}	1.324	1.078	1.202	3.350	0.982	0.894	0.831
F(000)	676.0	796.0	800.0	1216.0	852.0	956.0	1048.0
Crystal size/mm^3	0.355 × 0.247 × 0.142	0.213 × 0.15 × 0.078	0.287 × 0.134 × 0.127	0.05 × 0.02 × 0.01	0.253 × 0.154 × 0.05	0.253 × 0.169 × 0.078	0.216 × 0.169 × 0.088
Radiation/Å	MoKα, 0.71073	MoKα, 0.71073	MoKα, 0.71073	Synchrotron, 0.7749	MoKα, 0.71073	MoKα, 0.71073	MoKα, 0.71073
2Θ range for data collection/°	5.228 to 56.576	4.024 to 56.618	4.046 to 56.798	3.992 to 52.228	2.41 to 56.584	3.728 to 56.632	3.508 to 56.726
Index ranges	$-16 \le h \le 14$ $-13 \le k \le 13$ $-14 \le l \le 14$	$-10 \le h \le 10$ $-16 \le k \le 16$ $-24 \le l \le 24$	$-10 \le h \le 9$ $-15 \le k \le 15$ $-24 \le l \le 24$	$-10 \le h \le 10$ $-19 \le k \le 19$ $-16 \le l \le 16$	$-11 \le h \le 11$ $-17 \le k \le 17$ $-23 \le l \le 23$	$-11 \le h \le 11$ $-18 \le k \le 18$ $-24 \le l \le 24$	$-19 \le h \le 19$ $-11 \le k \le 11$ $-26 \le l \le 26$
Reflections collected	13204	31559	41891	37933	29491	49812	46814
Independent reflections	3268	4259	4239	3652	9415	5246	3344
R_{int}	0.0178	0.0520	0.0400	0.1904	0.0370	0.0340	0.0444
Data/restraints/ parameters	3268/2/167	4259/0/192	4239/0/192	3652/0/269	9415/41/437	5246/0/244	3344/235/201
Goodness-of-fit on F^2	1.042	1.180	1.064	1.037	1.005	1.089	1.031
Final R indexes [$I \ge 2\sigma$ (I)]	R_1 = 0.0131 wR_2 = 0.0319	R_1 = 0.0400 wR_2 = 0.0740	R_1 = 0.0385 wR_2 = 0.0920	R_1 = 0.0617 wR_2 = 0.1579	R_1 = 0.0307 wR_2 = 0.0591	R_1 = 0.0264 wR_2 = 0.0603	R_1 = 0.0284 wR_2 = 0.0652
Final R indexes [all data]	R_1 = 0.0133 wR_2 = 0.0320	R_1 = 0.0445 wR_2 = 0.0751	R_1 = 0.0401 wR_2 = 0.0931	R_1 = 0.0820 wR_2 = 0.1757	R_1 = 0.0499 wR_2 = 0.0645	R_1 = 0.0294 wR_2 = 0.0615	R_1 = 0.0372 wR_2 = 0.0693
Largest diff. peak/hole/e Å^{-3}	0.18/−0.12	0.37/−0.52	0.93/−0.57	1.03/−0.74	0.33/−0.51	0.39/−0.22	0.30/−0.23
Flack parameter	0.077(10)	0.32(3) *(racemic twin)	0.47(3) *(racemic twin)	none	none	none	none

[Mn**L1**Cl_2] crystallizes as well-formed colorless blocks in the non-centric space group *Cc* with a single molecule in the asymmetric unit. This space group was confirmed by the Flack parameter; the crystal structure determination was particularly good, with R = 1.33% and wR(F^2) = 3.20% for all observed data.

The four nitrogen atoms of the ligand all coordinate to the central Mn^{2+} ion (Figure 1). The coordination is completed by two chloride ions to give a rather distorted octahedral geometry at the metal. There is some variation in the Mn-N bond lengths; N1-Mn1 and N3-Mn1 are 2.2761(14) and 2.2935(13) Å, respectively, while those nitrogen atoms joined by the bridge form longer bonds: N2-Mn1 = 2.3255(14) Å and N4-Mn1 = 2.3401(13) Å. The N-Mn-N bond angles for adjacent nitrogen atoms lie in the range of 72.80(5) to 77.18(5)°, but the Mn ion lies somewhat outside the pocket of the ligand such that the angle of N1-Mn1-N3 is 142.60(5)°.

Table 2. Details of selected bond lengths and angles.

Structure	M-N1/Å M-N3/Å	M-N2/Å M-N4/Å	N1-M-N3/° N2-M-N4/°
[Mn(**L1**)Cl_2]	2.2761(14) 2.2935(13)	2.3255(14) 2.3041(13)	142.60(5) 73.49(5)
[Mn(**L3**)Cl_2]	2.392(3) 2.313(3)	2.302(3) 2.314(3)	142.48(10) 74.85(10)
[Fe(**L3**)Cl_2]	2.320(3) 2.257(3)	2.235(4) 2.233(4)	146.15(12) 77.20(11)
[Fe(**L5**)$(OAc)_2$] Ref: [50]	2.3024(10) 2.3024(9)	2.2091(10) 2.2228(11)	148.84(4) 77.91(4)
[Mn(**L9**)Cl_2]	**Mn1:** 2.3378(13) 2.3242(14) **Mn2:** 2.3510(14) 2.3786(15)	**Mn1:** 2.2706(14) 2.2981(13) **Mn2:** 2.3138(14) 2.2873(14)	**Mn1:** 144.69(5) 74.86(5) **Mn2:** 143.88(5) 74.15(5)
[Mn(**L11**)Cl_2]	2.4019(11) 2.3935(11)	2.3090(11) 2.2834(11)	143.56(4) 74.09(4)
[Fe(**L11**)Cl_2] · 2MeCN	2.2705(15) 2.262(2)	2.278(3) 2.169(3)	148.85(7) 77.38(12)

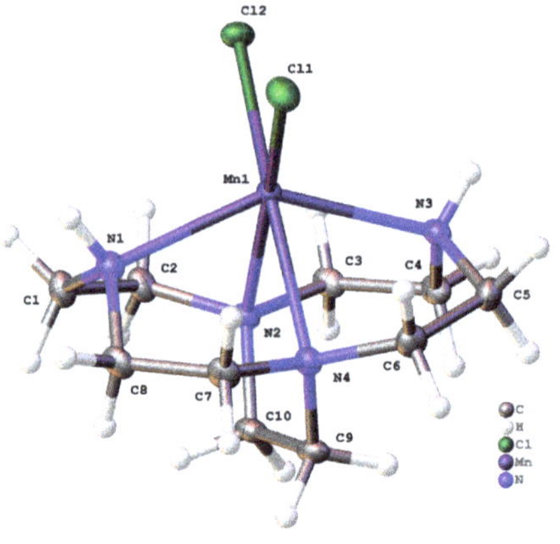

Figure 1. Asymmetric unit of [Mn**L1**Cl_2] with atoms drawn as 50% probability ellipsoids.

In the solid state, there are hydrogen bonds (N1-H1···Cl2 and N3-H3···Cl1) between adjacent molecules, which form hydrogen-bonded tapes that run parallel to the crystallographic *c* axis.

[Mn**L3**Cl_2] crystallizes in the chiral space group $P2_12_12_1$ with a single molecule in the asymmetric unit. This space group was identified from the systematic absences and the structure refined as a racemic twin with twin fractions 0.68 and 0.32(2). There was a very good fit to the observed data, with R = 4.45% and wR(F^2) = 7.51% for all data.

The coordination about the Mn^{2+} ion is very similar to that in [Mn**L1**Cl_2]; the central ion is coordinated by four nitrogen atoms from the ligand and two cis chlorides in a heavily distorted octahedral arrangement (Figure 2). However, there is a change in the relative lengths of the Mn-N bonds. In contrast to [Mn**L1**Cl_2], here, Mn1-N1 is 2.392(3) Å and

Mn1-N3 is 2.313(3) Å, but Mn1-N2 and Mn1-N4 are 2.302(3) and 2.314(3) Å, respectively. It appears that addition of substituents to the non-bridged nitrogen atoms extends the Mn-N bond lengths, presumably on steric grounds. It is notable that the methylene group (C11) and the methyl group (C14) are aligned so that H···Cl distances are approximately 3.1 Å, which may signify an interaction. The bond lengths of C11-C12 and C12-C13 in the allyl group are 1.507(5) and 1.316(5) Å, respectively, confirming the allyl group has indeed been added intact. C-H···Cl interactions are formed from the two bound chlorides to a pair of hydrogen atoms attached to one side of the ethylene bridge of an adjacent molecule related by a translation of ***a***. This interaction is repeated to generate tapes of molecules that run parallel to the crystallographic ***a*** axis. Between these tapes, there are further C-H···Cl interactions that assemble the molecules into a dense hydrogen-bonded array.

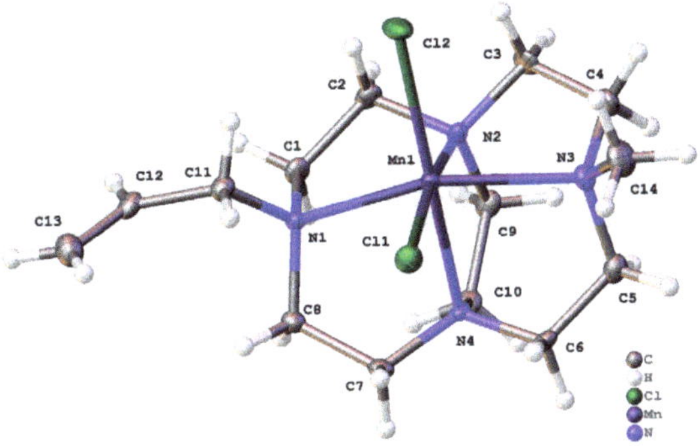

Figure 2. Asymmetric unit of [Mn**L3**Cl_2] with atoms drawn as 50% probability ellipsoids.

[Fe**L3**Cl_2] adopts the same structure as [Mn**L3**Cl_2] in space group $P2_12_12_1$. Each of the two structures is close to being centrosymmetric but is not. The systematic absences do not suggest any other space group, but some classes of data are systematically weak but not absent. It is possible to derive a centrosymmetric model in *Pnma* in which the whole molecule is disordered over a mirror plane. The fit is much poorer and there are a large number of systematic absence violations that preclude a centrosymmetric model. The final fit in $P2_12_12_1$ is good, with R = 4.01% and wR(F^2) = 9.30% for all data. The coordination about the metal is very similar to the Mn analogue, with Fe1-N1 and F1-N3 distances being 2.320(3) and 2.257(3) Å, respectively. The bonds lengths from nitrogen atoms at the ethylene bridge are 2.235(4) and 2.233(4) Å. The Fe1-N1 and Fe1-N3 bonds subtend an angle of 146.15(12)° at the iron center.

We have previously reported the structure of [Fe**L5**$(OAc)_2$] [50]. The coordination about the metal is very similar, except that the two chlorides are replaced by a pair of strictly mono-dentate acetate ligands. The iron to nitrogen bond lengths are 2.2091(10) and 2.2228(11) Å for the nitrogens at the ethylene bridge. The Fe-N bond length for the nitrogen atoms that have the benzyl substituents (N1 and N3) are 2.3024(10) and 2.3024(9) Å (*sic*), respectively. The Fe is slightly better contained within the pocket of the macrocycle so that the N1-Fe-N3 bond angle is 148.84(4)°.

A crystal structure of the known mono-methylated, mono-benzylated cyclam-glyoxal di-iodide [32] was obtained during the course of this work and is shown in Figure 3. The X-ray scattering was rather weak even though the data were collected using synchrotron radiation. However, the structure solves well in the centrosymmetric space group $P2_1/n$ with a single molecule in the asymmetric unit. The positive charge of the two tetraalkylammonium centers is balanced by two iodide ions. There is good evidence for disorder in the iodide positions and the single water molecule. In the solid state, the substituted cyclam ions are arranged in layers parallel to the (1 0 −1) plane and oriented so that they generate channels parallel to the crystallographic ***a*** direction. The water and iodide are found in these channels, between layers of ions.

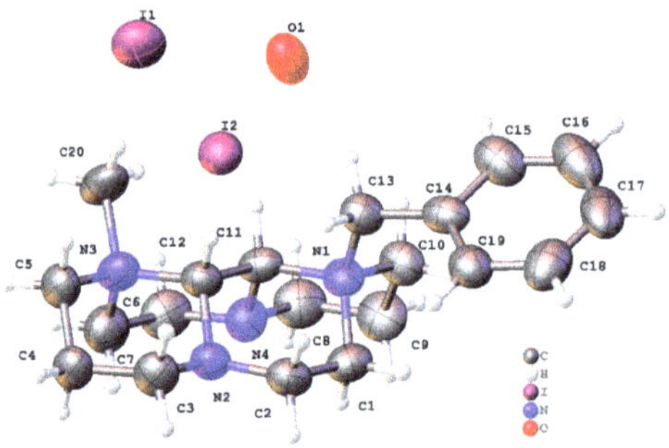

Figure 3. Asymmetric unit of **L8** precursor with minor disorder not shown. Atoms are represented as 50% probability ellipsoids.

[Mn**L9**Cl_2] crystallizes in the centrosymmetric space group *P*-1 with two unique molecules in the asymmetric unit (Z′ = 2). Refinement of the model was straightforward, and minor disorder in one of the pendant allyl groups was dealt with using normal methods. The final fit was very good: R = 4.99% and wR(F^2) = 6.45% for all data. Similar to the earlier examples, the metal ion is found in a distorted octahedral coordination geometry with a pair of cis chloride ions, as shown in Figure 4.

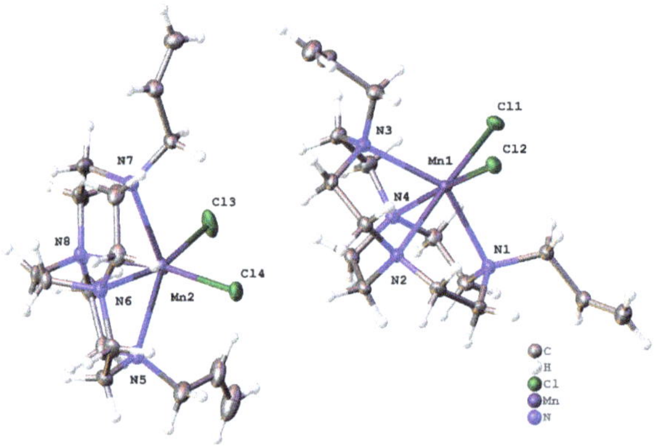

Figure 4. Asymmetric unit of [Mn**L9**Cl_2] with minor disorder not shown. Atoms are represented as 50% probability ellipsoids.

There are small differences in the orientation of the ligands around the metal ions in the two unique complexes, but the differences are not substantial. The familiar distorted octahedral coordination is observed, and, as in other cases, the nitrogen atoms joined by the ethylene bridge form shorter bonds to the metal than those bearing allyl substituents, as shown in Table 2, although the bonds about Mn2 are longer than those about Mn1.

The two unique complexes in the asymmetric unit form a hydrogen-bonded dimer through C-H···Cl interactions between bound chlorides of one complex and hydrogens from the ethylene bridge of the other. One of these hydrogen bonds is bifurcated such that the chloride is a hydrogen bond acceptor from two C-H groups. There is a further single trifurcated C-H···Cl hydrogen bond between pairs of these dimers, which assembles them into chains that run parallel to the (110) direction. Between these chains, additional C-H···Cl interactions generate a hydrogen-bonded three-dimensional network.

The crystal structure of [Mn**L11**Cl_2] is centrosymmetric (space group $P2_1/c$) but in other respects is similar to the other complexes described above. Once again, the fit of the model to observed data is excellent: R = 2.94% and wR(F^2) = 6.16% for all data. The asymmetric unit contains a single Mn^{2+} ion with four nitrogen atoms from the ligand bent away from two cis chloride ions, as shown in Figure 5. Here, the nitrogen atoms of the ethylene bridge (N2 and N4) are substantially closer to the metal than N1 and N3. Mn-N2 and Mn-N4 bonds are approximately 2.30 Å, but those for N1 and N3 are elongated with values at approximately 2.40 Å (Table 2).

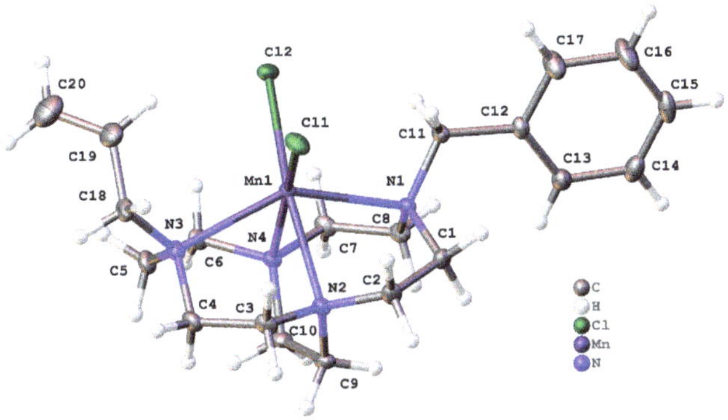

Figure 5. Asymmetric unit of [$MnL11Cl_2$] with atoms represented as 50% probability ellipsoids.

In the solid state, extensive hydrogen bonding between bound chloride and C-H groups produces hydrogen-bonded sheets in the *xy* plane. There are van der Waals interactions between layers, and close approach of C-H to the benzyl substituent may give rise to further weak C-H···π interactions.

[$FeL11Cl_2$] crystallizes with acetonitrile within the structure to give the formula of [$FeL11Cl_2$] · 2MeCN. The space group is *Pnma* (centrosymmetric), and the asymmetric unit contains one half complex that is disordered over a mirror plane. This added significant complication to the modeling of the observed electron density, but a satisfactory fit was possible using standard methods of disorder modeling: R(F) = 3.72% and wR(F^2) = 6.93% (all data). The asymmetric unit is shown in Figure 6, and a representation of the whole complex appears in Figure 7. The basic coordination about the Fe^{2+} ion is similar to that in the equivalent Mn^{2+} example, but the M-N bonds are shorter, although the trend of nitrogen atoms on the ethylene bridge having shorter bond lengths is preserved, as shown in Table 2.

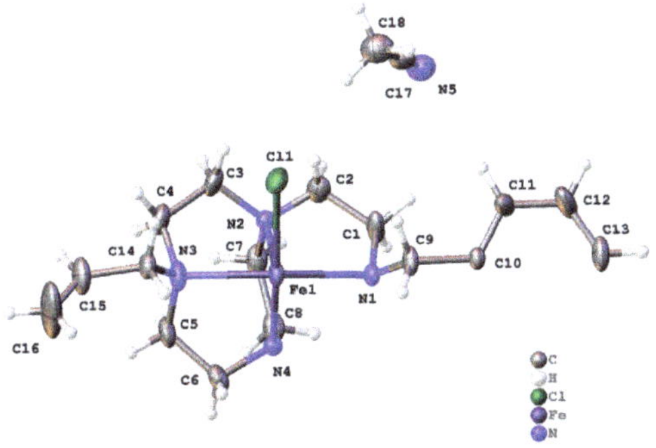

Figure 6. Asymmetric unit of [$FeL11Cl_2$] · 2MeCN with atoms represented as 50% probability ellipsoids.

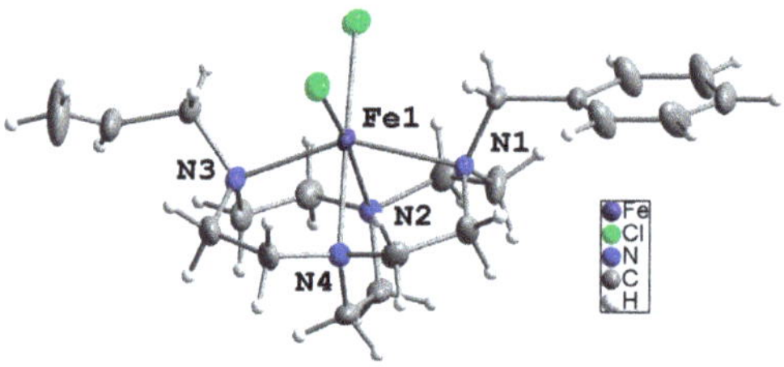

Figure 7. Complete [Fe**L11**Cl_2] complex with atoms represented as 50% probability ellipsoids.

Focusing on the metrics about the metal centers of the **L1** and **L11** ligands (Table 2), it is apparent that the addition of substituents to the ethylene-bridged cyclen changes the coordination behavior of the ligand. [Mn**L1**Cl_2] contains the unsubstituted ethylene-bridged cyclen, and the nitrogen atoms (N2 and N4) of the bridge are further from the metal than the two NH donors (N1 and N3). The bite angle of the two bridge nitrogen atoms is the smallest of these complexes described here. Adding substituents to N1 and N3 reverses the trend in bond lengths; in all of the other complexes, the mean bond lengths for M-N1 and M-N3 are longer than those for M-N2 and M-N4.

A trend in the bond lengths and angles is observed for the Mn^{2+} and Fe^{2+} complexes of ligands **L3** and **L11**. For **L3**, the Fe-N bond lengths are ~2.5% shorter than for Mn-N. Examination of equivalent M^{2+} structures in the CCDC [51] reproduces this trend, namely M-N bonds of the bridge are ~3.3% longer for Fe^{2+}, while for the other nitrogen atoms the bond lengths are 2.0% longer. Further, the ethylene bridge bite angle (N2-M-N4) is significantly smaller for Mn, while the N1-M-N3 angle is larger for Fe. The data in the CCDC confirm this as well; the N2-M-N4 and N1-M-N3 bite angles are 3.8% and 2.9% smaller for Mn^{2+}, respectively. These changes signify that the Fe^{2+} resides deeper into the cavity of the macrocycle than Mn, presumably as a consequence of the smaller size of Fe^{2+}. Similar trends are observed for **L11**; Fe-N bond lengths are substantially shorter than those of Mn-N. The bite angle of N2-M-N4 is larger for Fe (77.38(12)° vs. 74.09(4)° for Mn) and the N1-Fe-N3 angle is 148.85(7)° for Fe (143.56(4)° for Mn).

2.3. Acid Decomplexation Studies

Cu**L**Cl_2 or [Cu**L**Cl][PF_6] complexes of the novel ligands studied here were synthesized whenever possible (pure **L11** and **L12** complexes could not be synthesized and were excluded from this study). Decomplexation of such copper(II) complexes of cross-bridged tetraazamacrocycles under harsh conditions of strong acid and high temperature have become a standard method for examining the kinetic stability imparted by the ligand on metal complexes with these ligands [21–23,31,49,52–54]. Due to the proton sponge nature of these highly basic tetraamines, potentiometric titrations of metal complexation reactions in aqueous solution are generally not possible, as it is impossible to fully deprotonate the ligands in water. It is therefore necessary to conduct the kinetic evaluation under harsh acidic conditions (Table 3). Cu^{2+} is generally used, as the complexes are highly colored with only one d-d absorption band, which can easily be monitored using electronic spectroscopy, allowing the pseudo-first-order kinetic analysis and half-life calculations (see Experimental Section for details). Manganese(II) and iron(II) complexes themselves lack strong d-d absorption bands, making it impossible to perform the same experiments on them. Therefore, the copper(II) complex, with the constant being the Cu^{2+} ion itself, has been adopted as the standard used in order to compare the effect of each different cross-bridged ligand on stability. This is an imperfect system, but it gives relative ligand effects. It is not meant to imply that the Mn or Fe complexes would be as stable but to give the relative effects of the differing ligand structures on kinetic stability.

Table 3. Kinetic stability of $[CuLCl_2]$ and $[CuLCl]PF_6$ complexes of ligands **L1–L10** under harsh acidic conditions.

Ligand	Pseudo-First-Order Half-Life for Decomplexation of 1.000 mM Copper Complexes					
	30 °C 1 M HCl	30 °C 5 M HCl	50 °C 5 M HCl	70 °C 5 M HCl	90 °C 5 M HCl	References
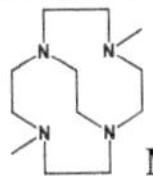 **Me_2Bcyclen**	36 min		<1 min			[55]
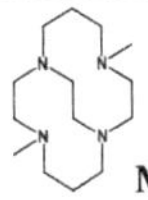 **Me_2EBC**			7.3 day		79 min	[22,55]
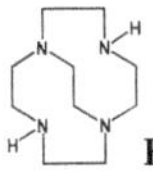 **H_2Bcyclen (L1)**	<1 min					[21]
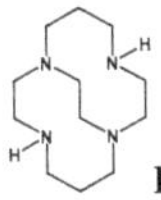 **H_2Bcyclam (L2)**	18.5 d		5.40 h [This work]		11.8 min	[54]
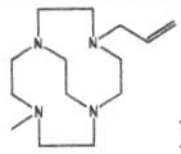 **Me_1Allyl_1Bcyclen (L3)**	25 min					This work
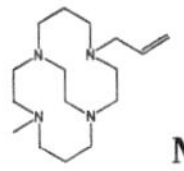 **Me_1Allyl_1Bcyclam (L4)**			2.33 h			This work
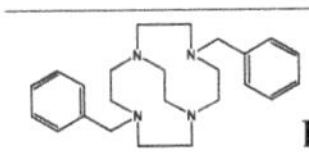 **Bn_2Bcyclen (L5)**	4.2 h	7.8 min				[33,56]
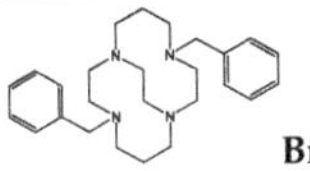 **Bn_2Bcylam (L6)**			2.38 h	23.4 min		[33,56]
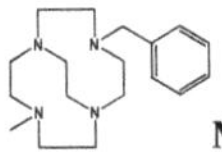 **Me_1Bn_1Bcyclen (L7)**	24.8 min					This work
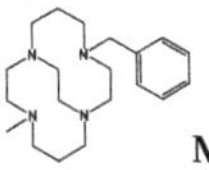 **Me_1Bn_1Bcyclam (L8)**			15.05 h	1.4 h		This work
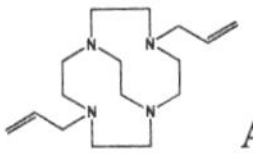 **$Allyl_2$Bcyclen (L9)**	51.7 min					This work
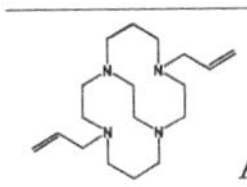 **$Allyl_2$Bcyclam (L10)**			1.86 h	15 min		This work

Cross-bridged cyclen copper complexes are not particularly stabilized by the topological complexity and rigidity imparted by the ethylene cross-bridge [56]. The small cavity does not allow for optimal complementarity between the small ligand and the Cu^{2+} cation [57]. Even though the cross-bridge does rigidify the ligand/complex, without good complementarity, the rigidity effects do not result in large kinetic stability gains. The small cyclen macrocycle is rather rigid itself, with only ethylene chains and all five-membered chelate rings in the complex. Adding the cross-bridge simply distorts the pseudo-octahedral or square pyramidal structure, leading to minimal stabilization, if any at all, over the unbridged complexes [56]. Even so, it is interesting to probe the pendant arm effects on the stability of the copper complexes. Data were obtained, or were already available, for all cyclen complexes at 30 °C in 1 M HCl, allowing for direct comparison.

The simplest ethylene cross-bridged cyclen H_2Bcyclen (**L1**) has a very short half-life of less than one minute, even in 1 M HCl acid conditions and at 30 °C [21]. The stability of the Cu^{2+} complexes increases in the following ligand order upon addition of substituents, making all tertiary nitrogen donors (Table 3) H_2Bcyclen (**L1**) < Bn_1Me_1Bcyclen (**L7**)~$Allyl_1Me_1$Bcyclen (**L3**) < **Me_2Bcyclen** < $Allyl_2$Bcyclen (**L9**) < Bn_2Bcyclen (**L5**). The trend(s) here are not completely transparent, but some comments can be made. It appears that a single Me group combined with one larger substituent is not optimal for stability, as **L7** and **L3** are only modestly stabilized. Two Me substituents (**Me_2Bcyclen**) give more stability, although the effect is not large. Finally, two larger allyl or Bn substituents on the ligand (**L9** and **L5**, respectively) impart even greater stability, with **L5** exhibiting the highest stability (4.2 h vs. 51.7 min).

Ethylene cross-bridged cyclam complexes are stabilized to a much greater extent by the topological complexity and rigidity of the ethylene cross-bridge. The complementarity match, (mostly a size effect) of the larger cyclam cavity for the Cu^{2+} ion allows stabilization through the additional constraint factors [56]. As such, the Cu-Bcyclam complexes are usually studied at higher acid concentration (5 M HCl) and at elevated temperature (50 °C, 70 °C, and 90 °C). For example, Cu(**Me_2Bcyclam**)Cl^+ has a half-life of 79 min even at 90 °C in 5 M HCl, conditions that quickly destroy the vast majority of transition metal complexes. Data were obtained, or were already available, for all cyclam complexes at 50 °C in 5 M HCl, allowing for direct comparison.

The stability of the Cu^{2+} cyclam complexes increases in the following ligand order (Table 3): $Allyl_2$Bcyclam (**L10**) < $Allyl_1Me_1$Bcyclam (**L4**)~Bn_2Bcyclam (**L6**) < H_2Bcyclam (**L2**) < Bn_1Me_1Bcyclam (**L8**) < **Me_2Bcyclam**. Interestingly, the bulkier substituents (Bn, allyl) actually appear to destabilize the copper complexes with the cyclam ring size, rather than stabilizing them as seen for the cyclen ligands. In fact, two of the three most stable complexes, with **L2** and **Me_2EBC**, have either only hydrogen or methyl substituents. Of the most stable cyclam complexes, **L8**, Bn_1Me_1Bcyclam contains a single large Bn group. Though only a hypothesis, the poor size match of the cyclen systems may gain stability as large allyl and Bn groups block access by H^+ to the otherwise exposed (due to the severely distorted octahedral geometry) Cu^{2+}-N bonds that must protonate in order for metal decomplexation to occur. On the other hand, the much better complementarity of the cyclam systems leads to less distortion of the octahedral geometries. Here, the smaller substituents (H, Me) distort the Cu^{2+}-N bonds to a lesser extent, while the larger allyl and Bn substituents may disrupt the complementarity to a greater extent, leading to faster decomplexation. Regardless, what is observed is that the ethylene cross-bridged cyclam complexes, in particular, are remarkably kinetically stable under harsh acidic conditions, which would suggest oxidation catalysts (in the case of Fe and Mn) that should survive to function for significant lengths of time during the proposed water purification processes.

2.4. Cyclic Voltammetry

Cyclic voltammetry studies on 1.000 mM solutions of the Fe and Mn complexes in MeCN were carried out in order to correlate available reversible redox processes with their reactivity (Figures S14–S21). The results of these studies are presented in Table 4. Previous work on the Fe and Mn complexes of cross-bridged tetraazamacrocycles is included and has generally shown that higher oxidation catalyst reactivity is related to higher oxidation potentials for reaching the higher oxidation states of these metals. Copper complexes were not included in this cyclic voltammetry study, as they are not active oxidation catalysts.

Table 4. Redox potentials (V) vs. SHE (0.00 V) and ΔV (mV) for 1.000 mM solutions of the Fe and Mn complexes of **L1–L12** in MeCN.

Complex	$E_{1/2}$ $M^{2+/3+}$ [V] (ΔV [mV])	$E_{1/2}$ $M^{3+/4+}$ [V] (ΔV [mV])	Reference
Fe(**Me_2Bcyclen**)Cl_2	+0.036 (64)	-----	[12]
Mn(**Me_2Bcyclen**)Cl_2	+0.466 (70)	+1.232 (102)	[12]
Fe(**Me_2Bcyclam**)Cl_2	+0.110 (63)	-----	[12]
Mn(**Me_2Bcyclam**)Cl_2	+0.585 (61)	1.343 (65)	[12]
[Fe**L1**Cl_2]	-0.055 (89)	-----	[13]
[Mn**L1**Cl_2]	+0.389 (280)	not accessible in DMF	[13]
[Fe**L2**Cl_2]	-0.113 (78)	-----	[13]
[Mn**L2**Cl_2]	+0.239 (79)	not accessible in DMF	[13]
[Fe**L3**Cl_2]	+0.094 (82)	-----	This work
[Mn**L3**Cl_2]	+0.545 (71)	+1.338 (86)	This work
[Fe**L4**C12]	+0.165 (96)	-----	This work
[Mn**L4**Cl_2]	+0.628 (59)	+1.374 (83)	This work
[Fe**L5**Cl_2]	+0.071 (85)	-----	[13]
[Mn**L5**Cl_2]	+0.400 (65)	not accessible in DMF	[13]
[Fe**L6**Cl_2]	+0.157 (85)	-----	[13]
[Mn**L6**Cl_2]	+0.577 (72)	not accessible in DMF	[13]
[Fe**L7**Cl_2]	+0.065 (64)	-----	This work
[Mn**L7**Cl_2]	+0.550 (76)	+1.355 (82)	This work
[Fe**L8**Cl_2]	+0.176 (86)	-----	This work
[Mn**L8**Cl_2]	+0.609 (93)	+1.404 (92)	This work
[Fe**L9**Cl_2]	+0.114 (92)	-----	This work
[Mn**L9**Cl_2]	+0.572 (68)	+1.385 (79)	This work
[Fe**L10**Cl_2]	+0.194 (59)	-----	This work
[Mn**L10**Cl_2]	+0.616 (72)	+1.422 (83)	This work
[Fe**L11**Cl_2]	+0.130 (78)	-----	This work
[Mn**L11**Cl_2]	+0.558 (72)	+1.390 (73)	This work
[Fe**L12**Cl_2]	+0.196 (72)	-----	This work
[Mn**L12**Cl_2]	+0.619 (85)	+1.433 (92)	This work

The cyclic voltammetry data are useful in evaluating the oxidizing ability of the higher-valent oxidation states of these catalysts with the reversibility generally desirable for catalytic species. These ligands support $Fe^{2+/3+}$ ions and $Mn^{2+/3+/4+}$ ions with good reversibility in acetonitrile (see ESI for CV curves). A few complexes, those of **L1**, **L2**, **L5**, and **L6**, were only soluble in DMF, thus their data may have slight solvent effects and lack the $Mn^{3+/4+}$ couple due to the limits of DMF solvent becoming oxidized itself. One trend that is quite consistent is that the smaller cyclen complex in a cyclen/cyclam pair having the same pendant arms is always easier to oxidize. This is likely a simple cavity size effect, with the smaller cyclen ring selecting for the smaller M^{3+} or M^{4+} cation and the larger cyclam ring selecting for the larger, lower-valent metal ion. Only the unsubstituted **L1** and **L2** show an inversion of this ring-size selectivity, which may be the result of less rigidity of these ligands due to having two secondary nitrogen donors, while all other ligands have all four nitrogen donors as more constrained tertiary nitrogen atoms.

Substituent effects are observed, although they are relatively small. For the best comparison, only those complexes examined in the same solvent (MeCN) are compared directly. For Fe complexes, replacing one methyl group with one benzyl group results in an increase in $\Delta E_{1/2}$ of ~0.03–0.06 V. For example, Fe(**Me₂Bcyclam**)Cl_2 and Fe**L8**Cl_2 have $E_{1/2}$ values of +0.110 V and +0.176 V, respectively. The change is a bit larger for Mn complexes, where $\Delta E_{1/2}$ of ~0.07–0.12 V is seen. For example, Mn(**Me₂Bcyclen**)Cl_2 and Me**L7**Cl_2 have $E_{1/2}$ $Mn^{3+/4+}$ values of +1.232 V and +1.355 V, respectively. Replacing successive methyl groups with allyl groups for Fe complexes results in an increase in $\Delta E_{1/2}$ of ~0.04–0.06 V per allyl group. Fe(**Me₂Bcyclam**)Cl_2, Fe**L4**Cl_2, and Fe**L10**Cl_2 have $E_{1/2}$ values of +0.110 V, +0.165 V, and +0.194 V, respectively. The change is again larger for Mn complexes, where $\Delta E_{1/2}$ of ~0.06–0.12 V is seen. For example, Mn(**Me₂Bcyclen**)Cl_2, Mn**L3**Cl_2, and Mn**L9**Cl_2 have $E_{1/2}$ $Mn^{3+/4+}$ values of +1.232 V, +1.338 V, and 1.385 V, respectively. These last two series show that the substituent effects are cumulative.

Interestingly, the highest oxidation potentials are found in the mixed pendant arm ligands **L11** and **L12**. The maximum $E_{1/2}$ value for an $Fe^{2+/3+}$ couple in a cyclen-based complex was found for Fe**L11**Cl_2, with +0.130 V. The maximum $E_{1/2}$ value for an $Fe^{2+/3+}$ couple in a cyclam-based complex was found for Fe**L12**Cl_2, with +0.196 V. The maximum $E_{1/2}$ value for an $Mn^{3+/4+}$ couple in a cyclen-based complex was found for Mn**L11**Cl_2, with +1.390 V. This value is actually higher than the oxidation potential for the well-known Mn(**EBC**)Cl_2 catalyst (+1.343 V), showing that substituent choices can serve to replace ring-size effects in order to modify oxidation potential. The maximum $E_{1/2}$ value for an $Mn^{3+/4+}$ couple in a cyclam-based complex was found for Mn**L12**Cl_2, with +1.433 V, a full 0.090 V higher than for the Mn(**Me₂Bcyclam**)Cl_2 catalyst (+1.343 V). Di-allyl ligands **L9** and **L10** similarly raise the respective oxidation potentials of their Fe and Mn complexes, though slightly less effectively than the $Allyl_1Bn_1$ ligands **L11** and **L12** (see Table 4). The main goal of adding these particular pendant arms was to investigate copolymerization reactions that could incorporate these catalysts into water-insoluble polymers for eventual use as heterogeneous water purification materials. However, increasing their oxidation potentials might be predicted to lead to more reactivity as oxidation catalysts, which is examined below.

Future electrochemical work will need to build off what has been learned: 1. the larger cyclam ligands support higher oxidations states in general, leading to (see below) higher oxidation catalysis reactivity and 2. pendant arm effects are minimal but maximized when both an allyl and benzyl pendant arm are present in the same complex. As we continues towards the goal of heterogeneous catalysts, we will need to discern if oxidation potentials are changed when the complexes are fixed into copolymeric materials, as has been observed previously [58,59]. Fortunately, many polymers are soluble in solvents such as DMF and THF in which cyclic voltammetry can be carried out [60]. We will endeavor to track the changes in cyclic voltammetry and how they track with oxidation efficiency for polymeric compounds in future work.

2.5. *Dye-Bleaching Studies*

The goal of our overall project is to develop catalysts capable of efficient removal of organic pollutants (contaminants of emerging concern, CEC) from wastewater via oxidation catalysis. Eventually, we hope to develop catalysts such as those disclosed here into heterogeneous catalytic materials via copolymerization using the allyl and benzyl pendant arms as sites of copolymerization to incorporate the Fe and Mn cross-bridge macrocycle complex cores into water-insoluble polymers for easy separation, or perhaps as cartridge-filling material for contaminated water to be passed through with purified water exiting the cartridge. In this work, we focus on screening a large number (24) of new complexes to determine their homogeneous oxidation catalytic ability prior to selecting lead compounds for copolymerization. We have selected three extensively used dyes [1,7,9,10] (Scheme 4), methylene blue (MB), methyl orange (MO), and Rhodamine B (RhB), as model organic pollutants for our degradation study using our metal–ligand oxidation catalysts (metal = Fe and Mn with **L1–L12**) (Figures S22–S33). The two cationic dyes (MB and RhB) and the anionic dye (MO) have diverse functional groups including NMe_2^+, $-SO_3^-$, -COOH, and $=NEt_2^+$. These oxidation catalysts were able to degrade all three of the dyes in the presence of hydrogen peroxide (H_2O_2) as the terminal oxidant. In a survey of many potential catalysts to identify lead compounds, it is not feasible to perform the dye-bleaching experiments for all complexes at a wide range of pHs due to time and resource constraints. We chose neutral pH 7 to conduct this screening study, as we knew the complexes and dyes were stable at this pH and that most natural and municipal water sources have pHs in the pH 6–8 range [61].

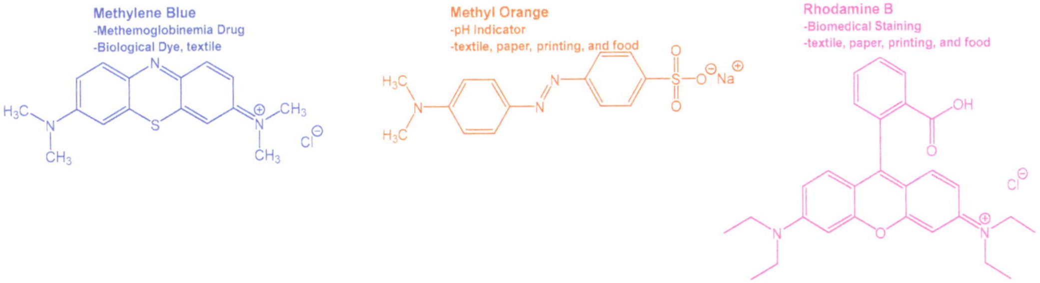

Scheme 4. Dyes used in the dye-bleaching studies.

Table 5 illustrates the average TOF values of control substances and oxidation catalysts for the degradation of methylene blue (MB), methyl orange (MO), and Rhodamine B (RhB). The eight complexes in red stand out among the rest as having at least one significantly higher-than-average TOF. Ranked order (1 being best) among these top eight catalysts are given in parentheses following each TOF value. The fifth column provides an overall ranked order based on an average rank order for the three dyes. For discussion purposes, the highest redox potential observed, and the kinetic stability half-life of the copper analogue (when available), are included in the table. Dye-bleaching plots, showing A_t/A_o vs. time for all catalysts, can be found in the ESI, along with narrative observations for each ligand's Fe/Mn pair of complexes. Here, we will present a broader discussion of general trends and observations about what factors lead to better dye bleaching.

Observation 1: Manganese catalysts generally outperform iron catalysts, which is the case for ligands **L3–L4**, **L6–L10**, and **L12**. Ligands **L5** and **L11** gave similar activity for both metals. Only iron catalysts with **L1** and **L2** gave clearly superior results. It should be noted that **L1** and **L2** have two secondary nitrogen donors, which sterically allows for dimerization to occur [13], forming LFe-O-FeL and LMn-$(O)_2$-Mn structures. Dimerization is sterically prevented by all alkyl substitutions of these hydrogen atoms, even methyl groups. It is likely that the Fe-O-Fe dimers are forming in the presence of H_2O_2, as has been observed [13], and that these dimers are the active catalysts, as they have one open coordination site per Fe to interact with the oxidant and substrate. Perhaps the coordinatively saturated Mn dimers, lacking an open coordination site, are unable to act as catalysts. All other ligands, having alkyl groups in place of these hydrogens, are prevented from dimerizing [13], which allows the monomeric complexes to persist in solution and act as the oxidation catalysts, where Mn is more efficient than Fe, as has been proposed for the well-known Mn**Me$_2$Bcyclam**Cl_2 catalyst [12].

Table 5. Turnover frequency (TOF) values, and other relevant parameters, for control and all 24 metal–complex catalysts for degradation of MB, MO, and RhB.

Catalyst	Average TOF for MB (h^{-1}) (Rank among Top 8)	Average TOF for MO (h^{-1}) (Rank among Top 8)	Average TOF for RhB (h^{-1}) (Rank among Top 8)	Catalyst Rank Based on Average Rank Order for Top 8 Catalysts for All Three Dye TOFs	Highest Redox Potential Observed (Fe^{3+}, Mn^{3+}, or Mn^{4+})	Half-Life of Cu^{2+} Complex Under Acid Conditions (Cyclen Complexes in 1 M HCl, 30 °C; Cyclam Complexes in 5 M HCl, 50 °C)
H_2O_2 only	0.018	0.034	0.002			
$FeCl_2$ only	0.111	0.926	0.055			
$MnCl_2$ only	0.003	0.518	0.047			
[Fe**L1**Cl_2]	0.54	1.33	0.31		−0.055 V	<1 min
[Mn**L1**Cl_2]	0.32	1.08	0.14		+0.389 V	<1 min
[Fe**L2**Cl_2]	1.21 (7)	31.35 (3)	3.62 (1)	**3 (tie) (Avg = 3.67)**	−0.113 V	5.40 h
[Mn**L2**Cl_2]	0.27	0.147	0.18		+0.239 V	5.40 h
[Fe**L3**Cl_2]	0.33	1.46	0.64		+0.94 V	25 min
[Mn**L3**Cl_2]	1.67 (5)	6.42 (8)	1.104 (7)	**6 (Avg = 6.67)**	+1.338 V	25 min
[Fe**L4**C12]	0.35	1.26	0.63		+0.165 V	2.33 h
[Mn**L4**Cl_2]	2.03 (4)	16.26 (4)	1.05 (8)	**4 (Avg = 5.33)**	+1.374 V	2.33 h
[Fe**L5**Cl_2]	0.13	4.08	0.21		+0.071 V	4.2 h
[Mn**L5**Cl_2]	0.16	0.60	0.11		+0.400 V	4.2 h
[Fe**L6**Cl_2]	0.45	3.41	0.62		+0.157 V	2.38 h
[Mn**L6**Cl_2]	0.74 (8)	16.14 (5)	1.27 (5)	**5 (tie) (Avg = 6)**	+0.577 V	2.38 h
[Fe**L7**Cl_2]	0.13	0.083	0.13		+0.065 V	24.8 min
[Mn**L7**Cl_2]	2.64 (3)	8.32 (7)	2.34 (2)	**3 (tie) (Avg = 3.67)**	+1.355 V	24.8 min
[Fe**L8**Cl_2]	0.57	2.82	1.54		+0.176 V	15.05 h
[Mn**L8**Cl_2]	2.09 (3)	108.9 (1)	1.46 (4)	**2 (Avg =2.67)**	+1.404 V	15.05 h
[Fe**L9**Cl_2]	0.35	0.77	0.92		+0.114 V	51.7 min
[Mn**L9**Cl_2]	0.53	3.13	0.35		+1.385 V	51.7 min
[Fe**L10**Cl_2]	0.171	2.2	0.35		+0.194 V	1.86 h
[Mn**L10**Cl_2]	1.23 (6)	9.40 (6)	1.16 (6)	**5 (tie) (Avg = 6)**	+1.422 V	1.86 h
[Fe**L11**Cl_2]	0.19	0.43	0.26		+0.130 V	n/a
[Mn**L11**Cl_2]	0.18	0.86	0.15		+1.390 V	n/a
[Fe**L12**Cl_2]	0.25	1.26	0.70		+0.196 V	n/a
[Mn**L12**Cl_2]	5.87 (1)	36.49 (2)	1.61 (3)	**1 (Avg = 2.0)**	+1.433 V	n/a

Observation 2: Cyclam-based ligands tend to produce more effective catalysts than cyclen-based ligands. Six of the eight catalysts identified as superior (red in Table 5) have cyclam ligands. Part of the reason may simply be stability, as we demonstrated that cyclam ligands were clearly superior to cyclen ligands in kinetic stability under

harsh acidic conditions. For reference, the last column of Table 5 shows the stability data for the copper complexes with these ligands. Clearly, there is some loose correlation between stability and catalytic efficiency. However, these dye-bleaching experiments were performed at pH = 7, so it is likely that acid-promoted ligand dissociation is not a significant concern here. From structural information gleaned from a vast number of crystal structures of ethylene cross-bridge tetraazamacrocycle complexes reported in the Cambridge Structural Database (CSD) [51], the cyclam ligands provide a much better complementarity match (size, geometry, and electronics matching between ligand and metal) for first row transition metal ions. Therefore, the less distorted, and likely better orbital-overlapped, M-N bonds may simply lead to more reactive, or more robust, catalysts under these oxidizing conditions. Mn and Fe complexes in water with oxidizers such as H_2O_2 generally form insoluble oxides, a thermodynamic sink that leads to inactivation of many aqueous catalysts. Such a mechanism is avoided here due to the cross-bridged ligand stabilization of the complexes for both cyclam and cyclen complexes.

Observation 3: As demonstrated in the cyclic voltammetry experiments, higher oxidation potentials for the Fe^{3+} and Mn^{4+} species appear to lead to more efficient catalysis. Most obviously, the Fe complexes cannot achieve the +4 state and the associated high oxidation potential that this would require and are as such clearly inferior catalysts as a group. Conversely, all of the Mn complexes that were examined in MeCN can reversibly obtain the Mn^{4+} oxidation state and generally are more efficient catalysts. It is telling that the two top catalysts (ranked by "average rank ordering" each dye TOF among the top eight identified catalysts) are Mn^{4+}-obtaining catalysts with observed oxidation potentials above 1.4 V: [Mn**L12**Cl_2] at +1.433 V and [Mn**L8**Cl_2] at 1.404 V. The other "top" Mn catalysts all achieve the Mn^{4+} oxidation state at or above 1.338 V. Clearly, catalysts that can achieve higher oxidation potentials are more active in the dye-bleaching tests we carried out. Interestingly, [Mn**L10**Cl_2] ranks only "6", even though its oxidation potential is 1.422 V. Thus, there are more factors at play here, even though oxidation potential appears to be a primary selector for high activity.

Observation 4: The three highest efficiency Mn catalysts all have a single benzyl pendant arm and either an allyl or methyl second pendant arm. This combination seems to provide the kinetic stability and oxidation potential for an efficient Mn catalyst. Perhaps the benzyl's hydrophobicity may also be important in allowing good interaction with the organic substrate dyes, all of which contain multiple aromatic groups (Scheme 4). We targeted benzyl pendant arms as possible copolymerization groups using Friedel–Crafts polymerization [36] reactions and have begun to work towards producing such copolymers containing these monomeric catalysts as potential heterogeneous water purification materials.

Future work will need to build off the spectroscopic tracking of dye-bleaching work and what has been learned, namely, that Mn cyclam complexes with high oxidation potentials are the most efficient catalyst. We will need to begin to incorporate other substrates besides dyes, which are straightforward to track spectroscopically but not the only type of contaminant of emerging concern [62]. Antimicrobials and antibiotics are increasingly concerning to health officials as contaminants in waste water [63,64] and will thus be future substrates for our catalysts to be tested against. We will incorporate mass spectra tracking of aliquots over time to help determine rate as well as to gain information on potential mechanisms and products formed [65–67].

3. Experimental

3.1. General

All reagents and solvents were purchased from commercial sources and used as received. All solvents were of reagent grade and anhydrous. Electrospray mass spectra were recorded on a Shimadzu LCMS 2020 Electrospray Mass Spectrometer (Shimadzu, Kyoto, Japan). NMR spectra were collected on a Varian Bruker AVANCE II 300 MHz NMR Spectrometer instrument (Varian Bruker, Billerica, MA, USA). Electrochemical experiments were completed on a BAS100B Electrochemical Analyzer (BASi, West Lafayette, IN, USA) at a scan rate of 200 mV/s under nitrogen atmosphere. A button Pt electrode, a Pt-wire electrode, and a Ag-wire electrode were used as working electrode, counter electrode, and pseudo-reference electrode, respectively. Acetonitrile solutions of the complexes (1 mM) with tetrabutylammonium hexafluorophosphate (0.1 M) were used as a supporting electrolyte. SHE using ferrocene (+0.400 V versus SHE) or acetylferrocene (+0.680 V versus SHE) was used as an internal standard for the measurements.

3.2. *Synthesis*

Ligands **L1** [45], **L2** [45], **L5** [45], **L6** [45], **L7** [33], and **L8 [32]** were synthesized by literature procedures. Mn**L1**Cl_2, Fe**L1**Cl_2, Mn**L2**Cl_2, Fe**L2**Cl_2, Mn**L5**Cl_2, Fe**L5**Cl_2, Mn**L6**Cl_2, Fe**L6**Cl_2 [13], and Fe**L8**Cl_2 [46] were prepared as previously published. Copper complexes were used only for kinetic decomplexation studies, for which the copper complexes of **L1**, **L2**, **L5**, and **L6** have already been published (see references in Table 3). [Cu**L11**Cl][PF_6] and [Cu**L12**Cl][PF_6] complexation never produced complexes with acceptable elemental analyses and are therefore not included in this study.

3.2.1. Synthesis of Novel Ligands and Precursors

Synthesis of **3**. A total of 15.606 g (0.0464 mol, 1 eq) of the mono-methylated cyclen-glyoxal condensate **1** [47,48] was suspended in 400 mL of CH_3CN in a 1 L RB flask. Then, 34.70 mL of allyl bromide (56.16 g, 0.464 mol, 10 eq) was added and the reaction was stoppered and stirred at room temperature for seven days. The product was filtered and washed (several times) with CH_3CN and ether and subsequently dried under vacuum, yielding 17.879 g (94%) of **3**. Elemental Analysis (%): calc $C_{14}H_{26}N_4Br_2 \cdot 0.9\ H_2O$ (426.41 g/mol): C 39.43, H 6.57, N 13.14; Found C 39.66, H 6.20, N 13.16. MS (ES) m/z = 125 (L-2Br^-)$^{2+}$.

Synthesis of **4**. A total of 9.935 g (0.0273 mol, 1 eq) of the mono-methylated cyclam-glyoxal condensate **2** [48] was suspended in 625 mL of CH_3CN in a 1 L RB flask. Then, 39.27 mL of allyl bromide (49.62 g, 0.410 mol, 15 eq) was added and the reaction was refluxed for 14 days under a nitrogen atmosphere using a water condenser and heating mantle. After the reaction was cooled to room temperature, the white solid product was filtered off using a glass frit then washed with CH_3CN and ether and dried under vacuum, yielding 11.058 g (92%) of **4**. Elemental Analysis (%): calc $C_{16}H_{30}N_4Br_2 \cdot 1.3\ H_2O$ (461.67 g/mol): C 41.63, H 7.12, N 12.14; Found C 41.90, H 7.49, N 12.16. MS (ES) m/z = 139 (L-2Br^-)$^{2+}$.

Synthesis of **L3**. A total of 17.879 g (0.0436 mol, 1 eq) of **3** was stirred into 1500 mL of 95% EtOH in a 2 L RB flask. Then, 16.494 g (0.436 mol, 10 eq) of $NaBH_4$ was slowly added to the reaction at room temperature and stirred under a nitrogen atmosphere for five days. Next, 6 M HCl was added slowly to the RB flask until the solution reached a pH of 2. Ethanol was evaporated off using a rotary evaporator. Next, the product was dissolved in 200 mL of 30% KOH solution and extracted with 5 × 175 mL fractions of benzene using a separatory funnel. The benzene layer was collected and dried with sodium sulfate overnight. The sodium sulfate was filtered off and the remaining solvent evaporated to give a pale yellow oily product, which was dried under vacuum to yield 8.808 g (80%) of **L3**. Elemental Analysis (%): calc $C_{14}H_{28}N_4 \cdot 0.4\ H_2O$ (259.61 g/mol): C 64.77, H 11.18, N 21.58; Found C 64.82, H 11.48, N 21.63. MS (ES) m/z = 253 $(LH)^+$.

Synthesis of **L4**. A total of 11.653 g (0.0266 mol, 1 eq) of **4** was stirred into 860 mL of 95% EtOH in a 2 L RB flask. Then, 10.059 g (0.266 mol, 10 eq) of $NaBH_4$ was slowly added to the reaction at room temperature and stirred under nitrogen atmosphere for five days. Next, 6 M HCl was added slowly to the RB flask until the solution reached a pH of 2. Ethanol was evaporated off using a rotary evaporator. Next, the product was dissolved in 200 mL of 30% KOH solution and extracted with 3 × 175 mL fractions of benzene using a separatory funnel. The benzene layer was collected and dried with sodium sulfate overnight. The sodium sulfate was filtered off and the remaining solvent evaporated to give a pale yellow oily product, which was dried under vacuum to yield 4.132 g (55%) of

L5. Elemental Analysis (%): calc $C_{16}H_{32}N_4 \cdot 0.6\ H_2O$ (291.26 g/mol): C 65.98, H 11.49, N 19.24; Found C 65.87, H 11.35, N 19.30. MS (ES) $m/z = 281\ (LH)^+$.

Synthesis of **7**. A total of 10.200 g (0.0592 mol, 1 eq) of **5** [48] was added to a 500 mL RB flask and dissolved in 100 mL of MeCN. Then, 71.384 g (50.99 mL, 0.592 mol, 10 eq) of allyl bromide was added, and the reaction was stoppered and stirred for five days. The white solid product was collected by filtration, washed with MeCN and ether, and dried under vacuum to yield 20.200 g (78%) of **7**. Elemental Analysis (%): calc $C_{16}H_{28}N_4Br \cdot 0.5\ H_2O$ (445.24 g/mol): C 43.16, H 6.57, N 12.58; Found C 43.37, H 6.72, N 12.61. MS (ES) $m/z = 139\ (L\text{-}2Br)^{2+}$.

Synthesis of **8**. A total of 7.214 g (0.0360 mol, 1 eq) of **6** [48] was added to a 500 mL RB flask and dissolved in 100 mL of MeCN. Then, 39.201 g (28.00 mL, 0.360 mol, 10 eq) of allyl bromide was added, and the reaction was stoppered and stirred for seven days. The white solid product was collected by filtration, washed with MeCN and ether, and dried under vacuum to yield 13.528 g (81%) of **7**. Elemental Analysis (%): calc $C_{16}H_{28}N_4Br \cdot 1.0\ CH_3CN.$ $0.75\ H_2O$ (518.85 g/mol): C 46.30, H 7.09, N 13.50; Found C 46.52, H 7.18, N 13.11. MS (ES) $m/z = 152\ (L\text{-}2Br)^{2+}$.

Synthesis of **L9**. A total of 20.201 g (0.0463 mol, 1 eq) of **7** was suspended in 1500 mL of 95% ethanol in a 2 L RB flask. Then, 17.515 g of sodium borohydride was slowly added to it over 10 min, and the reaction was stirred under nitrogen for seven days. Next, 6 M HCl was added to the reaction until the solution reached a pH of 2. The ethanol was evaporated, and the white residue was then dissolved into 50 mL of DI H_2O. Following this, 150 mL of 30% KOH solution was added, and the pH of the solution reached 14. Extraction was carried out with 4 × 100 mL fractions of benzene. The benzene layers were collected and dried over Na_2SO_4 overnight. After filtration of sodium sulfate, benzene was evaporated to give the pale yellow oily product, which was dried under vacuum to yield 8.555 g (66%) of **L9**. Elemental Analysis (%): calc $C_{16}H_{30}N_4 \cdot 0.4\ H_2O$ (285.65 g/mol): C 67.28, H 10.87, N 19.61; Found C 67.03, H 10.69, N 19.90. MS (ES) $m/z = 279\ (LH)^+$.

Synthesis of **L10**. A total of 13.50 g (0.0291 mol, 1 eq) of **8** was suspended in 1500 mL of 95% EtOH in a 2 L RB flask. Then, 10.999 g (0.291 mol, 10 eq) of sodium borohydride was added, and the reaction mixture was stirred under nitrogen atmosphere for five days. Next, 6 M HCl solution was added until the pH reached 2. The ethanol was evaporated, and the residual white solid was dissolved in 50 mL DI water followed by addition of 250 mL of 30% KOH solution. Extraction was carried out with 4 × 100 mL fractions of benzene. The benzene layers were collected and dried over Na_2SO_4 overnight. After filtration of sodium sulfate, benzene was evaporated to give the pale yellow oily product, which was dried under vacuum to yield 5.124 g (57%) of **L10**. Elemental Analysis (%): calc $C_{18}H_{34}N_4$ (285.65 g/mol): C 70.54, H 11.18, N 18.28; Found C 70.50, H 11.64, N 18.38. MS (ES) $m/z = 307\ (LH)^+$.

Synthesis of **11**. A total of 22.065 g (0.0604 mol, 1 eq) of **9** [48] was suspended in 300 mL of acetonitrile. Then, 73.072 g (0.604 mol, 10 eq) of allyl bromide was added, and the reaction was stoppered and stirred for three days. The white solid product was collected by filtration, washed with MeCN and ether, and dried under vacuum to yield 24.301 g (83%) of **9**. Elemental Analysis (%): calc $C_{20}H_{30}N_4Br_2 \cdot 2.0\ H_2O$ (522.33 g/mol): C 45.99, H 6.56, N 10.73; Found C 45.85, H 6.39, N 10.72. MS (ES) $m/z = 407\ (L\text{-}Br)^+$; $m/z = 163\ (L\text{-}2Br)^{2+}$.

Synthesis of **12**. A total of 4.379 g (0.0111 mol, 1 eq) of **10** [48] was suspended in 50 mL of acetonitrile. Then, 13.429 g (0.111 mol, 10 eq) of allyl bromide was added, and the reaction was stoppered and stirred for three days. The white solid product was collected by filtration, washed with MeCN and ether, and dried under vacuum to yield 1.835 g (32%) of **12**. Elemental Analysis (%): calc $C_{22}H_{34}N_4Br_2 \cdot 1.3\ H_2O$ (537.77 g/mol): C 45.14, H 6.86, N 10.42; Found C 49.35, H 6.94, N 10.11. MS (ES) $m/z = 433\ (L\text{-}Br)^+$.

Synthesis of **L11**. A total of 14.591 g (0.0300 mol, 1 eq) of **11** was suspended in 1000 mL of 95% ethanol in an RB flask. Then, 11.351 g (0.300 mol, 10 eq) of sodium borohydride was slowly added to it over 10 min, and the reaction was stirred under nitrogen for nine days. Next, 6 M HCl was added to the reaction until the solution reached a pH of 2.

The ethanol was evaporated, and the white residue was then dissolved into 50 mL of DI H_2O. Following this, 200 mL of 30% KOH solution was added, and the pH of the solution reached 14. Extraction was carried out with 5 × 175 mL fractions of benzene. The benzene layers were collected and dried over Na_2SO_4 overnight. After filtration of sodium sulfate, benzene was evaporated to give the pale yellow oily product, which was dried under vacuum to yield 6.432 g (65%) of **L11**. Elemental Analysis (%): calc $C_{20}H_{32}N_4 \cdot 0.1\ C_6H_6 \cdot 1.6\ H_2O$ (365.13 g/mol): C 67.76, H 9.88, N 15.34; Found C 67.61, H 9.69, N 14.99. MS (ES) m/z = 329 $(LH)^+$.

Synthesis of **L12**. A total of 9.243 g (0.0180 mol, 1 eq) of **12** was suspended in 1000 mL of 95% ethanol in a 2 L RB flask. Then, 6.798 g (0.180 mol, 10 eq) of sodium borohydride was slowly added to it over 10 min, and the reaction was stirred under nitrogen for eight days. Next, 6 M HCl was added to the reaction until the solution reached a pH of 2. The ethanol was evaporated, and the white residue was then dissolved into 50 mL of DI H_2O. Following this, 200 mL of 30% KOH solution was added, and the pH of the solution reached 14. Extraction was carried out with 5 × 175 mL fractions of benzene. The benzene layers were collected and dried over Na_2SO_4 overnight. After filtration of sodium sulfate, benzene was evaporated to give the pale yellow oily product, which was dried under vacuum to yield 6.044 g (94%) of **L12**. Elemental Analysis (%): calc $C_{22}H_{36}N_4 \cdot 1.2\ H_2O$ (378.17 g/mol): C 69.87, H 10.23, N 14.82; Found C 69.79, H 9.84, N 15.11. MS (ES) m/z = 357 $(LH)^+$.

3.2.2. Synthesis of Novel Metal Complexes

General Procedure for Complexation of $CuCl_2$, $MnCl_2$, or $FeCl_2$

In an inert atmosphere glove box, 0.001 mol of ligand and 0.001 mol of anhydrous MCl_2 were stirred in 15 mL anhydrous aprotic solvent if M = Fe/Mn (either MeCN, DMF), or MeOH of M = Cu, for 24 h. Products, blue (Cu), white (Mn), or light brown (Fe), may precipitate and be filtered from the reaction solution (Fraction A) and be washed with MeCN and allowed to air dry in the glovebox. Alternatively, products may remain soluble in the reaction mixture (Fraction B). For "Fraction B" complexes, 200 mL of anhydrous ether were added to the filtrate, which resulted in the precipitation of the product, which was collected on a fine frit, washed with ether, allowed to dry while it was open to the glovebox atmosphere, and then scraped from the frit with a spatula into a vial for storage. For Cu, "Fraction C" complexes were necessary in some cases to achieve pure products: the reaction mixture was removed from the glovebox, filtered to remove trace solids, and evaporated to dryness. The residue was then dissolved in a minimum of methanol, and a methanol solution of 5 eq of NH_4PF_6 was added to precipitate the blue solid products, which were filtered off, washed with methanol and ether, and dried under vacuum.

[Cu**L3**Cl_2]: MeCN. Fraction B. Yield: 0.213 g (55%). Elemental Analysis (%): calc $[Cu(C_{14}H_{28}N_4)Cl_2] \cdot 1.5\ H_2O$ (386.85 g/mol): C 40.63, H 7.55, N 13.54; Found C 40.86, H 7.21, N 13.70. MS (ES) m/z = 352 $(CuLCl)^+$.

Mn**L3**Cl_2: MeCN. Fraction B. Yield: 0.172 g (45%). Elemental Analysis (%): calc $[Mn(C_{14}H_{28}N_4)Cl_2] \cdot 0.8\ H_2O$ (392.66 g/mol): C 42.82, H 7.60, N 14.27; Found C 42.82, H 7.44, N 14.39. MS (ES) m/z = 342 $(MnLCl)^+$.

Fe**L3**Cl_2: MeCN. Fraction B. Yield: 0.157 g (41%). Elemental Analysis (%): calc $[Fe(C_{14}H_{28}N_4)Cl_2] \cdot 0.3\ H_2O$ (384.55 g/mol): C 43.73, H 7.50, N 14.57; Found C 43.63, H 7.58, N 14.55. MS (ES) m/z = 343 $(FeLCl)^+$.

[Cu**L4**Cl][PF_6]: MeCN. Fraction C. Yield: 0.286 g (49%). Elemental Analysis (%): calc $[Cu(C_{16}H_{32}N_4)Cl_{0.5}][PF_6]_{1.5}$ (579.18 g/mol): C 33.18, H 5.57, N 9.67; Found C 33.15, H 5.62, N 9.65. MS (ES) m/z = 379 $(CuLCl)^+$.

Mn**L4**Cl_2: MeCN. Fraction B. Yield: 0.185 g (45%). Elemental Analysis (%): calc $[Mn(C_{16}H_{32}N_4)Cl_2] \cdot 1.0\ H_2O$ (424.31 g/mol): C 45.29, H 8.08, N 13.20; Found C 45.18, H 8.00, N 13.37. MS (ES) m/z = 370 $(MnLCl)^+$.

Fe**L4**Cl_2: MeCN. Fraction B. Yield: 0.163 g (40%). Elemental Analysis (%): calc $[Fe(C_{16}H_{32}N_4)Cl_2] \cdot 1.5\ H_2O$ (424.31 g/mol): C 44.26, H 8.12, N 12.90; Found C 43.87, H 7.66, N 13.27. MS (ES) m/z = 371 $(FeLCl)^+$.

[Cu**L7**Cl][PF_6]: MeOH. Fraction B. Yield: 0.289 g (79%). Elemental Analysis (%): calc [Cu($C_{18}H_{30}N_4$)Cl] [PF_6] · 2.5 H_2O (481.95 g/mol): C 44.86, H 7.32, N 11.63; Found C 44.51, H 7.26, N 11.89. MS (ES) m/z = 400 (CuLCl)$^+$.

Mn**L7**Cl_2: DMF. Fraction A. Yield: 0.295 g (69%). Elemental Analysis (%): calc [Mn($C_{18}H_{30}N_4$)Cl_2] (428.30 g/mol): C 50.48, H 7.06, N 13.08; Found C 50.38, H 7.24, N 12.92. MS (ES) m/z = 393 (MnLCl)$^+$.

Fe**L7**Cl_2: DMF. Fraction A. Yield: 0.308 g (72%). Elemental Analysis (%): calc [Mn($C_{18}H_{30}N_4$)Cl_2] · 1.0 DMF. 1.0 H_2O (511.31 g/mol): C 49.33, H 7.49, N 13.70; Found C 49.41, H 7.66, N 13.53. MS (ES) m/z = 394 (FeLCl)$^+$.

[Cu**L8**Cl]PF_6: MeOH. Fraction C. Yield: 0.523g (84%). Elemental Analysis (%): calc [Cu($C_{20}H_{34}N_4$)Cl][PF_6] · 0.3 NH_4PF_6 (623.38 g/mol): C 38.54, H 5.69, N 9.66; Found C 38.22, H 5.75, N 9.72. MS (ES) m/z = 428 (CuLCl)$^+$.

[Cu**L9**Cl]PF_6: MeOH. Fraction C. Yield: 0.169 g (31%). Elemental Analysis (%): calc [Cu($C_{16}H_{30}N_4$)Cl][PF_6] · 1.0 H_2O (540.41 g/mol): C 35.56, H 5.97, N 10.37; Found C 35.67, H 5.84, N 10.57. MS (ES) m/z = 376 (CuLCl)$^+$.

Mn**L9**Cl_2: MeCN. Fraction B. Yield: 0.238 g (59%). Elemental Analysis (%): calc [Mn($C_{16}H_{30}N_4$)Cl_2] · 0.5 H_2O (413.29 g/mol): C 46.50, H 7.56, N 13.56; Found C 46.73, H 7.69, N 13.85. MS (ES) m/z = 369 (MnLCl)$^+$.

Fe**L9**Cl_2: MeCN. Fraction B. Yield: 0.188 g (46%). Elemental Analysis (%): calc [Fe($C_{16}H_{30}N_4$)Cl_2] · 0.5 H_2O (414.19 g/mol): C 46.40, H 7.54, N 13.53; Found C 46.33, H 7.63, N 13.58. MS (ES) m/z = 370 (FeLCl)$^+$.

[Cu**L10**Cl][PF_6]: MeOH. Fraction C. Yield: 0.226 g (41%). Elemental Analysis (%): calc [Cu($C_{18}H_{34}N_4$)Cl][PF_6] · 1.5 H_2O (540.41 g/mol): C 37.44, H 6.46, N 9.70; Found C 37.52, H 6.11, N 9.43. MS (ES) m/z = 404 (CuLCl)$^+$.

Mn**L10**Cl_2: MeCN. Fraction B. Yield: 0.217 g (50%). Elemental Analysis (%): calc [Mn($C_{18}H_{34}N_4$)Cl_2] · 1.0 H_2O (450.35 g/mol): C 48.01, H 8.06, N 12.44; Found C 48.30, H 8.31, N 12.45. MS (ES) m/z = 397 (MnLCl)$^+$.

Fe**L10**Cl_2: MeCN. Fraction B. Yield: 0.195 g (45%). Elemental Analysis (%): calc [Fe($C_{18}H_{34}N_4$)Cl_2] · 1.5 H_2O (460.26 g/mol): C 46.97, H 8.10, N 12.17; Found C 46.93, H 7.98, N 12.06. MS (ES) m/z = 398 (FeLCl)$^+$.

Mn**L11**Cl_2: MeCN. Fraction B. Yield: 0.253 g (56%). Elemental Analysis (%): calc [Mn($C_{20}H_{32}N_4$)Cl_2] · 0.5 H_2O (424.31 g/mol): C 51.84, H 7.18, N 12.09; Found C 51.59, H 7.19, N 11.94. MS (ES) m/z = 419 (MnLCl)$^+$.

Fe**L11**Cl_2: MeCN. Fraction B. Yield: 0.386 g (85%). Elemental Analysis (%): calc [Fe($C_{20}H_{32}N_4$)Cl_2] · 3.5 H_2O (518.30 g/mol): C 46.35, H 7.58, N 10.81; Found C 46.64, H 7.24, N 10.50. MS (ES) m/z = 420 (FeLCl)$^+$.

Mn**L12**Cl_2: MeCN. Fraction A. Yield: 0.147 g (30%). Elemental Analysis (%): calc [Mn($C_{22}H_{36}N_4$)Cl_2] · 0.3 H_2O (487.80 g/mol): C 54.17, H 7.56, N 11.49; Found C 54.02, H 7.55, N 11.50. MS (ES) m/z = 446 (MnLCl)$^+$.

Fe**L12**Cl_2: MeCN. Fraction A. Yield: 0.162 g (34%). Elemental Analysis (%): calc [Fe($C_{22}H_{36}N_4$)Cl_2] · 0.6 H_2O (494.11 g/mol): C 53.48, H 7.59, N 11.34; Found C 53.21, H 7.55, N 11.35. MS (ES) m/z = 447 (FeLCl)$^+$.

3.3. *Characterization*

3.3.1. Acid Decomplexation Studies

This experiment is a typical technique for comparing the kinetic stability of new cross-bridged ligands [21–23,31,49,52–54]. The d-d absorption band (generally near 600 nm) of the respective copper complexes (1 mM concentration) was studied on a Shimadzu UV-3600 spectrophotometer (Shimadzu, Kyoto, Japan) in the presence of acidic solutions of varying concentration. The decomposition behaviors of the different copper complexes in acidic conditions were monitored at increasing temperatures until they decomposed (in a matter of hours/days). As the absorption decreased over time, appearance of an isosbestic point on each spectrum confirmed the presence of a single decomposition product, likely $CuCl_{2(aq)}$. The decomposition of copper complexes followed pseudo-first-order kinetics.

The half-lives of the respective complexes were calculated from the plot of ln absorbance (absorbance at maxima of the spectra) vs. time.

3.3.2. X-ray Crystallography Studies

The crystal structure data for all of the metal complexes were collected using a Bruker diffractometer employing Mo Kα radiation (λ = 0.71073 Å), a TRIUMPH curved-graphite monochromator, and either an APEX-II or PHOTON-II detector. Data were corrected for the effects of absorption using a multi-scan method within SADABS. The sample was held at 120 K in an Oxford Cryosystems nitrogen gas cryostream.

L8 precursor: Data were collected through the SCrALS (Service Crystallography at the Advanced Light Source) program at Beamline 11.3.1 at the Advanced Light Source (ALS), Lawrence Berkeley National Laboratory. The instrument employed is based on a commercial Bruker diffractometer with a PHOTON 100 detector and an operating wavelength of 0.7749 Å. The data were corrected for the effects of beam decay and absorption using SADABS. The sample was held at 150 K in an Oxford Cryosystems nitrogen gas cryostream.

CCDC-2268469-2268475 contain the supplementary crystallographic data (Tables S1–S52) for this paper. These data can be obtained free of charge from The Cambridge Crystallographic Data Centre via www.ccdc.cam.ac.uk/data_request/cif.

3.4. Dye-Bleaching Studies

All of the dye-decolorization studies were performed in water at pH 7, following the procedure of Yin et al. [68]. The three dyes, methylene blue (MB), methylene orange (MO), and Rhodamine B (RhB), had concentrations of 3.35 mM, 3.83 mM, and 2.60 mM, respectively, whereas the metal complex stock solutions had a 1.00 mM concentration. For each experiment, 10.00 mL of dye solution was added to 10.00 mL of metal complex stock solution. pH was kept neutral with the usage of either 0.100 M HCl or 0.100 M NaOH. Further, water was added to each set to adjust the volume up to 25 mL. Accordingly, the final concentrations of the metal complex and three dyes, namely, MB, MO, and RhB, were calculated to be 0.40 mM, 1.34 mM, 1.53 mM, and 1.04 mM, respectively. For these studies, 0.25 mL of 30% H_2O_2 was added to each set to trigger the dye-bleaching process. To monitor the progress of the reaction, 1 mL reaction mixture aliquots were taken out at certain time intervals. After removal, each aliquot set was diluted to 100 mL with water in order to quench the bleaching reaction. Next, absorbance spectra were recorded, and the absorbance maxima values at 664 nm, 465 nm and 553 nm were monitored for reaction mixture containing MB, MO, and, RhB, respectively. Time zero was defined as the time of H_2O_2 addition. Further, the absorbance was monitored at 5, 10, 20, 30, 60, 120, 180, 240, 300 min, or more (until the dye reached ~10% of its actual absorbance). For each set, at least two trials were performed, and all the absorbance values reported are an average value of the trials.

4. Conclusions

Potentially copolymerizable allyl and benzyl pendant-armed ethylene cross-bridged tetraazamacrocycles were synthesized, and twenty of their novel Mn, Fe, and Cu complexes were produced and characterized. Two cis coordination cites are available for interaction with substrate and oxidant due to the cis-folded pseudo-octahedral geometry forced by the rigidifying ethylene cross-bridge, as demonstrated by multiple X-ray crystal structures. The cyclam-based ligands, in particular, provide superior stabilization with half-lives of the Cu complexes of many minutes or even hours in 5 M HCl at 50–90 °C, as determined by kinetic stability experiments under harsh acidic and high-temperature conditions. Reversible redox processes indicated stabilization of Fe^{2+}/Fe^{3+} and $Mn^{2+}/Mn^{3+}/Mn^{4+}$ oxidation states, supporting the likelihood of catalytic oxidation for these complexes. Finally, Mn cross-bridged cyclam complexes with at least one benzyl pendant arm performed best in dye-bleaching experiments with methylene blue, methyl orange, and rhodamine B. Future work will include selection of the most successful monomeric catalysts for copolymerization to produce future heterogeneous water purification materials.

Supplementary Materials: The following supporting information can be downloaded at: https://www.mdpi.com/article/10.3390/molecules28186466/s1. Figures S1–S12: ^{1}H and ^{13}C NMR spectra for novel organic compounds. Figure S13: Example of kinetic decomplexation plot to determine $t_{1/2}$ for copper complexes. Figures S14–S21: Cyclic voltammograms for novel Mn and Fe complexes. Figures S22–S33: Dye bleaching plots of catalyst, H_2O_2, and dye (MB, MO, RhB). Tables S1–S52: Crystallographic details for all new structures [69–71].

Author Contributions: Conceptualization, T.J.H.; methodology, T.J.H., L.G. and M.R.K. validation, L.G. and M.R.K.; formal analysis, T.J.H., S.M., J.A.K., A.G.O. and T.J.P.; investigation, T.J.H., M.R.K., L.G., J.T.P., W.A.T., E.M.A.A., A.G.M., D.J.C.-S., N.L.F., T.H., I.K., D.J.H., J.A.K., A.G.O. and T.J.P.; resources, T.J.H., J.A.K., A.G.O. and T.J.P.; data curation, T.J.H., S.M., J.A.K., A.G.O. and T.J.P.; writing—original draft preparation, S.M., T.J.H. and T.J.P.; writing—review and editing, T.J.H., J.A.K., A.G.O. and T.J.P.; visualization, T.J.H., S.M. and T.J.P.; supervision, T.J.H.; project administration, T.J.H.; funding acquisition, T.J.H., J.A.K. and A.G.O. All authors have read and agreed to the published version of the manuscript.

Funding: This material is based on work supported by the National Science Foundation under Grant No. OIA-1946093. Any opinions, findings, conclusions, or recommendations expressed in this material are those of the author(s) and do not necessarily reflect the views of the National Science Foundation. Acknowledgement is made to the donors of the American Chemical Society Petroleum Research Fund for partial support of this work. T.J.H., E.M.A.A., A.G.M., D.J.C.-S. and L.G. also acknowledge OK-LSAMP and NSF for support.

Data Availability Statement: Data presented in this study are available in the Supplementary Materials or upon request from the corresponding authors.

Acknowledgments: Samples for synchrotron crystallographic analysis were submitted through the SCrALS (Service Crystallography at Advanced Light Source) program. Crystallographic data were collected at Beamline 11.3.1 at the Advanced Light Source (ALS), Lawrence Berkeley National Laboratory. The ALS is supported by the U.S. Department of Energy, Office of Energy Sciences, under contract DE-AC02-05CH11231.

Conflicts of Interest: The authors declare no conflict of interest.

Sample Availability: Samples of the compounds are not available.

References

1. Schäfer, F.P.; Schmidt, W.; Volze, J. Organic dye solution laser. *Appl. Phys. Lett.* **1966**, *9*, 306–309. [CrossRef]
2. Mustroph, H.; Stollenwerk, M.; Bressau, V. Current Developments in Optical Data Storage with Organic Dyes. *Angew. Chem.* **2006**, *45*, 2016–2035. [CrossRef]
3. Chiu, Y.-H.; Chang, T.-F.M.; Chen, C.-Y.; Sone, M.; Hsu, Y.-J. Mechanistic Insights into Photodegradation of Organic Dyes Using Heterostructure Photocatalysts. *Catalysts* **2019**, *9*, 430. [CrossRef]
4. Tkaczyk, A.; Mitrowska, K.; Posyniak, A. Synthetic organic dyes as contaminants of the aquatic environment and their implications for ecosystems: A review. *Sci. Total. Environ.* **2020**, *717*, 137–222. [CrossRef]
5. Epling, G.A.; Lin, C. Photoassisted bleaching of dyes utilizing TiO2 and visible light. *Chemosphere* **2002**, *46*, 561–570. [CrossRef] [PubMed]
6. Wang, Z.; Yuan, R.; Guo, Y.; Xu, L.; Liu, J. Effects of chloride ions on bleaching of azo dyes by Co^{2+}/oxone regent: Kinetic analysis. *J. Hazard. Mater.* **2011**, *190*, 1083–1087. [CrossRef] [PubMed]
7. Zucca, P.; Neves, C.M.B.; Simões, M.M.Q.; Neves, M.D.G.P.M.S.; Cocco, G.S.; Sanjust, E. Immobilized Lignin Peroxidase-Like Metalloporphyrins as Reusable Catalysts in Oxidative Bleaching of Industrial Dyes. *Molecules* **2016**, *21*, 964. [CrossRef]
8. Santos, M.C.; Antonin, V.S.; Souza, F.M.; Aveiro, L.R.; Pinheiro, V.S.; Gentil, T.C.; Lima, T.S.; Moura, J.P.C.; Silva, C.R.; Lucchetti, L.E.B.; et al. Decontamination of wastewater containing contaminants of emerging concern by electrooxidation and Fenton-based processes—A review on the relevance of materials and methods. *Chemosphere* **2022**, *307*, 135763. [CrossRef]
9. Kiselev, V.M.; Evstrop'ev, S.K.; Starodubtsev, A.M. Photocatalytic degradation and sorption of methylene blue on the surface of metal oxides in aqueous solutions of the dye. *Opt. Spectrosc.* **2017**, *123*, 809–815. [CrossRef]
10. da Rocha, M.G.; Nakagaki, S.; Ucoski, G.M.; Wypych, F.; Machado, G.S. Comparison between catalytic activities of two zinc layered hydroxide salts in brilliant green organic dye bleaching. *J. Colloid Interface Sci.* **2019**, *541*, 425–433. [CrossRef]

11. Zhang, Y.H.G.; Zhang, G. A novel mixed-phase TiO_2/kaolinite composites and their photocatalytic activity for degradation of organic contaminants. *Chem. Eng. J.* **2011**, *172*, 936–943. [CrossRef]
12. Hubin, T.J.; McCormick, J.M.; Collinson, S.R.; Buchalova, M.; Perkins, C.M.; Alcock, N.W.; Kahol, P.K.; Raghunathan, A.; Busch, D.H. New iron(II) and manganese(II) complexes of two ultra-rigid, cross-bridged tetraazamacrocycles for catalysis and biomimicry. *J. Am. Chem. Soc.* **2000**, *122*, 2512–2522. [CrossRef]
13. Hubin, T.J.; McCormick, J.M.; Collinson, S.R.; Alcock, N.W.; Clase, H.J.; Busch, D.H. Synthesis and X-ray crystal structures of iron(II) and manganese(II) complexes of unsubstituted and benzyl substituted cross-bridged tetraazamacrocycles. *Inorg. Chim. Acta* **2003**, *346*, 76–86. [CrossRef]
14. Busch, D.H.; Collinson, S.R.; Hubin, T. Catalysts and Methods for Catalytic Oxidation. U.S. Patent 6,906,189, 14 June 2005.
15. Yin, G.; Danby, A.M.; Kitko, D.; Carter, J.D.; Scheper, W.M.; Busch, D.H. Olefin Epoxidation by Alkyl Hydroperoxide with a Novel Cross-Bridged Cyclam Manganese Complex: Demonstration of Oxygenation by Two Distinct Reactive Intermediates. *Inorg. Chem.* **2007**, *46*, 2173–2180. [CrossRef] [PubMed]
16. Yin, G.C.; Danby, A.M.; Kitko, D.; Carter, J.D.; Scheper, W.M.; Busch, D.H. Oxidative reactivity difference among the metal oxo and metal hydroxo moieties: pH dependent hydrogen abstraction by a manganese(IV) complex having two hydroxide ligands. *J. Am. Chem. Soc.* **2008**, *130*, 16245–16253. [CrossRef] [PubMed]
17. Chattopadhyay, S.; Geiger, R.A.; Yin, G.; Busch, D.H.; Jackson, T.A. Oxo- and Hydroxomanganese(IV) Adducts: A Comparative Spectroscopic and Computational Study. *Inorg. Chem.* **2010**, *49*, 7530–7535. [CrossRef]
18. Dong, L.; Wang, Y.; Lu, Y.; Chen, Z.; Mei, F.; Xiong, H.; Yin, G. Lewis-Acid-Promoted Stoichiometric and Catalytic Oxidations by Manganese Complexes Having Cross-Bridged Cyclam Ligand: A Comprehensive Study. *Inorg. Chem.* **2013**, *52*, 5418–5427. [CrossRef]
19. Tan, P.; Kwong, H.-K.; Lau, T.-C. Catalytic oxidation of water and alcohols by a robust iron(iii) complex bearing a cross-bridged cyclam ligand. *Chem. Commun.* **2015**, *51*, 12189–12192. [CrossRef]
20. Annunziata, A.; Esposito, R.; Gatto, G.; Cucciolito, M.E.; Tuzi, A.; Macchioni, A.; Ruffo, F. Iron(III) Complexes with Cross-Bridged Cyclams: Synthesis and Use in Alcohol and Water Oxidation Catalysis. *Eur. J. Inorg. Chem.* **2018**, *2018*, 3304–3311. [CrossRef]
21. Matz, D.L.; Jones, D.G.; Roewe, K.D.; Gorbet, M.J.; Zhang, Z.; Chen, Z.Q.; Prior, T.J.; Archibald, S.J.; Yin, G.C.; Hubin, T.J. Synthesis, structural studies, kinetic stability, and oxidation catalysis of the late first row transition metal complexes of 4,10-dimethyl-1,4,7,10-tetraazabicyclo [6.5.2]pentadecane. *Dalton Trans.* **2015**, *44*, 12210–12224. [CrossRef]
22. Jones, D.G.; Wilson, K.R.; Cannon-Smith, D.J.; Shircliff, A.D.; Zhang, Z.; Chen, Z.; Prior, T.J.; Yin, T.J.; Hubin, T.J. Synthesis, structural studies, and oxidation catalysis of the late-first-row-transition-metal complexes of a 2-pyridylmethyl pendant-armed ethylene cross-bridged cyclam. *Inorg. Chem.* **2015**, *54*, 2221–2234. [CrossRef] [PubMed]
23. Odendaal, A.Y.; Fiamengo, A.L.; Ferdani, R.; Wadas, T.J.; Hill, D.C.; Peng, Y.; Heroux, K.J.; Golen, J.A.; Rheingold, A.L.; Anderson, C.J.; et al. Isomeric Trimethylene and Ethylene Pendant-armed Cross-bridged Tetraazamacrocycles and in Vitro/in Vivo Comparisions of their Copper(II) Complexes. *Inorg. Chem.* **2011**, *50*, 3078–3086. [CrossRef] [PubMed]
24. Sun, X.; Wuest, M.; Weisman, G.R.; Wong, E.H.; Reed, D.P.; Boswell, C.A.; Motekaitis, R.; Martell, A.E.; Welch, M.J.; Anderson, C.J. Radiolabeling and in vivo behavior of copper-64-labeled cross-bridged cyclam ligands. *J. Med. Chem.* **2002**, *45*, 469–477. [CrossRef] [PubMed]
25. Boswell, C.A.; Sun, X.; Niu, W.; Weisman, G.R.; Wong, E.H.; Rheingold, A.L.; Anderson, C.J. Comparative in vivo stability of copper-64-labeled cross-bridged and conventional tetraazamacrocyclic complexes. *J. Med. Chem.* **2004**, *47*, 1465–1474. [CrossRef] [PubMed]
26. Heroux, K.J.; Woodin, K.S.; Tranchemontagne, D.J.; Widger, P.C.B.; Southwick, E.; Wong, E.H.; Weisman, G.R.; Tomellini, S.A.; Wadas, T.J.; Anderson, C.J.; et al. The long and short of it: The influence of N-carboxyethyl versus N-carboxymethyl pendant arms on in vitro and in vivo behavior of copper complexes of cross-bridged tetraamine macrocycles. *Dalton Trans.* **2007**, *21*, 2150–2162. [CrossRef]
27. Sprague, J.E.; Peng, Y.; Fiamengo, A.L.; Woodin, K.S.; Southwick, E.A.; Weisman, G.R.; Wong, E.H.; Golen, J.A.; Rheingold, A.L.; Anderson, C.J. Synthesis, Characterization and In Vivo Studies of Cu(II)-64-Labeled Cross-Bridged Tetraazamacrocycle-amide Complexes as Models of Peptide Conjugate Imaging Agents. *J. Med. Chem.* **2007**, *50*, 2527–2535. [CrossRef]
28. Anderson, C.J.; Wadas, T.J.; Wong, E.H.; Weisman, G.R. Cross-bridged Macrocyclic Chelators for Stable Complexation of Copper Radionuclides for PET Imaging. *Q. J. Nucl. Med. Mol. Imaging* **2008**, *52*, 185–192.
29. Ferdani, R.; Stigers, D.J.; Fiamengo, A.L.; Wei, L.; Li, B.T.Y.; Golen, J.A.; Rheingold, A.L.; Wiesman, G.R.; Wong, E.H.; Anderson, C.J. Syntheis, Cu(II) complexation, 64Cu-labeling and biological evaluation of cross-bridged cyclam chelators with phosphonate pendant arms. *Dalton Trans.* **2012**, *41*, 1938–1950. [CrossRef]
30. Burke, B.P.; Miranda, C.S.; Lee, R.E.; Renard, I.; Nigam, S.; Clemente, G.S.; D'Huys, T.; Ruest, T.; Domarkas, J.; Thompson, J.A.; et al. Cu PET Imaging of the CXCR4 Chemokine Receptor Using a Cross-Bridged Cyclam Bis-Tetraazamacrocyclic Antagonist. *J. Nucl. Med.* **2020**, *61*, 123–128. [CrossRef]
31. Wadas, T.J.; Wong, E.H.; Weisman, G.R.; Anderson, C.J. Coordinating Radiometals of Copper, Gallium, Indium, Yttrium, and Zirconium for PET and SPECT Imaging of Disease. *Chem. Rev.* **2010**, *110*, 2858–2902. [CrossRef]
32. Khan, A.; Nicholson, G.; Greenman, J.; Madden, L.; McRobbie, G.; Pannecouque, C.; De Clercq, E.; Ullom, R.; Maples, D.L.; Maples, R.D.; et al. Binding Optimization through Coordination Chemistry: CXCR4 Chemokine Receptor Antagonists from Ultrarigid Metal Complexes. *J. Am. Chem. Soc.* **2009**, *131*, 3416–3417. [CrossRef] [PubMed]

33. Maples, R.D.; Cain, A.N.; Burke, B.P.; Silversides, J.D.; Mewis, R.; D'huys, T.; Schols, D.; Linder, D.P.; Archibald, S.J.; Hubin, T.J. Aspartate-based CXCR4 chemokine receptor binding of cross-bridged tetraazamacrocyclic copper(II) and zinc(II) complexes. *Chem. A Eur. J.* **2016**, *22*, 12916–12930. [CrossRef] [PubMed]
34. Hoang, T.; Mondal, S.; Allen, M.B.; Garcia, L.; Krause, J.A.; Oliver, A.G.; Prior, T.J.; Hubin, T.J. Synthesis and Characterization of Late Transition Metal Complexes of Mono-Acetate Pendant Armed Ethylene Cross-Bridged Tetraazamacrocycles with Promise as Oxidation Catalysts for Dye Bleaching. *Molecules* **2023**, *28*, 232. [CrossRef] [PubMed]
35. Munnik, P.; De Jongh, P.E.; De Jong, K.P. Recent Developments in the Synthesis of Supported Catalysts. *Chem. Rev.* **2015**, *115*, 6687–6718. [CrossRef]
36. Ganesan, V.; Yoon, S. Direct Heterogenization of Salphen Coordination Complexes to Porous Organic Polymers: Catalysts for Ring-Expansion Carbonylation of Epoxides. *Inorg. Chem.* **2020**, *59*, 2881–2889. [CrossRef]
37. Huang, Y.; Liu, T.; Lin, J.; Lu, J.; Lin, Z.; Cao, R. Homochiral NIckel Coordination Polymers Based on Salen(Ni) Metalloligands: Synthesis, Structure, and Catalytic Alkene Epoxidation. *Inorg. Chem.* **2011**, *50*, 2191–2198. [CrossRef]
38. Yuan, G.; Jiang, H.; Zhang, L.; Liu, Y.; Cui, Y. Metallosalen-based crystalline pourous materials: Synthesis and property. *Coord. Chem. Rev.* **2019**, *378*, 483–499. [CrossRef]
39. Liu, T.-T.; Liang, J.; Xu, R.; Huang, Y.-B.; Cao, R. Salen-Co(III) insertion in multivariate cationic metal-organic frameworks for the enhanced cycloaddition reaction of carbon dioxide. *Chem. Commun.* **2019**, *55*, 4063–4066. [CrossRef]
40. Tan, L.; Tan, B. Hypercrosslinked porous polymer materials: Design, synthesis, and applications. *Chem. Soc. Rev.* **2017**, *46*, 3322–3356. [CrossRef]
41. Friedel, C.; Crafts, J.M. Sur une nouvelle méthode générale de synthèse d'hydrocarbures, d'acétones, etc. *Compte. Rend.* **1877**, *84*, 1392.
42. Begni, F.; Gullo, F.; Paul, G.; Rea, R.; Ferrari, M.-C.; Mangano, E.; Cossi, M.; Gatti, G.; Marchese, L. Optimization of the Friedel–Crafts Alkylation for the Synthesis of Hyper-Cross-Linked Polymers. *ACS Appl. Polym. Mater.* **2022**, *4*, 5281–5286. [CrossRef]
43. Leeuwen, P.W.N.M. Alkene Polymerisation. In *Homogeneous Catalysis*, 1st ed.; Springer: Dordrecht, The Netherlands, 2004; pp. 191–228. [CrossRef]
44. Zuo, X.; Mosha, D.; Archibald, S.J.; McCasland, A.K.; Hassan, A.M.; Givens, R.S.; Busch, D.H. Towards the soil poultice and a new separations methodology: Rebinding of macrocyclic metal complexes to molecularly imprinted polymers specifically templated via noncovalent interactions. *J. Coord. Chem.* **2005**, *58*, 21–39. [CrossRef]
45. Weisman, G.R.; Wong, E.H.; Hill, D.C.; Rogers, M.E.; Reed, D.P.; Calabrese, J.C. Synthesis and transition-metal complexes of new cross-bridged tetraamine ligands. *Chem. Commun.* **1996**, *8*, 947–948. [CrossRef]
46. Khan, M.O.F.; Keiser, J.; Amoyay, P.N.A.; Hossain, M.F.; Vargas, M.; Le, J.G.; Simpson, N.C.; Roewe, K.D.; Carder Freeman, T.N.; Hasley, T.R.; et al. Discovery of Antischistosomal Drug Leads Based on Tetraazamacrocyclic Derivatives and Their Metal Complexes. *Antimicr. Agents Chemoth.* **2016**, *60*, 5331–5336. [CrossRef] [PubMed]
47. Rohovec, J.; Gyepes, R.; Cisarova, I.; Rudovsky, J.; Lukes, I. Nucleophilic reactivity of perhydro-3,6,9,12-tetraazacyclopenteno[1,3-*f,g*]acenaphthylene. A unified approach to *N*-monosubstituted and *N,N''*-disubstituted cyclen derivatives. *Tetrahedron Lett.* **2000**, *41*, 1249–1253. [CrossRef]
48. Le Baccon, M.; Chuburu, F.; Toupet, L.; Handel, H.; Soibinet, M.; Deschamps-Olivier, I.; Barbier, J.-P.; Aplincourt, M. Bis-aminals: Efficient tools for bis-macrocycle synthesis. *New J. Chem.* **2001**, *25*, 1168–1174. [CrossRef]
49. Hubin, T.J.; McCormick, J.M.; Collinson, S.R.; Alcock, N.W.; Busch, D.H. Ultra rigid cross-bridged tetraazamacrocycles as ligands—The challenge and the solution. *Chem. Commun.* **1998**, *16*, 1675–1676. [CrossRef]
50. Hubin, T.J.; Walker, A.N.; Davilla, D.J.; Carder Freeman, T.N.; Epley, B.M.; Hasley, T.R.; Amoyaw, P.N.A.; Jain, S.; Archibald, S.J.; Prior, T.J.; et al. Tetraazamacrocyclic derivatives and their metal complexes as antileishmanial leads. *Polyhedron* **2019**, *163*, 42–53. [CrossRef]
51. Groom, C.R.; Bruno, I.J.; Lightfoot, M.P.; Ward, S.C. The Cambridge Structural Database. *Acta Cryst.* **2016**, *B72*, 171–179. [CrossRef]
52. Hubin, T.J.; McCormick, J.M.; Alcock, N.W.; Clase, H.J.; Busch, D.H. Crystallographic characterization of stepwise changes in ligand conformations as their internal topology changes and two novel cross-bridged tetraazamacrocyclic Copper(II) complexes. *Inorg. Chem.* **1999**, *38*, 4435–4446. [CrossRef]
53. AlHaddad, N.; Lelong, E.; Suh, J.-M.; Cordier, M.; Lim, M.H.; Royal, G.; Platas-Iglesias, C.; Bernard, H.; Tripier, R. Copper(II) and zinc(II) complexation with N-ethylene hydroxycyclams and consequences on the macrocyclic backbone configuration. *Dalton Trans.* **2022**, *51*, 8640–8656. [CrossRef] [PubMed]
54. Woodin, K.S.; Heroux, K.J.; Boswell, C.A.; Wong, E.H.; Weisman, G.R.; Niu, W.; Tomellini, S.A.; Anderson, C.J.; Zakharov, L.N.; Rheingold, A.L. Kinetic inertness and electrochemical behavior of Cu(II) tetraazamacrocyclic complexes: Possible implications for in vivo stability. *Eur. J. Inorg. Chem.* **2005**, *7*, 4829–4833. [CrossRef]
55. Hubin, T.J.; Alcock, N.W.; Morton, M.D.; Busch, D.H. Synthesis, structure, and stability in acid of copper(II) and zinc(II) complexes of cross-bridged tetraazamacrocycles. *Inorg. Chim. Acta* **2003**, *348*, 33–40. [CrossRef]
56. Walker, A.N.; Ayala, M.A.; Mondal, S.; Bergagnini, M.C.; Bui, P.J.D.; Chidester, S.N.; Doeden, C.I.; Esjornson, L.; Sweany, B.R.; Garcia, L.; et al. A Bridge Too Far? Comparison of Transition Metal Complexes of Dibenzyltetraazamacrocycles with and without Ethylene Cross-Bridges: X-ray Crystal Structures, Kinetic Stability, and Electronic Properties. *Molecules* **2023**, *28*, 895. [CrossRef]
57. Busch, D.H. The Compleat Coordination Chemistry--One Practioner's Perspective. *Chem. Rev.* **1993**, *93*, 847–860. [CrossRef]

58. Schubert, U.S.; Andres, P.R. New functional polymers and materials based on 2,2′:6′2″-terpyridine metal complexes. *Adv. Mater.* **2004**, *16*, 1043–1068.
59. Park, J.; Pasupathy, A.N.; Goldsmith, J.I.; Chang, C.; Yaish, Y.; Petta, J.R.; Rinkoski, M.; Sethna, J.P.; Abruna, H.D.; McEuen, P.L.; et al. Coulomb blockade and the Kondo effect in single-atom transistors. *Nature* **2002**, *417*, 722–725. [CrossRef]
60. Ledwon, P.; Ovsiannikova, D.; Jarosz, T.; Gogoc, S.; Nitschke, P.; Domagala, W. Insight into the properties and redox states of n-dopable conjugated polymers based on naphtalene diimide units. *Electrochim. Acta* **2019**, *307*, 525–535. [CrossRef]
61. Kulthanan, K.; Nuchkull, P.; Varothai, S. The pH of water from various sources: An overview for recommendation for patients with atopic dermatitis. *Asia Pacific. Allergy* **2013**, *3*, 155–160. [CrossRef]
62. Bayabil, H.K.; Teshome, T.; Li, Y.C. Emerging Contaminants in Soil and Water. *Front. Environ. Sci.* **2022**, *10*, 873499. [CrossRef]
63. Wang, J.; Chu, L.; Wojnárovits, L.; Takács, E. Occurrence and fate of antibiotics, antibiotic resistant genes (ARGs) and antibiotic resistant bacteria (ARB) in municipal wastewater treatment plant: An overview. *Sci. Total Environ.* **2020**, *744*, 140997. [CrossRef] [PubMed]
64. Niang, M.; Reichard, J.F.; Maier, A.; Talaska, G.; Ying, J.; Domingo, J.S.; Varughese, E.; Boczek, L.; Huff, E.; Reponen, T. Ciprofloxacin- and azithromycin-resistant bacteria in a wastewater treatment plant. *J. Occup. Environ. Hyg.* **2023**, *20*, 219–225. [CrossRef] [PubMed]
65. Mangalgiri, K.P.; Blaney, L. Elucidating the Stimulatory and Inhibitory Effects of Dissolved Organic Matter from Poultry Litter on Photodegradation of Antibiotics. *Environ. Sci. Technol.* **2017**, *51*, 12310–12320. [CrossRef] [PubMed]
66. Mangalgiri, K.P.; Timko, S.A.; Gonsior, M.; Blaney, L. PARAFAC Modeling of Irradiation- and Oxidation-Induced Changes in Fluorescent Dissolved Organic Matter Extracted from Poultry Litter. *Environ. Sci. Technol.* **2017**, *51*, 8036–8047. [CrossRef]
67. Mangalgiri, K.P.; Patton, S.; Wu, L.; Xu, S.; Ishida, K.P.; Liu, H. Optimizing Potable Water Reuse Systems: Chloramines or Hydrogen Peroxide for UV-Based Advanced Oxidation Process? *Environ. Sci. Technol.* **2019**, *53*, 13323–13331. [CrossRef]
68. Xu, A.; Xiong, H.; Yin, G. Decolorization of Dye Pollutants by Manganse Complexes with Rigid Cross-Bridged Cyclam Ligands and Its Mechanistic Investigations. *J. Phys. Chem. A* **2009**, *113*, 12243–12248. [CrossRef]
69. Dolomanov, O.V.; Bourhis, L.J.; Gildea, R.J.; Howard, J.A.K.; Puschmann, H.X. OLEX2: A Complete Structure Solution, Refinement and Analysis Program. *J. Appl. Cryst.* **2009**, *42*, 339–341. [CrossRef]
70. Sheldrick, G.M. SHELXT-Integrated Space-Group and Crystal-Structure Determination. *Acta Cryst.* **2015**, *A71*, 3–8. [CrossRef]
71. Sheldrick, G.M. Crystal Structure Refinement with SHELXL. *Acta Cryst.* **2015**, *C71*, 3–8. [CrossRef]

separations MDPI

Review

Adsorption-Based Pretreatment of Irrigation Water to Prevent Water Quality Issues

Tamás Kucserka [1,*], Gábor István Németh [1], Ivett Pálfi [1], Zsolt L. Kiss [1,2], Etelka Tombácz [1,*] and Ildikó Galambos [1]

[1] Soós Research and Development Center, University of Pannonia, Zrínyi M. str. 18., H-8800 Nagykanizsa, Hungary; nemeth.gabor@pen.uni-pannon.hu (G.I.N.); palfi.ivett@pen.uni-pannon.hu (I.P.); zsoltkiss027@gmail.com (Z.L.K.); galambos.ildiko@pen.uni-pannon.hu (I.G.)

[2] Department of Civil and Architectural Engineering, University of Wyoming, 1000 E. University Ave., Laramie, WY 82071, USA

* Correspondence: kucserka.tamas@pen.uni-pannon.hu (T.K.); tombacz.etelka@pen.uni-pannon.hu (E.T.)

Abstract: Access to and the use of irrigation water of adequate quality and targeted nutrient supplementation have become more widespread in recent years. Both crop quality and the irrigation system lifetime are affected by the quality of the water used for irrigation. Micro-irrigation (e.g., drip and sprinkle) is becoming increasingly common alongside the more typical irrigation methods, but it requires expertise and pre-treatment to ensure a proper water supply. The most significant problem is clogging, which can greatly reduce irrigation efficiency. Treatment for irrigation purposes mainly depends on the contaminants that are present in the water. The main treatment options available are biological, electromagnetic and electrostatic treatments, but these have a wide range of effectiveness levels compared to membrane separation technologies. In addition, adsorption treatments are also available, which, depending on the adsorbent used, can greatly improve the pre-treatment of irrigation water. This work provides an overview of adsorbents suitable for the treatment of irrigation water and their effectiveness. The separation of interfering components via adsorption is effective and promising for future application as the expected irrigation demands increase.

Keywords: irrigation; water treatment; adsorbents; circular economy

1. Introduction

1.1. The Role of Climate Change in the Necessity of Irrigation

One of the regional consequences of climate change is that weather extremes are becoming more frequent [1], so our water management needs to be reconsidered. The main problem in itself is not that the annual amount of precipitation decreases, but that the precipitation does not arrive evenly [2] and summer drought periods become more frequent [3]. Climate change and the consequent increasing dryness, forest fires, the lowering of the ground water level and the beginning of soil erosion all contribute to the process of desertification, but many factors directly related to humans are also to blame [4]. Industry and agriculture are primarily responsible for water consumption. The latter accounts for about half of the total water consumption [5]. To the best of our ability, we must do everything we can to prevent and reduce the damage caused by drought, so that we can supply water at the right time and in the right amount. Water supply plays a key role in the conditions for competitive agriculture, and the efficient use of irrigation water of adequate quality can be a key element of this supply.

1.2. The Role of Irrigation Systems

One of the biggest challenges of the 21st century is determining how to provide enough clean water for everyone (agriculture, industry and the entire living world) and not just the population [6]. Irrigation is the artificial application of water to the surface of the soil and then to deeper layers, using a variety of specialized equipment (e.g., pipes, pumps and sprinkler, dripping systems). Irregular rainfall distribution and drought make irrigation necessary in certain areas. A wide range of irrigation systems are available to distribute water evenly over the surface to be irrigated. The water used for irrigation may originate from ground water, spring water, well water, surface water, rivers, lakes, reservoirs, or even treated wastewater or desalinated water [7]. The use of irrigation and targeted nutrient supply in quality and intensive vegetable and fruit cultivation has been spreading more and more in recent years [8]. The quality of the crops, their aesthetic appearance and the life of the irrigation system are both influenced by the quality of the water used for irrigation [9]. There are many options available for irrigation water treatment. Biological treatment methods could be promising for organic and heavy metal compounds; however, they require precise operation and control. For salt removal, many studies have reported electromagnetic and electrostatic treatment as fast and quite easier treatment methods; however, the efficiency of these techniques could be lower since they are a pressure-driven membrane separation process. The use of adsorbents can be a good way to remove certain undesirable components of irrigation water. Two large groups of adsorbents are known: conventional and non-conventional adsorbents, which can achieve different efficiencies in water purification. Conventional adsorbents generally have a higher adsorption capacity compared to non-traditional adsorbents, which are readily available and therefore inexpensive, but have a lower adsorption capacity.

2. Overview of the Requirements for Irrigation Water

2.1. Legal Requirements

Nowadays, it is no longer enough to provide the physical conditions for irrigation; there are also an increasing number of legal requirements to be met. As environmental requirements are becoming more stringent due to the increasing value of clean water, no water-based activities can escape the scrutiny of the legal regulatory regime. In the following, the most important aspects and principles will be summarized; these may influence professional conclusions.

The regulatory background for irrigation covers a wide range of legislation, depending on the source, quality and agricultural area of the irrigation water.

In Hungary, the concept of water management and the related responsibilities serves to comply with the relevant EU and national legislation, i.e., it includes compliance with the requirements of Directive 91/271/EEC [10] on urban wastewater treatment, Directive 98/83/EC [11] on the quality of water intended for human consumption, the Water Framework Directive, Act LIII of 1995 [12] on the protection of the environment, Act LIII of 1996 [13] on the protection of nature, Act CLXXXV of 2012 [14] on waste, and Directive 2007/60/EC [15] on water damage.

The European Union Directive 2000/60/EC [16] is the most important and community-wide guideline for the protection of the good quality of the water bodies covered by the community.

The EU's Water Framework Directive (WFD) aims to achieve and maintain the good status of surface, coastal, transitional and groundwater, thereby contributing to the protection of ecosystems. An important objective of the Directive is to prevent the pollution of waters, and it sets limits on anthropogenic emissions. The measures taken to achieve and maintain the objectives set out in the WFD are summarized in the river basin management plan at a national level.

Decree 90/2008 (18 July 2008) [17] of the Ministry of Agriculture and Rural Development stipulates that a soil protection plan must be drawn up for, inter alia, irrigation for agricultural purposes, on the basis of which the justification for the measure must be presented and a proposal for protection against erosion must be made after soil sampling in an accredited laboratory.

The suitability of irrigation water should also be assessed in terms of abstraction. The requirements for the suitability of irrigation water are set out in the standard MSZ-10-640:1989 [18].

In response to the challenges of climate change, to improve the competitiveness and adaptability of agriculture, to promote the wider use of irrigated agriculture and to establish irrigation communities, the National Assembly has enacted Act CXIII of 2019 [19] on Irrigation Agriculture. According to the law, "In order to carry out irrigation farming, the holder of the water right of the water facility is entitled to an irrigation easement, under which the owner or user of the property served must tolerate the establishment and operation of a linear water facility for irrigation on his property by the farmer carrying out irrigation farming, to carry out the necessary water works for the purpose of irrigation for the continuation of his agricultural activity, and to transfer the irrigation equipment, provided that it does not preclude the proper use of the property".

It can be seen from the above that the Government is seeking to promote the competitiveness of agriculture and the protection of land and water through the appropriate use and quality of the land, in accordance with the objectives of the European Union. More modern, but more expensive irrigation technologies, can achieve the more sustainable use of water. These aspects will become increasingly important in the future.

2.2. *What Water Sources Can We Work with?*

The most common sources of water available for cultivation are surface water (water from lakes, canals, possibly rivers) and underground water [20]. For subsurface waters, we must separate groundwater and aquifers from a quality point of view. Groundwater is the water of the first water-tight layer, and water bodies shallower than 20 m (dug and drilled wells). Aquifers are located in layers deeper than 20 m, depending on the geological and hydrological conditions. These waters may only be extracted with pre-planned and authorized wells. The professional basis for this is to protect it from possible pollution from the surface, since it is our most valuable, cleanest, long-term water reserve [21].

From the perspective of producers and water users, the quality assessment of the three types of water is quite different. The differences are to be found in the constancy of the composition and the effects of environmental pollutants. Surface waters are constantly exposed to quality-affecting influences: inflows, water withdrawals, precipitation, washing in, dilution, and concentration, which constantly modify the composition. When using surface water for horticulture, constant caution is a good idea; at the very least, we should measure the salinity (EC) on a weekly basis, which draws our attention to major changes. Regarding groundwater wells, we can say that their waters can only be recommended for horticultural crops with greater compromises, because they can contain not only a significant amount of nutrients, but also harmful substances accumulated to a toxic level [22].

2.3. *Quality Requirements of Irrigation Water*

The quality of irrigation water varies from country to country. For irrigation, natural water quality would be the most suitable; however, we cannot usually ensure rainwater quality from the available water sources. Therefore, in addition to the positive effects, irrigation can also have negative effects (Table 1), which are partly due to the water quality [23].

Table 1. The positive and negative effects of irrigation [23].

Positive Effects	Negative Effects
better plant water supply	structural damage
increased nutrient access	deterioration of topsoil water management
increased nutrient intake	possible leaching of nutrients
leaching of harmful salts	salinization
protection against erosion and deflation	in case of overwatering reduction, rise in groundwater level, waterlogging

The quality of irrigation water refers to the range of properties of the water to be used that affect the soil, the life processes of the plant and the technology of water distribution.

Therefore, when determining the quality indicators of irrigation water, it is advisable to examine the triple interaction of water–soil–plant [7,24]. Methods for ensuring water quality have developed continuously and basically set certain limit values based on two aspects, namely the effects on the soil–plant combination and the effects on the operation of irrigation equipment. The first group basically includes chemical parameters, and the second group includes physical, chemical and biological parameters [25].

Among the chemical requirements, the most important characteristics are as follows:

- salinity,
- indicators expressing the effect of hydrocarbons and carbonates,
- the amount of sodium ions compared to calcium–magnesium ions,
- the relative ratio of magnesium,
- electrical conductivity,
- chloride, iron and manganese content [26].

2.3.1. pH

Changes in the pH of the water during cultivation will affect the quality of the plants. A low pH inhibits the absorption of vital nutrients such as calcium, potassium, magnesium and molybdenum [27]. At the same time, a low pH increases toxicity because some trace elements are absorbed too easily [28]. At too high a pH, other substances such as phosphates and other important trace elements become unavailable to the plant [29].

2.3.2. Salinity

The sodium adsorption ratio (SAR) is the most important characteristic. It means a value representing the relative amount of sodium ions to the combined amount of calcium and magnesium ions in water using the following formula: $SAR = [Na]/(([Ca] + [Mg])/2)^{1/2}$, where all concentrations are expressed as milliequivalents of charge per liter [30]. Salinity is generally dangerous for the soil, but it can also be harmful to plants. Salinity is expressed by the amount of salt dissolved in a volume unit (g/L) or by the specific electrical conductivity (EC) of the solution. When establishing the limit values, the primary consideration is that the irrigation water should not cause a level of salt accumulation in the soil that could be harmful. Therefore, it is not possible to establish a general rule regarding this, because it depends on many properties (e.g., water permeability, rainfall, irrigation method, salt tolerance of the plant [31]. It is generally accepted that if the salt content of the irrigation water is less than 500 mg/L ($EC < 0.78$ mS/cm), then the salt pool of the irrigated soil usually does not increase significantly [32].

As a result of certain environmental changes (e.g., concentration, dilution, increase in pH), some of the Ca and Mg ions precipitate out of waters with a high carbonate–hydrocarbonate content [33], thus increasing the proportion of Na ions in the solution, and thereby making the water salinizing [31].

2.3.3. The Amount of Sodium Ions Compared to Calcium-Magnesium Ions

Various parameters have been proposed to estimate the amount of carbonate and hydrocarbonate ions and their expected chemical interactions, e.g., phenolphthalein alkalinity, soda equivalent, saturation index, effective Ca and Mg concentration, relative hydrocarbonate and carbonate ratio [34,35]. The principle of certification in this area is that high-quality irrigation water must not contain free soda at all and must not show phenolphthalein alkalinity [36].

In terms of the cation composition of the irrigation water, a low sodium level is favorable [37]. However, in terms of the salinizing effect of water, the most important thing is not the absolute amount of Na-ions, but their proportion compared to other cations [38]. Sodium is a natural component and enters the water through the dissolution of minerals, but it can also be of municipal origin. It is the biggest problem in soil cultivation, where it has a harmful soil-destroying and salinizing effect; this is because it replaces other cations (e.g., calcium ion) on soil colloids and thus significantly changes the structure, functioning, nutrient and water management characteristics of the soil [39].

2.3.4. Manganese Content

A significant amount of magnesium ions from irrigation water with a high magnesium content can bind to soil colloids, which adversely affects the physical and water management properties of the soil. Therefore, it is advisable to take the Mg/(Ca + Mg) ratio or its percentage into account when classifying water [34,40].

3. Physical Structure of Irrigation Systems and the Arising Problems

For the irrigation of different crops, there are different watering options, ranging from methods that have been used for decades to modern precision options that are more efficient. Agriculture is the largest water-consuming sector, and therefore the focus of attention when water management is an important point of discussion. This sector has recently been confronted with trends in water supply restrictions, making the efficient use of optimized water resources through the use of multiple irrigation systems essential. Compared to conventional irrigation methods, drip irrigation systems provide frequent and efficient water supply and also control the amount of water used. Therefore, the impact of different irrigation systems should be studied to find the best option for each case [41].

3.1. Surface Irrigation Systems

The most common irrigation method is surface irrigation, which is mainly used for crops covering large areas [42]. This type of irrigation works via gravity, using pre-dug trenches without the need for a pump. The most commonly used surface irrigation systems are border and furrow irrigations. The latter is more commonly used for the water supply of row crops and orchards; however, it is more demanding and, due to the technology, results in more losses and lower efficiency in water use. This method has different design, management factors and elements that have been thoroughly analyzed [43]. Furrow irrigation is the oldest and most widely used irrigation system. Pressure irrigation methods remain less popular due to their high energy costs and the use of automation [44]. Some researchers have made comparisons between different irrigation systems in the Jaizan Plain based on soil and land characteristics. The results showed that drip and sprinkler irrigation methods are more efficient and effective than conventional surface techniques, where drainage is the most limiting factor. The research concludes that due to the rapid decrease in and shortage of water resources, current gravity irrigation methods need to be replaced by modern (drip and sprinkler) technologies as soon as possible [45].

3.1.1. Sprinkler Irrigation

The method of using sprinkler irrigation is similar to natural rainfall. Water is delivered through a pressurized system to the pipes that release it. Eventually, sprinklers will spray it into the air, where droplets are fragmented, and fall to the ground. The pump-fed system should be designed so that each sprinkler distributes an equal amount of water [46].

The micro sprinkler technology operates at low pressure and can be used in areas in which drip irrigation is not recommended or is not efficient enough to keep the plant foliage constantly moist, or in which overhead irrigation would be required. Micro sprinklers are categorized according to several aspects, such as flow rate, wetted diameter and the method of spraying, such as producing a mist or a spray [47].

Sprinkler irrigation is preferred over subsurface drip irrigation for cotton production, but this technique is not widespread due to climatic conditions and salt accumulation [48]. Some studies compare drip and sprinkler irrigation and show that for some crops, such as grapes, sprinkler irrigation results in a greater root surface area as you move away from the trunk [49]. The study in the Jaizan Plain also showed that drip and sprinkler irrigation is more effective than conventional surface irrigation. The major limiting factor was found to be the gravel soil structure, drainage and calcium carbonate when using sprinkler irrigation [45].

3.1.2. Drip Irrigation

Drip irrigation (also known as trickle irrigation) is a very slow dripping technique (2–20 L per hour) using a small-diameter plastic system of tiny emitters placed on the outside of the pipe, which are called drippers. The water is discharged close to the plant so only the soil near the root is wetted, instead of the whole soil profile, like in the case of surface irrigation and sprinkler irrigation. With drip irrigation, treatment is much more frequent (usually every 1–3 days) than other techniques; therefore, it provides a more favorable moisture level for the plants [50]. Under-foil drip irrigation is a complex water-efficient irrigation method that can directly improve the soil water–salt conditions in the root zone of plants and affects physiological characteristics such as plant emergence, root spread and above-ground growth, therefore improving plant growth, development and yield. Numerous studies have been carried out on the characteristics of soil water and salt movement during under-film drip irrigation [51].

Some of the most convincing reasons for using drip systems are cost efficiency, easy set-up and operation, as well as low maintenance costs [52]. The drip method provides micro irrigation, which helps to reduce water use and avoid unnecessary nutrient excess. Its operation is based on direct irrigation for roots, which provides a constant flow with equal distribution while minimizing evaporation. After proper pretreatment, water is delivered through the system's network of pipes to drippers placed close to the plants, which can be either holes in the pipes or units placed on top of them to deliver the water directly to the roots through capillary tubes. The accuracy of the system may depend on its design goals and its maintenance and operation technology [53]. The drip irrigation system will also include a slow-release device to control and optimize the flow of water to the soil. There are two different types of drip irrigation, which are subsurface and surface methods. The main difference is that the former is applied to the soil, while the latter near the roots. It is important to note that a maximum of two emitters per plant is sufficient for proper irrigation [53].

3.2. General Structure of Irrigation Systems

Traditionally used irrigation (border, basin, furrow) only requires a sufficient water surface and sometimes a pump for periodic irrigation. In contrast, precision irrigation types require the construction of a complete irrigation system. When using sprinklers or drip irrigation, the system design is almost the same, and only the method of water discharge varies.

The main components used in drip irrigation include a pressure regulation valve, a filter, a system controller, injectors, gauges, flow pipes and emitters [54]. Eventually, the water is delivered to the plants via polyethylene pipes, which have smaller holes at certain distances. In the installation, it is important that the distances are properly measured so that the pressure and water flow are nearly the same even at the farthest parts of the pipes [55].

From the relevant reservoir, a pump delivers the water through filters (mostly gravel) to the pressure regulator and then to the screen filter [55]. The water is then fed to the main branch, from where it is sent to several sub-branches (Figure 1). Fertilizers can be applied through this type of system. This can result in a reduction in the fertilizer amount and costs [56].

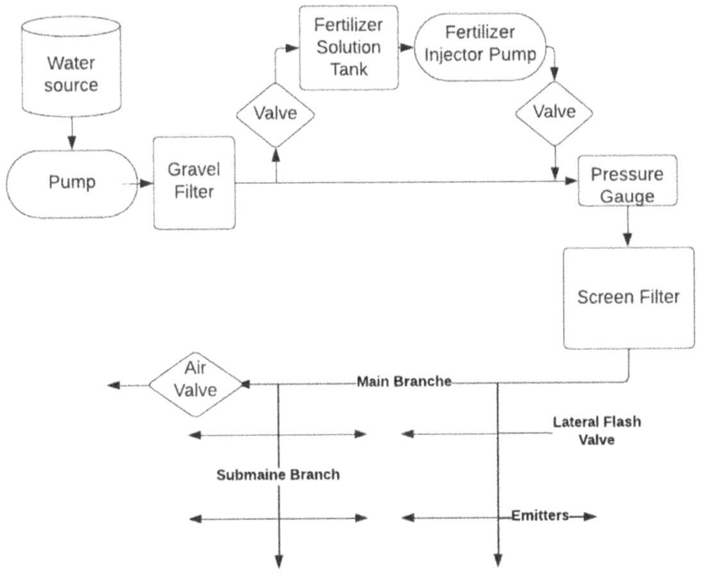

Figure 1. The general structure of drip irrigation systems.

3.3. *Clogging*

Emitter clogging is the biggest limiting factor affecting the use of drip irrigation. The use of poor-quality water in micro irrigation systems greatly increases the potential for clogging, reducing the possible areas of use and increasing maintenance costs. Studying the main elements of clogging and their effects and mechanisms of accumulation have become indispensable [57]. Fluid flow in narrow pipes and emitters can easily become clogged by solid particles, chemical constituents and microorganisms during irrigation [46].

3.3.1. Physical Clogging

Physical clogging is mainly caused by various organic and inorganic materials suspended in the water [58]. These particles of clay, sand, plastic and other residues cannot be filtered out and can therefore cause the clogging of the emitter. Physical clogging caused by solid particles accounts for nearly 31% of the clogging potential, making it one of the most common causes of clogging [46].

3.3.2. Chemical Clogging

Chemical clogging is caused by substances dissolved in water, such as carbonates, phosphates, Fe^{2+}, Ca^{2+} andMg^{2+}, which under certain conditions precipitate on the inner wall of the pipe. This deposition is greatly influenced by temperature, pressure, pH and the concentration of ions [59]. In addition, the nutrients added to the water also influence clogging; for example, phosphorus added to the water greatly increases the precipitation of carbonate, quartz and silicate [60]. Fe^{2+} can cause serious problems in the irrigation system if its concentration exceeds 1.5 g m^{-3} [46], and a water hardness above 300 mg L^{-1} also can threaten the irrigation system [61].

3.3.3. Biological Clogging

The drip system provides a suitable habitat for algae, bacteria and other organic organisms that can form slime. This material, in combination with other contaminants, can cause biofouling [62]. However, in underground drip systems, the effect of roots can also be considered as biological clogging [63].

4. Overview of Current Irrigation Water Treatment Methods

Nowadays, irrigation water treatment methods, in which natural water sources are extensively used, are becoming more focused and based on water pollution and hectic rainfalls. The most often reported problem is the high salinity of this water, which could negatively effect agriculture production even if it is not reduced to a level that is acceptable for the different plants. The effectiveness of treatment methods is influenced by the background quality of the irrigation water and includes physical, chemical, and biological parameters. Most of those factors could impact which water treatment is efficient to achieve the required water quality needs.

4.1. Overview of Contaminants

Organic contaminants are not desirable in water for irrigation based on infection risk, clogging and fouling issues. Organic matter is an organic substance that is broken down from plants, animals, microorganisms and from geological origins, and that can be found in water sources, especially in surface water in many cases [64]. These organic substances could increase the turbidity of the water supply and could affect disinfection methods, especially for UV sources. Different types of microbes may require different treatment methods, disinfection concentrations and contact times, with longer disinfection times possibly also helping to avoid infection. The pH of the irrigation water also affects the efficiency of certain treatments. For example, ozone treatments require different dosages based on pH, and chlorination is more effective in water with a lower pH [64,65].

Excess calcium and magnesium in a water source contribute to the hardness of water and can lead to the scaling of irrigation lines. This build up can decrease the effectiveness of water treatment processes such as UV, making it important to minimize scale deposits in the irrigation system [64,66].

4.2. Biological Treatment

Biological treatment is an effective method via which to reduce the organic content of irrigation water; however, one of its main challenges is keeping the microorganism alive in the treatment system and providing an equal water quality without harming the food safety or yield and the fruit or derivatives' quality. A membrane bioreactor was tested by treating polluted surface water. The results showed that up to 99.99% of the inlet *Escherichia coli* and 98.52% of the total coliforms were removable, as well as some harmful compounds and heavy metals in particular, including 82% of arsenic, 82% of cadmium, 97% of chromium, 93% of copper and 99% of lead; however, boron and manganese were not removed from the permeate. The sole gravel filter and heavy metal removal device (HMR) system were not able to remove the pathogen microorganisms. However, connecting the UV treatment could solve this issue [67]. In another study, a semi-closed horizontal tubular photobioreactor (PBR) at the pilot scale was tested to remove a total of 35 target compounds, including benzotriazoles, benzophenones, antibiotics and different pharmaceuticals present in irrigation water in a peri-urban rural area. This water running through an open channel and was a mixture of reclaimed wastewater from a nearby wastewater treatment plant (WWTP) and run-off from the different agricultural fields in the area. It has been reported that the compounds studied are usually not fully eliminated during conventional wastewater treatment, which justifies the need to investigate alternative treatment strategies. A total of 21 of these compounds were detected in the irrigation water. Benzotriazoles were only partially removed after the microalgae treatment, with an elimination efficiency of conventional WWTPs. The UV filter benzophenone-3 (BP3) showed variable removals, ranging from no elimination to 51%, whereas 4-methylbenzilidenecamphor (4MBC) was completely eliminated. Regarding pharmaceuticals, the average removals were higher and in the range of 60–100%; this is with the exception of the antibiotics sulfamethoxazole (46%) and sulphapyridine, which were not removed [68]. Removing target compounds from irrigation water has a key role in protecting plants and products from unwanted changes and from accumulating these compounds in the agricultural products.

4.3. Desalination Processes

Saline groundwater irrigation is a crucial method by which to reduce the shortage of freshwater resources. However, saline water could have impact on the yield of plants and on the soil.

4.3.1. Ultrasound Treatment

Ultrasound (US) treatment is one of the most promising treatment methods and has been evaluated in long-term studies with various plant species, including some highly salinity-sensitive plants, under greenhouse conditions. The ultra sonication system provided encouraging results when used to irrigate a number of vegetable species with high-electrical-conductivity irrigation water. The results from the long-term use of ultrasound treatment increased the yield of lettuce and celery plants [69].

4.3.2. Magnetic Treatment

Another low-cost and promising operating method is applying permanent magnets for irrigation water treatment [70]. Surendran and coworkers evaluated the effects of magnetic treatment on irrigation water and the growth and yield parameters of cow pea using pot and field experiments. During their experiment, normal water and hard water at 150 and 300 ppm, respectively, and saline water at 500, 1000 and 2000 ppm of both the control and magnetic-treated solutions were used, respectively. Permanent magnets with the capacity of 1800–2000 G were applied. Their investigation indicated that the magnetic treatment of irrigation water types achieved an improvement in the crop growth and yield parameters of cow pea as well. Magnetic treatments have the potential to decrease the electrical conductivity, total dissolved solids and salinity levels of all solutions. The increase in pH was observed in all the treatments. Magnetized irrigation water increased the soil moisture compared with the control sample [71].

Furthermore, of permanent magnet electromagnetic treatment was also investigated during the irrigation of corn and potato plants. The results showed a significant increase in the germination rate of corn seedlings watered with electromagnetic-treated saline water, particularly when the water was exposed to electromagnetic fields for 15 min. A significant reduction in the soil salinity (ECe) and Na^+ and Cl^- contents of the soils irrigated with electromagnetic-treated saline water was detected compared to the soils irrigated with non-treated saline water. The electromagnetic treatment of saline water caused significantly higher K, N and P adsorption values in all the potato samples [72]. Wiedenfeld studied the sugarcane yield and juice quality by using an electrostatic precipitator. The study was conducted to compare the effect of water irrigation with 1.3 dS m^{-1} vs. 3.4 dS m^{-1} on sugarcane and to evaluate whether the electrostatic conditioning treatment of the water influenced the salt effects. The results showed that the cane and sugar yields were reduced approximately 17% by the 3.4 dS m^{-1} water compared to the 1.3 dS m^{-1} water, but the juice quality parameters were not affected by the soil salinity levels [73].

4.3.3. Ionization

The ability of ionizing brackish water to prevent salinity damage to cotton plant growth with an improving irrigation water productivity was investigated by Wei and coworkers, who applied deep irrigation treatments. As the results indicated, the ionization effectively reduced the mass of salt from the late seedling stage to harvest by 12.8–65%, especially for the treatment using 562.5 mm of ionized brackish water. The maximum amount of cotton was obtained in the treatment using 487.5 mm of ionized brackish water, which increased to an average of 19% compared to 487.5 mm in the non-ionized brackish water treatment experiment. The water use efficiency improved under the ionized brackish water treatments [74].

4.3.4. Membrane Separation

The membrane separation process is well known in many water and wastewater treatment applications; however, its operation and maintenance costs could decrease its popularity for use in irrigation water treatment. In special cases, geothermal water treatment is a possible economically worthwhile treatment method. In many areas of Hungary and Europe, geothermal water is used for heating purposes. However, the used/spent geothermal water rarely gets injected back into the ground. Some amount of used geothermal water ends up in rivers or lakes and causes increasing salinity levels in surface water bodies; this could have an effect on the natural vegetation and fauna. Jarma and co-workers investigated the use of treated geothermal water for the irrigation of tomato plants. Membranes (NF8040-70 as NF membrane and TM720D-400 as RO membrane) were selected for this experiment. According to the results, almost all of the minerals that are needed for plant growth were rejected by the tested membranes. Therefore, for remineralization purposes, mixed water was also tested. In the case of the NF membrane, the permeate was mixed 1:1 with non-treated geothermal water; this ratio was 2:3 in the case of the RO permeate and geothermal water, according to the higher level of salt rejected by the RO membrane. The costs of the product water were found to be 0.76 and 1.56 $\$/m^3$ for the NF and RO processes, respectively [75]. The management of saline water through desalination or blending with low-salinity canal water may be a viable strategy for potato irrigation in semi-arid regions that have saline groundwater and scarce canal water. A 4-year study was conducted to investigate the effects of four water qualities [canal water (CW), desalinated water (DSW), saline groundwater (GW) and mixed water (MW) (CW + GW)], three N–P–K fertigation levels [F100 = 100% NPK (190, 28.4 and 53.9 kg ha^{-1} of N, P and K, respectively), F80 = 80% NPK and F60 = 60% NPK] and their interaction on the performance of two potato varieties grown in a sandy loam alluvial soil of a semi-arid region. The electrical conductivity (EC) of CW, DSW, GW and MW ranged from 0.3 to 0.4, 0.2–0.3, 3.2–4.2 and 1.8–2.2 dS m^{-1}, respectively. Drip irrigation with GW resulted in a 21–44% reduction in the total tuber yield compared with CW. Irrigation with MW increased the total tuber yield by 9–37% compared with GW. Irrigation with DSW effected a 41% and 14% increase in the total tuber yield compared with GW and MW, respectively. The plant growth and yield parameters reached their highest amount at F100, which reduced dramatically with a decrease in the fertigation level for all water qualities. However, these differences were slight between GW and MW, and more moderate between CW and DSW. That is the reason why, in drip-fertigated potatoes using saline or mixed water in a sandy loam soil, it is suggested that 80% NPK is used to avoid a reduction in the tuber yield and to reduce the expense of cultivation and environmental pollution [76].

4.4. *Effects of Salinity*

Saline water could have an effect on soil nitrogen transport, organic and inorganic content of the soil, and yield of the products. Therefore, when irrigating with saline water, all of its effects need to be considered. In order to utilize underground saltwater resources, and furthermore reduce the amount of irrigation water and nitrogen fertilizer used, Wang and coworkers carried out a field experiment in Xinjiang. The study was conducted to explore the effects of the irrigation water's quantity, salinity, and nitrogen application rate on nitrogen fertilizer absorption and cotton yield. Three experimental factors and four levels were investigated in the experiment. Irrigation amounts of 0.7, 1.0, 1.3 and 1.6 Etc, irrigation water salinity levels of 10.9, 7.8, 4.7, 1.5 dS m^{-1} and nitrogen applications of 100, 200, 300 and 400 kg N ha^{-1} were set in the experiment. The results indicated that the salt content of the irrigation water had no effect on the soil nitrogen transport, and the soil nitrogen content only increased when more nitrogen fertilizer was used [77]. The experiment was performed by using fresh water and saline water at the seedling, jointing, heading, and filling stages. The salinity of the soil, growth, and amount of summer maize were measured by Sang and coworkers. The results showed that alternating fresh and saline water irrigation increased the salinity of the soil. The maximum quantum yield, effective quantum yield of photochemical energy conversion, photochemical quenching, and non-photochemical quenching varied greatly at the jointing stage and heading stage. Furthermore, the yield of maize that was irrigated with fresh water at the heading stage (8.53 t ha^{-1}) was greater than that at the jointing (7.69 t ha^{-1}) and filling stages (7.45 t ha^{-1}). Therefore, these findings indicate that in areas in which fresh water is scarce, priority should be given to the application of fresh water at the heading stages for summer maize irrigation [78]. Saline water irrigation's effect on the soil organic carbon (SOC) content was smaller compared to its effect on the soil inorganic carbon (SIC) content. Dong et al. irrigated soil with 1 g L^{-1} and 8 g L^{-1} of saline water in a winter wheat and summer maize rotation system. After 14 years of irrigation, the soils were analyzed in terms of their soil water, soil salt, SOC, and SIC contents. The results showed that, compared with 1 g L^{-1} of water irrigation, 8 g L^{-1} of saline water irrigation significantly increased the soil water and salt contents. Moreover, 8 g L^{-1} of saline water irrigation significantly decreased the SOC and SIC contents in the 0–20 cm soil layer ($p < 0.05$) and mainly decreased the SOC content in the >1 mm aggregates and the wheat-derived SOC content in the bulk soil. In comparison, 4 g L^{-1} of saline water had no significant effect on the soil water, soil salt, SOC, and SIC contents. These results indicated that a high concentration of saline water irrigation is harmful for soil carbon sequestration, while a low concentration of saline water does not affect soil carbon sequestration. Thus, using no more than 4 g L^{-1} of saline water irrigation for 14 years can maintain soil carbon storage in areas experiencing a water shortage [79].

Overall, several treatment methods are available and have been proven to reduce natural or anthropogenic organic and inorganic contaminants, microorganisms, and salt to achieve the safe irrigation of water; these do not reduce the yield of the products and do not change the soil and groundwater characteristics. However, some of the above-mentioned target compounds and contaminants could have long-term effects on the environment, but these effects are not yet well known.

4.5. Adsorbents in Irrigation Water Treatment

The use of wastewater for irrigation purposes is becoming widespread, but this raises new problems, such as the presence of heavy metals [80]. Although the heavy metal content in wastewater is low, it can accumulate in the soil and reach crops [81]. Adsorption is a suitable treatment technology for removing metal (for example Cd and Pb), or for reducing salinity, depending on the material used [82]. The use of adsorbents can be implemented using a flow-through system (similar to a sand or gravel filter) or by mixing them in a tank for a suitable period of time. Combinations of certain adsorbents are able to remove chemical oxygen demand and the total dissolved solids, and also reduce the turbidity very efficiently [83]. Adsorbents for the pre-treatment of irrigation water and their effects are discussed in more detail below.

5. Overview of the Most Important, Easily Available Adsorbents Based on Their Effectiveness

In water treatment, adsorption is an effective process for the removal of contaminants, in which ions and molecules are removed from the liquid phase via adsorption to the solid phase [84,85]. The adsorbent material can be a mineral, organic or biological material [86]. The selection of a suitable adsorbent for the removal of pollutants in water depends on the concentration and type of the pollutant, the adsorbent's adsorbing capacity, and its efficiency [85]. The adsorption capacity of adsorbents is influenced by temperature, pH, the adsorbent dose, adsorbent particle size, agitation speed, and contact time [87]. A direct comparison of the results obtained using different adsorbents is therefore not possible since the experimental conditions are not the same during the different tests. Adsorbents can be divided into two major categories: conventional adsorbents and non-conventional adsorbents (Figure 2, Table 2) [86].

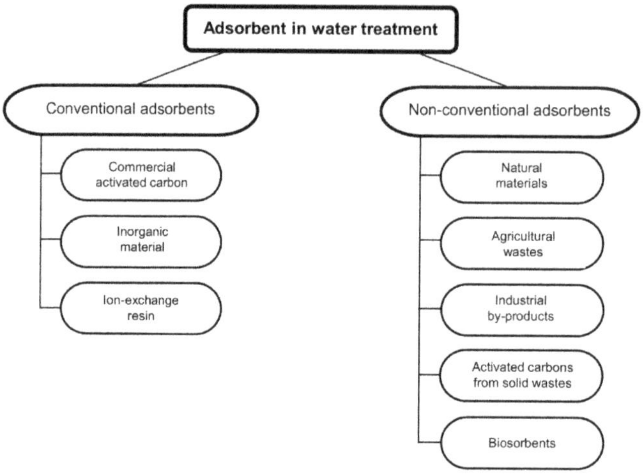

Figure 2. Different types of adsorbents for the removal of pollutants from water [86].

Table 2. Some examples of the most used adsorbents in water treatment, conditions and their efficiency.

Type of Adsorbent	Type of Pollutants	pH	Adsorption Efficiency (mgg^{-1})	References
	Conventional adsorbents			
Activated carbon + cation resin	K^+ Na^+ Ca^{2+} Mg^{2+} Cl^- HCO_3^-	4.50 4.50 4.50 4.50 4.50 4.50	0.26 13.80 2.50 2.73 11.50 18.30	[88]
Activated carbon + zeolite	K^+ Na^+ Ca^{2+} Mg^{2+} Cl^- HCO_3^-	7.20 7.20 7.20 7.20 7.20 7.20	0.05 2.10 1.00 0.12 0.90 4.00	[88]
Activated carbon	Hardness		0.86	[89]
Zeolite	Mn^{2+}	8.70	30.89	[90]
Thermally activated natural zeolite (NZ 200)	Fe^{2+} Mn^{2+}	8.95 8.95	6.13 0.86	[91]
Modified zeolite	Na^+	7.87 ± 0.25	N/A	[92]
TiO_2@Zeolite	Mn^{2+} Fe^{3+} Hardness	7.00 7.00 7.00	94.10 150.10 131.8 (Ca^{2+}) 703.6 (T.H.) *	[93]
Activated carbon + nano zero-valent iron + natural zeolite	COD	6.50–8.50	35.64	[83]
Zeolite clinoptilolite	Hardness		7.48	[89]
Activated alumina	Hardness		44954.00	[89]
	Non-conventional adsorbents			
Sawdust and barley husk	Fe^{2+}	6.50	N/A	[94]
Manganese oxide-coated sand1 (MOCS1) Manganese oxide-coated sand2 (MOCS2) Iron oxide-coated sand1 (IOCS2) Iron oxide-coated sand2 (IOCS2) Manganese and iron oxide-coated sand (MIOCS)	Mn^{2+}	8.00	2.61 0.83 0.40 0.55 0.88	[95]
Saccharomyces cerevisiae	Mn^{2+}	4.50–9.00	N/A	[96]
Pleurotus mushroom compost	Mn^{2+}	6.00	N/A	[97]

* T.H.: Total Hardness.

5.1. Conventional Adsorbents

5.1.1. Activated Carbon

One of the most common adsorbents currently in the field of water treatment is activated carbon [85,98]. Carbon-based materials such as activated carbon (AC) can be used well as adsorbents, proving to be highly effective in removing pollutants from water [83,85,99]. Activated carbon serves as a versatile material for the removal of heavy metals and organic compounds [100–102] and desalination [83,88,103]. The term 'activated carbon' is used for carbon-containing materials with a high micropores volume, favorable pore size distribution, extensive surface area (600–4300 m^2g^{-1}) [85,104] and high adsorption capacity [83,85,98,105,106]. Almost any carbonaceous material can be used to produce activated carbon, such as coconut shells, lignite, wood, sawdust, coal, and peat [98,103,105,107,108], but its properties are influenced by the raw material and the activation method [85]. Activated carbons are generally very-broad-spectrum adsorbents [86] and their low price makes them favorable [83]. Furthermore, their favorable features include their high thermal, mechanical, and light stability [89]. AC is available in two forms: powdered activated carbon (PAC) and granular activated carbon (GAC) [98,105]. PAC generally comprises particles with a diameter of 0.2 mm, and it has a large external surface area and a high adsorption rate [105]. GAC comprises particles with a maximum diameter of 5 mm, and its external surface area is smaller than that of PAC [105]. It is generally used in fixed-bed filter systems to remove pollutants from water because the granular form is better suited to continuous contact [98,105]. The adsorption efficiency of GAC can be affected by the carbon source material, carbon particle size, influent contaminant concentration, and contact time [109]. Grant et al. [109] conducted a study using a small-scale Granular Activated Carbon (GAC) system to effectively remove a range of agrichemicals, including acephate, bifenthrin, chlorpyrifos, flurprimidol, glyphosate, hydrogen peroxide + peracetic acid, imidacloprid, paclobutrazol, didecyldimethylammonium chloride (DDAC), triclopyr, and uniconazole, from water. In laboratory conditions, a decrease in the concentration of plant protection agents was observed in all cases, and a greater percentage decrease was observed in 64 s of contact time than in 12 s. At the same time, it was pointed out that further studies are needed using the high flow rate, water pressure, volume, and level of suspended solids typically found in recirculated irrigation systems.

Ghanbarizadeh et al. [89] compared the adsorption capacity of unmodified activated carbon with activated carbons modified using HCl, NaCl and HCl + NaCl. The modified activated carbon, which was first treated with salt and then with acid, was the most effective in its total hardness removal.

Activated carbon can prove to be even more effective in water purification in combination with other adsorbents. Aghakhani et al. [88] conducted experiments on the desalination of irrigation water by combining activated carbon with other adsorbents, zeolite, peat, anion exchange resin, and cation exchange resin. The combination of AC with cationic resin worked most effectively in desalination.

5.1.2. Carbon Nanotubes

Nano-adsorbents show great promise in water treatment and water purification compared to traditional adsorbents, as they can provide greater efficiency and a faster adsorption speed due to their large surface area [85,110]. Activated carbon nanotubes are one of the most commonly used nano adsorbents in wastewater treatment [110].

Carbon nanotubes (CNTs) possess unique structural, electronic, optoelectronic, semiconductor, mechanical, chemical, and physical properties. CNTs are well suited to adsorbing heavy metals from water [110].

Stafiej and Pyrzynska [111] studied the adsorption of heavy metals on CNT as a function of the pH change. In the case of CTNs, pH plays an important influencing role in the adsorption of certain ions. According to Stafiej and Pyrzynska [111], the adsorption of cations on carbon nanotubes (CNTs) is more favorable at a pH higher than the point of zero charge of the adsorbent. Conversely, a decrease in pH results in a decrease in the adsorption of cations.

5.1.3. Zeolite

Zeolite is one of the most widely used adsorbents, and is a microporous, hydrated aluminosilicate material that has an infinite, open, three-dimensional structure [85,112–115]. Ion exchangers are widely utilized to enhance water quality due to their capacity to exchange and adsorb cations present in water. This property enables them to effectively remove undesirable cations and improve the overall quality of the water [89,93,115]. The advantage is that it is available in large quantities and is therefore cheap and easily accessible [85,116].

Natural zeolite is suitable for removing heavy metal elements and for the desalination of water; this is supported by several research studies [92,93,114,117]. Furthermore, natural zeolite and its modified forms are suitable for removing anions and organic substances from water [115]. The pH of the solution affects the cation adsorption capacity of the zeolite, as strong acids and alkalis dissolve the zeolite aluminosilicate framework. For this reason, the highest cation exchange capacity is achieved at a neutral to slightly acidic pH [113]. He et al. [90] used zeolite synthesized from fly ash to remove heavy metals. The heavy metal removal efficiency increased by increasing the adsorbent dose and increasing the pH from 1.0 to 5.0.

Ghanbarizadeh et al. [89] compared the adsorbent capacity of natural and modified zeolite for hardness removal and studied the optimal contact times. The modified zeolite showed a higher efficiency when removing hardness, and most ions were removed with a contact time of 90 min.

In their study, Neag et al. [91] employed thermally activated natural zeolite for the removal of iron (Fe) and manganese (Mn) from groundwater. They found that the zeolite activated at 200 °C demonstrated the highest effectiveness in simultaneously removing both Fe and Mn. Fe removal was more efficient than Mn removal.

Natural zeolites can be modified using various methods, such as acid treatment, ion exchange, and the use of surfactants [115]; thus, they are also suitable for removing anions from water, and a higher adsorption capacity can be achieved when adsorbing organic substances and anions [113,115]. Paul et al. [92] used modified zeolite treated with $Ca(OH)_2$ for desalination. They found that it is suitable for removing Na from salt water, while also neutralizing the acidic desalinated water. However, compared to untreated zeolite, it shows a lower efficiency during sodium removal.

Mubarak et al. [93] used TiO_2-modified zeolite adsorbent to remove iron, manganese, and the total and calcium hardness from groundwater. The presence of a new active functional group resulted in the improved efficiency of the modified zeolite adsorbent compared to zeolite. This improvement was observed in terms of the removal of Mn(II), Fe(III), the total hardness, and calcium hardness. In addition, it was found that the contact time can be reduced by 29–50% if a zeolite adsorbent modified with TiO_2 is used.

Gibb et al. [118] used sequentially calcined layered double hydroxide (CLDH) and acid-treated zeolite as adsorbents for aqueous solution desalination. CLDH first adsorbs Cl^- and temporarily raises the pH, then the acid-treated zeolite removes Na^+ and neutralizes the pH. The combined use of CLDH and acid-treated zeolite removes cations and anions (Na^+, K^+, Ca^{2+}, Mg^{2+} and Cl^-) and results in a neutral pH.

In their study, Amiri et al. [83] employed a combination of natural zeolite, activated carbon, and nano zero-valent iron to treat the gray water from a student dormitory. The treated water was subsequently utilized for irrigation purposes. This triple combination of adsorbents was found to be the most suitable for reducing COD and TDS, and almost completely removed turbidity. It proved to be more effective than using natural zeolite. The optimal contact time for the removal of COD, TDS, and turbidity was 4 h, after which no significant changes were observed. According to their research, when the three materials were used independently as adsorbents, zeolite was the most effective for reducing

TDS and removing turbidity, while activated carbon proved to be the most effective for reducing COD.

5.1.4. Activated Alumina

In addition, Ghanbarizadeh et al. [89] tested activated alumina to reduce the hardness of groundwater. They found that a contact time of 90 min was the most effective; further increasing the contact time did not significantly improve the removal efficiency of the treatment process. The modification of activated alumina with salt did not increase the adsorption efficiency, while modification with acid and salt acid reduced it.

5.2. Non-Conventional Adsorbents

There are substances in nature that are available in large quantities and can remove pollutants from water and be used as adsorbents [85,119]. The raw materials of low-cost adsorbents usually do not require complicated pretreatment, which makes them economical [119].

5.2.1. Clay

Among the natural adsorbents, the use of clay is widespread, especially clays consisting of montmorillonite, bentonite, kaolinite, and zeolite [85]. Its positive properties include a high sorption capacity, high chemical and mechanical stability, a high uptake capacity, and a large surface [85,114]. Another advantage is that there are a large number of cations (Ca^{2+}, Mg^{2+}, H^+, K^+, $NH_4{}^+$, Na^+) on their surface, which can easily be replaced by adsorbates [85].

Bentonite can also be used as an adsorbent; it is a sedimentary rock containing the clay mineral montmorillonite, a mineral mixture [114]. It is suitable for adsorbing many toxic substances and cations in water, but shows a lower affinity towards negative groups (e.g., phosphates) [114].

5.2.2. Bentonite

Bastamy et al. [114] utilized bentonite in pilot-scale bed columns constructed from various materials to investigate its effectiveness in removing pollutants from wastewater. In the experimental setup, the column was divided into three sections. The top and bottom sections were filled with a layer of gravel, each measuring 10 cm in thickness. The middle section of the column was dedicated to ion exchange beds, which were composed of a mixture of sand and bentonite. The sand-to-bentonite ratio used in the beds was 1:2. They were able to achieve a high removal efficiency, so the water could immediately be used for irrigation. The optimal removal efficiency for the tested parameters was achieved when the contact time was set at 4 h. Subsequently, the removal efficiency reached a stable state, indicating that further increasing the contact time would not result in any significant improvements in the removal process.

5.2.3. Sand

Sand is also a substance that occurs in large quantities in nature and can be used for the adsorption of heavy metals from water. Kan et al. [95] attempted to remove Mn from water using Fe and Mn oxide-coated sand. The primary objective of the study was to examine the influence of pH and temperature on the removal efficiency of manganese (Mn) from wastewater. The pH significantly affects the adsorption of Mn, and increasing the pH from 5.5 to 8 caused a significant increase in the adsorption capacity. Furthermore, the temperature is also a significant factor, and the adsorption of Mn increased as the temperature increased from 25 °C to 45 °C. In general, manganese oxides have a higher adsorption capacity for Mn than iron oxides.

5.2.4. Agricultural Wastes

Agricultural wastes can be utilized as alternative adsorbents due to their unique chemical properties and distinctive structure. The polymer chains present in these wastes often contain specific functional groups, including alcohol, phenol, aldehyde, carboxyl, and ketone. These functional groups contribute to their adsorption capabilities and make them suitable for various applications [85,120]. They are economical as they are abundantly available, renewable, environmentally friendly, and low-cost [121].

Peels of fruits and vegetables, among the agricultural waste materials, are significant as potential adsorbents in water treatment. Many of these peels are typically discarded as waste without further use, but they possess the potential to be utilized as effective adsorbents. By repurposing these peels, their adsorption properties can be harnessed to remove contaminants and improve water quality, making them a valuable resource in sustainable water treatment strategies [121].

Agricultural wastes are mostly not considered effective adsorbents, but their adsorption efficiency can be significantly increased by physical and chemical methods [122]. In bioadsorption, a simple chemically modified adsorbent can be developed from certain plant materials such as bark, leaves, husks, stems, branches, and pods [120].

Biosorbents have a relatively high biosorption capacity, strong mechanical strength, good thermal stability, and chemical resistance [123]. In recent years, biomass has mainly been used as a biosorbent, since it is difficult to meet the needs of practical applications using a single type of biological adsorbent. For the removal of heavy metals, biomass has already been combined with nanoparticles in several experiments, thereby achieving excellent adsorption properties. The creation of multi-biological adsorbents is also an effective modification, as their use can have a symbiotic and synergistic effect. Biosorbents show a lower adsorption efficiency compared to traditional adsorbents, which is why they are currently less in demand [124].

Cork is obtained from the trunk and main branches of *Quercus suber* L. in the form of a half-tube. Sorption using natural cork or cork modified to biochar or activated carbon can be used for a wide range of pollutants (heavy metals, organic pollutants) and in different states (liquid, gas). Cork is more selectively adsorptive for Hg than for Cd, with minimal interference of the latter, while in seawater the adsorption of Hg is much more reduced due to the formation of Hg–chlorine complexes. Anionic metals are more easily adsorbed at acidic pH levels, as electrostatic attraction may play a role in their uptake [125].

In their study, Yan et al. produced biochar from corn stalks, which can be conveniently and inexpensively used as a tool for removing various metal contaminants from wastewater. During their experiments, they investigated the removal efficiency of Cd and Pb contaminants. They conducted batch experiments while varying different factors such as pH, temperature, and background ionic strength. Increasing the pH and temperature was found to enhance the adsorption capacity, while the ionic strength showed negative effects [126].

In their study, Murtaza et al. conducted a comprehensive analysis of the mechanisms involved in the immobilization and adsorption of heavy metals and organic pollutants using biochar under varying environmental conditions. They concluded that the efficiency of biochar is influenced by factors such as pyrolysis temperature, feedstock type, residence time, and application rate. Increasing the pyrolysis temperature facilitates the production of biochars with higher porosity and a specific surface area, thereby enhancing their adsorption potential. It is important to note that the pyrolysis process generates impurities like volatile organic compounds (VOCs), dioxins, and polycyclic aromatic hydrocarbons (PAHs), which have adverse environmental effects. Hence, using unpolluted feedstock as the starting material and employing appropriate carbonization technology are imperative measures [127].

Raheem et al. [94] attempted to remove iron from groundwater using an adsorbent made from a mixture of sawdust and barley husk in different proportions. They found that a mixing ratio of sawdust and barley husk of 0.5:1 resulted in the highest adsorption efficiency at pH 6.5.

Microorganisms could be a new alternative for removing metal ions from water. *Saccharomyces cerevisiae* is one of the most promising bioadsorbents for the removal of metal ions from water.

Fadel et al. [96] investigated the possibility of removing manganese ions from water with *S. cerevisiae*. They concluded that it can be used well for this purpose, and that the pH has a significant role in the adsorption efficiency. Increasing the pH from 5 to 9 significantly increased the adsorption efficiency. Regarding the contact time, the microorganism proved to be the most effective during a contact time of 30 to 45 min.

Kamarudzaman et al. [97] used mushroom compost in a fixed-bed column as a biosorbent to remove Mn from an aqueous solution. Pleurotus mushroom compost contains lignocellulosic, which binds metal ions, and is also available in large quantities, which makes it an inexpensive bioorganic.

Peat is present in large quantities in nature and can also be used as a bioadsorbent in water purification [85]. It has a large surface area and a high adsorption capacity [85]. The main components of peat are lignin, cellulose, and humic acid [85].

6. Conclusions

Traditional surface irrigation is outdated due to scarce water reserves, but the many micro irrigation techniques that are considered modern, in addition to their efficiency and proven results, require a high level of knowledge and attention, both in their construction and application. The quality of irrigation water is affected by both chemical and physical pollutants. Chemical pollutants are important for both the soil and the irrigation system, while mechanical pollutants can mainly endanger the operational safety. With regular irrigation, special attention must be paid to water quality, as the harmful effects of irrigation often occur cumulatively (salt accumulation, leaching, etc.). The chemical treatment of irrigation water is problematic, so the best method is to choose a water source that ensures good quality. The mechanical treatment of irrigation water is determined by the sensitivity of the irrigation equipment. It is not worth filtering the water unnecessarily, because it involves a loss of energy. However, it must be taken into account that the nutrients introduced during irrigation can easily colonize microorganisms (algae, bacteria), which can cause the build-up of slime, making it easier for particles to accumulate in the system.

The use of adsorbents can be a good method for water purification, which makes it suitable for irrigation. Conventional adsorbents usually have a higher adsorption capacity but are expensive. On the other hand, non-conventional adsorbents are easily accessible and therefore cheap, but have a lower adsorption capacity. Otherwise, biological treatment methods are reported to be useful for organic contaminants and the removal of heavy metals; however, it is quite challenging to keep the microorganism alive and handle many other factors.

For salt and to avoid the formation of scaling, a couple of desalination methods have been used; these include electromagnetic field, electrostatic precipitator and even membrane separation techniques, which have a couple of limitations, such as a lower effectivity at a higher salt concentration, scaling, membrane fouling and a high energy consumption for running it. The use of adsorbents for purification could help solve many of the issues of using water for irrigation that would otherwise not be used for other purposes and become wastewater. The specific advantages of using adsorbents for wastewater treatment could be their low operation cost on the energy side and specific removal of many organic and inorganic pollutants. Therefore, the separation of interfering components via adsorption is effective and promising for future application as expected irrigation demands increase.

Author Contributions: Conceptualization, T.K. and E.T.; writing—review and editing, T.K., G.I.N., I.P. and Z.L.K.; supervision, I.G.; funding acquisition and resources: I.G. All authors have read and agreed to the published version of the manuscript.

Funding: This work was implemented by the TKP2021-NKTA-21 project with the support provided by the Ministry of Culture and Innovation of Hungary from the National Research, Development and Innovation Fund, financed under the 2021 Thematic Excellence Programme funding scheme.

Data Availability Statement: No new data were created or analyzed in this study.

Acknowledgments: We would like to thank all of the participants in our review, who generously shared their time, experiences, and insights with us. Their willingness to engage with our research was essential to the success of this project, and we are deeply grateful for their participation. Overall, this research project would not have been possible without the support and contributions of our institute. We are deeply grateful to all of those who helped to make this review a reality, and we hope that our findings will make a meaningful contribution to the field.

Conflicts of Interest: The authors declare no conflict of interest. The funders had no role in the design of the study; in the collection, analyses, or interpretation of data; in the writing of the manuscript; or in the decision to publish the results.

References

1. Zittis, G.; Almazroui, M.; Alpert, P.; Ciais, P.; Cramer, W.; Dahdal, Y.; Fnais, M.; Francis, D.; Hadjinicolaou, P.; Howari, F.; et al. Climate Change and Weather Extremes in the Eastern Mediterranean and Middle East. *Rev. Geophys.* **2022**, *60*, e2021RG000762. [CrossRef]
2. Malhi, G.S.; Kaur, M.; Kaushik, P. Impact of Climate Change on Agriculture and Its Mitigation Strategies: A Review. *Sustainability* **2021**, *13*, 1318. [CrossRef]
3. Tabari, H. Climate Change Impact on Flood and Extreme Precipitation Increases with Water Availability. *Sci. Rep.* **2020**, *10*, 13768. [CrossRef] [PubMed]
4. *Climate Change and Land: An IPCC Special Report on Climate Change, Desertification, Land Degradation, Sustainable Land Management, Food Security, and Greenhouse Gas Fluxes in Terrestrial Ecosystems*; 2019; pp. 1–864.
5. Akhtar, N.; Syakir Ishak, M.I.; Bhawani, S.A.; Umar, K. Various Natural and Anthropogenic Factors Responsible for Water Quality Degradation: A Review. *Water* **2021**, *13*, 2660. [CrossRef]
6. Khan, S.; Tariq, R.; Yuanlai, C.; Blackwell, J. Can Irrigation Be Sustainable? *Agric. Water Manag.* **2006**, *80*, 87–99. [CrossRef]
7. Hagan, R.M. Water-Soil-Plant Relations. *Calif. Agric.* **1957**, *11*, 9–12.
8. Ampim, P.A.Y.; Obeng, E.; Olvera-Gonzalez, E. Indoor Vegetable Production: An Alternative Approach to Increasing Cultivation. *Plants* **2022**, *11*, 2843. [CrossRef] [PubMed]
9. Gerdes, M.E.; Cruz-Cano, R.; Solaiman, S.; Ammons, S.; Allard, S.M.; Sapkota, A.R.; Micallef, S.A.; Goldstein, R.E.R. Impact of Irrigation Water Type and Sampling Frequency on Microbial Water Quality Profiles Required for Compliance with U.S. Food Safety Modernization Act Produce Safety Rule Standards. *Environ. Res.* **2022**, *205*, 112480. [CrossRef]
10. European Council. 91/271/EEC Directive Concerning Urban Waste Water Treatment. *Off. J. Eur. Communities II* **1991**, *135*, 40–52.
11. Council of European Communities. 98/83/EC Directive on the Quality of Water Intended for Human Consumption. *Off. J. Eur. Communities* **1998**, *330*, 32–54. [CrossRef]
12. *Act No. LIII of 1995 on the General Rules of Environmental Protection*; Budapest, Hungary, 1995.
13. *Act No. LIII of 1996 on Nature Conservation*; Budapest, Hungary, 1996.
14. *Act No. CLXXXV of 2012 on Waste*; Budapest, Hungary, 2012.
15. European Parliament; European Council. 2007/60/EC Directive on the Assessment and Management of Flood Risks. *Off. J. Eur. Union* **2007**, *288*, 27–34.
16. European Parliament; European Council. 2000/60/EC Directive Establishing a Framework for Community Action in the Field of Water Policy. *Off. J. Eur. Communities* **2000**, 1–72.
17. Decree of the Ministry of Agriculture No. 90/2008. (VII. 18.) on Detailed Rules for Drawing Up the Soil Protection Plan. Budapest, Hungary, 2008. Available online: https://www.ecolex.org/details/legislation/decree-no-90-of-2008-vii-18-fvm-of-the-ministry-of-agriculture-and-rural-development-laying-down-detailed-rules-of-elaboration-of-soil-conservation-plans-lex-faoc124169/ (accessed on 24 July 2023).
18. MSZ-10-640:1989 Water Management. Water Quality. Requirements for the Quality of Irrigation Water (Hungarian Standard). Budapest, Hungary, 1989. Available online: https://oszkdk.oszk.hu/storage/00/02/38/12/dd/1/Alkalmazkod___v__zgazd__lkod__s_ABSZTRAKT.pdf (accessed on 24 July 2023).
19. *Act No. CXIII on Farming Irrigation*; Budapest, Hungary, 2019.
20. Paudel, K.P.; Pandit, M.; Roger, H. Irrigation Water Sources and Irrigation Application Methods Used by U.S. Plant Nursery Producers. *Water Resour. Res. Res.* **2016**, *52*, 698–712. [CrossRef]
21. Ma, L.; Liu, Y.; Zhang, J.; Yang, Q.; Li, G.; Zhang, D. Impacts of Irrigation Water Sources and Geochemical Conditions on Vertical Distribution of Pharmaceutical and Personal Care Products (PPCPs) in the Vadose Zone Soils. *Sci. Total Environ.* **2018**, *626*, 1148–1156. [CrossRef] [PubMed]
22. Malakar, A.; Snow, D.D.; Ray, C. Irrigation Water Quality—A Contemporary Perspective. *Water* **2019**, *11*, 1482. [CrossRef]

23. Stefanovics, P.; Filep, G.; Füleky, G. *Talajtan*; Mezőgazda Kiadó: Budapest, Hungary, 2010.
24. Darab, K.; Ferencz, K. *Öntözött Területek Talajtérképezése*; Országos Mezőgazdasági Minőségvizsgáló Intézet: Budapest, Hungary, 1969.
25. George, P.R. *Agricultural Water Quality Criteria: Irrigation Aspects*; Department of Agriculture: Merredin, Australia, 1983.
26. Zaman, M.; Shahid, S.A.; Heng, L. *Guideline for Salinity Assessment, Mitigation and Adaptation Using Nuclear and Related Techniques*; Springer: Berlin, Germany, 2018; ISBN 9783319961897.
27. Zhao, D.; Hao, Z.; Wang, J.; Tao, J. Effects of PH in Irrigation Water on Plant Growth and Flower Quality in Herbaceous Peony (*Paeonia lactiflora* Pall.). *Sci. Hortic.* **2013**, *154*, 45–53. [CrossRef]
28. Guimarães, J. de J.; Sousa, F.G.G. de; Román, R.M.S.; Dal Pai, A.; Rodrigues, S.A.; Sarnighausen, V.C.R. Effect of Irrigation Water PH on the Agronomic Development of Hops in Protected Cultivation. *Agric. Water Manag.* **2021**, *253*, 106924. [CrossRef]
29. Adamu, G.K. Quality of Irrigation Water and Soil Characteristics of Watari Irrigation Project. *Am. J. Eng. Res.* **2013**, *02*, 59–68.
30. Seilsepour, M.; Rashidi, M.; Khabbaz, B.G. Prediction of Soil Exchangeable Sodium Ratio Based on Soil Sodium Adsorption Ratio. *World Acad. Sci. Eng. Technol.* **2010**, *46*, 255–257.
31. Machado, R.M.A.; Serralheiro, R.P. Soil Salinity: Effect on Vegetable Crop Growth. Management Practices to Prevent and Mitigate Soil Salinization. *Horticulturae* **2017**, *3*, 30. [CrossRef]
32. Yuan, C.; Feng, S.; Huo, Z.; Ji, Q. Effects of Deficit Irrigation with Saline Water on Soil Water-Salt Distribution and Water Use Efficiency of Maize for Seed Production in Arid Northwest China. *Agric. Water Manag.* **2019**, *212*, 424–432. [CrossRef]
33. Ajay, S. Soil Salinization Management for Sustainable Development: A Review. *J. Environ. Manag.* **2021**, *277*, 111383. [CrossRef]
34. Smith, C.J.; Oster, J.D.; Sposito, G. Potassium and Magnesium in Irrigation Water Quality Assessment. *Agric. Water Manag.* **2015**, *157*, 59–64. [CrossRef]
35. Rácz Istvánné Nemes, Z.J. *Öntözővizek Minősége a Hazai Vízkultúrás Zöldségtermesztésben*; Corvinus University of Budapest: Budapest, Hungary, 2007.
36. Sarda, P.; Sadgir, P. Assessment of Multi Parameters of Water Quality in Surface Water Bodies—A Review. *Int. J. Res. Appl. Sci. Eng. Technol.* **2015**, *3*, 331–336.
37. Rengasamy, P.; Olsson, K.A. Irrigation and Sodicity. *Aust. J. Soil Res.* **1993**, *31*, 821–837. [CrossRef]
38. Halliwell, D.J.; Barlow, K.M.D.M.; Nash, D.M. A Review of the Effects of Wastewater Sodium on Soil Physical Properties and Their Implications for Irrigation Systems. *Aust. J. Soil Res.* **2001**, *39*, 1259–1267. [CrossRef]
39. Erel, R.; Ben-Gal, A.; Dag, A.; Schwartz, A.; Yermiyahu, U. Sodium Replacement of Potassium in Physiological Processes of Olive Trees (Var. Barnea) as Affected by Drought. *Tree Physiol.* **2014**, *34*, 1102–1117. [CrossRef]
40. Qadir, M.; Schubert, S.; Oster, J.D.; Sposito, G.; Minhas, P.S.; Cheraghi, S.A.M.; Murtaza, G.; Mirzabaev, A.; Saqib, M. High-magnesium Waters and Soils: Emerging Environmental and Food Security Constraints. *Sci. Total Environ.* **2018**, *642*, 1108–1117. [CrossRef]
41. Martínez-Gimeno, M.A.; Bonet, L.; Provenzano, G.; Badal, E.; Intrigliolo, D.S.; Ballester, C. Assessment of Yield and Water Productivity of Clementine Trees under Surface and Subsurface Drip Irrigation. *Agric. Water Manag.* **2018**, *206*, 209–216. [CrossRef]
42. Sable, R.; Kolekar, S.; Gawde, A.; Takle, S.; Pednekar, A. A Review on Different Irrigation Methods. *Int. J. Appl. Agric. Res.* **2019**, *14*, 49–60.
43. Holzapfel, E.A.; Pannunzio, A.; Lorite, I.; de Oliveira, A.S.S.; Farkas, I. Design and Management of Irrigation Systems. *Chil. J. Agric. Res.* **2009**, *69*, 17–25. [CrossRef]
44. Pereira, L.S. Surface Irrigation Systems. *Sustain. Irrig. Agric.* **1996**, 269–289. [CrossRef]
45. Albaji, M.; Golabi, M.; Boroomand Nasab, S.; Zadeh, F.N. Investigation of Surface, Sprinkler and Drip Irrigation Methods Based on the Parametric Evaluation Approach in Jaizan Plain. *J. Saudi Soc. Agric. Sci.* **2015**, *14*, 1–10. [CrossRef]
46. Nakayama, F.S.; Bucks, D.A. Water Quality in Drip/Trickle Irrigation: A Review. *Irrig. Sci.* **1991**, *12*, 187–192. [CrossRef]
47. Yan, H.; Hui, X.; Li, M.; Xu, Y. Development in Sprinkler Irrigation Technology in China. *Irrig. Drain.* **2020**, *69*, 75–87. [CrossRef]
48. Bansal, G.; Mahajan, A.; Verma, A.; Bandhu Singh, D. A Review on Materialistic Approach to Drip Irrigation System. *Mater. Today Proc.* **2021**, *46*, 10712–10717. [CrossRef]
49. Henrique Bassoi, L.; Hopmans, J.W.; André de Castro Jorge, L.; Miranda de Alencar, C.; Antonio Moura Silva, J. Grapevine Root Distribution in drip and microsprinkler irrigation. *Sci. Agric.* **2003**, *60*, 377–387. [CrossRef]
50. Goldberg, D.; Gornat, B.; Rimon, D. *Drip Irrigation. Principles, Design and Agricultural Practices*; Drip Irrigation Scientific Publications: Kfar Shmaryahu, Israel, 1976.
51. Dong, R.; Liu, W.; Qu, J.; Cao, W. Accumulation of Na+ in Cotton Field under Mulched Drip Irrigation of Brackish Water in Arid Areas. *Separations* **2023**, *10*, 180. [CrossRef]
52. Zapata, N.; Playán, E.; Castillo, R.; Gimeno, Y.; Oliván, I.; Jiménez, A.; Carbonell, X.; Fábregas, M.; López-Pardo, J.R.; Vicente, L.M.; et al. A Methodology to Classify Irrigated Areas: Application to the Central Ebro River Basin in Aragón (Spain). *Agric. Water Manag.* **2020**, *241*, 106365. [CrossRef]
53. Arshad, I. Importance of Drip Irrigation System Installation and Management—A Review. *PSM Biol. Res.* **2020**, *5*, 22–29.
54. Flores, J.H.N.; Faria, L.C.; Neto, O.R.; Diotto, A.V.; Colombo, A. Methodology for Determining the Emitter Local Head Loss in Drip Irrigation Systems. *J. Irrig. Drain. Eng.* **2020**, *147*, 06020014. [CrossRef]
55. Shi, K.; Lu, T.; Zheng, W.; Zhang, X.; Zhangzhong, L. A Review of the Category, Mechanism, and Controlling Methods of Chemical Clogging in Drip Irrigation System. *Agriculture* **2022**, *12*, 1–20. [CrossRef]

56. Prathyusha, K.; Chaitanya Suman, M. Design of Embedded Systems for the Automation of Drip Irrigation. *Int. J. Appl. Innov. Eng. Manag.* **2012**, *1*, 245–258.
57. Li, Y.; Pan, J.; Chen, X.; Xue, S.; Feng, J.; Muhammad, T.; Zhou, B. Dynamic Effects of Chemical Precipitates on Drip Irrigation System Clogging Using Water with High Sediment and Salt Loads. *Agric. Water Manag.* **2019**, *213*, 833–842. [CrossRef]
58. Bucks, D.A.; Nakayama, F.S.; Gilbert, R.G. Trickle Irrigation Water Quality and Preventive Maintenance. *Agric. Water Manag.* **1979**, *2*, 149–162. [CrossRef]
59. Pitts, D.J.; Haman, D.Z.; Smajstria, A.G. *Causes and Prevention of Emitter Plugging in Micro Irrigation Systems*; Bulletin 2; Institute of Food and Agricultural Sciences, University of Florida: Gainesville, FL, USA, 1990.
60. Muhammad, T.; Zhou, B.; Liu, Z.; Chen, X.; Li, Y. Effects of Phosphorus-Fertigation on Emitter Clogging in Drip Irrigation System with Saline Water. *Agric. Water Manag.* **2021**, *243*, 106392. [CrossRef]
61. El Housse, M.; Hadfi, A.; Karmal, I.; Ben-Aazza, S.; Belattar, M.; Errami, M.; Mohareb, S.; Driouiche, A. Study of the Effect of Inorganic Inhibitor on the Calcium Carbonate Precipitation in the Localized Irrigation Systems. *Nanotechnol. Environ. Eng.* **2021**, *6*, 2–9. [CrossRef]
62. Yan, D.; Bai, Z.; Rowan, M.; Gu, L.; Shumei, R.; Yang, P. Biofilm Structure and Its Influence on Clogging in Drip Irrigation Emitters Distributing Reclaimed Wastewater. *J. Environ. Sci.* **2009**, *21*, 834–841. [CrossRef]
63. Ravina, I.; Paz, E.; Sofer, Z.; Marcu, A.; Schischa, A.; Sagi, G.; Yechialy, Z.; Lev, Y. Control of Clogging in Drip Irrigation with Stored Treated Municipal Sewage Effluent. *Agric. Water Manag.* **1997**, *33*, 127–137. [CrossRef]
64. Dery, J.L.; Brassill, N.; Rock, C.M. Minimizing Risks: Use of Surface Water in Pre-Harvest Agricultural Irrigation Part I: Understanding Water Quality & Treatment Options Why Treat Agricultural Irrigation Water? *Univ. Arizona Coop. Ext.* **2019**, *1793*, 1–4.
65. Lim, K.Y.; Foo, K.Y. Hazard Identification and Risk Assessment of the Organic, Inorganic and Microbial Contaminants in the Surface Water after the High Magnitude of Flood Event. *Environ. Int.* **2021**, *157*, 106851. [CrossRef] [PubMed]
66. Jiang, D.; Ao, C.; Bailey, R.T.; Zeng, W.; Huang, J. Simulation of Water and Salt Transport in the Kaidu River Irrigation District Using the Modified SWAT-Salt. *Agric. Water Manag.* **2022**, *272*, 107845. [CrossRef]
67. Battilani, A.; Steiner, M.; Andersen, M.; Back, S.N.; Lorenzen, J.; Schweitzer, A.; Dalsgaard, A.; Forslund, A.; Gola, S.; Klopmann, W.; et al. Decentralised Water and Wastewater Treatment Technologies to Produce Functional Water for Irrigation. *Agric. Water Manag.* **2010**, *98*, 385–402. [CrossRef]
68. Vassalle, L.; Sunyer-Caldú, A.; Uggetti, E.; Díez-Montero, R.; Díaz-Cruz, M.S.; García, J.; García-Galán, M.J. Bioremediation of Emerging Micropollutants in Irrigation Water. The Alternative of Microalgae-Based Treatments. *J. Environ. Manag.* **2020**, *274*. [CrossRef] [PubMed]
69. Gertsis, A.; Zoukidis, K. Irrigation with Highly Saline Water: A New Innovative Water Treatment System Evaluated for Vegetable Production in Greenhouse. *Eur. Water* **2017**, *59*, 331–337.
70. Seyyedi, M. Magnetic Fields and Manipulation of Water Behaviors: A Viable Method for Reducing RO Specific Energy Consumption. In Proceedings of the 2022 Membrane Technology Conference and Exposition, Las Vegas, NV, USA, 21–25 February 2022.
71. Surendran, U.; Sandeep, O.; Joseph, E.J. The Impacts of Magnetic Treatment of Irrigation Water on Plant, Water and Soil Characteristics. *Agric. Water Manag.* **2016**, *178*, 21–29. [CrossRef]
72. Hachicha, M.; Kahlaoui, B.; Khamassi, N.; Misle, E.; Jouzdan, O. Effect of Electromagnetic Treatment of Saline Water on Soil and Crops. *J. Saudi Soc. Agric. Sci.* **2018**, *17*, 154–162. [CrossRef]
73. Wiedenfeld, B. Effects of Irrigation Water Salinity and Electrostatic Water Treatment for Sugarcane Production. *Agric. Water Manag.* **2008**, *95*, 85–88. [CrossRef]
74. Wei, K.; Zhang, J.; Wang, Q.; Guo, Y.; Mu, W. Irrigation with Ionized Brackish Water Affects Cotton Yield and Water Use Efficiency. *Ind. Crops Prod.* **2022**, *175*, 114244. [CrossRef]
75. Jarma, Y.A.; Karaoğlu, A.; Senan, I.R.A.; Meriç, M.K.; Kukul, Y.S.; Özçakal, E.; Barlas, N.T.; Çakıcı, H.; Baba, A.; Kabay, N. Utilization of Membrane Separation Processes for Reclamation and Reuse of Geothermal Water in Agricultural Irrigation of Tomato Plants-Pilot Membrane Tests and Economic Analysis. *Desalination* **2022**, *528*, 115608. [CrossRef]
76. Garg, N.; Choudhary, O.P.; Thaman, S.; Sharma, V.; Singh, H.; Vashistha, M.; Sekhon, K.S.; Sharda, R.; Dhaliwal, M.S. Effects of Irrigation Water Quality and NPK-Fertigation Levels on Plant Growth, Yield and Tuber Size of Potatoes in a Sandy Loam Alluvial Soil of Semi-Arid Region of Indian Punjab. *Agric. Water Manag.* **2022**, *266*, 107604. [CrossRef]
77. Wang, T.; Wang, Z.; Zhang, J.; Ma, K. An Optimum Combination of Irrigation Amount, Irrigation Water Salinity and Nitrogen Application Rate Can Improve Cotton (for Fiber) Nitrogen Uptake and Final Yield. *Ind. Crops Prod.* **2022**, *187*, 115386. [CrossRef]
78. Sang, H.; Guo, W.; Gao, Y.; Jiao, X.; Pan, X. Effects of Alternating Fresh and Saline Water Irrigation on Soil Salinity and Chlorophyll Fluorescence of Summer Maize. *Water* **2020**, *12*, 3054. [CrossRef]
79. Dong, X.; Wang, J.; Zhang, X.; Dang, H.; Singh, B.P.; Liu, X.; Sun, H. Long-Term Saline Water Irrigation Decreased Soil Organic Carbon and Inorganic Carbon Contents. *Agric. Water Manag.* **2022**, *270*, 107760. [CrossRef]
80. Ouoba, N.; Phung, L.D.; Sasaki, A.; Pham, D.V.; Watanabe, T. Drip Fertigation with Treated Municipal Wastewater and Soil Amendment with Composted Sewage Sludge for Sustainable Protein-Rich Rice Cultivation. *Environ. Technol. Innov.* **2022**, *28*, 102569. [CrossRef]
81. Yang, S.; Feng, W.; Wang, S.; Chen, L.; Zheng, X.; Li, X.; Zhou, D. Farmland Heavy Metals Can Migrate to Deep Soil at a Regional Scale: A Case Study on a Wastewater-Irrigated Area in China. *Environ. Pollut.* **2021**, *281*, 116977. [CrossRef] [PubMed]

82. Lin, Y.; Zhang, X.; Wang, Y.; Shi, E.; Lin, H.; Chen, G. Removal of Pb^{2+} and Cd^{2+} from Irrigation Water and Replenishment of Mineral Nutrients Using a Low-Cost Mineral Adsorbent Derived from Potassium-Rich Aluminum Silicates. *J. Environ. Chem. Eng.* **2023**, *11*, 109282. [CrossRef]
83. Amiri, M.J.; Bahrami, M.; Badkouby, M.; Kalavrouziotis, I.K. Greywater Treatment Using Single and Combined Adsorbents for Landscape Irrigation. *Environ. Process.* **2019**, *6*, 43–63. [CrossRef]
84. Hameed, B.H.; Tan, I.A.W.; Ahmad, A.L. Adsorption Isotherm, Kinetic Modeling and Mechanism of 2,4,6-Trichlorophenol on Coconut Husk-Based Activated Carbon. *Chem. Eng. J.* **2008**, *144*, 235–244. [CrossRef]
85. Singh, N.B.; Nagpal, G.; Agrawal, S. Rachna Water Purification by Using Adsorbents: A Review. *Environ. Technol. Innov.* **2018**, *11*, 187–240. [CrossRef]
86. Crini, G.; Lichtfouse, E.; Wilson, L.D.; Morin-Crini, N. Conventional and Non-Conventional Adsorbents for Wastewater Treatment. *Environ. Chem. Lett.* **2019**, *17*, 195–213. [CrossRef]
87. Jain, C.K.; Malik, D.S.; Yadav, A.K. Applicability of Plant Based Biosorbents in the Removal of Heavy Metals: A Review. *Environ. Process.* **2016**, *3*, 495–523. [CrossRef]
88. Aghakhani, A.; Mousavi, S.F.; Mostafazadeh-Fard, B.; Rostamian, R.; Seraji, M. Application of Some Combined Adsorbents to Remove Salinity Parameters from Drainage Water. *Desalination* **2011**, *275*, 217–223. [CrossRef]
89. Ghanbarizadeh, P.; Parivazh, M.M.; Abbasi, M.; Osfouri, S.; Dianat, M.J.; Rostami, A.; Dibaj, M.; Akrami, M. Performance Enhancement of Specific Adsorbents for Hardness Reduction of Drinking Water and Groundwater. *Water* **2022**, *14*, 2749. [CrossRef]
90. He, K.; Chen, Y.; Tang, Z.; Hu, Y. Removal of Heavy Metal Ions from Aqueous Solution by Zeolite Synthesized from Fly Ash. *Environ. Sci. Pollut. Res.* **2016**, *23*, 2778–2788. [CrossRef] [PubMed]
91. Neag, E.; Török, A.I.; Tanaselia, C.; Aschilean, I.; Senila, M. Kinetics and Equilibrium Studies for the Removal of Mn and Fe from Binary Metal Solution Systems Using a Romanian Thermally Activated Natural Zeolite. *Water* **2020**, *12*, 1614. [CrossRef]
92. Paul, B.; Dynes, J.J.; Chang, W. Modified Zeolite Adsorbents for the Remediation of Potash Brine-Impacted Groundwater: Built-in Dual Functions for Desalination and PH Neutralization. *Desalination* **2017**, *419*, 141–151. [CrossRef]
93. Mubarak, M.F.; Mohamed, A.M.G.; Keshawy, M.; elMoghny, T.A.; Shehata, N. Adsorption of Heavy Metals and Hardness Ions from Groundwater onto Modified Zeolite: Batch and Column Studies. *Alex. Eng. J.* **2022**, *61*, 4189–4207. [CrossRef]
94. Raheem, S.A.; Kadhim, E.J.; Abdulhasan, M.J. Comparative Study of Iron Removal from Groundwater Using Low Cost Adsorbents. *J. Ecol. Eng.* **2022**, *23*, 18–23. [CrossRef]
95. Kan, C.C.; Aganon, M.C.; Futalan, C.M.; Dalida, M.L.P. Adsorption of Mn2+ from Aqueous Solution Using Fe and Mn Oxide-Coated Sand. *J. Environ. Sci.* **2013**, *25*, 1483–1491. [CrossRef]
96. Fadel, M.; Hassanein, N.M.; Elshafei, M.M.; Mostafa, A.H.; Ahmed, M.A.; Khater, H.M. Biosorption of Manganese from Groundwater by Biomass of Saccharomyces Cerevisiae. *HBRC J.* **2017**, *13*, 106–113. [CrossRef]
97. Kamarudzaman, A.N.; Chay, T.C.; Amir, A.; Talib, S.A. Biosorption of Mn(II) Ions from Aqueous Solution by Pleurotus Spent Mushroom Compost in a Fixed-Bed Column. *Procedia—Soc. Behav. Sci.* **2015**, *195*, 2709–2716. [CrossRef]
98. Bhatnagar, A.; Minocha, A.K. Conventional and Non-Conventional Adsorbents for Removal of Pollutants from Water—A Review. *Indian J. Chem. Technol.* **2006**, *13*, 203–217.
99. McKay, G.; Bino, M.J.; Altamemi, A.R. The Adsorption of Various Pollutants from Aqueous Solutions on to Activated Carbon. *Water Res.* **1985**, *19*, 491–495. [CrossRef]
100. Leboda, R. Carbon-Mineral Adsorbents—New Type of Sorbents Part II. Surface Properties and Methods of Their Modification. *Mater. Chem. Phys.* **1993**, *34*, 123–141. [CrossRef]
101. Gabaldón, C.; Marzal, P.; Seco, A.; Gonzalez, J.A. Cadmium and Copper Removal by a Granular Activated Carbon in Laboratory Column Systems. *Sep. Sci. Technol.* **2000**, *35*, 1039–1053. [CrossRef]
102. Bhatnagar, A.; Sillanpää, M. A Review of Emerging Adsorbents for Nitrate Removal from Water. *Chem. Eng. J.* **2011**, *168*, 493–504. [CrossRef]
103. Aghakhani, A.; Mousavi, S.F.; Mostafazadeh-Fard, B. Desalination of Saline Water with Single and Combined Adsorbents. *Desalin. Water Treat.* **2013**, *51*, 1928–1935. [CrossRef]
104. Ahmed, A.S.; Alsultan, M.; Sabah, A.A.; Swiegers, G.F. Carbon Dioxide Adsorption by a High-Surface-Area Activated Charcoal. *J. Compos. Sci.* **2023**, *7*, 179. [CrossRef]
105. Carrott, P.J.M.; Carrott, M.R. Lignin—From Natural Adsorbent to Activated Carbon: A Review. *Bioresour. Technol.* **2007**, *98*, 2301–2312. [CrossRef]
106. Ahmed, M.; Chin, Y.H.; Guo, X.; Zhao, X.M. Microwave Assisted Digestion Followed by ICP-MS for Determination of Trace Metals in Atmospheric and Lake Ecosystem. *J. Environ. Sci.* **2017**, *55*, 1–10. [CrossRef] [PubMed]
107. Mattson, J.S.; Mark, H.B. *Activated Carbon: Surface Chemistry and Adsorption from Solution*; M. Dekker: New York, NY, USA, 1971; ISBN 0824714431.
108. Chiang, Y.C.; Chiang, P.C.; Chiang, H.L.; Chang, E.E. Determination of Chemical Characterizations of Activated Carbons from Various Raw Materials. *Toxicol. Environ. Chem.* **1998**, *67*, 275–291. [CrossRef]
109. Grant, G.A.; Fisher, P.R.; Barrett, J.E.; Wilson, P.C. Removal of Agrichemicals from Water Using Granular Activated Carbon Filtration. *Water. Air. Soil Pollut.* **2019**, *230*, 7. [CrossRef]

110. Sadegh, H.; Ali, G.A.M.; Gupta, V.K.; Makhlouf, A.S.H.; Shahryari-ghoshekandi, R.; Nadagouda, M.N.; Sillanpää, M.; Megiel, E. The Role of Nanomaterials as Effective Adsorbents and Their Applications in Wastewater Treatment. *J. Nanostruct. Chem.* **2017**, *7*, 1–14. [CrossRef]
111. Stafiej, A.; Pyrzynska, K. Adsorption of Heavy Metal Ions with Carbon Nanotubes. *Sep. Purif. Technol.* **2007**, *58*, 49–52. [CrossRef]
112. Mumpton, F.A. La Roca Magica: Uses of Natural Zeolites in Agriculture and Industry. *Proc. Natl. Acad. Sci. USA* **1999**, *96*, 3463–3470. [CrossRef] [PubMed]
113. Wen, J.; Dong, H.; Zeng, G. Application of Zeolite in Removing Salinity/Sodicity from Wastewater: A Review of Mechanisms, Challenges and Opportunities. *J. Clean. Prod.* **2018**, *197*, 1435–1446. [CrossRef]
114. El Bastamy, E.; Ibrahim, L.A.; Ghandour, A.; Zelenakova, M.; Vranayova, Z.; Abu-Hashim, M. Efficiency of Natural Clay Mineral Adsorbent Filtration Systems in Wastewater Treatment for Potential Irrigation Purposes. *Sustainability* **2021**, *13*, 5738. [CrossRef]
115. Wang, S.; Peng, Y. Natural Zeolites as Effective Adsorbents in Water and Wastewater Treatment. *Chem. Eng. J.* **2010**, *156*, 11–24. [CrossRef]
116. Babel, S.; Kurniawan, T.A. Low-Cost Adsorbents for Heavy Metals Uptake from Contaminated Water: A Review. *J. Hazard. Mater.* **2003**, *97*, 219–243. [CrossRef]
117. Gibb, N.P.; Dynes, J.J.; Chang, W. Synergistic Desalination of Potash Brine-Impacted Groundwater Using a Dual Adsorbent. *Sci. Total Environ.* **2017**, *593–594*, 99–108. [CrossRef] [PubMed]
118. Gibb, N.P.; Dynes, J.J.; Chang, W. A Recyclable Adsorbent for Salinized Groundwater: Dual-Adsorbent Desalination and Potassium-Exchanged Zeolite Production. *Chemosphere* **2018**, *209*, 721–729. [CrossRef] [PubMed]
119. Lim, A.P.; Aris, A.Z. A Review on Economically Adsorbents on Heavy Metals Removal in Water and Wastewater. *Rev. Environ. Sci. Biotechnol.* **2014**, *13*, 163–181. [CrossRef]
120. Bilal, M.; Ihsanullah, I.; Younas, M.; Ul Hassan Shah, M. Recent Advances in Applications of Low-Cost Adsorbents for the Removal of Heavy Metals from Water: A Critical Review. *Sep. Purif. Technol.* **2022**, *278*, 119510. [CrossRef]
121. Bhatnagar, A.; Sillanpää, M.; Witek-Krowiak, A. Agricultural Waste Peels as Versatile Biomass for Water Purification—A Review. *Chem. Eng. J.* **2015**, *270*, 244–271. [CrossRef]
122. Younas, H.; Younas, F. Wastewater Application in Agriculture—A Review. *Water Air Soil Pollut.* **2022**, *233*, 329. [CrossRef]
123. Beni, A.A.; Esmaeili, A. Biosorption, an Efficient Method for Removing Heavy Metals from Industrial Effluents: A Review. *Environ. Technol. Innov.* **2020**, *17*, 100503. [CrossRef]
124. Qin, H.; Hu, T.; Zhai, Y.; Lu, N.; Aliyeva, J. The Improved Methods of Heavy Metals Removal by Biosorbents: A Review. *Environ. Pollut.* **2020**, *258*, 113777. [CrossRef]
125. Jesus, J.; Nunes da Silva, R.; Pintor, A. Advances in Cork Use in Adsorption Applications: An Overview of the Last Decade of Research. *Separations* **2023**, *10*, 1–18. [CrossRef]
126. Yan, S.; Yu, W.; Yang, T.; Li, Q.; Guo, J. The Adsorption of Corn Stalk Biochar for Pb and Cd: Preparation, Characterization, and Batch Adsorption Study. *Separations* **2022**, *9*, 22. [CrossRef]
127. Murtaza, G.; Ahmed, Z.; Eldin, S.M.; Ali, I.; Usman, M.; Iqbal, R.; Rizwan, M.; Abdel-Hameed, U.K.; Haider, A.A.; Tariq, A. Biochar as a Green Sorbent for Remediation of Polluted Soils and Associated Toxicity Risks: A Critical Review. *Separations* **2023**, *10*, 197. [CrossRef]

water

MDPI

Article

Stream Water Quality Control and Odor Reduction through a Multistage Vortex Aerator: A Novel In Situ Remediation Technology

Arnab Ghosh [1,2], Mijeong Choi [1], Dain Yoon [3], Sunghoon Kim [4], Jaebum Kim [1], Jurng-Jae Yee [5] and Sunghyuk Park [1,5,*]

1 Department of Civil Engineering, Dong-A University, Busan 49315, Republic of Korea; arnab@dau.ac.kr (A.G.)
2 University Core Research Center for Disaster-Free Safe Ocean City Construction, Dong-A University, Busan 49315, Republic of Korea
3 Busan Port Authority, Port Construction Department, Busan 48943, Republic of Korea
4 AI Research Laboratory, Korea Water Resource Corporation Research Institute, Daejeon 34045, Republic of Korea
5 Department of ICT Integrated Safe Ocean Smart Cities Engineering, Dong-A University, Busan 49315, Republic of Korea
* Correspondence: shpark@dau.ac.kr

Abstract: In this work, we report the restoration of a polluted urban stream by employing the multistage vortex aerator (MVA), an in-line mixer device that improves the dissolved oxygen concentration of polluted streams and accelerates the water purification rate. It was observed during the field experiment that the dissolved oxygen was enhanced up to 7.05 mg/L and the water quality was improved to a good grade. As a result, the complex odor was successfully eliminated and reduced by up to 71.9%, while the water quality grade was also improved by more than two grades on average. Stream water quality indicators monitored for twelve months revealed high removal rates of total phosphorous (56.4%) and suspended solids (61%). The study demonstrated MVA as a promising eco-friendly technology for significant improvement in urban stream water quality. Moreover, the MVA process creates no secondary pollution and is believed to be a sustainable treatment option for odorous water bodies. Overall, the MVA process is technically feasible for implementation, and this study provides a specific reference as a basis for the treatment of polluted water bodies in urban settings.

Keywords: multistage vortex aerator; dissolved oxygen; odor reduction; water quality control

1. Introduction

Water is not only necessary for all life on Earth, but it is also a valuable resource for human civilization [1]. Urban growth and transitional development have worsened the environment as industry and urbanization have accelerated rapidly. Large amounts of industrially polluted wastewater and domestic sewage are discharged arbitrarily during urban construction and development [2]. Excessive discharge significantly reduces the self-purification capacity of river water bodies and contributes to eutrophication issues in lakes and rivers. Contaminated odorous water bodies generate substantial environmental concerns around the globe today, threatening ecological integrity, compromising the image of the metropolitan landscape, and increasing people's health risk index [3]. Odorous water bodies develop due to dynamic water factors, metal element pollution, organic pollution, sediment pollution, and other variables [4]. The problem is primarily caused by a combination of physical (such as temperature), chemical (such as FeS and/or MnS in enveloping water formed by the interaction of Fe^{2+} or Mn^{2+} with S^{2-}), and biological (such as algae) reactions of pollutants in the water under hypoxic or anaerobic conditions [5–7]. Rivers and lakes in cities and towns serve several purposes, including water supply, temperature regulation, and water storage [8]. They play an essential role in ensuring the stability of the human ecological environment. Therefore, treating odorous water bodies in today's urban water environment is of utmost importance.

Treatment approaches for odorous water bodies are classified into four categories: physical, chemical, biological, and ecological restoration technologies [9]. The physical, chemical, and biological treatment measures are in situ treatment methods, while restoration is an ex situ treatment technique. Some examples of standard physical treatment technologies include pollution interception and source management, sediment dredging, aeration, and oxygenation [10,11]. Physical methods, in general, may remove pollutants from water bodies without generating secondary pollution, but the expense is often high, and the workload is significant.

On the other hand, chemical treatment methods include flocculation and sedimentation, algae-removing agents, adsorbents, and so forth [12]. Chemical treatment processes have immediate effects and are highly relevant; nevertheless, an excessive chemical injection results in serious secondary contamination [6]. Biological methods, in principle, are natural and environmentally benign, with high stability, although they require a more extended treatment cycle. Biofilm technology and aquatic plant purification are two examples of biological treatments [13]. These techniques, however, are prone to secondary pollution and are not appropriate for the continuous and effective purification of water under complex conditions. Ex situ treatment by restoration entails separating and transferring the polluted water and sediments to water treatment facilities and returning the treated water to the initial water body. However, external treatment and restoration technology demand a sizable investment of material and monetary resources. Dealing with the extracted sediment from the water body presents another challenging issue [14]. In this context, pure oxygen injection employing a multistage vortex aerator (MVA) can be a promising in situ treatment alternative to alleviate the possibility of secondary pollution. Additionally, there are no reports on stream water quality improvement employing an MVA system.

An adequate amount of oxygen is critical in both streams and sewers. In the absence of oxygen, volatile fatty acids and hydrogen sulfide are produced, resulting in an undesirable odor [15,16]. Hydrogen sulfide, produced by sustained anoxic conditions has a highly unpleasant odor and can be fatal in high concentrations. It has been proposed that such high concentrations of hydrogen sulfide can be reduced and controlled by employing oxygen injection [17]. Odor measurement in Korea is based on designated odor substances and complex odors. It is evaluated using the results of measurement at the boundary line and the emission limit value specified in the Odor Prevention Act. Among the 22 designated odor substances, hydrogen sulfide is the representative odor material, has a high odor index, occurs in anaerobic or anoxic conditions, and is also generated by sulfur-reducing bacteria, representative microorganisms that reduce sulfate ions to hydrogen sulfide.

Pollutants from non-point sources and sewage sludge are partially flowing and deposited into the Goejeongcheon stream, the target site of our research. Because of this, the stream's dissolved oxygen levels were near to those of anaerobic conditions. In this study, the dissolved oxygen concentration was augmented by dissolving oxygen in the target stream using the multistage vortex aerator (MVA). We evaluated the feasibility of using pure oxygen and a vortex aerator for stream restoration to suppress river anaerobicization and promote aerobic conditions. The study investigated MVA as an eco-friendly technology for the improvement of urban stream water quality parameters in a sustainable way.

2. Materials and Research Methods

2.1. Research Site Location

Goejeongcheon stream, located in Saha-gu, Busan, is a local stream. The stream flows through the downtown area and connects to the Nakdong-gang estuary. The length of the open area is 0.634 km, and the length of the concrete-covered area is 4.735 km, with a total length of 5.37 km. The stream was covered to alleviate traffic congestion, and the covered section is now used as a road. The covered stream runs through the downtown area, where residential areas and markets are closely located. Considering the land use status of the Goejeongcheon watershed, the forest area is the largest portion, covering 4.78 km^2 of the total area of 5.36 km^2, followed by residential areas and roads. Because Goejeongcheon is a low-tide river, the water level fluctuates significantly over time. As a result, when the water level is low, the bottom of the river becomes visible. According to the annual meteorological report of the Busan Meteorological Administration, the annual rain is 1495.60 mm, the lowest temperature is −12.6 °C, the highest temperature is 36.7 °C, the average yearly temperature is 14.5 °C, and the average annual humidity is 65.4%. The target site of this study is the open section of Goejeongcheon stream (Figure 1), with a total length of 0.634 km and an area of 8200 m^2. The stream connects to the estuary of the Nakdonggang river, which has the characteristics of a tidal river. The average water level is at least 31 cm, and the maximum is 127 cm. Since 2005, residents have complained about Goejeongcheon's stench, citing headaches and disruption to their living environment due to the foul odor.

Figure 1. Location of Goejeongcheon research site.

2.2. Multistage Vortex Aerator (MVA)

The multistage vortex aerator (Figure 2) is an oxygen-dissolving system that uses an in-line mixing method. This device effectively and quickly transfers the oxygen required in processes such as livestock wastewater treatment and allied fields. It was developed to accelerate wastewater purification by boosting the dissolved oxygen concentration in the wastewater and activating aerobic bacteria, as well as to reduce the occurrence of odors by decomposing various organic compounds. The multistage vortex aerator (MVA) system comprises a pure oxygen generator and an oxygen dissolver. A submerged pump transports water into the MVA, injects oxygen into the inlet, and delivers efficient mixing of oxygen and liquid via a vortex flow inside the dissolver's multistage partition walls. The absorption of gas depends on the film thickness between the gas and liquid, and the absorption rate can be increased by factors that can reduce this thickness [18]. By increasing the time and number of collisions between the two phases, the thickness of the film between the gas and liquid is decreased, and the efficiency of dissolving oxygen in the liquid increases. The detailed structure of the device is illustrated in our earlier reported work [19]. Since the MVA circulates and supplies pure oxygen, it was confirmed that the MVA increases dissolved oxygen concentrations to higher levels in a shorter time than other existing aeration devices, such as fine pore diffusers. Fine pore diffusers suffer from the drawbacks of inorganic scaling and biological fouling, which can reduce the transfer efficiency and cause excessive head loss [20].

Figure 2. Pictorial representation of a multistage vortex aerator. Adapted with permission from [19].

In addition, although the conventionally used oxygenators use the 21% of oxygen in the air as an oxygen source, there is a limit to the movement speed of oxygen, whereas the multistage vortex aerator uses high-purity oxygen (about 90% oxygen). Therefore, the oxygen saturation concentration of the aqueous solution is about four times higher than that of the air diffuser, and it is observed that the dissolved oxygen in water can be rapidly increased. The increased number of collisions inside the MVA reduces the gas–liquid interfacial film thickness, thereby improving oxygen transfer efficiency.

By maintaining high dissolved oxygen (DO) levels in the river using the MVA, the aerobic microorganisms in the river can actively degrade organic debris, and high DO concentration inhibits the growth of odor-causing strictly anaerobic SRB (sulfate-reducing bacteria) [21]. The multistage vortex aerator suppresses the occurrence of anaerobic conditions, which are considered the origin of odors, enabling it to remove and prevent odors in a short period of time.

2.3. Field Experiment

The basic operation of the MVA system is depicted in Figure 3. Briefly, the river water was pumped up at a flow rate of 1 m^3/min using a submersible pump, and oxygen (~90% pure oxygen) generated through an oxygen generator was injected and dissolved into the MVA inlet. Thereafter, the oxygen-enriched water was discharged into the river. The dissolved oxygen concentration at the MVA outlet was maintained at an average of 22 mg/L. The dissolved oxygen concentration of the oxygenated water was measured by sampling through a sample collection valve located at the outlet of the MVA. The cycle was repeated for 16 h a day, and the system was operated for 12 months. The hydraulic retention time (HRT) of the water inside the MVA is 1.4 min. The oxygen generator was installed 115 m from the main public access, which is an upstream point, and the MVA outlet was installed at 90 m (point A) (Figure 4). The multistage vortex aerator consists of one oxygen generator, one dissolver, one pump, and a water level sensor.

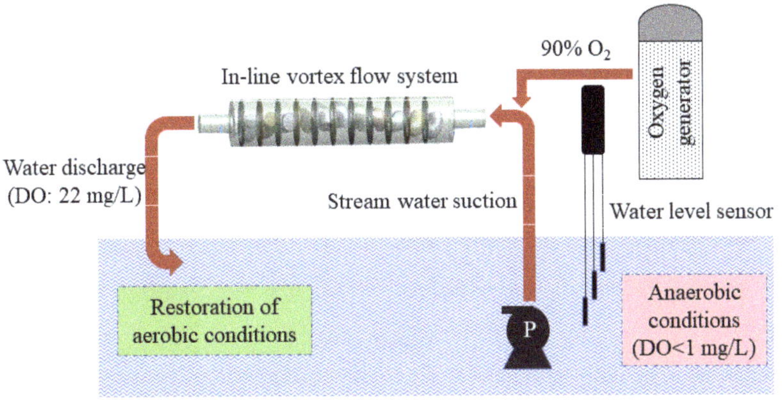

Figure 3. Schematic diagram of stream water treatment using multistage vortex aerator (MVA).

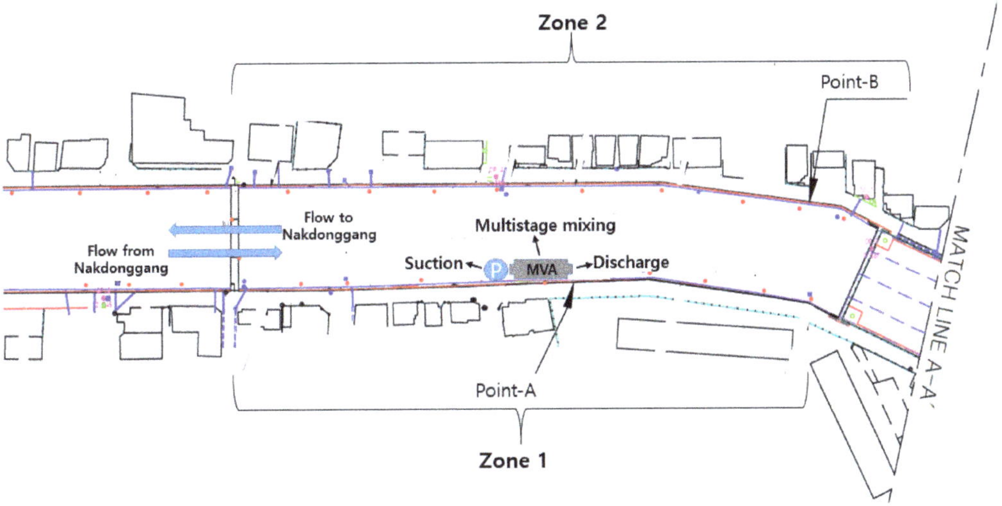

Figure 4. Schematic diagram for field experimental sampling points for water quality and odor analysis (Blue arrows illustrate the direction of water flow, and the squares represent the residential buildings).

A mechanical water level sensor was installed to enable steady system operation and prevent idling, such that the system only operated whenever the river's water level was 35 cm or higher. The mechanical water level sensor consists of three distinct sensors, each with a different length. If two or more of the three sensors are exposed to the atmosphere, the device shuts down. Furthermore, due to the device's noise, it was operated from 6 a.m. to 10 p.m.

2.4. Odor and Water Quality Analysis

2.4.1. Odor Analysis

The complex odor strength level was measured using a handheld odor meter, OMX-SRM (Shinyei, Japan). In addition, the intensity of the odor of the water samples was measured using the threshold odor number (TON) based on the olfactometric evaluation method. In this method, the sample is diluted with odor-free water until the least discernible odor is obtained.

2.4.2. Water Quality Measurement

A multi-parameter water quality meter from YSI was used to measure dissolved oxygen. The dissolved oxygen was determined by directly inserting the probe into the stream water. The measurement of suspended substances was based on the standard water quality test method, and the chemical oxygen demand (COD), total nitrogen (T-N), and total phosphorus (T-P) were performed according to the HUMAS water quality analysis KIT manual and placed into the water quality analyzer for subsequent analysis. The water quality analyzer used was the HS-2300 PLUS model, which performs measurements by the absorbance method using a color reagent.

3. Experimental Results and Discussion

3.1. Field Experiment

At the field-testing location, point A represents the section of the stream influenced by the MVA, while point B represents the section without the MVA installation (Figure 4). The section with the MVA installation is called 'zone 1', and 'zone 2' represents the control section without the MVA installation. Goejeongcheon has the characteristics of both a lake and a river. Notably, the Water Resources Management Information System (WAMIS) standards in Korea are more stringent for lake water quality, and therefore, we adopted these criteria for our work. The effectiveness of the MVA installation was confirmed by comparing the obtained water quality data with the lake water quality and environmental standards according to WAMIS (Table 1) [22].

Table 1. Water quality parameters and environmental standards for classification.

Classification	Parameters				
	COD (mg/L)	Suspended Solids (mg/L)	Dissolved Oxygen (mg/L)	Total Phosphorous (mg/L)	Total Nitrogen (mg/L)
Very good (Ia)	2 or less	1 or less	7.5 or more	0.01 or less	0.2 or less
Good (Ib)	3 or less	5 or less	5.0 or more	0.02 or less	0.3 or less
Slightly good (II)	4 or less	5 or less	5.0 or more	0.03 or less	0.4 or less
Fair (III)	5 or less	15 or less	5.0 or more	0.05 or less	0.6 or less
Slightly bad (IV)	8 or less	15 or less	2.0 or more	0.10 or less	1.0 or less
Poor (V)	10 or less	Floating suspended matter	2.0 or more	0.15 or less	1.5 or less
Very bad (VI)	>10	–	<2.0	>0.15	>1.5

3.1.1. Dissolved Oxygen (DO)

Data for dissolved oxygen by season were recorded for one year, with the average of the data from spring to autumn being reported (Figure 5). Data analysis for each season was required to examine the effectiveness of the MVA in dissolved oxygen improvement. In Figure 5, it can be observed that the seasonal variation in dissolved oxygen concentration followed a similar pattern at points A and B, respectively. At point B, the average dissolved oxygen was 0.98 mg/L in summer, 2.55 mg/L in spring, and 1.81 mg/L in autumn, wherein spring and autumn exhibited higher dissolved oxygen concentrations compared with summer because the dissolved oxygen concentration is inversely correlated to the stream temperature [23]. Compared with spring, the DO in summer was 1.57 mg/L lower, while the DO increased by 0.83 mg/L in the autumn. Likewise, at point A, the average DO in summer was the lowest at 4.62 mg/L, while in spring, the DO was measured at 7.05 mg/L, with an increment of 2.44 mg/L compared with summer. The average DO measured in autumn was 6.45 mg/L, a 1.83 mg/L increase over the summer DO level. The seasonal comparison of the DO concentration at points A and B revealed a 63.8% DO increment from

2.55 mg/L to 7.05 mg/L in spring and a 78.6% DO increment from 0.98 mg/L to 4.62 mg/L in summer. In autumn, it was observed that the DO increased by 71.8%, from 1.81 mg/L to 6.45 mg/L. The reason for the seasonal variation in the dissolved oxygen concentration is attributed to the increase in bacterial diversity in warmer months [24] and the decline in oxygen solubility due to an increase in the temperature of the stream.

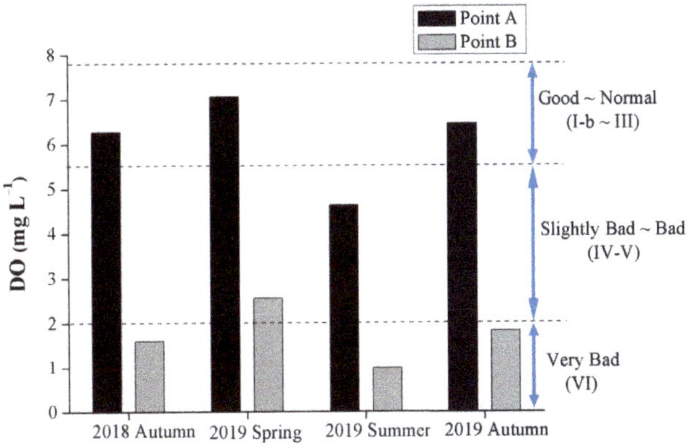

Figure 5. Variation in dissolved oxygen (DO) content over different seasons in the field experiment.

Furthermore, although there was seasonal variation, it is important to observe that dissolved oxygen could be significantly improved from the very bad to good grade when the stream water was cycled through the installed MVA system at point A. It is evident from the plot in Figure 5 that compared with point B, the average DO concentration increased by 3.5 times at point A in the presence of the MVA system. The results indicate that when the MVA is employed, the increase in dissolved oxygen concentration is significant in contrast with point B.

3.1.2. Complex Odor

The data for variation in seasonal complex odor were recorded for a period of one year (Figure 6a). At point A, the average of the complex odors measured in the autumn of 2018 was the highest at 10.19, and the average of the complex odors measured in the spring of 2019 was 5.67, suggesting a 44.7% decrease from the autumn of 2018. In summer, the average odor was 8.51, which was reduced by 24.8% in the autumn of 2019. In addition, comparing the autumn of 2018 and the autumn of 2019, the complex odor in the autumn of 2019 was 6.46, indicating a 36.6% reduction.

At point B, where the MVA system was not installed, the average complex odor was 23.91 in autumn 2018, 17.74 in spring 2019, 33.22 in summer 2019, and 22.98 in autumn 2019, with the highest value observed in summer. Overall, a negligible reduction of 3.87% in complex odor was noted when comparing the autumn values of 2018 and 2019.

It is imperative to compare the complex odor between points A and B to determine the efficiency of the MVA system in complex odor reduction. Comparing the complex odor by season, in the autumn of 2018, there was a 57.4% decrease, and in the spring of 2019, there was an almost 68% reduction. In the summer of 2019, there was a 74.4% reduction in complex odors, and in the autumn of 2019, a reduction of 71.9% was observed. In addition, it was observed that the complex odor varied by season, and comparing the values for 2019 alone, the complex odor was observed to be lower in spring and autumn than in summer. Increased temperatures in summer can further degrade water quality and increase the degree of black odorous water by influencing the release of phosphorous from sediments [25,26]. Therefore, the findings highlight that the odor was very strong at point B in all the seasons, particularly in summer, and the effect of odor reduction was investigated. Comparing the autumn of 2018 and the autumn of 2019 (after the installation of MVA system), it is apparent that the decrease in complex odor was greater after one year of system operation. Notably, it was confirmed that TON and complex odor decreased as dissolved oxygen increased (Figure 6b,c). A correlation analysis was performed using SPSS v. 22.0 to estimate the degree of association among multiple variables, including DO, TON, and complex odor, and a correlation matrix was computed (Table 2). The table shows the degree of linear association between any two of the parameters, which is measured by the degree of correlation as a coefficient (R). The R-values indicate the highly correlated inter-relationships between DO, TON and complex odor. The confidence interval was 95% and the significance level was $p < 0.05$. It was observed that the TON and complex odor are negatively correlated with DO because with an increase in DO, the complex odor as well as TON decreases. The circulation of the water through the MVA system provides enough dissolved oxygen (~22 mg/L) to the aerobic bacteria so they can actively break down the odorous compounds, resulting in odor reduction [27].

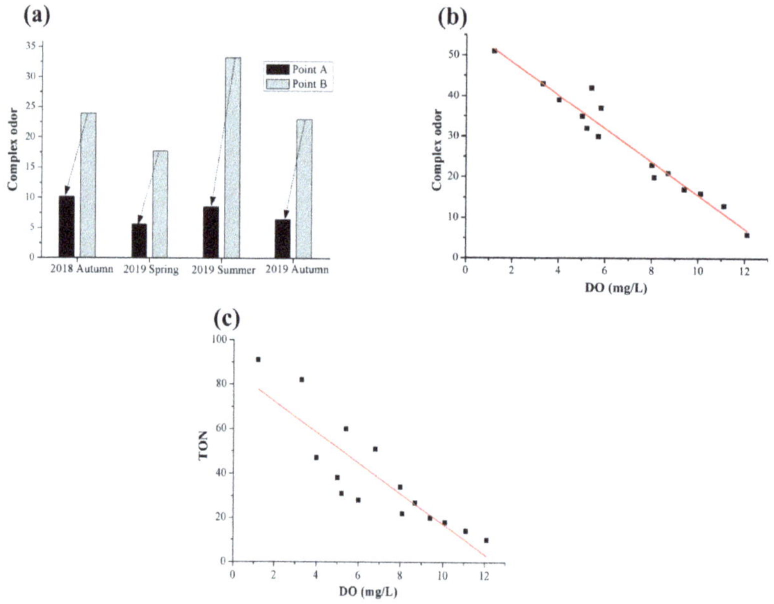

Figure 6. (**a**) Complex odor at point A and point B according to seasonal variation; (**b**) influence of dissolved oxygen (DO) on complex odor; and (**c**) effect of DO on TON.

Table 2. Correlation matrix of the important parameters.

	DO	TON	Complex Odor
DO	1	−0.8744 **	−0.9649 **
TON	0.8744 **	1	0.9186 **
Complex odor	0.9649 **	0.9186 **	1

Note: **: *p*-value < 0.05.

3.1.3. Water Quality Analysis

The descriptive statistics of the experimental data of water quality parameters at points A and B over the given period are presented in Table 3.

Table 3. Descriptive statistics of the water quality parameters.

	Min	Max	Average	STD
T-N at point A	0.300	4.510	2.10552	1.216509
T-N at point B	0.530	4.810	2.52828	1.189926
T-P at point A	0.005	0.045	0.01529	0.009180
T-P at point B	0.020	0.060	0.03510	0.011119
COD at point A	5.150	19.140	14.67156	2.835519
COD at point B	6.220	19.960	16.63875	3.201404
SS at point A	2.667	25.000	14.25950	6.083765
SS at point B	8.000	136.700	36.44611	28.228620

At point A of zone 1, the average total phosphorus (TP) concentration was measured at 0.015 mg/L and it was 0.035 mg/L at point B of zone 2, indicating a 56.4% reduction. The significant removal of TP could be attributed to the enhanced DO concentration due to the presence of the MVA system, which can promote aerobic microbial growth, resulting in accelerated phosphorous consumption [28]. There is a strong possibility that the product of the decomposition of the organic matter facilitated the deposit and subsequent adsorption of the phosphorous onto the sediments [29,30]. Urban streams are more susceptible to eutrophication because of their lower self-purification capacity than rivers. Therefore, from the environmental perspective, it is importance to prevent the accumulation of nutrients causing eutrophication. During the field experiment period, the water quality grade, based on the average TP concentration, gradually improved from bad to normal. Overall, it improved from a very bad to good grade at the end of the treatment period (Figure 7a).

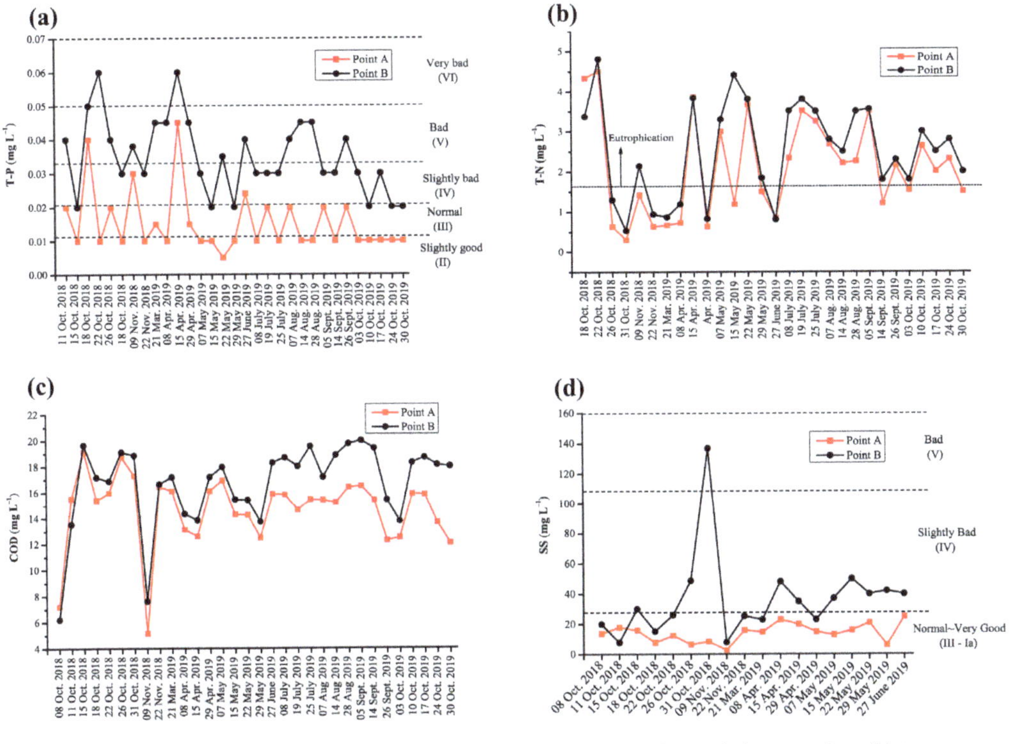

Figure 7. (**a**) Field experiment total phosphorous (T-P) concentration variation over time; (**b**) total nitrogen (T-N) concentration variation over time; (**c**) COD concentration variation over time; (**d**) suspended solids (SS) content variation over time.

Total nitrogen (TN) was also compared and analyzed at point B of zone 2 and point A of zone 1, where the MVA was installed. The TN concentration over a period of one year ranged from 0.3 mg/L to 4.81 mg/L, depending on the seasonal variation (Figure 7b). Total nitrogen was reduced by up to 17% on average at point A compared with point B. The aerobic conditions promoted by the application of the MVA aid the microorganisms in nitrogen removal by oxidation of anaerobic ammonia.

The chemical oxygen demand (COD) was compared at point A (having high DO) of zone 1 with point B of zone 2 (Figure 7c). The COD was influenced by the tide and displayed a range of concentrations. Based on the analysis of a year of data, it was observed that the average chemical oxygen demand at point A of zone 1 was reduced by up to 12% compared with that at point B in zone 2. The increase in dissolved oxygen stimulates the microbial activity, where the microbes decompose and consume the particulate organic matter (POM), thereby lowering the chemical oxygen demand [31].

In the presence of the MVA system at point A, the suspended solid (SS) concentration was reduced by 61% on average compared with point B (Figure 7d). Overall, the water quality improved from a poor to moderate grade and from a moderate to very good grade after circulation through the MVA system over a period of twelve months. Pure oxygen (~90%) injection and multistage mixing through the MVA enhanced the dissolved oxygen content and augmented the biological decomposition of organic matter, thereby reducing the suspended solids concentration.

The above findings demonstrate that the MVA system enhances the dissolved oxygen and thereby reduces the concentration of pollutants in odorous water by stimulating the growth of microorganisms.

4. Conclusions

In this research, we evaluated the feasibility of installing and operating the MVA in Goejeongcheon stream, Saha-gu, Busan, from September to November 2018 and March to November 2019. The novel MVA system is an in situ remediation technology that efficiently transfers and increases the dissolved oxygen in the water body, thereby promoting aerobic microbial activity. The proposed MVA process was monitored for nearly twelve months and demonstrated promising results, with a 3.5-times increase in DO content and maximum removal rates of TP and SS of 56.4% and 61%, respectively. Several water quality indicators met the Class Ia (very good) standard, while others achieved the Class Ib (good) surface water standard.

The present work demonstrates that by increasing the dissolved oxygen concentration, it was possible to reduce odor sources and improve the water quality of the polluted research site. The MVA system can be operated in a continuous mode devoid of any fouling problems. It can be a viable alternative to treat odorous water bodies through pure oxygen injection. It is reported as the first case in Korea and worldwide of improving the odor and water quality of a stream by employing a multistage vortex aerator (MVA). Overall, the effects of increased dissolved oxygen on odor reduction and water quality parameters (T-N, T-P, COD, S-S) are documented and discussed in this paper. Based on the results, the proposed system can be developed into a full-scale river purification system in the future. The findings of this work provide valuable insights that can be applied in scaling up the system in future studies. However, for the large-scale application of this technology, the production cost of pure oxygen and the durability of the equipment need to be examined in detail.

Author Contributions: Conceptualization, S.P.; methodology, M.C.; software, J.K., S.K. and A.G.; validation, J.K.; formal analysis, A.G. and D.Y.; investigation, M.C., S.K. and D.Y.; resources, J.-J.Y.; data curation, M.C.; writing—original draft preparation, A.G.; writing—review and editing, S.P.; visualization, A.G. and S.K.; supervision, S.P.; project administration, J.-J.Y.; funding acquisition, J.-J.Y. and S.P. All authors have read and agreed to the published version of the manuscript.

Funding: This work was supported by a National Research Foundation of Korea (NRF) grant funded by the Korea government (MSIT) (2021R1I1A3060770) and the Basic Science Research Program through the National Research Foundation of Korea (NRF) funded by the Ministry of Education (2016R1A6A1A03012812).

Data Availability Statement: The data presented in this study are available on request from the corresponding author.

Acknowledgments: The authors are very grateful to all anonymous reviewers and proofreaders for their insightful comments which helped us in improving the overall quality of this work.

Conflicts of Interest: The authors declare no conflict of interest.

References

1. Qu, X.; Alvarez, P.J.J.; Li, Q. Applications of nanotechnology in water and wastewater treatment. *Water Res.* **2013**, *47*, 3931–3946. [CrossRef] [PubMed]
2. Hsu, A.; Yeo, Z.Y.; Weinfurter, A. Emerging digital environmental governance in China: The case of black and smelly waters in China. *J. Environ. Plann. Man.* **2020**, *63*, 14–31. [CrossRef]
3. Song, C.; Liu, X.; Song, Y.; Liu, R.; Gao, H.; Han, L.; Peng, J. Key blackening and stinking pollutants in Dongsha River of Beijing: Spatial distribution and source identification. *J. Environ. Manag.* **2017**, *200*, 335–346. [CrossRef] [PubMed]
4. Peng, W.; Yisong, W.; Pingfang, Z.; Mei, L.; Jingya, S.; Fusheng, Z. Analysis of formation and mechanisms of black and smelly river water in island cities. *Meteorol. Environ. Res.* **2018**, *9*, 42–48. [CrossRef]
5. Gao, J.; Jia, J.; Xing, F.; Wang, Y.; Xu, X.; Yang, Y.; Zou, X.; Gao, S.; Qi, S. Changes in water and sediment exchange between the Changjiang River and Poyang Lake under natural and anthropogenic conditions. *China. Sci. Total Environ.* **2014**, *481*, 542–553. [CrossRef]
6. Perera, E.D.P.; Jinno, K.; Hiroshiro, Y. Bacteria-mediated Reduction and Precipitation of $Fe(OH)_3$ and FeS in the Subsurface of a Coastal Aquifer: A Numerical Investigation. *Water Qual. Expo. Health* **2010**, *2*, 15–30. [CrossRef]
7. Negoro, T.; Ando, M.; Ichikawa, N. Blue-green algae in Lake Biwa which produce earthy-musty odors. *Water Sci. Technol.* **1988**, *20*, 117–123. [CrossRef]
8. Liang, Z.; Siegert, M.; Fang, W.; Sun, Y.; Jiang, F.; Lu, H.; Chen, G.; Wang, S. Blackening and odorization of urban rivers: A biogeochemical process. *Fems Microbiol. Ecol.* **2018**, *94*, fix180. [CrossRef]
9. Cao, J.; Sun, Q.; Zhao, D.; Xu, M.; Shen, Q.; Wang, D.; Wang, Y.; Ding, S. A critical review of the appearance of black-odorous waterbodies in China and treatment methods. *J. Hazard. Mater.* **2020**, *385*, 121511. [CrossRef]
10. Toffolon, M.; Ragazzi, M.; Righetti, M.; Teodoru, C.R.; Tubino, M.; Defrancesco, C.; Pozzi, S. Effects of artificial hypolimnetic oxygenation in a shallow lake. Part 1: Phenomenological description and management. *J. Environ. Manag.* **2013**, *114*, 520–529. [CrossRef]
11. Liu, C.; Shao, S.; Shen, Q.; Fan, C.; Zhou, Q.; Yin, H.; Xu, F. Use of multi-objective dredging for remediation of contaminated sediments: A case study of a typical heavily polluted confluence area in China. *Environ. Sci. Pollut. R.* **2015**, *22*, 17839–17849. [CrossRef] [PubMed]
12. He, H.; Yu, Q.; Lai, C.; Zhang, C.; Liu, M.; Huang, B.; Pu, H.; Pan, X. The treatment of black-odorous water using tower bipolar electro-flocculation including the removal of phosphorus, turbidity, sulfion, and oxygen enrichment. *Front. Environ. Sci. Eng.* **2021**, *15*, 18. [CrossRef]
13. Zhang, Y.; He, F.; Xia, S.; Zhou, Q.; Wu, Z. Studies on the treatment efficiency of sediment phosphorus with a combined technology of PCFM and submerged macrophytes. *Environ. Pollut.* **2015**, *206*, 705–711. [CrossRef]
14. Zhu, G.; Chen, J.; Zhang, S.; Zhao, Z.; Luo, H.; Hursthouse, A.S.; Wan, P.; Fan, G. High removal of nitrogen and phosphorus from black-odorous water using a novel aeration-adsorption system. *Environ. Chem. Lett.* **2022**, *20*, 2243–2251. [CrossRef]
15. Huisman, J.L.; Weber, N.; Gujer, W. Reaeration in sewers. *Water Res.* **2004**, *38*, 1089–1100. [CrossRef] [PubMed]
16. Yoon, D.; Choi, M.; Park, S. Urban Streams' Water Quality and Odor Control Using Pure Oxygen and Vortex Aerator. *KSCE J. CEER* **2021**, *41*, 493–504.
17. Jiang, G.; Sun, J.; Sharma, K.; Yuan, Z. Corrosion and odor management in sewer system. *Curr. Opin. Biotechnol.* **2015**, *33*, 192–197. [CrossRef]
18. Lewis, W.K.; Whitman, W.G. Principles of gas absorption. *J. Ind. Eng. Chem.* **1924**, *16*, 1215–1220. [CrossRef]
19. Park, S.H.; Batchelor, B.; Ghosh, A. Gas transfer model for a multistage vortex aerator: A novel oxygen transfer system for dissolved oxygen improvement. *J. Environ. Manag.* **2022**, *319*, 115704. [CrossRef]
20. Wang, M.; Mo, H.; Liu, G.; Qi, L.; Yu, Y.; Fan, H.; Xu, X.; Luo, T.; Shao, Y.; Wang, H. Impact of scaling on aeration performance of fine-pore membrane diffusers based on a pilot-scale study. *Sci. Rep.* **2020**, *10*, 4902. [CrossRef]
21. Mara, D.; Horan, N. *Handbook of Water and Wastewater Microbiology*; Academic Press: Cambridge, MA, USA, 2003. [CrossRef]
22. River Water Quality Environmental Standards (2018.5 Revision), Water Resources Management Information System (WAMIS), Republic of Korea. 2018. Available online: http://www.wamis.go.kr/wke/wke_wqbase_lst.do (accessed on 15 March 2023).

23. Walczyńska, A.; Sobczyk, L. The underestimated role of temperature–oxygen relationship in large-scale studies on size-to-temperature response. *Ecol. Evol.* **2017**, *17*, 7434–7441. [CrossRef] [PubMed]
24. Mestre, M.; Höfer, M.; Sala, M.M.; Gasol, J.M. Seasonal Variation of Bacterial Diversity Along the Marine Particulate Matter Continuum. *Front. Microbiol.* **2020**, *11*, 1590. [CrossRef]
25. Gibbons, K.J.; Bridgeman, T.B. Effect of temperature on phosphorus flux from anoxic western Lake Erie sediments. *Water Res.* **2020**, *182*, 116022. [CrossRef] [PubMed]
26. Kong, M.; Han, T.; Chen, M.; Zhao, D.; Chao, J.; Zhang, Y. High mobilization of phosphorus in black-odor river sediments with the increase of temperature. *Sci. Total Environ.* **2021**, *775*, 145595. [CrossRef]
27. Zhu, J. A review of microbiology in swine manure odor control. *Agric. Ecosyst. Environ.* **2000**, *78*, 93–106. [CrossRef]
28. Yu, S.; Huang, C.; Huang, Y.; Li, Z.; Li, Y.; Xiao, Y.; Xu, J.; Ni, S. Case study of the in-situ restoration of black-odorous water by combined process of forced aeration and biological contact oxidation. *Water Sci. Technol.* **2022**, *85*, 827–838. [CrossRef]
29. Zhang, D.; Yang, H.; Lan, S.; Wang, C.; Li, X.; Xing, Y.; Yue, H.; Li, Q.; Wang, L.; Xie, Y. Evolution of urban black and odorous water: The characteristics of microbial community and driving-factors. *J. Environ. Sci.* **2022**, *112*, 94–105. [CrossRef]
30. Ilyas, H.; Masih, I. The effects of different aeration strategies on the performance of constructed wetlands for phosphorus removal. *Environ. Sci. Pollut. Res. Int.* **2018**, *25*, 5318–5335. [CrossRef]
31. Siuda, W.; Chróst, R.J. Decomposition and utilization of particulate organic matter by bacteria in lakes of different trophic status. *Pol. J. Environ. Stud.* **2002**, *11*, 53–66.

water

MDPI

Article

Analysis of Desalination Performance with a Thermal Vapor Compression System

Zineb Fergani [1], Zakaria Triki [1], Rabah Menasri [1], Hichem Tahraoui [1,2,*], Mohammed Kebir [3], Abdeltif Amrane [4,*], Nassim Moula [5], Jie Zhang [6] and Lotfi Mouni [7]

1 Laboratory of Biomaterials and Transport Phenomena, University of Medea, Medea 26000, Algeria
2 Laboratoire de Génie des Procédés Chimiques, Department of Process Engineering, University of Ferhat Abbas, Setif 19000, Algeria
3 Research Unit on Analysis and Technological Development in Environment (URADTE CRAPC), BP 384, Bou-Ismail, Tipaza 42000, Algeria
4 Univ Rennes, Ecole Nationale Supérieure de Chimie de Rennes, CNRS, ISCR—UMR6226, F-35000 Rennes, France
5 Fundamental and Applied Research in Animal and Health (FARAH), Department of Veterinary Management of Animal Resources, Faculty of Veterinary Medicine, University of Liege, 4000 Liege, Belgium
6 School of Engineering, Merz Court, Newcastle University, Newcastle upon Tyne NE1 7RU, UK
7 Laboratory of Management and Valorization of Natural Resources and Quality Assurance, SNVST Faculty, Akli Mohand Oulhadj University, Bouira 10000, Algeria
* Correspondence: hichemm.tahraouii@gmail.com (H.T.); abdeltif.amrane@univ-rennes1.fr (A.A.)

Abstract: Multi-effect distillation with thermal vapor compression (MED-TVC) is a highly energy-efficient desalination technology that can provide a reliable and sustainable source of high-quality water, particularly in areas with limited energy infrastructure and water resources. In this study, a numerical model based on exergoeconomic approach is developed to analyze the economic performance of a MED-TVC system for seawater desalination. A parallel/cross feed configuration is considered because of its high energy efficiency. In addition, a parametric study is performed to evaluate the effects of some operational parameters on the total water price, such as the top brine temperature, seawater temperature, motive steam flow rate, and number of effects. The obtained results indicate that the total water price is in the range of 1.73 USD/m^3 for a distilled water production of 55.20 kg/s. Furthermore, the exergy destructions in the effects account for 45.8% of the total exergy destruction. The MED effects are also identified to be the most relevant component from an exergoeconomic viewpoint. Careful attention should be paid to these components. Of the total cost associated with the effects, 75.1% is due to its high thermodynamic inefficiency. Finally, the parametric study indicates that adjusting the top brine temperature, the cooling seawater temperature, the motive steam flow rate, and the number of effects has a significant impact on the TWP, which varies between 1.42 USD/m^3 and 2.85 USD/m^3.

Keywords: water desalination; multi-effect distillation; thermal vapor compression; exergoeconomic analysis; total water price

1. Introduction

Desalination technologies have become an important pillar of the world's freshwater supplies; they are effective in supporting countries and industries to satisfy their water needs. There are many desalination plants around the world on which communities depend for their economic activity and portable water supply. As desalination systems become a common application in the world, different desalination technologies are available, such as membrane techniques and thermally driven techniques [1]. In this regard, desalination systems are generally designed and optimized from an energy and economic perspective. The exergy-based method has been successfully used for the analysis and optimization of energy and the economic performance of desalination systems [2–6].

Multi-effect desalination is among the thermal desalination systems to receive significant attention for its high efficiency, operational and maintenance simplicity, the feasibility of using low-grade heat, and economic sustainability [7,8]. MED integrated in a thermal vapor compressor system has recently received much attention in research; the major advantages of MED-TVC desalination systems are high thermal efficiency with low energy requirement, a small condenser size, and reasonable production costs [9].

A number of studies have been conducted to analyze the energetic performance of MED-TVC systems using the exergy approach [10–17], in which exergy analysis revealed that the thermo-compressor and effects are the major sources of exergy destruction in a MED-TVC system. Hamed et al. [10] applied an exergy analysis to evaluate and compare the performance of a TVC system with those of the conventional MEB and MVC desalination systems. It was observed that the TVC system generates the lowest exergy destruction among the three systems. In addition, a major part of the exergy destruction in the TVC system occurs in the first effect and in the thermo-compressor. In [11], the thermo-compressor and the evaporator were also identified as the main sources of exergy destruction for single- and multi-effect thermo-vapor compression systems.

Al Asfour et al. [12] conducted a parametric study using energy and exergy analysis for three configurations of a multi-effect thermal vapor compression desalination system. The results revealed that the steam ejector and evaporators are the major sources of exergy destruction in the three configurations. The decrease in the top brine temperature leads to a decrease in the specific heat consumption and exergy destruction. Choi et al. [13] presented an energy and exergy analysis for a MED-TVC system developed by Hyundai Heavy Industry for identifying the system's efficiency improvement potential. It was found that the largest specific exergy losses in the MED-TVC desalination system occur in the TVC and the effects. Comprehensive thermal and exergy analyses for MED and MED–TVC systems were conducted to reveal the mechanism of the system's internal exergy loss [14].

The exergy consumption of the MED TVC system is evidently higher than for the MED system, and the exergy efficiency of the system improves as the number of suction effects increases. Binamer [15] developed a mathematical model of a MED-TVC desalination system to evaluate the system performance using energy and exergy analysis. It was shown that the first effect accounted for about 31% of the total destruction of the exergy effects. The specific exergy destruction is significantly reduced when increasing the number of effects as well as operating at lower top brine temperatures. Sadri et al. [16] also developed a mathematical model to predict and optimize the exergy efficiency of a MED-TVC system. The model could predict the effects of various variables on the performance of the system. The authors indicated that careful attention should be paid to evaporators, since they are the main sources of exergy destruction. Menasri et al. [17] presented a MED–TVC system model based on energy and exergy analyses. The results revealed that the main sources of exergy destruction in the MED-TVC system are the ejector and the evaporators, while a parametric study revealed that operating with a lower brine temperature and lower motive steam flow rate is recommended to improve the overall efficiency.

In recent studies, the economic performance of MED-TVC desalination systems has been analyzed by using the exergoeconomic method. Elsayed et al. [9] conducted an exergoeconomic analysis on four different feed configurations for a MED-TVC desalination system. The effect of the economic variables, such as cost index, interest rate, and electricity cost, on the total water price was investigated. Their results showed that the parallel cross feed (PCF) configuration has the minimum production cost (2.09 USD/m^3). In another study, Elsayed et al. [1] also developed an exergoeconomic model to analyze the performance of a multi-effect desalination plant integrated to a mechanical vapor compressor unit (MED-MVC). The results indicated that the total water price was 1.63 USD/m^3. Samaké et al. [18] performed an exergoeconomic analysis of a multiple-effect desalination system with ejector vapor compression. Both forward and parallel/cross configurations were considered. It was found that the required total capital investment and the unit cost of the produced potable water always increase when the number of effects increases.

In recent studies, Javadi et al. [19–21] have explored the use of different combinations, including gas and steam, parabolic solar collectors, MED desalination units, and reverse osmosis, to generate water and electricity. They worked on the optimization of these cogeneration systems through performing exergy, economic, and environmental analyses. Ghasemiasl et al. [22] conducted an economic optimization for a cogeneration system consisting of gas and steam turbines, a MED system, and a solar farm; the results showed that the interest rate of freshwater production increased from 6 to 12 due to the increase in the number of effects.

Based on the review mentioned above, it is obvious that MED-TVC desalination systems have attracted the attention of many researchers. Most studies focus on the exergy analysis of MED-TVC desalination systems; however, few focus on their economic analysis. The present study will add more value to the existing studies by providing a detailed economic analysis using an exergoeconomic approach for a MED-TVC desalination system, in order to estimate the total water price and identify the most relevant components of the system from an exergoeconomic point of view. By analyzing the costs and benefits of different design options, operational strategies, and maintenance practices, decision makers can identify ways to improve the efficiency and effectiveness of the MED-TVC system. In addition, a parametric study is conducted here to evaluate the effect of some process parameters on the total water price, such as top brine temperature, seawater temperature, motive steam flow rate, and number of effects.

2. MED-TVC System Description

A schematic diagram of the MED-TVC desalination system is presented in Figure 1. The MED-TVC system mainly consists of four effects (evaporators); a steam ejector (thermo-compressor); a condenser; and pumps for seawater, brine, and distillate. As shown in this figure, the feed configuration of MED integrated with the TVC unit considered in the present study is PCF. The cooling seawater flow (stream 1) enters the condenser to cool the distillate water and increase the seawater feed temperature (stream 2). The seawater leaves the condenser (stream 3) and is divided into two streams. One stream (stream 4) returns to the tank and the other stream (stream 5) is directed to the effects and split into four parts (streams 25–28). Motive steam (stream 7) is supplied to the steam ejector and is mixed with entrained steam (stream 6), which was separated from the steam generated from the last effect (stream 12).

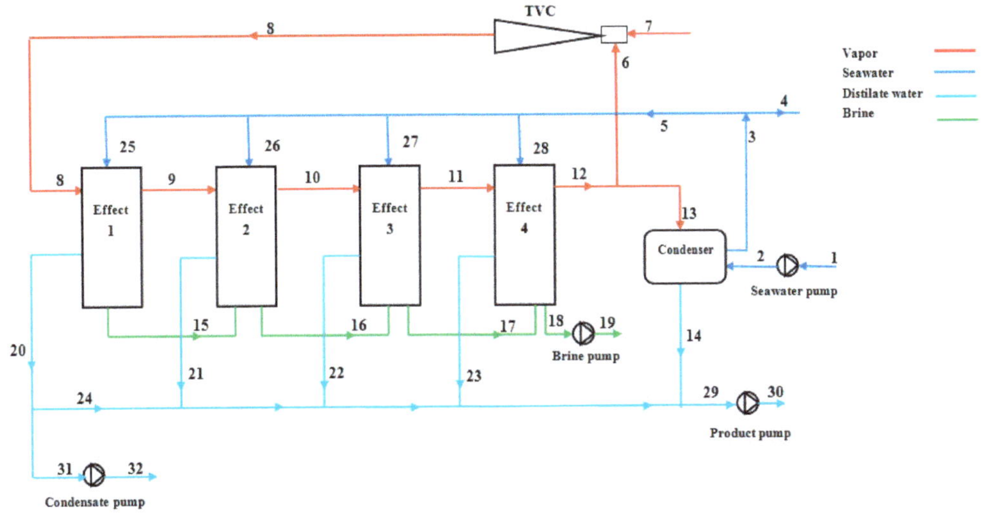

Figure 1. Schematic diagram of the MED-TVC desalination system with parallel cross feed configuration.

The combined vapor streams are then compressed in the ejector to meet the first thermal requirement (stream 8). The other part (stream 13) enters the condenser and heats the seawater to a higher temperature. Thereby, it condenses as distillate water and passes to the product line (stream 14). The vapor generated from the first effect (stream 9) is passed through demisters and enters effect number two (stream 10), and so on into the next effect (stream 11). The brine remaining in the feed water of the first effect (stream 15) is directed to the second effect and so on to the other effects (streams 16–18). In the last effect, raw water is discharged as brine (stream 19). At the same time, the condensate steam leaving the first effect (stream 20) is divided into two streams. One stream (stream 31) is fed to the water to the steam cycle of the steam power plant, while the other stream (stream 24) flows with all streams collected from the four effects (streams 21–23) in the fresh water line (stream 29).

3. Mathematical Modeling of the System

3.1. Thermodynamic Analysis

The thermodynamic model of the proposed MED-TVC desalination system was developed in MATLAB 7.9 software using mass and energy conservation equations according to the following assumptions:

- The desalination plant operates in a steady state condition.
- The effects have a similar temperature difference and the feed-flow rate is constant.
- The distillate and vapor are salt-free.
- A pump average efficiency of 75% is considered.

In addition, the model includes the equations for calculating the physical properties of the seawater and the heat transfer coefficients. A detailed description of seawater properties' (enthalpy, entropy, chemical potential and density) calculation and the thermodynamic model developed for a combined cycle for the MED-TVC desalination system were previously presented by the authors [17,23].

3.2. Exergy Analysis

Exergy analysis involves calculation of the exergy flow of each stream and the exergy destruction of each component of the system. Neglecting the kinetic and potential energies, the exergy flow rate of each stream is given as follows:

$$\dot{E}_{ph} = \dot{m}[(h - h_0) - T_0(s - s_0)] \tag{1}$$

$$\dot{E}_{ch} = \dot{m} \sum \omega_k \left(\mu_k^s - \mu_{0k}^0 \right) \tag{2}$$

The global exergy of each flow is defined as follows:

$$\dot{E} = \dot{E}_{ph} + \dot{E}_{ch} \tag{3}$$

The exergy destruction is defined as the irreversibility rate of the system. For each component, the exergy destruction rate can be obtained by the following equation:

$$\dot{E}_{D,k} = \dot{E}_{F,k} - \dot{E}_{P,k} \tag{4}$$

3.3. Exergoeconomic Analysis

The method of exergoeconomic analysis was used to evaluate the economic performance of the MED-TVC system presented in this study. There are several exergoeconomic approaches reported in the literature [24]; the specific exergy costing method (SPECO), which is one of the most widely applied methods for energy systems, is used in this study.

The SPECO method consists of three basic steps, including the definition of control volumes for all components, the development of auxiliary equations according to the fuel and product rules, and the formulation of the cost balance equations [25].

In an exergoeconomic analysis, the cost balance for each system component can be written as:

$$\sum \dot{C}_{out,k} = \sum \dot{C}_{in,k} + \dot{Z}_k \tag{5}$$

$$\dot{C}_i = c_i \dot{E}_i \tag{6}$$

The term $\dot{Z}$ represents the total cost rate associated with the capital investment of the component, as defined by [26,27]:

$$\dot{Z}_k = \frac{CRF \varphi Z_k}{N} \tag{7}$$

CRF is the capital recovery factor, which is given as follows:

$$CRF = \frac{i(1+i)^n}{(1+i)^n - 1} \tag{8}$$

Z_k represents the purchase cost of the component that is calculated based on the cost formulas in the literature [9,18]. It is noted that the purchase cost of each component has been updated to the year 2020 by using the chemical engineering plant cost index (CEPCI) [28].

$$\text{Cost at reference year} = \text{original cost} \times \frac{\text{cost index for the refernce year}}{\text{cost index for the original year}} \tag{9}$$

Combining Equation (5) for all the MED-TVC system components with the appropriate auxiliary equations, a linear equation system is developed to obtain the unknown flows costs.

$$\left[\dot{E}_k\right] \times c_k = \left[\dot{Z}_k\right] \tag{10}$$

The remaining cost balances and corresponding auxiliary equations for each component of the system are given in Table 1 [9,18].

Table 1. Exergy and cost balance equations with corresponding auxiliary equations for the MED-TVC system [9,18].

Item	Fuel Exergy	Product Exergy	Cost Balances	Auxiliary Equations
TVC	$\dot{m}_7(e_7 - e_8)$	$\dot{m}_6(e_8 - e_6)$	$\dot{C}_6 + \dot{C}_7 + \dot{Z}_{TVC} = \dot{C}_8$ $Z_{TVC} = 15962.46\, \dot{m}_8 \left(\frac{T_6}{P_6}\right)^{0.55} P_8{}^{-0.75}$	$\dot{C}_7 = SHC \times D$
Effect 1	$\dot{E}_8 - \dot{E}_{20}$	$\dot{E}_9 + \dot{E}_{15} - \dot{E}_{25}$	$\dot{C}_8 + \dot{C}_{25} + \dot{Z}_{Ef1} = \dot{C}_9 + \dot{C}_{15} + \dot{C}_{20}$ $Z_{Ef1} = 201.67\, UAdP_t{}^{0.15} dP_s{}^{-0.15}$	$\frac{\dot{C}_8}{\dot{E}_8} = \frac{\dot{C}_{20}}{\dot{E}_{20}}$ $\frac{\dot{C}_{15}}{\dot{E}_{15}} = \frac{\dot{C}_{25}}{\dot{E}_{25}}$
Effect 2	$\dot{E}_9 - \dot{E}_{21}$	$\dot{E}_{10} + \dot{E}_{16} - \dot{E}_{15} - \dot{E}_{26}$	$\dot{C}_9 + \dot{C}_{15} + \dot{C}_{26} + \dot{Z}_{Ef2} = \dot{C}_{10} + \dot{C}_{16} + \dot{C}_{21}$ $Z_{Ef2} = 201.67\, UAdP_t{}^{0.15} dP_s{}^{-0.15}$	$\frac{\dot{C}_9}{\dot{E}_9} = \frac{\dot{C}_{21}}{\dot{E}_{21}}$ $\frac{\dot{C}_{16}}{\dot{E}_{16}} = \frac{\dot{C}_{26}}{\dot{E}_{26}}$
Effect 3	$\dot{E}_{10} - \dot{E}_{22}$	$\dot{E}_{11} + \dot{E}_{17} - \dot{E}_{16} - \dot{E}_{27}$	$\dot{C}_{10} + \dot{C}_{16} + \dot{C}_{27} + \dot{Z}_{Ef3} = \dot{C}_{11} + \dot{C}_{17} + \dot{C}_{22}$ $Z_{Ef3} = 201.67\, UAdP_t{}^{0.15} dP_s{}^{-0.15}$	$\frac{\dot{C}_{10}}{\dot{E}_{10}} = \frac{\dot{C}_{22}}{\dot{E}_{22}}$ $\frac{\dot{C}_{17}}{\dot{E}_{17}} = \frac{\dot{C}_{27}}{\dot{E}_{27}}$
Effect 4	$\dot{E}_{11} - \dot{E}_{23}$	$\dot{E}_{12} + \dot{E}_{18} - \dot{E}_{17} - \dot{E}_{28}$	$\dot{C}_{11} + \dot{C}_{17} + \dot{C}_{28} + \dot{Z}_{Ef4} = \dot{C}_{12} + \dot{C}_{18} + \dot{C}_{23}$ $Z_{Ef4} = 201.67\, UAdP_t{}^{0.15} dP_s{}^{-0.15}$	$\frac{\dot{C}_{11}}{\dot{E}_{11}} = \frac{\dot{C}_{23}}{\dot{E}_{23}}$ $\frac{\dot{C}_{18}}{\dot{E}_{18}} = \frac{\dot{C}_{28}}{\dot{E}_{28}}$
Condenser	$\dot{E}_{13} - \dot{E}_{14}$	$\dot{E}_3 + \dot{E}_2$	$\dot{C}_2 + \dot{C}_{13} + \dot{Z}_{Con} = \dot{C}_3 + \dot{C}_{14}$ $Z_{Con} = 201.67\, UAdP_t{}^{-0.15} dP_s{}^{-0.15}$	$\frac{\dot{C}_{13}}{\dot{E}x_{13}} = \frac{\dot{C}_{14}}{\dot{E}x_{14}}$

Table 1. *Cont.*

Item	Fuel Exergy	Product Exergy	Cost Balances	Auxiliary Equations
Pump$_{SW}$	$\dot{W}_{sw}$	$\dot{E}_2 - \dot{E}_1$	$\dot{C}_1 + \dot{C}_{w,1} + \dot{Z}_{swp} = \dot{C}_2$ $Z_{swp} = 13.92\,(m_2 * \Delta P)^{0.55} * \left(\frac{\eta_P}{1-\eta_P}\right)^{1.05}$	$\dot{C}_1 = (SLC + SCC) \times D$ $\dot{C}_{w,1} = \dot{C}_{elec}$
Pump$_P$	$\dot{W}_p$	$\dot{E}_{30} - \dot{E}_{29}$	$\dot{C}_{29} + \dot{C}_{w,2} + \dot{Z}_{pp} = \dot{C}_{30}$ $Z_{pp} = 13.92\,(m_{30} * \Delta P)^{0.55} * \left(\frac{\eta_P}{1-\eta_P}\right)^{1.05}$	$\dot{C}_{w,2} = \dot{C}_{elec}$
Pump$_{Cond}$	$\dot{W}_{Cond}$	$\dot{E}_{32} - \dot{E}_{31}$	$\dot{C}_{31} + \dot{C}_{w,3} + \dot{Z}_{Distp} = \dot{C}_{32}$ $Z_{Cond} = 13.92\,(m_{32} * \Delta P)^{0.55} * \left(\frac{\eta_P}{1-\eta_P}\right)^{1.05}$	$\dot{C}_{w,3} = \dot{C}_{elec}$
Pump$_B$	$\dot{W}_B$	$\dot{E}_{19} - \dot{E}_{18}$	$\dot{C}_{18} + \dot{C}_{w,4} + \dot{Z}_{Bp} = \dot{C}_{19}$ $Z_{Bp} = 13.92\,(m_{19} * \Delta P)^{0.55} * \left(\frac{\eta_P}{1-\eta_P}\right)^{1.05}$	$\dot{C}_{w,4} = \dot{C}_{elec}$

3.4. Performance Indicators

- Total water price (TWP), which is the ratio between the cost streams of water exiting the desalination unit to the amount of the desalinated water [29].

$$TWP = \frac{\dot{C}_4 + \dot{C}_{19} + \dot{C}_{30} + \dot{C}_{32}}{\dot{m}_{30} \times 3.6} \tag{11}$$

- Exergy destruction cost rate ($\dot{C}_{D,k}$), defined as [30]

$$\dot{C}_{D,k} = c_{F,k}\dot{E}_{D,k} \tag{12}$$

- Exergoeconomic factor calculated by:

$$f_k = \frac{\dot{Z}_k}{\dot{Z}_k + \dot{C}_{D,k}} \tag{13}$$

4. Results and Discussion

First, it should be noticed that the developed thermodynamic model of the MED-TVC system was validated using the reported data of a model developed by Elsayed et al. [9]. The input values for validation and the validation results are given in Table 2. As shown in this table, there is good agreement between the current results and the available data in [9].

The results of the exergy and exergoeconomic analyses and the parametric study are presented in the next sections. The main initial parameters used to simulate the MED-TVC system are provided in Table 3. The simulation results, including the mass flow rate, temperature, specific exergy, and cost rate of all of the matter streams shown in Figure 1, are summarized in Figure 2. The fresh water production for the evaluated system is 55.2 kg/s (198.7 m^3/h), and the total water price (TWP) is found to be around 1.73 USD/m^3. This value is of the same range as those reported in the literature for other MED-TVC units, varying between 0.41 USD/m^3 and 2.09 USD/m^3 [1,9–30].The difference can be attributed to the MED-TVC unit configuration and the cost index used to update the equipment purchase cost.

Table 2. Validation results of MED-TVC model.

Parameter	Reference [9]	Present Model
Inputs		
Number of effects	4	4
Motive pressure (kPa)	2300	2300
Top brine temperature (°C)	60.1	60.1
Last brine temperature (°C)	45.4	45.4
Seawater temperature (°C)	31.5	31.5
Feed temperature (°C)	41.5	41.5
Feed content (g/kg)	35	35
Brine content (g/kg)	53	53
Motive steam flow rate (kg/s)	8.89	8.89
Outputs		
Distillate production, D (kg/s)	**56.12**	**59.9**
Gain output ratio (GOR)	**6.31**	**6.73**
Specific cooling seawater flow ($sM_{sw} = \dot{m}_{sw}/D$)	**6.7**	**7.4**
Specific heat consumption (SHC = 2330/GOR)	**369.3**	**346.2**

Table 3. Parameters assumed for the simulations.

Parameter	Value
Cooling seawater temperature (°C)	31.5
Feed seawater temperature (°C)	41.5
Top brine temperature (°C)	60.5
Motive pressure (kPa)	2300
Motive steam flow rate (kg/s)	8.8
Number of effects	4
Interest rate (%)	5
System lifetime (yr)	20
Annual system operating hours (h)	7884
Specific labor cost (USD/m^3)	0.05 [a]
specific chemical cost (USD/m^3)	0.025 [a]
Specific steam cost (USD/m^3)	1.28 [a]
electrical energy cost (USD/kWh)	0.08 [a]

Note: [a] Adapted from [9].

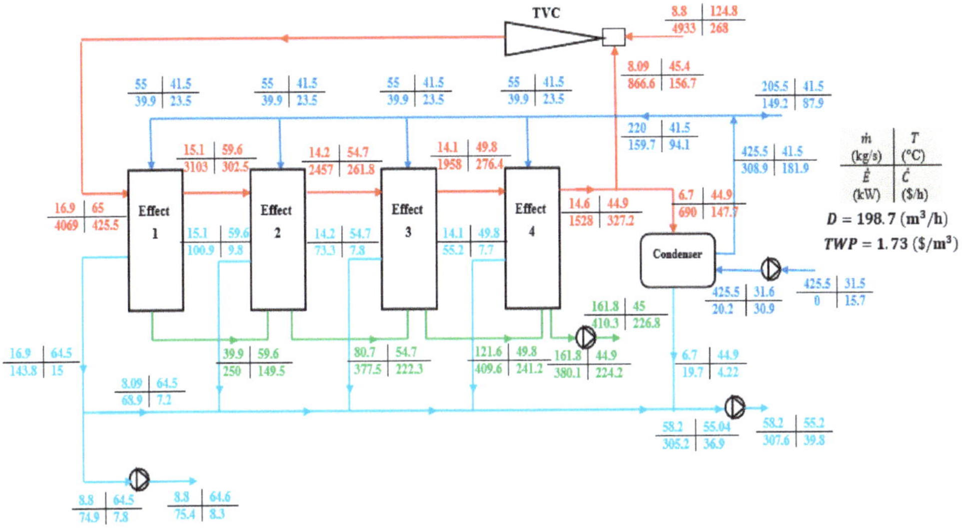

Figure 2. Cost flow diagram of the MED-TVC system.

4.1. Exergy Analysis

Table 4 contains the results obtained from the exergetic analysis, i.e., exergy rates associated with fuel, product, exergy destruction and exergy destruction ratios for each component. It can be seen that the components that should be improved first are the evaporators and the TVC unit. The exergy destruction of the evaporators is 1946.3 kW, which accounts for 45.8% of the system exergy destruction, and the exergy destruction of the TVC unit is 1730.8 kW, which accounts for 40.7% of the system exergy destruction. This is mainly due to the heat transfer process in the effects associated with phase change and to the high pressure of motive steam and the temperature difference through the TVC unit.

Table 4. Exergetic analysis results.

Item	$\dot{E}_{F,k}$ (kW)	$\dot{E}_{P,k}$ (kW)	$\dot{E}_{D,k}$ (kW)	$\dot{E}_{D,k}$ (%)
TVC	2813.3	1082.5	1730.8	40.7
Effect	3925.5	3314.1	1946.3	45.8
Condenser	670.2	288.7	381.5	8.9
$Pump_{SW}$	170.2	20.2	150.0	3.5
$Pump_{P}$	34.9	2.3	32.6	0.8
$Pump_{B}$	32.5	29.7	2.8	0.1
$Pump_{Cond}$	4.4	0.9	3.5	0.1

The second priority for improvement is the condenser; the exergy destruction for this component is 381.5 kW. While pumps (except the feed pump) have the smaller share of exergy destruction, ranging from 0.1% to 3.5%, the improvement is less important than the improvement of the above-mentioned components. Therefore, designers should focus on evaporators, TVC unit, and condenser to improve the exergetic performance of a desalination plant. It should be noted that the exergy destruction in the effects can be reduced by increasing the number of effects [17]. On the other hand, seawater preheater before the condenser inlet can be used to improve the exergy performance of the condenser by recovering energy from the fresh water outlets to increase the temperature of the seawater at the inlet of the condenser.

4.2. Exergoeconomic Analysis

Table 5 and Figure 3 show the results of the exergoeconomic analysis of the components of the MED-TVC system. Different exergoeconomic variables were calculated for each component, such as the investment cost rate ($\dot{Z}_k$), the cost rates associated with exergy destruction ($\dot{C}_{D,k}$), the exergoeconomic factor (f), and the sum of ($\dot{Z}_k + \dot{C}_{D,k}$). According to ($\dot{Z}_k$), values, the cheapest component is the TVC unit (0.7 USD/h) and pumps (0.71 USD/h), and the most expensive component is the effects (72.4 USD/h) and condenser (7.7 USD/h), due to the required large heat transfer area. In addition, for all components, low values of the exergoeconomic factor (f) were observed, except for the pump (60%). A low value of the exergoeconomic factor suggests a decrease in the cost rates associated with the exergy destruction of a component.

Table 5. Exergoeconomic analysis results.

Item	$\dot{Z}_k$ (USD/h)	$\dot{C}_{D,k}$ (USD/h)	f (%)
TVC	0.7	28.6	2.4
Effects	72.4	217.3	24.9
Condenser	7.7	81.7	8.6
$Pump_{SW}$	0.2	13.2	1.5
$Pump_{P}$	0.2	2.5	7.4
$Pump_{B}$	0.3	0.2	60
$Pump_{Cond}$	0.01	0.3	3.2

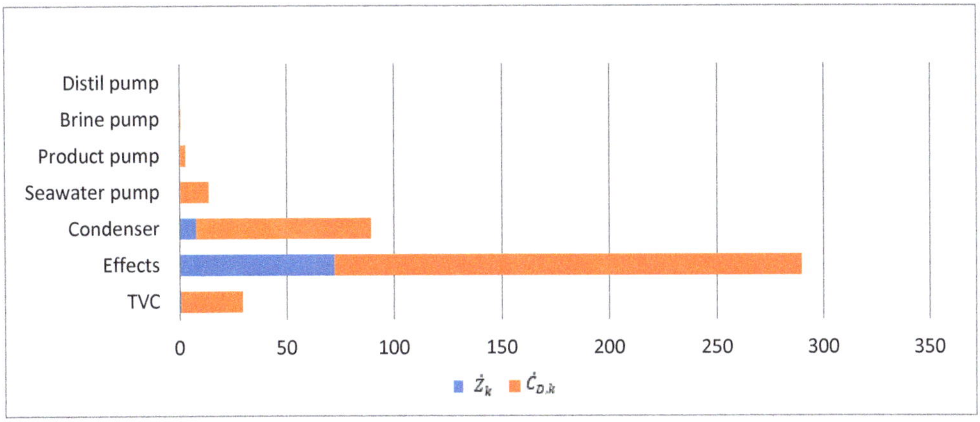

Figure 3. The sum of $(\dot{Z}_k + \dot{C}_{D,k})$ for the MED-TVC system components.

It can be seen from Figure 3 that the highest sum $(\dot{Z}_k + \dot{C}_{D,k})$ is for the effects, followed by the condenser. The exergoeconomic factor of the effects is 24.9%, suggesting that the capital cost has a relatively much lower rate than the cost of exergy destruction. Therefore, 75.1% of the total cost associated with the effects is due to its high thermodynamic inefficiency (high exergy destruction). The second cost ineffective component in the system is the condenser; its exergoeconomic factor is observed to be 8.6%, meaning that only 8.6% of the cost is owing to component capital cost and 91.4% of the cost is because of high exergy destruction. The third cost ineffective component in the system is found to be the TVC unit, and its exergoeconomic factor is obtained to be approximately 2.4%, which means that only 2.4% of the cost is owing to component capital cost and 97.6% of the cost is owing to high exergy destruction.

4.3. Parametric Study

A parametric study was carried out in order to investigate the effect of certain parameters on the total water price of the MED-TVC system. The key parameters of the study were the top brine temperature, seawater temperature, motive steam flow rate, and number of effects.

4.3.1. Effect of the Top Brine Temperature

Figure 4 shows the variations in the total water price of the MED-TVC system as the top brine temperature changes. It can be seen that the TWP monotonically decreases over the given range of TBT, but its rate of decrease is lower (from 1.8 USD/m^3 to 1.42 USD/m^3) as a higher TBT is approached. This trend in the TWP is explained by the nature of variation of the cost streams of water exiting the desalination unit with an increase in TBT. Although the distillate production increases linearly with TBT, the variation of the cost streams of water with an increase in TBT is not linear, and there is a small increase in the slope of increase in the cost streams with TBT. This feature is attributed to the variation of LMTD with an increase in TBT.

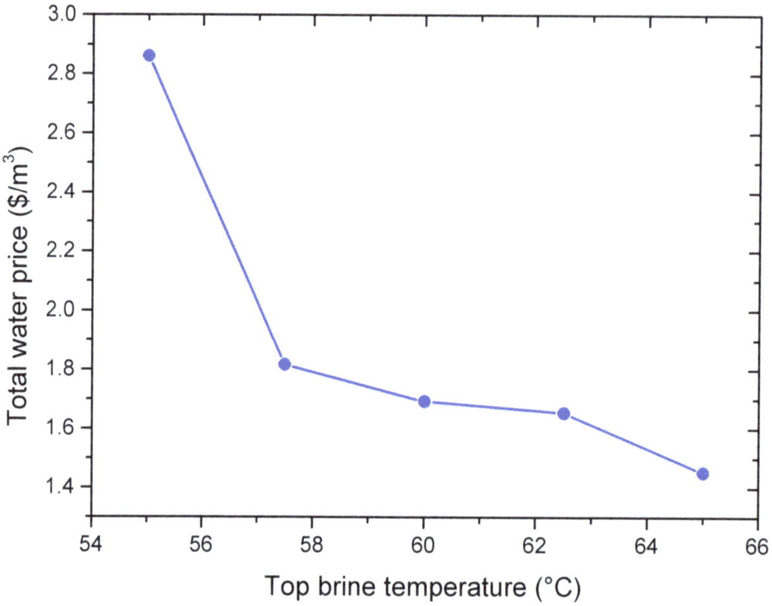

Figure 4. Effect of top brine temperature on the TWP.

4.3.2. Effect of the Seawater Temperature

Figure 5 shows the variation of total water price with the seawater temperature. As it can be seen from this figure, the increase in the cooling seawater temperature leads to an increase in the TWP of the system. The TWP is increased by 12%, from 1.6 USD/m^3 to 1.79 USD/m^3. The reason is that the increase in the seawater temperature results in an increase in the feed water temperature, and feed water with higher temperatures needs lower heat to be preheated. Therefore, the lower the specific heat consumption, the lower the TWP. This is because the energy required to desalinate seawater depends on the difference in temperature between the feed seawater and the heated evaporator surface. Warmer seawater requires less energy to heat to the boiling point, which results in lower energy consumption and operating costs. However, there are some limitations to the benefits of warmer seawater. For example, if the temperature of the seawater is too high, it may cause scaling or fouling of the heat exchangers, which can decrease the efficiency of the desalination process and increase maintenance costs. The obtained results indicate that the increase in the seawater temperature from 25°C to 33°C leads to a 10% increase in the TWP of the MED-TVC system.

4.3.3. Effect of the Motive Steam Flow Rate

The influence of the motive steam flow variation on the total water price is shown in Figure 6. It is observed that an increase in the motive steam flow rate from 6 to 10 kg/s leads to a slight increase in the TWP. The increase in the TWP is only 2.3% when the motive steam flow rate rises from 6 to 9 kg/s and then increases by 4.4% when the motive steam flow increases to 10 kg/s. This can be explained by the higher influence of the motive steam flow on the specific heat consumption because the addition of steam flow rate leads to an increase in the temperature and the pressure of compressed vapor. This will lead to the generation of more vapors and, consequently, more distillate water, but higher energy will be needed to evaporate the seawater in all effects. Therefore, the amount of freshwater production is more sensitive to the motive steam flow rate of the MED-TVC system than the total water price.

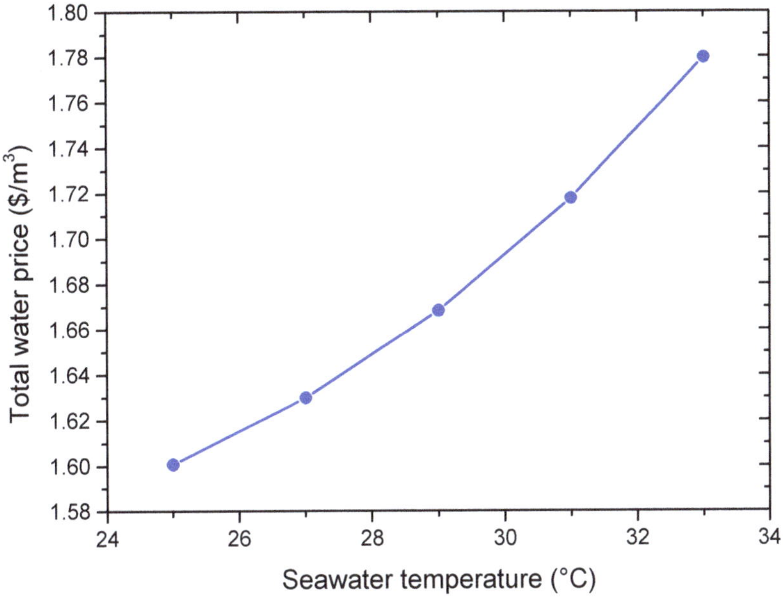

Figure 5. Effect of cooling seawater temperature on the TWP.

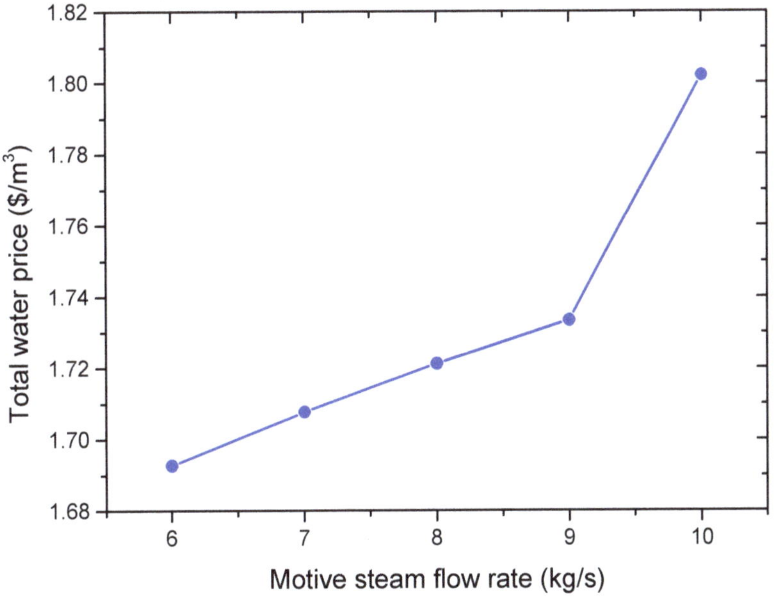

Figure 6. Effect of motive steam flow rate on the TWP.

4.3.4. Effect of the Number of Effects

The number of effects used in a MED-TVC process is an important factor in determining the total water price. Figure 7 shows the variation of the total water price with the number of effects. As shown in this figure, the TWP monotonically rises with the addition of more effects to the MED-TVC system, due to the increase in the capital and operating costs of the desalination process. By adding more effects, freshwater production and energy efficiency increase, but this also increases the capital cost, maintenance cost, and downtime, which can lead to a higher total water price. It is estimated that the increase in the total water price is about 26% as the number of effects increases from 4 to 10. The minimum TWP is obtained at four effects. Therefore, using four effects is preferable for the MED-TVC system to obtain the lowest price of desalinated water.

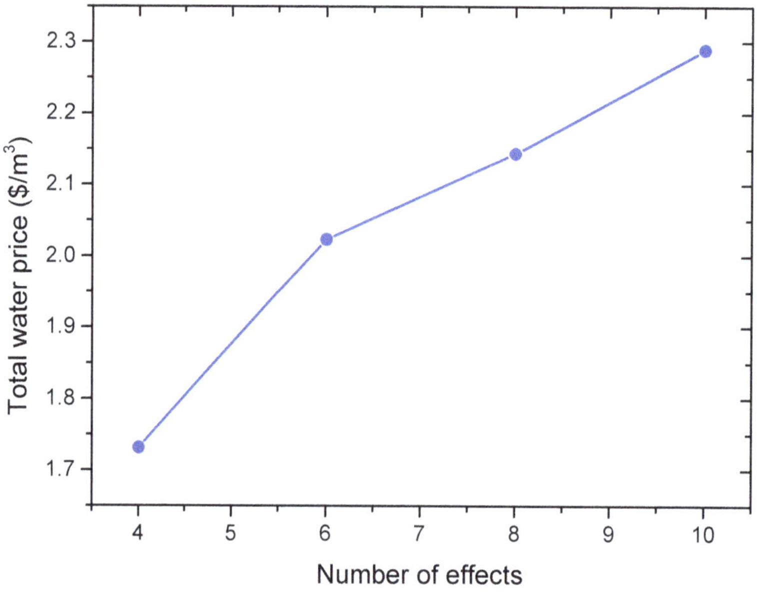

Figure 7. Influence of the number of effects on the TWP.

5. Conclusions

In this paper, a parallel/cross MED-TVC desalination system with four effects was investigated, and its performance was evaluated using exergy and exergoeconomic approaches. In addition, a parametric study was performed to analyze the effects of several operating parameters on the system economic performance. The main conclusions arising from the present study can be summarized as follows:

- The total water price of the MED-TVC system is estimated to be 1.73 USD/m^3 for a freshwater production of 16.16 kg/s.
- Among all system components, the effects, TVC unit, and condenser are the most relevant components from an exergetic and an exergoeconomic point of view.
- The top brine temperature, cooling seawater temperature, motive steam flow rate, and number of effects have significant effects on the economic performance of the MED-TVC system.

Finally, an exergoenvironmental analysis using life cycle assessment is needed and will be performed in a future study to evaluate the environmental impact of the MED-TVC desalination system and improve its ecological performance.

Author Contributions: Conceptualization, Z.T., H.T., M.K., A.A., N.M. and L.M.; Methodology, Z.F., Z.T., R.M., H.T., A.A., J.Z. and L.M.; Software, Z.F., Z.T., H.T. and J.Z.; Validation, Z.F., Z.T., R.M., H.T., M.K., A.A., N.M., J.Z. and L.M.; Formal analysis, Z.F., Z.T., R.M., H.T., M.K., A.A., J.Z. and L.M.; Investigation, Z.F., Z.T., H.T., M.K., A.A., N.M. and L.M.; Resources, Z.F., Z.T., R.M., A.A. and L.M.; Data curation, Z.F., Z.T., R.M., A.A. and J.Z.; Writing—original draft, Z.F.; Writing—review & editing, Z.T., H.T., M.K., A.A., N.M., J.Z. and L.M.; Visualization, Z.T., R.M., H.T., M.K., A.A., N.M., J.Z. and L.M.; Supervision, Z.T., H.T. and A.A.; Project administration, Z.T., A.A., N.M., J.Z. and L.M. All authors have read and agreed to the published version of the manuscript.

Funding: This research received no external funding.

Data Availability Statement: Not applicable.

Conflicts of Interest: The authors declare no conflict of interest.

Nomenclature

Symbols	Description
A	heat transfer area (m^2)
$\dot{C}$	cost rate (USD/h)
$\dot{C}_{elec}$	electricity cost (USD/kWh)
C	cost per exergy unit (USD/GJ)
$c_{F,k}$	cost rate of the exergy of fuel (USD/GJ)
D	distillate production (m^3/h)
$\dot{E}$	exergy rate (kW)
$\dot{E}_{ch}$	chemical exergy(kW)
$\dot{E}_{D,k}$	exergy destruction (kW)
$\dot{E}_{F,k}$	exergy rate of the fuel (kW)
$\dot{E}_{ph}$	physical exergy (kW)
$\dot{E}_{P,k}$	exergy rate of the product (kW)
h	specific enthalpy (kJ/kg)
h_0	specific enthalpy at ambient conditions (kJ/kg)
I	interest rate (%)
K	kth component
$\dot{m}$	mass flow rate (kg/s)
N	operating hours(h)
P	pressure (kPa)
N	lifetime of the unit desalination (yr)
s	specific entropy (kJ/kg K)
s_0	specific entropy at ambient conditions (kJ/kg K)
T	temperature (°C)
SCC	specific chemical cost (USD/m^3)
SHC	specific heat consumption (USD/m^3)
SLC	specific labor cost (USD/m^3)
U	heat transfer coefficient (W/m^2°C)
$\dot{W}_B$	brine pump power (kW)
$\dot{W}_{Cond}$	condensate pump power(kW)
$\dot{W}_p$	product pump power (kW)
$\dot{W}_{sw}$	feed pump power (kW)
Z	component purchasing cost (USD)
$\dot{Z}$	capital investment cost rate (USD/h)
φ	maintenance factor (%)
ω_k	mass fraction (kg/kg)
μ	chemical potential (kJ/kg)
η_P	pump efficiency (%)

Abbreviations

LMDT	Logarithmic Mean Temperature Difference
MEB	Multi-Effect Boiling
MED	Multiple-Effect Desalination
MVC	Mechanical Vapor Compression
PCF	Parallel Cross Feed
SPECO	Specific Exergy Costing Method
TBT	Top Brine Temperature
TVC	Thermal Vapor Compression
TWP	Total Water Price

References

1. Elsayed, M.L.; Mesalhy, O.; Mohammed, R.H.; Chow, L.C. Transient and thermo-economic analysis of MED-MVC desalination system. *Energy* **2019**, *16*, 283–296. [CrossRef]
2. Macedonio, F.; Drioli, E. An exergetic analysis of a membrane desalination system. *Desalination* **2010**, *261*, 293–299. [CrossRef]
3. Fitzsimons, L.; Corcoran, B.; Young, P.; Foley, G. Exergy analysis of water purification and desalination: A study of exergy model approaches. *Desalination* **2015**, *359*, 212–224. [CrossRef]
4. Miladi, R.; Frikha, N.; Gabsi, S. Exergy analysis of a solar-powered vacuum membrane distillation unit using two models. *Energy* **2017**, *120*, 872–883. [CrossRef]
5. Askari, I.B.; Ameri, M.; Calise, F. Energy, exergy and exergo-economic analysis of different water desalination technologies powered by Linear Fresnel solar field. *Desalination* **2018**, *425*, 37–67. [CrossRef]
6. Triki, Z.; Fergani, Z.; Bouaziz, M.N. Exergoeconomic and exergoenvironmental evaluation of a solar-energy-integrated vacuum membrane distillation system for seawater desalination. *Desalin. Water Treat.* **2021**, *225*, 380–391. [CrossRef]
7. Mistry, K.H.; Antar, M.A.; Lienhard, V.J.H. An improved model for multiple effect distillation. *Deasalin. Water Treat.* **2013**, *51*, 807–821. [CrossRef]
8. Ophir, A.; Lokiec, F. Advanced MED process for most economical sea water desalination. *Desalination* **2005**, *182*, 187–198. [CrossRef]
9. Elsayed, M.L.; Mesalhy, O.; Mohammed, R.H.; Chow, L.C. Exergy and thermo-economic analysis for MED-TVC desalination systems. *Desalination* **2018**, *447*, 29–42. [CrossRef]
10. Hamed, O.A.; Zamamiri, A.M.; Aly, S.; Lior, N. Thermal performance and exergy analysis of a thermal vapour compression desalination system. *Energy Convers. Manag.* **1996**, *37*, 379–387. [CrossRef]
11. Al-Najem, N.M.; Darwish, M.A.; Youssef, F.A. Thermovapor compression desalters: Energy and availability—Analysis of single-and multi-effect systems. *Desalination* **1997**, *110*, 223–238. [CrossRef]
12. Alasfour, F.N.; Darwish, M.A.; Amer, A.B. Thermal analysis of ME—TVC+MEE desalination systems. *Desalination* **2005**, *174*, 39–61. [CrossRef]
13. Choi, H.S.; Lee, T.J.; Kim, Y.G.; Song, S.L. Performance improvement of multiple-effect distiller with thermal vapour compression system by exergy analysis. *Desalination* **2005**, *182*, 239–249. [CrossRef]
14. Cao, C.; Xie, L.; Xu, S.; Du, Y. Exergy analysis and optimization of MED–TVC system with different effect group divisions. *Desalination* **2021**, *500*, 114891. [CrossRef]
15. Binamer, A. Second law and sensitivity analysis of large ME-TVC desalination units. *Desalin. Water Treat.* **2015**, *53*, 1234–1245. [CrossRef]
16. Sadri, S.; HaghighiKhoshkhoo, R.; Ameri, M. Multi objective optimization of the MED-TVC system with exergetic and heat transfer analysis. *Energy Equip. System.* **2017**, *5*, 419–430.
17. Menasri, R.; Triki, Z.; Bouaziz, M.N.; Hamrouni, B. Energy and exergy analyses of a novel multi-effect distillation system with thermal vapour compression for seawater desalination. *Desalin. Water Treat.* **2022**, *246*, 54–67. [CrossRef]
18. Samaké, O.; Galanis, N.; Sorin, M. Thermo-economic analysis of a multiple-effect desalination system with ejector vapour compression. *Energy* **2018**, *144*, 1037–1051. [CrossRef]
19. Javadi, M.A.; Khalaji, M.; Ghasemiasl, R. Exergoeconomic and environmental analysis of a combined power and water desalination plant with parabolic solar collector. *Desalin. Water Treat.* **2020**, *193*, 212–223. [CrossRef]
20. Hemmatpour, K.; Ghasemiasl, R.; Javadi, M.A. Time-Transient Optimization of Electricity and Fresh Water Cogeneration Cycle Using Gas Fuel and Solar Energy. *Mathematics* **2023**, *11*, 571. [CrossRef]
21. Javadi, M.A.; Khodabakhshi, S.; Ghasemiasl, R.; Jabery, R. Sensivity analysis of a multi-generation system based on a gas/hydrogen-fueled gas turbine for producing hydrogen, electricity and fresh water. *Energy Convers. Manag.* **2022**, *252*, 115085. [CrossRef]
22. Ghasemiasl, R.; Javadi, M.A.; Nezamabadi, M.; Sharifpur, M. Exergetic and economic optimization of a solar-based cogeneration system applicable for desalination and power production. *J. Therm. Anal. Calorim.* **2021**, *145*, 993–1003. [CrossRef]
23. Sharqawy, M.H.; Lienhard, J.H.; Zubair, S.M. Thermophysical properties of seawater: A review of existing correlations and data. *Desalin. Water Treat.* **2010**, *16*, 354–380. [CrossRef]

24. Abusoglu, A.; Kanoglu, M. Exergoeconomic analysis and optimization of combined heat and power production: A review. *Renew. Sust. Energ. Rev.* **2009**, *13*, 2295–2308. [CrossRef]
25. Lazzaretto, A.; Tsatsaronis, G. SPECO: A systematic and general methodology for calculating efficiencies and costs in thermal systems. *Energy* **2006**, *31*, 1257–1289. [CrossRef]
26. Mohammadkhani, F.; Khalilarya, S.; Mirzaee, I. Exergy and exergoeconomic analysis and optimization of diesel engine based Combined Heat and Power (CHP) system using genetic algorithm. *Int. J. Exergy* **2013**, *12*, 139–161. [CrossRef]
27. Habibi, H.; Zoghi, M.; Chitsaz, A.; Shamsaiee, M. Thermo-economic performance evaluation and multi-objective optimization of a screw expander-based cascade Rankine cycle integrated with parabolic trough solar collector. *Appl. Therm. Eng.* **2020**, *180*, 115827. [CrossRef]
28. Jenkins, S. Chemical Engineering Plant Cost Index Annual Average 2020. Available online: http://www.chemengonline.com/pci-home (accessed on 18 April 2022).
29. Mohammed, R.H.; Ibrahim, M.M.; Abu-Heiba, A. Exergoeconomic and multi-objective optimization analyses of an organic Rankine cycle integrated with multi-effect desalination for electricity, cooling, heating power, and fresh water production. *Energy Convers. Manag.* **2021**, *231*, 113826. [CrossRef]
30. Bejan, A.; Tsatsaronis, G.; Moran, M.J. *Thermal Design and Optimization*; John Wiley & Sons: Hoboken, NJ, USA, 1996.

separations

MDPI

Communication

Conceptual Design of a Compact Water Purification Unit Using Reed Bed Filtration

Elias Farah [1,*], Maria Khalil [2], Manuella Richa [2] and Chantal Abou Harb [1]

1 Department of Civil Engineering, School of Engineering, Holy Spirit University of Kaslik (USEK), Jounieh P.O. Box 446, Lebanon; abouharbchantal@gmail.com
2 Environmental Engineering and Sustainability Management, École Polytechnique, CEDEX, 91128 Palaiseau, France; maria.khalil@polytechnique.edu (M.K.); manuella.richa@polytechnique.edu (M.R.)
* Correspondence: eliasafarah@usek.edu.lb

Abstract: One of the most widespread global challenges is the insufficient provision of potable water, which affects individuals across diverse geographical regions. It is anticipated that issues related to water scarcity and quality will escalate in tandem with the expanding human population and the rapid pace of global development. Water sources are massively polluted hence, not safe for drinking nor irrigation. As a consequence, it is very important to have a water purification treatment plant to provide good water quality. Given the pressing need to ensure universal access to safe and clean drinking water, this investigation aims to engineer a compact and space-efficient apparatus that can expeditiously produce purified water. The proposed system seeks to optimize water purification performance while minimizing spatial requirements and operational duration. Its size is minimized by combining the three processes: coagulation, flocculation, and clarification together in one tank. Following to the aforementioned reservoir, an integrated natural system is employed to reduce the usage of chemicals and establish an ecologically sustainable platform. A hydraulic study is conducted to obtain the dimensioning of the several units which can be later scaled according to the flowrate. The latter was assumed in this study to be 2 L/s, then the compact unit can serve up to 800 persons by scaling the model and adjusting it.

Keywords: disinfection; filtration; purification; reed bed; water

1. Introduction

Access to clean and safe water is essential for human survival and well-being. However, in many parts of the world, access to clean water is limited or non-existent, particularly in developing countries and rural areas. The scarcity of unpolluted water can engender far-reaching consequences for human and animal well-being, ecological balance, fishing, leisure, transportation, and various commercial endeavors [1,2]. Inadequate drinking water quality has been linked to several recognized diseases such as cholera, arsenic, and dental/skeletal fluorosis [3,4]. Consequently, the global community has endeavored to seek viable, economical, and effective interventions to facilitate universal access to safe drinking water over several decades [2,5–9]. One of the main solutions is the use of water purification treatment units.

Various studies have used several treatment techniques. The utilization of vetiver grass, ordinary sand, rice husk, coconut fibers, activated carbon, oysters, and moringa seeds for the remediation of lake water has demonstrated effective removal of a significant proportion of impurities and pollutants [10]. Other treatments have been used Slow Sand Filtration (SSF) and Rapid Sand Filtration (RSF). Studies have shown that slow sand filtration (SSF) is highly proficient in eliminating suspended particles that exceed the filter's pore size, whereas rapid sand filtration (RSF) is an efficient method for eliminating turbidity. However, RSF alone may not be sufficient to eliminate viruses, fluorides, arsenic, salts, and other contaminants without pre- and post-treatment measures [11]. A biological sand filter (BSF) supplemented with metallic iron Fe^0 has been employed for decentralized safe drinking water treatment. The findings indicate that BSF, operating on gravity and affordability, is a suitable and effective technology for decentralized water supply, successfully removing particles and pathogens through a synergistic interplay of biological, chemical, and physical processes [12].

Water treatment is a process that demands substantial energy inputs, with the level of energy consumption dependent on a multitude of factors, including raw water quality, source of water, age of water delivery infrastructure, conveyance distance, water storage, and elevation variations [13,14]. The Drinking Water Treatment Plant (DWTP) in Jamshoro Pakistan consumed 7.4 Wh/m^3 to treat 0.017 m^3/day of river water [15]. An existing DWTP using solar photovoltaics (PVs) to reduce carbon emission showed that the largest consumers of energy were the processes of coagulation (1.95 Wh/m^3) and flocculation (1.93 Wh/m^3) and a 500 kW PV system was found to be sufficient to offset the energy consumption of the water treatment only operations, for a net present value of $0.24 million [14].

The expenses of water treatment hinge on several variables, such as the quality of raw water, the selected treatment technologies, governing regulations, energy sources, and the volume of water processed [16]. Several research studies reveal that the enhancement of source water quality correlates with a reduction in water treatment costs [17]. But water quality is a worldwide problem, and Lebanon is one of the top countries facing a similar issue, with little awareness about the importance of solving it, thus the absence of funding for such projects. More people die every year from the consequences of unsafe water than the greatest impacts are on children under the age of five. Every 8 s a child dies from water-related diseases [18]. In order to protect human life, an eco-friendly compact size of DWTP can be a great solution that serves small communities and can be movable, thus helpful in times of natural disasters where infrastructure may be destroyed. This paper has the potential to achieve this goal.

In this study, we present a conceptual design for a small-scale water purification treatment unit using reed bed eco-friendly filtration that can be used in remote and underserved communities. Our design is based on a combination of different treatment methods and technologies, including filtration, disinfection, and adsorption. The goal of this paper is to provide a detailed description of our conceptual design, including the treatment processes, materials, and equipment used, as well as the expected performance of the unit.

2. Compact Water Purification Treatment Unit

2.1. Process Flow Diagram

A process flow diagram for the DWTP is shown in Figure 1. The solid arrows denote the sequential advancement of water treatment, the dashed lines indicate the locations for chemical injection, and the dotted arrows represent the discharge of sediments. The surface water was extracted from rivers or lakes and then screened to remove large solids, rags, and debris of 6 mm or larger. This water is pumped and channeled to the drinking water reduction plant through an inlet pipe leading to a tank. During this stage, a coagulant is added to the water. Therefore, the tank is used in order to carry out the following processes: coagulation, in which chemicals are introduced to the water to initiate the formation of gelatinous hydracids through the attraction of particles together; flocculation, where chemicals are gradually mixed with the water to facilitate the aggregation of small suspended solids and the formation of flocs; and clarification, which reduces the solid load and removes approximately 85% to 95% of suspended solids and colloidal matter by controlling the speed of the propeller and introducing the necessary coagulants. Subsequently, a reed bed system replaces the filtration unit, with water passing through a porous medium that removes the remaining suspended solids, color, taste, odor, iron, manganese, and up to 90% of bacteria, creating a nature-based solution. Depending on the intended application of the water, an optional disinfection tank may be situated downstream before the water is either stored in a separate tank or distributed directly.

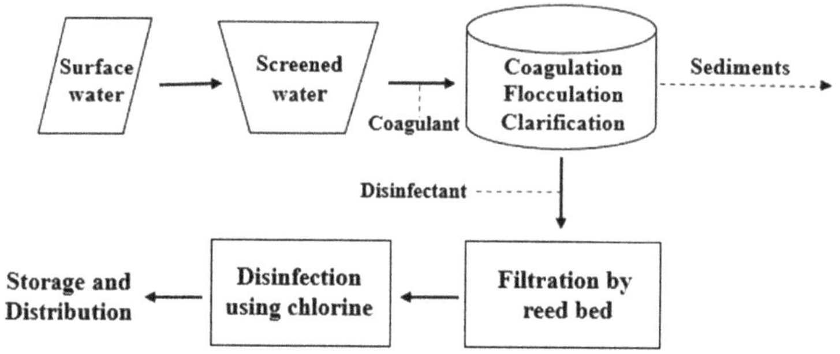

Figure 1. Process flow diagram for a proposed drinking water treatment plant.

2.2. *Conceptual Design*

The compact unit will operate at a flowrate is equal to 2 L/s. This process can serve up to 800 persons by scaling the model and adjusting it. The current study focused on sizing an eco-friendly concept of DWTP by designing the various unit processes involved. This section includes the conceptual design for all stages of the DWTP.

2.2.1. Inlet Pipe Design

Referring to Table 1, and taking the water velocity in the inlet pipe equal to 1 m/s, the required area of a PVC circular inlet pipe is determined by dividing the flowrate by the fluid velocity, therefore the diameter of the inlet pipe should be equal to 50 mm.

Table 1. Intake conduit design criteria [19].

Number of Conduits	One Minimum, Two or More Preferred
Velocity	1.2–2.0 ft/s (46–61 cm/s) at design flow; 3–4 ft/s (90–120 cm/s) maximum
Type of construction	Tunnel or pipeline
Slope or grade	Continuous to drain or to an air release valve

2.2.2. Coagulant Tank Design

Generally, to determine the quantity of the required coagulant "Alum" for this process, a jar test must be conducted to know the water quality. A minimum alum dosage of 25 mg/L is used in our design [20], with a density of 1340 kg/m^3 and a velocity of 0.8 m/s. The alum coagulant is active at 100% and its commercial strength is 50%. To determine the required volume of the alum (Equation (1)), the required mass should be divided by the density of the fluid. Therefore, using Equations (2) and (4), the required volume of alum must be 0.006 m^3/day. Taking a depth of the alum tank equal to 25 cm, the proposed alum tank dimension used in this conceptual design is 20 cm × 15 cm × 25 cm.

$$V_{alum} = \frac{M_{r,alum}}{\rho} \tag{1}$$

$$M_{r,alum} = \frac{M_{alum}}{CS} \tag{2}$$

$$M_{alum} = C \times Q \tag{3}$$

where V_{alum} is the required volume of alum (m^3/day), $M_{r,alum}$ is the required mass of alum (kg/day), ρ is the alum density (kg/m^3), CS is the commercial strength (%), C is the alum dosage (kg/m^3), and Q is the flowrate (m^3/day).

2.2.3. Circular Tank Design

Since the circular tank is dedicated to include three processes as mentioned before: coagulation, flocculation, and clarification, therefore its volume is determined by the largest volume required among them.

- For rapid mixing during coagulation, the detention time should be $10 < DT < 60$ s [18]. Taking DT = 20 s, the required volume will be 0.04 m^3 based on Equation (4).

$$V = Q_{avg} \times DT \tag{4}$$

where V is the circular tank volume (m^3), Q_{avg} is the average flowrate (m^3/s), and DT is the detention time (s).
- For the clariflocculation, the detention time should be between 2 and 4 h. Assuming that the time it takes to complete the flocculation and clarification respectively, is 30 and 150 min [18], therefore the total detention time in this phase is 180 min. Using Equation (4), the required volume will be 21.6 m^3.

The governed value that determines the required volume of the circular tank is 21.6 m^3. Assuming a depth of 2 m, the calculated diameter will be 3.71 m. Therefore, a diameter of 4 m for the circular tank is used in the DWTP design.

2.2.4. Propeller Design

During the coagulation, a static rapid mixer is used in order to ensure a uniform dispersion of the coagulant throughout the basin. After coagulation, the flocculation process assists in the formation of flocs, which involves slow mixing to enhance the contact between the coagulant and the feed water and for the subsequent formation of flocs. For both processes, a propeller mixer is used which is placed at 1/3 of the tank depth; at 0.65 m from its bottom.

The velocity gradient (G) remains a significant parameter for the design of the mixer. The equation to compute the G value [21] is as follows:

$$G = \sqrt{\frac{P}{\mu V}} \tag{5}$$

where G is root-mean-square velocity gradient (m/s/m), P is the power imparted to water (N.m/s), V is the mixer volume (m^3), and μ is the dynamic viscosity (N.s/m^2).

Using the actual volume of the circular tank that has a value of 25.13 m^3, a dynamic viscosity of the fluid (μ) equal to 1.0087 N.s/m^2, and a specific value of G that depends on the duration of the two processes [18], the value of the power of the propeller can be determined as follow:

- For rapid mixing operating for 20 s, the rate of change of velocity should be 600 m/s/m $< G <$ 1000 m/s/m, taking G equal to 1000 m/s/m the corresponding power of the mixer during this phase will be 25,349 W.
- For flocculation operation for 30 min, the velocity gradient will be taken 50 m/s/m then the propeller power will be 64 W.

Hence the propeller will operate at a power of 25,349 W for 20 s during the rapid mixing phase, then operates at a power of 64 W for 30 min to form the flocs needed during the flocculation after which it stops for two and a half hours during the sedimentation to let the flocs deposit and quit through a sludge pipe, then restarts its cycle again.

In order to determine the mixer diameter, Table 2 can be used. Having a radial flow, the corresponding range is 0.14–0.5. For a tank diameter (T) of 4 m, and taking D/T equal to 0.25, the required impeller diameter (D) will be 1 m.

Table 2. Tank and impeller geometries for mixing [20].

Geometric Ratio	Range
D/T (radial)	0.14–0.5
D/T (radial)	0.17–0.4
H/D (either)	2.0–4.0
H/T (axial)	0.34–1.6
H/T (radial)	0.28–2.0
B/D (either)	0.7–1.6

2.2.5. Sludge Removal Pipe Design

Assuming that the sludge will accumulate during the flocculation in the last 10 cm of the tank, the volume of the deposits and its corresponding flowrate will be respectively 1.26 m^3 and 0.7 L/s. The velocity in the pipe should be between 1.2 and 1.8 m/s. Taking a velocity of 1.5 m/s, the PVC pipe diameter should be equal to 25 mm.

The sludge should be treated first and then disposed of; the treatment process is as follows: Sludge Pasteurization, Mesophilic Anaerobic Digestion, Thermophilic Aerobic Digestion, Composting (Windrows or Aerated Piles), Lime Stabilization of Liquid Sludge, Liquid Storage, Dewatering and Storage. According to the Waste Framework Directive, the sludge could undergo preparation for reuse, recycling, disposal, or other recovery processes. From these, two groups of disposal methods can be derived: organic recycling and energy and material recycling.

2.2.6. Filtration

For the filtration unit, the filtration rate should be between 5 and 7 m/h. Assuming its value equal to 7 m/h, the required area of the filtration system should be 1.03 m^2. Taking a length of 1 m, a width of 1.5 m and a depth of 0.8 m for the filtration system, the real volume of the reed bed system will be 1.2 m^3.

The filtration unit will include three layers:

- Bottom layer: made of gravel of size 25 to 40 mm with a 25 cm thickness.
- Middle layer: made of gravel of size 10 to 25 mm with a 25 cm thickness.
- Top layer: made of silty sand with a 10 cm thickness.

On top of all the layers, a reed bed is planted as shown in Figure 2 to ensure the consumption of organic matter in water by the bacteria present in the roots of the plants [22]. Finally, a PVC perforated pipe is placed above the bottom ground by 15 cm to let the drainage occur and collect the clean water. To build a prototype a sieve analysis test is conducted to place the three different layers, followed by the plantation of the reeds which requires six to eight months to fully grow and give the desired outcome. The species of reeds used include *Phragmites australis* and *Typha latifolia* with a density of 4 to 8 plants per m^2. They have long, thin stems that can grow up to several meters in height. Their roots are shallow and spread out, allowing the reeds to grow in wet soils.

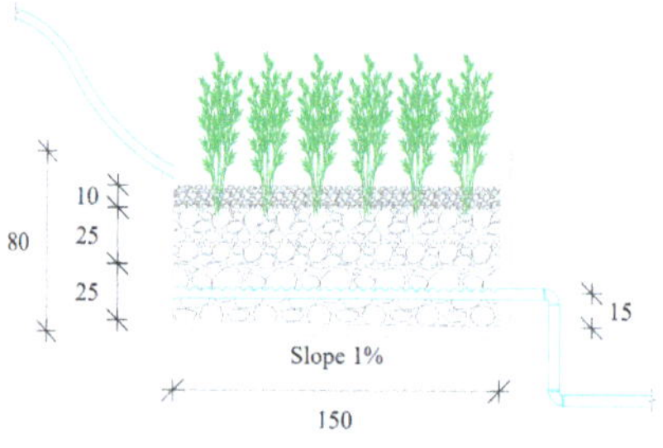

Figure 2. Reed bed filtration unit.

2.2.7. Disinfection

For the DWTP, filtered water is disinfected using sodium hypochlorite (NaClO), which is a chemical compound that contains chlorine. The chlorination design is achieved using the design criteria provided by [23]. Taking a chlorine dosage of 50 mg/L, its consumed mass will be 8.64 kg/day. Referring to the chlorine institute, the density of chlorine in its liquid state at a temperature of 15.6 °C is 1422 kg/m^3, then the required chlorine volume should be 0.006 m^3/day. Taking a 25 cm chlorine tank depth, its required area should be 0.024 m^2. Therefore, the proposed chlorine tank dimension is 20 cm × 15 cm × 25 cm.

Disinfection storage time is 30 min, this gives a stored volume of 3.6 m^3. Assuming a disinfection tank depth of 1.25 m, the minimum required area should be 2.88 m^2. Using 4 cylindrical tanks with the same depth, the required diameter should be 0.96 m. Therefore, in the disinfection DTWP phase, 4 cylindrical tanks with a depth of 1.25 m and a diameter of 1 m are used. To ensure that these dimensions are optimal, the detention time should be checked. The actual time is the total volume of tanks divided by the average flowrate. The calculated time is 32.7 min greater than 30 min which indicates that the disinfection tank dimensions are acceptable.

After the disinfection process, the chlorine level in water is monitored using amperometric meters to ensure that disinfection level is safe and efficient while also avoiding the risk of secondary pollution.

3. Impact

In September 2015, the United Nations (UN) adopted the Sustainable Development Goals (SDGs) aiming to significantly improve human well-being by 2030 [24]. This conceptual unit shown in Figure 3 is designed to address several SDGs in an eco-friendly way by employing recycled materials, resulting in up to a 50% reduction in cost. Its environmental and social impacts include:

- SDG number 3: "Good health and well-being", by eliminating water borne diseases that cause serious health issues.
- SDG number 6: "Clean water and sanitation", achieved by providing water that is palatable, odorless, and visually appealing
- SDG number 13: "Climate action", achieved by ensuring the safe and easy disposal of used materials.

This concept optimizes the size of the unit by integrating the first three processes in a single tank, followed by a nature-integrated system that relies on reed plants to act as chemical removers of bacteria and sand layers to facilitate natural filtration processes. Additionally, this approach encourages small communities, particularly those in small villages that lack the resources to construct an entire treatment plant.

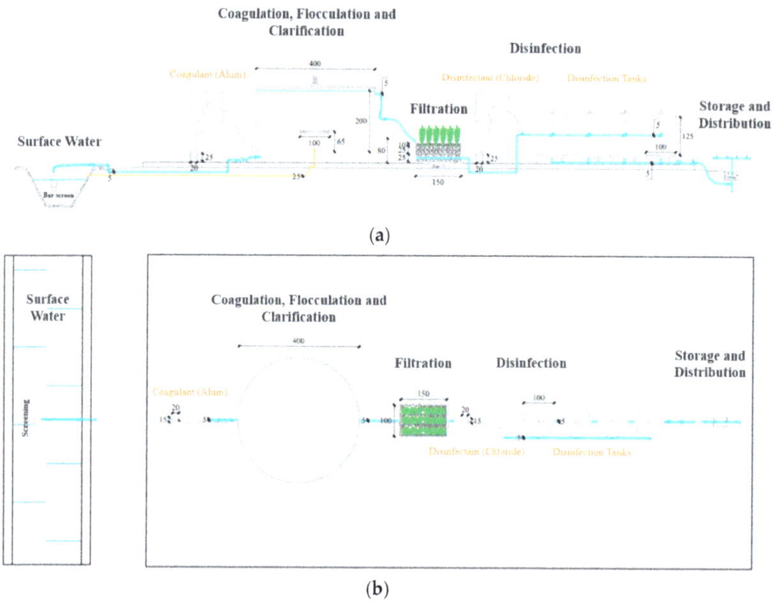

Figure 3. (**a**) Side view and (**b**) top view of the conceptual design of DWTP (all dimensions are expressed in cm).

4. Conclusions

An overview of the design of the main processes for drinking water was presented in this work. Despite the availability of many filtration processes, this paper proposed a new filtration concept using a reed bed system. Fixed bed filtration is widely regarded as the most significant filtration process due to its versatility and ease of operation. Among the various types of filtration, adsorptive filtration is considered the most cost-effective technology, owing to the availability of diverse adsorbents. Reed plants, which serve as natural adsorbents, not only augment pollutant removal but also enable us to achieve the long-sought-after sustainability goals. These techniques aid in eliminating waterborne illnesses, which pose a significant threat to public health.

Author Contributions: Conceptualization, E.F.; Formal analysis, M.K., M.R. and C.A.H.; Methodology, E.F.; Software, M.K., M.R. and C.A.H.; Supervision, E.F.; Validation, E.F.; Writing—original draft, M.K., M.R. and C.A.H.; Writing—review & editing, E.F. All authors have read and agreed to the published version of the manuscript.

Funding: This research received no external funding.

Data Availability Statement: Not applicable.

Conflicts of Interest: The authors declare no conflict of interest.

References

1. Ali, I. Water treatment by adsorption columns: Evaluation at ground level. *Sep. Purif. Rev.* **2014**, *43*, 175–205. [CrossRef]
2. Shannon, M.A.; Bohn, P.W.; Elimelech, M.; Georgiadis, J.G.; Marinas, B.J.; Mayes, A.M. Science and technology for water purification in the coming decades. *Nature* **2008**, *452*, 301–310. [CrossRef] [PubMed]
3. World Health Organization. *Guidelines for Drinking-Water Quality*; World Health Organization: Geneva, Switzerland, 2004; Volume 1.
4. Jain, S.K.; Agarwal, P.K.; Singh, V.P.; Jain, S.K.; Agarwal, P.K.; Singh, V.P. Water quality and related aspects. *Hydrol. Water Resour. India* **2007**, *57*, 997–1033.
5. Tong, Y.; McNamara, P.J.; Mayer, B.K. Adsorption of organic micropollutants onto biochar: A review of relevant kinetics, mechanisms and equilibrium. *Environ. Sci. Water Res. Technol.* **2019**, *5*, 821–838. [CrossRef]

6. Noubactep, C.; Caré, S.; Crane, R. Nanoscale metallic iron for environmental remediation: Prospects and limitations. *Water Air Soil Pollut.* **2012**, *223*, 1363–1382. [CrossRef] [PubMed]
7. Gonzalez-Perez, A.; Persson, K.M.; Lipnizki, F. Functional channel membranes for drinking water production. *Water* **2018**, *10*, 859. [CrossRef]
8. Ali, I. New generation adsorbents for water treatment. *Chem. Rev.* **2012**, *112*, 5073–5091. [CrossRef] [PubMed]
9. Chen, W.; Mo, J.; Du, X.; Zhang, Z.; Zhang, W. Biomimetic dynamic membrane for aquatic dye removal. *Water Res.* **2019**, *151*, 243–251. [CrossRef] [PubMed]
10. Patel, A.; Shah, A. Sustainable solution for lake water purification in rural and urban areas. *Mater. Today Proc.* **2020**, *32*, 740–745. [CrossRef]
11. Shabiimam, M.; Kazi, T.; Anas, P.; Shifa, S. Treatment of water using various filtration techniques: A review study. In Proceedings of the 3rd International Conference on Construction, Real Estate, Infrastructure and Project (CRIP) Management, Pune, India, 23–25 November 2018; pp. 229–235.
12. Yang, H.; Hu, R.; Ndé-Tchoupé, A.I.; Gwenzi, W.; Ruppert, H.; Noubactep, C. Designing the next generation of Fe0-based filters for decentralized safe drinking water treatment: A conceptual framework. *Processes* **2020**, *8*, 745. [CrossRef]
13. Goldstein, R.; Smith, W. *Water & Sustainability (Volume 4): US Electricity Consumption for Water Supply & Treatment—The Next Half Century*; Electric Power Research Institute: Washington, DC, USA, 2002.
14. Bukhary, S.; Batista, J.; Ahmad, S. Design aspects, energy consumption evaluation, and offset for drinking water treatment operation. *Water* **2020**, *12*, 1772. [CrossRef]
15. Bukhary, S.; Weidhaas, J.; Ansari, K.; Mahar, R.B.; Pomeroy, C.; VanDerslice, J.A.; Burian, S.; Ahmad, S. Using Distributed Solar for Treatment of Drinking Water in Developing Countries. In Proceedings of the World Environmental and Water Resources Congress 2017, Sacramento, CA, USA, 21–25 May 2017; pp. 264–276.
16. Plappally, A.K.; Lienhard, J.H. Costs for water supply, treatment, end-use and reclamation. *Desalination Water Treat.* **2013**, *51*, 200–232. [CrossRef]
17. Price, J.I.; Heberling, M.T.; Nietch, C.T. Economic support for decisions on source water protection. *J. -Am. Water Work. Assoc.* **2018**, *110*, 56. [CrossRef] [PubMed]
18. Jagadeesh, A. Safe drinking water for all through solar disinfection. *J. Rural Dev.* **2010**, *58*, 11–13.
19. American Water Works Association. *Water Treatment Plant Design*; McGraw-Hill Education: New York, NY, USA, 2012.
20. Davis, M.L. *Water and Wastewater Engineering: Design Principles and Practice*; McGraw-Hill Education: New York, NY, USA, 2010.
21. Hendricks, D.W. *Water Treatment Unit Processes: Physical and Chemical*; CRC Press: Boca Raton, FL, USA, 2018.
22. Prasad, R.K.; Chatterjee, S.; Sn, D.; Sharma, S.; Vairale, M.; Veer, V. Reed Bed Plants & Their Role in Waste Water Treatment & Reclamation. In Proceedings of the National Conference on Recent Advances in Biodegradation of Human Wastes, Tezpur, India, 16–17 December 2014.
23. Lee, C.; Lin, S.D. *Handbook of Environmental Engineering Calculations*; McGraw-Hill Education: New York, NY, USA, 2007.
24. Hering, J.G.; Maag, S.; Schnoor, J.L. A call for synthesis of water research to achieve the sustainable development goals by 2030. *Environ. Sci. Technol.* **2016**, *50*, 6122–6123. [CrossRef] [PubMed]

water

MDPI

Review

A Review of Bubble Aeration in Biofilter to Reduce Total Ammonia Nitrogen of Recirculating Aquaculture System

Putu Ayustin Suriasni [1], Ferry Faizal [1,2], Camellia Panatarani [1,2], Wawan Hermawan [2,3] and I Made Joni [1,2,*]

1 Department of Physics, Faculty of Mathematics and Natural Science, Padjadjaran University, Jl. Raya Bandung-Sumedang KM 21, Sumedang, West Java 45363, Indonesia
2 Functional Nano Powder (FiNder) University Center of Excellence, Padjadjaran University, Jl. Raya Bandung-Sumedang KM 21, Sumedang, West Java 45363, Indonesia
3 Department of Biology, Faculty of Mathematics and Natural Science, Padjadjaran University, Jl. Raya Bandung-Sumedang KM 21, Sumedang, West Java 45363, Indonesia
* Correspondence: imadejoni@phys.unpad.ac.id

Abstract: Aeration becomes an essential aspect of biofilter performance to reduce ammonia nitrogen in the Recirculating Aquaculture System (RAS). Efficient aeration introduces air into water media and offers an aerobic environment in the biofilter for microbial degradation of organic matter and ammonia nitrogen. The efficiency of the bubble aeration depends on the size of the bubbles; these include coarse bubble, microbubble, fine bubble, and ultrafine bubble or nanobubble. This review highlights an overview of bubble aeration features in a biofilter to reduce ammonia nitrogen. Moreover, key aspects responsible for the ammonia nitrogen removal efficiencies, such as oxygen transfer, microbial community, and biofilm thickness, are evaluated in this review. In conclusion, the bubble size of aeration affects the microbial community of nitrifying bacteria, consequently determining the growth and thickness of biofilm to improve ammonia removal efficiency. It is emphasized that fine bubble and nanobubble aeration have very positive prospects on improving biofilter performance, though they are currently not widely used in RAS.

Keywords: aeration; biofilter; RAS; bubble

1. Introduction

In recent years, the RAS has become extensively applied in aquaculture production to decrease the impact of wastewater pollution on the environment [1]. The RAS is a closed aquaculture system that reuses water by replacing water volume by less than 10% [2]. Water replacement is conducted through recirculation and implementing various treatments to reduce contaminants in the wastewater. Wastewater is polluted water generated from industrial processes, including aquaculture waste. Therefore, wastewater treatment must remove the contaminants and leave an effluent with a less negative impact on aquaculture and the environment. The stages broadly incorporate physical, biological, and chemical processes that take place in that order. The first stage is primary treatment using a physical process such as sedimentation and flotation to remove organic and inorganic solids. The second stage is secondary treatment using a biological process to transform pollutants into stable forms through oxidation or nitrification. Finally, the third stage is tertiary treatment using a chemical process to remove specific wastewater pollutants that are unresolved from secondary treatment [3,4].

Ammonia nitrogen is a critical process parameter to be controlled in the RAS, which strongly affects the health and growth of aquaculture species [5]. Therefore, introducing effective and efficient wastewater treatments is very crucial in RASs. A biofilter is one of the methods in RAS wastewater treatments. Efficient aeration enables the introduction of air into the water media, offering an aerobic environment as a biofilter for microbial degradation of contaminants such as ammonia nitrogen. Therefore, aeration becomes essential in aerobic biofilters where enhanced dissolved oxygen probably improves organic degradation performance. However, when selecting an aeration system, it is also needed to consider its energy usage. The advantages of advanced aeration technology offer multiple economic and environmental benefits, correspondingly reducing operating costs and energy savings. Therefore, recent aeration technology needs to consider implementing low-carbon technology in line with the demand of every nation to meet net-zero carbon emissions goals [6]. It means that a reliable biofilter with efficient aeration contributes to energy-saving solutions and cutting greenhouse gas emissions. Thus creating affordable, sustainable, long life-cycle and high-performance biofilters is the next urgently needed and significant challenge.

Based on the growth of the microorganisms, biofilters classify as suspended growth or fixed film [7]. In the suspended growth system, heterotrophic bacteria remain suspended in the water through stirring [8]. On the other hand, in an attached growth system or fixed film, autotrophic bacteria attach to a solid surface, such as rock, sand, metal, and plastic [9]. One of the most common fixed film bioreactors in an RAS is a Moving Bed Biofilm Bioreactor (MBBR). An MBBR utilizes small floating plastic media to serve as a surface for bacterial attachment and biofilm growth [10]. Furthermore, under aerobic conditions, nitrification occurs in biofilm when Total Ammonia Nitrogen (TAN) oxidizes into nitrite (NO_2^-) and nitrate (NO_3^-) [11]. Another example of the bioreactor in the RAS is the Biological Aerated Filter (BAF). The biofilm formed in the BAF reactor contains aerobic and anaerobic layers, which guarantee biomass structure stability and contaminant degradation [12].

The essential factors that influence biofilter performance are mass transfer and oxygen distribution. First, dissolved oxygen (DO) is responsible for enhancing biofilm growth and, consequently, TAN degradation through nitrification [13]. The mass or oxygen transfer from DO in the water to biofilm depends on many factors, such as biomass concentration, aeration rate, hydrodynamic condition, and biofilm characteristics [14]. Therefore, an aeration system with natural or mechanical aeration is usually utilized to provide a sufficient amount of DO for the biofilter. One of the mechanical aeration types used to provide DO is a bubble aerator. A bubble aerator with a finer size leads to a higher interfacial area and induces higher DO transfer [15,16]. As a result, the TAN removal efficiency of an MBBR with a similar reactor utilizing a fine bubble aerator was higher (45%) compared to a larger or coarse bubble size (33%) [17,18].

A study was done to review the properties of microbubbles and size measurement from available bubble generation technologies and provides the perspective of future major applications in water and wastewater treatment [19]. Another review focuses on applying Micro–Nano-Bubble (MNBs) generators for domestic and industrial wastewater treatment [20]. The MNBs technology considers an effective method to improve the adhesion of suspended particles due to the unique characteristics of bubbles, such as small-sized bubbles, higher surface area, slow rising velocity, negatively charged zeta potential, generation of free radicals, and stability. The MBBR applies through conventional nitrification and simultaneous nitrification–denitrification pathways for residual water of the RAS [21]. However, there is no study on the effect of the bubble size and oxygen volumetric transfer on the biofilm growth and microbial communities concerning TAN removal efficiency in treating RAS residual water. Therefore, the objective of this review is to provide an updated evaluation of bubble size and oxygen volumetric transfer of the aerator on TAN removal efficiency and the effect of bubble size on the growth and microbial communities of biofilms in biofilters applied for the treatment of residual water from an RAS.

2. Materials and Methods

This review aims to understand the effect of bubble aeration on ammonia removal efficiency and biofilm growth in biofilters applied in RASs. A systematic review with four steps was conducted to achieve this goal. The first step was identifying the database and then screening the result from the title and abstract. Furthermore, the selected papers in the full text were analyzed. The last step was choosing a group of documents to include. In this review, the PRISMA guideline was also adopted.

2.1. Objective

The main objective of this study was summarized in two research questions (RQ).

RQ1: How does bubble size affect oxygen transfer and ammonia removal efficiency in various types of biofilters in RASs?

RQ2: What is the relation between aeration and biofilm growth to remove ammonia in various types of biofilters in RASs?

Several keywords were determined from the research questions. The keywords, which consist of bubbles (and synonyms such as nanobubble), aeration, ammonia, biofilter (and correlated expressions such as MBBR), biofilm, and RAS, were included in the screening process. The Boolean operators were combined with these keywords and used as a search expression [22]. The search was carried out in ScienceDirect and Google Scholar. From the search, 1899 documents were found, as shown in Table 1, and proceeded to the screening step.

Table 1. Identification step results.

Search Local	Search Expression	Search Result	Types of Documents
ScienceDirect	(bubble! OR nanobubble!) AND (aeration) AND (ammonia) AND (biofilter! OR MBBR!) AND (biofilm!)	445	• 55 Review articles • 239 Research articles • 19 Encyclopedia • 79 Book Chapters • 10 Conference abstracts • 2 News • 3 Short communication • 38 Other
Google Scholar	(bubble! OR nanobubble!) AND (aeration) AND (ammonia) AND (biofilter! OR MBBR!) AND (biofilm!) AND (RAS!)	454	• 27 Patents • 43 Review articles

2.2. Screening

This step determined the year of publication screening from the total identified documents. The years of publication were limited to 2011–2022 in ScienceDirect and 2011–2021 for Google Scholar. As a result, 324 documents from ScienceDirect and 320 documents from Google Scholar were selected. Then, the type of paper screening was utilized to exclude all documents except for research papers. When the filtering considered only "research article" in ScienceDirect, 170 articles were received.

In contrast, in Google Scholar, the received articles totaled 251. Then, screening was performed by checking the title and abstract. If the title and abstract contained one or more keywords or search queries, it was included for full-text analysis. Consequently, 369 articles were excluded by this step. Furthermore, the screening step was done in November–December 2021.

2.3. Eligibility

Eligibility assessment was started by downloading the 52 articles from 2 search engines. Then, the 52 articles were analyzed, seeking the effect of bubble aeration on reducing ammonia nitrogen and its effect on the biofilm growth of biofilters. Out of 52 articles, five were excluded due to duplication, and 28 were excluded because the aeration/biofilter was not implemented in an RAS.

2.4. Inclusion

After full-text analysis, 19 articles were included in this study. These articles described the effect of bubble aeration on biofilm growth and reduced ammonia in the biofilter at the lab or a pilot of an RAS. The articles searched were published in the time range between 2011 and 2022. The analysis steps introduced in the PRISMA procedure are shown in Figure 1.

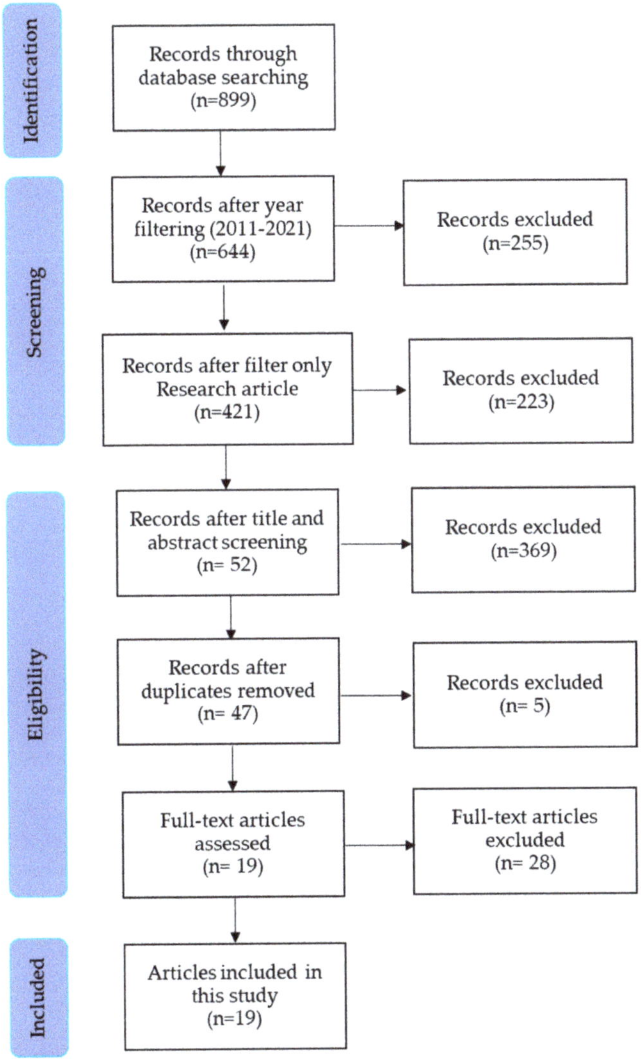

Figure 1. Document selection based on PRISMA workflow.

3. Types of Biofilter in RASs and Role of Bubble Aeration in Biofilter

There are two well-known biofiltration types based on the growth of microorganisms, i.e., attached and suspended growth, as shown in Figure 2. A fixed or attached film system is generally used in the RAS since it is more stable than a suspended growth system. The fixed film is classified as the emergent filter or the submerged filter. The emergent filter uses a water cascade that directly flows over biofilter media to maximize oxygen transfer [7].

The trickling filter is one emergent filter that uses wastewater infiltration sprayed from the top of the filter towards the bottom drain to form a biofilm on the surface of the packing media. Biofilm formation degrades the organic matter in wastewater passing through the packing media [23]. In comparison, the rotating biological contactors treat the wastewater by rotating the media in and out of the wastewater [24].

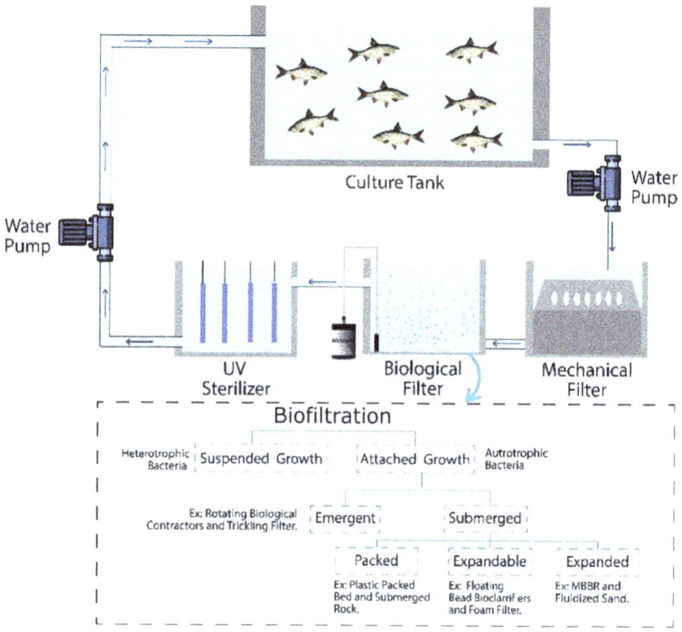

Figure 2. Types of biofilters in RASs.

In the submerged filter, surfaces of submerged biofilter media are used to support the attachment and growth of microbes, subsequently increasing biofilm growth [25]. Three types of submerged filter media are based on dynamic behavior: packed bed, expandable, and expanded. The packed bed filter utilizes a compact and immobile stack of media that behaves as porous media [26]. This biofilter creates a biofilm that is more stable and is not easily damaged in a small amount of oxygen [27]. The expandable filter utilizes bead media simultaneously to stimulate the growth of desirable microbes and to remove waste from the water [28]. The filtration process occurs by capturing suspended solids of the recirculating wastewater that passes through the packed beads media and subsequently causes pollutant degradation [29]. Thus, periodic cleaning of the packed beads is necessary to maintain the filtration performance and is achieved by mechanical, pneumatic, or hydraulic means [28,30]. In contrast, the expanded filter of an MBBR utilizes the motion of media; in this condition, the cyclic formation and erosion of biofilm occurs because of the hydraulic and pneumatic behavior of media [7]. However, an efficient biofilm for the MBBR needs a high specific surface area and to be provided a large growth zone. Therefore, precaution is needed to control the hydrodynamic behavior of aeration.

Bubble aeration is critical in determining biofilter efficiency in reducing ammonia nitrogen, as described in Figure 3. Bubble aeration utilizes air or gas in the form of bubbles to provide dissolved oxygen to the media of biofilm growth. The rising gas bubbles transfer oxygen into the liquid phase as dissolved oxygen. The dissolved oxygen is diffused into the biofilm towards the oxidative-phosphorylation sites. The biofilm is attached to the media as a solid phase. The oxygen transport suffers various resistances from the bubbles, liquid, and solid phases, as seen in Figure 3e. The resistance of the bubble comes from the liquid film resistance (k_L) around the bubble, which usually controls the overall transfer rate from the bubble to the liquid phase and the final destination of the biofilm, as seen in Figure 3e [31].

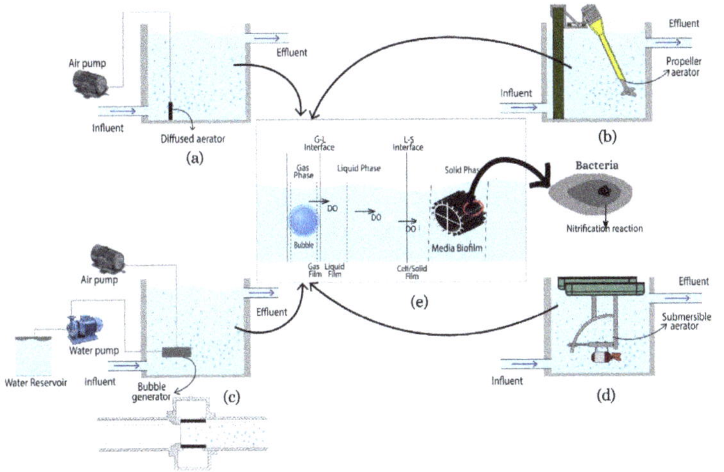

Figure 3. Oxygen transfer at various types of bubble aerators(**a**) diffused aerator; (**b**) propeller aerator; (**c**) bubble generator; (**d**) submersible aerator, and (**e**) transfer oxygen process in the biofilter.

Bubble aerators are classified as a diffused, propeller, and submersible, as shown in Figure 3a–d. The characteristics of generated bubbles depends on the mechanism of bubble formation, the location, and the interaction of the generator with water. For example, pressurized air or oxygen gas from a blower or compressor introduced in the diffuser method generated bubbles with the size and distribution of the bubbles depending on the pore size and morphology of the diffuser [32]. In comparison, the high-velocity agitated propeller-generated bubbles create highly dynamic behavior causing turbulence in the water, and the bubbles mix easily with the water. The submersible aerator consists of a pump that draws atmospheric air through a hollow suction pipe connected to the propeller at the other end submerged in water. The propeller rotates at a high-speed rpm, enough to cause air suction in the hollow pipe to generate bubbles due to the turbulent water and air mixture created by the propeller [33].

In contrast, the Bernoulli principle of fluid mechanics generates smaller and uniformly distributed bubble sizes, where the free air enters the pipes due to pressure differences in the pipes and the atmosphere [34]. In this review, the application of diffused aerators and bubble generators is highlighted in terms of their bubble size and oxygen transfer concerning the biofilm growth or biofiltration performances.

3.1. Bubble Size

Many researchers classified bubbles as nanobubbles (diameter <200 nm), fine bubbles (diameter 200−10 μm), microbubbles (diameter ≤50 μm), and macro bubbles (diameter 2−5 mm) [35,36]. However, ISO 20480-1 classifies a bubble with a volume equivalent to diameter. Thus, bubble terms based on their size are defined as micro, fine, and ultrafine bubbles/nanobubbles ranging from 1 to 100 μm, less than 100 μm, and less than 1 μm, respectively [37]. In conventional aerators, there are medium bubbles with a diameter of 1.5 to 3 mm, and giant bubbles or coarse bubbles with a diameter of over 3 mm.

Fine bubbles can be generated from a porous diffuser or membrane with tiny holes, while medium bubbles from an orifice, and coarse bubbles through pipes or plates [38]. The relation between the bubble size (d_b) and pore size (d_o) can be written in Equation (1), where g is gravitational acceleration, ρ is the density of the liquid, and σ is surface tension [39].

$$\frac{d_b}{d_o} = 1.8 \left[\frac{\sigma}{\rho g {d_o}^2}\right]^{1/3} \quad (1)$$

3.2. Biofilm Growth

Biofilms are groups of microbial cells attached to a solid surface or a media and encased with a matrix. The matrix contains microbial biopolymers composed of extracellular DNA, exopolysaccharides, and proteins [40]. Figure 4 shows the growth phase of the biofilm from the attachment stage to the release stage and its correlation with bubble size. The smaller size of the bubble, the more the microbial community is increased. The first stage of biofilm growth is reversible attachment. In this stage, free-floating (planktonic) bacteria begin to adhere to media acting as biocarrier surfaces. However, the bacteria are not fully attached and are possibly separated from the formation before being wholly attached. The second stage is irreversible bacteria binding. Bacteria begin to adhere perfectly and grow into mature biofilms due to the presence of an extracellular polymeric substance (EPS). The production of a slime-like layer from the EPS supports the structure of the biofilm. EPS is composed of proteins, polysaccharides, and nucleic acids. Then in the third stage, or the maturation stage, bacteria have a complex architecture, pores, and charges far from the substrate. At this stage, there is more protein, consequently increasing the bacterial number and diversity of the biofilm. When smaller bubbles are applied, more efficient oxygen transfer is generated to create more protein to promote an enhanced number and diversity of bacteria. In the fourth stage, strong bacterial adherence occurs due to matured biofilm formation. The last stage is detachment, where bacteria are detached from the biofilm due to the reduction of the EPS or lack food supply [41,42]. Biofilm growth rate (μ in h^{-1}) is calculated using Equation (2) where X_2 (mg/cm^3) is the density of biomass at the end of the experiment, X_1 (mg/cm^3) is the density of biomass at the first period of the experiment, and θ (h) is the experiment period [43].

$$\mu = \frac{(X_2 - X_1)}{X_1 \, x \, \theta} \quad (2)$$

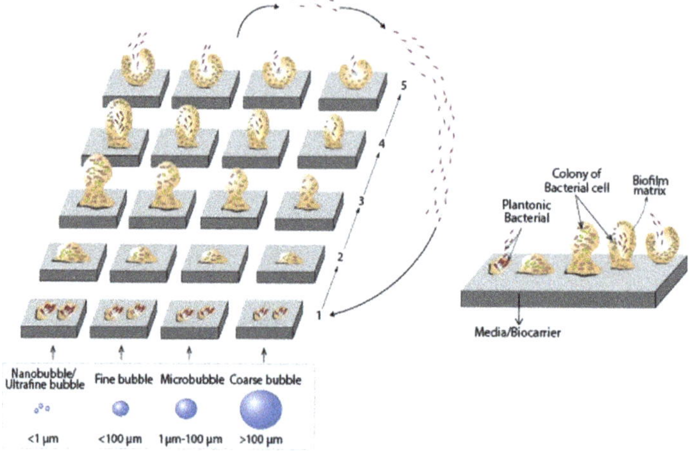

Figure 4. Effect of bubble size on the phase of biofilm growth: (1) Adhesion stage, (2) Extracellular Polymeric Substance (EPS) production, maturation stage, (3) formation of complex structure, (4) strong-adherence bacterial stage, and (5) cell detachment from the biofilm.

3.3. Nitrification in Biofilm

Characteristics of bubbles affect the oxygen transfer efficiency from gas through water into the biofilm [44]. Efficient oxygen transfer ensures sufficient DO for the nitrification process inside the biofilm. Nitrification is extremely important in reducing harmful ammonia in the RAS. Nitrification is one of the biological processes where NH_3 and NH_4^+ are oxidized to nitrate in two steps with the help of nitrifying bacteria with a reaction, as seen in Equations (3) and (4) [45]. Ammonium oxidizing bacteria (AOB) such as *Nitrosomonas* and *comammox* are responsible for converting ammonium nitrogen into nitrite (NO_2^-). Then, nitrite is converted into nitrate with the help of nitrite-oxidizing bacteria (NOB) such *Nitrospira* and *Nitrobacter*.

$$NH_4^+ + \frac{3}{2}O_2 \overset{AOB}{\rightarrow} NO_2^- + 2\,H^+\ H_2O \quad (3)$$

$$NO_2^- + \frac{1}{2}O_2 \overset{NOB}{\rightarrow} NO_3^- \quad (4)$$

3.4. Oxygen Transfer

An effective nitrification process depends on having enough dissolved oxygen obtained from the aerator with a small bubble size. Bubbles with smaller sizes have higher oxygen mass transfer [15]. Furthermore, small bubbles facilitate oxygen distribution through biofilm for the nitrification process. Thus, a smaller bubble size potentially produces higher ammonia degradation than a coarse bubble size. The relation between bubble size and oxygen transfer can be associated with specific interfacial area a. This value is related to fractional gas hold up ε_g and Sauter mean diameter of bubble d_b as written in Equation (5) [16,46]. Based on this equation, a smaller bubble diameter increases the gas–liquid interfacial area.

$$a = \frac{6\varepsilon_g}{d_b(1-\varepsilon_g)} \quad (5)$$

The value of a influences the mass transfer rate, which can be described with *dC/dt* as written in Equation (6) [47] where *dC* is the change of oxygen concentration with time *dt*, $k_La_{(T)}$ is the oxygen volumetric transfer coefficient at temperature T (the temperature at the time of the experiment), which is determined from K_L (liquid film coefficient) and $a_{(T)}$ (unit interfacial media at a certain temperature), C* is the oxygen saturation, and C is the DO concentration at time t. The $k_La_{(T)}(C^* - C)$ is the Oxygen Transfer Rate (OTR). The value of $k_La_{(T)}$ is an indicator of DO mass transfer in biofilters which is one of the critical factors in the aeration rate [14,48]. Furthermore, $k_La_{(T)}$ is calculated using Equation (7), which can be integrated from Equation (6) [49].

$$dCdt = k_La_{(T)}(C^* - C) \quad (6)$$

$$\ln\left(1 - \frac{C}{C*}\right) = k_La_{(T)}t \quad (7)$$

4. Results and Discussion

4.1. Biofilm Performance with Bubble Aeration

4.1.1. Effect of Medium and Coarse Bubble Aeration on the Biofilter Performance

The recent development of bubble aeration applied for fixed film biofilters is summarized in Table 2. The larger size of bubble aeration affected the media in the MBBR to maintain biofilm thickness by shear forces and allow diffusion transport of DO and ammonia ions to the biofilm. However, aggressive aeration on the media caused excessive biofilm erosion [50]. Medium bubble aeration was used to aerate five types of media with different surface areas and percentages of voidage in a pilot plant MBBR. The surface area of various media with spherical shapes, such as media 1 (Biofil), 2 (Bioball), and

3 (Biomarble), were 135, 220, and 310 m^2/m^3 with voidage of 95, 92, and 90%, respectively. In contrast, the surface area of media 4 (Biopipe) and 5 (Biotube) with cylindrical shapes were 600 and 1000 m^2/m^3 with voidage of 82.5 and 80%, respectively. The airflow rate was controlled at 3.6–18.7 m^3/m^2.h to keep stable DO levels. The ammonia removal efficiencies have reached 50 ± 13%, 64 ± 13%, and 63 ± 7% for media 1, 2, and 3, respectively after 30, 22, and 17 days. Media 4 and 5 have achieved ammonia removal efficiency percentages of, respectively, 32 ± 17% after 46 days and 34 ± 5% after 47 days. The highest removal efficiency has been achieved by media 2 due to the open spherical structure, higher voidage compared to media 3, 4, and 5, and even higher surface area than media 1. These factors caused an increase in oxygen distribution all over the media, and they offered larger surface areas for biofilm to attach to the media surface [17].

A coarse bubble from a disc diffuser was tested in the MBBR to enhance the media movement and degassing. Three media (MB3, K1 Kaldness, AMB) were used as MBBR media for water treatment in the red drum fingerlings hatchery. The highest TAN removal efficiency for those three media was generated when the DO was 6.9 ± 0.7 mg/L in high feed load rates (8.2 kg feed/m^3 media). The TAN removal efficiency reached 15.6 ± 0.4%, 15.4 ± 0.2%, and 11.6 ± 0.3% for MB3, K1 Kaldness, and AMB, respectively. MB3 media resulted in the highest TAN removal efficiency because it had a larger surface area (604 m^2/m^3 surface area) than the other media. The firm structure and larger size of MB3 media were easily aerated and provided enhanced TAN removal rates [50].

Meanwhile, coarse bubble diffusers were also tested on a lab-scale MBBR by maintaining DO concentration at 5.5–7.5 ± 0.5 mg/L [51]. The novel sponge carrier (SB) and Kaldness K5 were used as biofilm media. The TAN removal efficiency of SB media was 91.65 ± 1.3% which was higher than Kaldness media (86.67 ± 2.4%) for 6 h of hydraulic retention time (HRT). Higher removal efficiency in SB media was due to the enhancement of aerobic conditions, which led to a complete nitrification process. In addition, the porous structure of the SB media provided sufficient protection, so the bacteria were not easy to wash away [51].

Another study compared a coarse bubble diffuser and biofilm chip media (900 m^2/m^3 surface area) to control the water quality in an RAS for Atlantic Salmon smolt. The system consisted of three MBBRs with three nominal alkalinity treatments: 10 mg/L, 70 mg/L, and 200 mg/L. The DO was controlled between 85% and 90% of saturation in the experiment. The TAN concentration of inlet MBBR based on the 10 mg/L, 70 mg/L, and 200 mg/L alkalinity treatments were 0.65 ± 0.08, 0.43 ± 0.04, and 0.39 ± 0.05 mg/L, respectively. While the TAN generated from the outlet of MBBR based on the 10 mg/L, 70 mg/L, and 200 mg/L alkalinity treatments were 0.39 ± 0.06, 0.22 ± 0.03, and 0.23 ± 0.04 mg/L, respectively. The 70 mg/L alkalinity treatment contributed the highest mean TAN removal efficiency (around 50%) in the 20 week experiment. This efficiency occurred due to the higher inlet TAN of the 10 mg/L alkalinity treatment causing higher removal rates based on kinetic studies [52]. Similarly, a coarse bubble diffuser with a high flow rate (3.17 L/min) was applied to enhance the media movement of K1 Kaldness (500 m^2/m^3 surface area) in an MBBR for rearing Florida Pompano. The DO concentration was 8.2 ± 1.0 mg/L in the culture tanks. During the experiment, the average TAN concentration was 0.46 ± 0.27 mg/L, and TAN removal per pass was 33.0 ± 11.9% with a range of 8.3–62.4% [53]. At the same time, a coarse bubble was applied as the aeration system to ensure oxygen saturation of >100% for the MBBR in RAS culturing of European sea bass (*Dicentrarchus labrax*). Sea bass culture was exposed to four different nitrates: control (C) 0 mg/L, low nitrate (LN) 125 mg/L, medium nitrate (MN) 250 mg/L, and high nitrate (HN) 500 mg/L. Mean values (±SD) of TAN for different nitrate C, LN, MN, and HN had reached 0.10 ± 0.09 mg/L, 0.11 ± 0.10 mg/L, 0.11 ± 0.08 mg/L, and 0.11 ± 0.08 mg/L, respectively [54]. The MBBR with coarse bubbles showed that the TAN concentration meets the safety range for culturing sea bass (<2–6 mg/L) [55,56].

Table 2. Summary of biofilter performance at various types of aerator and biofilter types in RASs.

Type of Aerator	Biofilter Type (Volume in L)	RAS Culture Species	Media Type (Surface Area in m^2/m^3)	DO (mg/L)	$k_L a_{(T)} t$	Performance: Ammonia Removal (%)	Performance: Ammonia Concentration (mg/L)	Reference
Medium bubble diffuser	MBBR (2000)	N/A	Bioball (220)	6.1 ± 0.9	N/A	64 ± 13	N/A	[17]
Coarse Diffuser	MBBR (6360)	Red drum fingerlings (*Sciaenops ocellatus*)	MB3 (604)	9.9 ± 1.9	N/A	15.6 ± 0.4	1.055	[50]
Coarse bubble diffuser	MBBR (5.5)	Not specified	Novel sponge biocarrier (N/A)	5.5–7.5 ± 0.5	N/A	91.65 ± 2.4	0.83 ± 0.31	[51]
Coarse bubble diffuser	MBBR (7000)	Atlantic salmon parr (*Salmo salar*)	N/A	7.3	1.89	50 ± 3	0.22 ± 0.03	[52]
Coarse Bubble diffuser	MBBR (410)	Florida pompano (*Trachinotus carolinus*)	K1 Kaldness (N/A)	8.2 ± 1.0	1.39	62.4	0.46 ± 0.27	[53]
Coarse bubble diffuser	MBBR (N/A)	European sea bass (*Dicentrarchus labrax*)	N/A	10 ± 0.9	N/A	N/A	0.11 ± 0.08	[54]
Coarse bubble diffuser	Membrane Filter (N/A)	N/A	Commercial mixed-cellulose (MC) membranes (N/A)	6.75	1.54	38	7.81	[12]
Microbubble diffuser	airlift fluidized bed reactor (115)	Nile Tilapia (*Oreochromis niloticus*)	Granular activated carbon (N/A)	4.16	N/A	27	0.104	[2]
Fine bubble diffuser	BAF (1.6)	N/A	Puzzolane material (1740)	>2.1	0.31	50-80	N/A	[57]
Fine bubble diffuser	Fixed bed reactor (3.5)	N/A	Polycaprolactone 114 (PCL) (N/A)	5.63 ± 0.57	1.73	57.14	1.93 ± 0.37	[58]
Nanobubble generator and combination with coarse bubble diffuser	BAF (4.75)	N/A	Volcanic rock filter (N/A)	6.52	2.06	98.02	N/A	[59]
Nanobubbles generator	Membrane Filter (N/A)	N/A	Commercial mixed-cellulose (MC) membranes (N/A)	8.55	1.92	75	3.15	[12]

4.1.2. Effect of Micro, Fine, and Nanobubble Aeration on the Biofilter Performance

Recently, aeration systems with a finer bubble for RASs have gained much interest because of their high mass transfer and long-term existence in an aqueous environment. The RAS with cultured Nile Tilapia used microbubble aeration aimed to maintain the correct concentration of DO. This RAS used the aeration system of the airlift fluidized bed reactor. By maintaining the air flow rate at 10 L/min, the aerator guaranteed the needed quantity of DO concentration (4.61 mg/L). Thus, the mean removal efficiencies of BOD, COD, phosphorous, TAN, and total nitrogen had reached 47, 77, 38, 27, and 24%, respectively [2].

A fine bubble diffuser generating micro- and nanobubble was evaluated for the aeration performance in a BAF reactor intended to reduce the concentration of ammonia nitrogen. The DO concentration was maintained >2 mg/L, and the aeration rate was kept at 0.33 L/min. The sampling points were made in some places based on the biofilter's depth, 3, 4.5, 7.5, 12.5, 22.5, 32.5, and 40 cm. This investigation aimed to know the effect of hydraulic retention time (HRT) in correlation to biofilter performance. The average DO concentrations in the long term decreased along with biofilter depth. Ammonia removal efficiency between 50% and 80% received from sections between 3–7.5 cm depth was associated with high availability of DO (over 2.1 mg/L), and formed higher biofilm thickness (0.4–0.6 mm) [57]. The ammonia concentration was higher at a higher depth (12.5–40 cm). Higher ammonia concentrations occurred due to the effect of lower HRT causing high DO demand and consequently lowering the biofilm thickness because of a decrease in the availability of DO compared to the depth of 3–7.5 cm.

Similarly, a fine bubble diffuser was also used to study the effect of aeration in a fixed-bed reactor [58]. The study consisted of three treatment groups: anoxic, low DO, and aeration, producing DO concentrations ranging from 0.12 to 0.55 mg/L, 1.82 to 3.99 mg/L, and 5.29 to 6.56 mg/L, respectively. All treatments gave a DO concentration below 0.02 mg/L in the outlet. Reduced DO indicated oxygen consumption due to bacterial growth under aerobic conditions. The removal efficiency of NO_3-N after the start-up period of 11–18 days was maintained at 90% in the anoxic and low DO treatments. However, it decreased to 75% for the aeration treatment at the end of the experiment. Although aeration treatment showed a low NO_3-N removal efficiency, Total Nitrogen (TN) removal efficiency was the highest (90%) among other treatments. This efficiency in TAN removal was possibly due to complete nitrate removal, designating the effective nitrate transformation into nitrogen gas. Conversely, in the anoxic and low DO treatments, NO_3-N might be transformed into ammonia or nitrite (NO_2), and they remained in the system. In addition, the average TAN accumulation in all three treatments was observed at 1.93 ± 0.37, 1.37 ± 0.15, and 1.72 ± 0.46 mg/L during the steady-state period [58]. Therefore, the aeration efficiency in aerobic systems enhanced the total removal efficiency, especially in removing nitrite.

An aerator made from a combination of nanobubbles (420 nm) and air bubbles (coarse bubbles) has also been tried and compared to just nanobubbles used for biologically aerated filters (BAF) in a laboratory-scale wastewater treatment plant (WWTP) [59]. Both treatments used the same filter size and eight variations of aeration treatment with different oxygen resources (air/water ratio). The application of only nanobubbles enabled obtaining $NH_4^+ - N$ and TN removal efficiency at the maximum levels of 74.89 and 70.69%, respectively, if the DO concentration was 5.59 mg/L. When the combination of nanobubbles and coarse bubbles was used in BAF, the highest $NH_4^+ - N$ and TN removal efficiency were obtained at 98.02 and 54.98%, respectively, if the DO concentration was 6.52 mg/L. The DO was obtained at 6.52 mg/L when the air/water ratios were set up to be 0.2 and 0.5 for nanobubble and coarse bubbles, respectively. The highest removal efficiency of $NH_4^+ - N$ and TN were due to the dynamic interaction between the bubbles and media and the roles of bubbles size in oxygen mass transfer efficiency from bubbles to media at a particular air/water ratio. In addition, the coarse bubbles played an essential role in reinforcing the shear stress to the biofilm in correlation to the cyclic growth of biofilm formation. In contrast, nanobubbles gave effective mass transfer to provide high dissolved oxygen in the media. The allowable shear stress of the coarse bubble on the biofilm enabled film detachment. In addition, it maintained the thickness to effectively transport nutrients and oxygen into the biofilm [59]. Thus, higher dissolved oxygen generated in nanobubble aeration could provide higher TAN removal than coarse bubble aeration systems.

In contrast, nanobubbles with an aeration rate of 0.2 L/min was used in the lab-scale experiment, and coarse bubble aeration was applied to the system as a comparison [12]. A membrane filter of 1.2 μm thickness (ø =50 mm) commercial mixed-cellulose (MC) membrane was used as the biofilm carrier. The higher DO peak (10.012 mg/L) in biofilm has resulted in using nanobubble aeration compared with coarse bubble aeration (8.581 mg/L). The dissolved oxygen generated from nanobubble aeration remained higher than the coarse bubbles after 200 s. Consequently, higher ammonia removal (75%) has been achieved using nanobubble aeration. Thus, it also remarked that higher transfer oxygen in nanobubble than coarse bubble was considered responsible for the higher removal efficiency [12].

4.2. Biofilm Growth

4.2.1. Microbial Community

In wastewater treatment, the microbial community is an essential factor affecting biofilter performance to reduce ammonia through nitrification. One way to determine biofilm diversity is using the Shannon diversity index [60]. The Shannon index was used to describe variation at multiple levels of genetic organization from single nucleotide polymorphisms (SNP) through whole species or larger taxonomic units to ecosystems [61]. The Shannon index was calculated from the operational taxonomic unit (OTU). The OTU is a collection of 16S rRNA sequences with specific sequence divergence percentages [62]. We can know the number of necessary bacteria such as AOB and NOB in biofilm from the OTU. The diversity in the biofilm, which correlated with the bubble aeration, is summarized in Table 3.

Table 3. Summary of microbial community in biofilter.

Aeration Type	Biofilter Type (Volume in L)	Biofilter Volume (L)	Culture Species	Shanon	Performance		Reference
					Ammonia Removal (%)	TN Removal (%)	
Coarse bubble diffuser	MBBR	0.7	N/A	1.30	97.1	N/A	[63]
Coarse bubble diffuser	MBBR	N/A	Atlantic salmon post-smolts	2.87	50	N/A	[64]
Coarse bubble diffuser	BAF	10,850	Seawater RAS	5.04	41.7	41.3	[65]
Coarse bubble diffuser	BAF	10,850	Seawater RAS	8.36	85	N/A	[66]
Coarse bubble diffuser	MBBR	5	N/A	2.30	96	N/A	[13]
Coarse bubble diffuser	MBBR	0.4	N/A	2.85	36.04	N/A	[11]
Fine bubble diffuser	MBBR	0.4	N/A	2.88	51	N/A	[11]

Coarse bubble aeration was applied to investigate microbial communities in three MBBRs with different salinity [63]. The system used an air stone aerator to keep oxygen concentration at 7.2–7.6 mg O_2 L^{-1} in three MBBRs with K1 AnoxKaldness media. Each MBBR was fed with basal synthetic wastewater medium (freshwater reactor or FR), a mix of tap water and seawater (brackish reactor or BR), or only seawater (seawater reactor or SR). Nitrifying bacteria were classified in the family of *Nitrosomonadacea* (AOB) with a composition of 6 ± 0.4%, 34.7 ± 1.9%, and 14.7 ± 2.3% for FR, BR, and SR, respectively. NOB consisted of *Nitrospiraceae*, *Nitrospira*, and *Nitrobacter*. The OTU of NOB was 27 ± 1.0%, 8.1 ± 1.3%, and 13.4 ± 1.7% for FR, BR, and SR, respectively. The average ammonia removal efficiency of FR, BR, and SR reached 97.1%, 95%, and 95.4%, respectively. The other nitrification performances of the three reactors might be related to the ratio of NOB and AOB. The higher ammonia removal efficiency was attained from FR because of a higher population of NOB than AOB (NOB:AOB ratio of 4:1). The dominance of NOB over AOB correlated with good nitrification. Nevertheless, FR resulted in a lower Shannon diversity value (1.3 ± 0.08) than SR (1.4 ± 0.56) [63]. Similarly to coarse bubbles, this was also used in another study to investigate microbial communities in the biofilm of the MBBR. Three salinities, 12, 22, and 32 ppt, were used in a semi-commercial RAS for salmon post-smolt treatment. From the study, sampling was done in the MBBR inlet, outlet, and biofilm. Biofilm communities are higher regarding their Bray–Curtis similarities in comparison between time points. Biofilm communities seemed less influenced by salinity than water communities and more stable over time. The nitrifying biofilm was adapted to increase salinity over 2−4 weeks. The highest removal TAN was received when the average Bray–Curtis similarity was 0.75, and the Shannon index was 2.87 [64].

A coarse bubble was used as aeration in an experiment with four different microbial agents (DW, CES-1, BM-S-1, KBM-1) for a marine finfish RAS [65]. This study investigated four bioreactors: packed, aerated, floating, and mesh biofilters. Aeration was provided with a flow rate of 800 L/min in a floating bed biofilter. The ammonia nitrogen degradation rate reached 13.4, 41.0, 33.9, and 41.7% for DW, CES-1, BM-S-1, and KBM-1 over eight days of the experiment. The result showed that on the fourth day of the experiment, the KBM-1 community was more diverse than BMS-1, with 575 OTU and a 5.04 Shannon value. In all treatment groups, *proteobacteria* were dominant in 60–70% at the phylum level. Another abundance in the system consisted of *Bacteroidetes*, *Firmicutes*, *Fusobacteria*, *Actinobacteria*, and others [65]. A similar plant was also used to investigate bacterial diversity in a seawater RAS. The system used a packed bed biofilter, aerated biofilter, floating bed biofilter, mesh biofilter, and maturation tank [66]. Bubble aeration with a flow rate of 1000 L/min was applied as an aeration system in the aerated biofilter. The result of the experiment showed that the packed bed and mesh biofilter contained a high number of nitrifying bacteria. *Phylotypes* related to *Nitrosomonadaceae*, which contained most of the AOB occupied 0.06–3.56% (average 1.22%) of the packed bed biofilter and 1.19–6.41% (average 2.80%) of the mesh biofilter. The *Nitrospirae* group was also detected at values 0.02–10.56% (average 2.55%) and 0.32–9.45% (average 3.57%) in the packed bed and mesh biofilter, respectively. *Nitrospina*, a nitrite-oxidizing group belonging to *Deltaproteobacteria*, was detected up to 9.0%. This result showed that packed bed and mesh biofilter types had affected ammonia removal in the RAS. Whereas in the aerated biofilter, the value of *Nitrosomonadaceae* was about 0.1–1% and *Nitrospirae* about 0.5–1.5%. The highest Shannon Index was 8.36 for the packed bed biofilter [66].

Coarse bubble aeration with a 48 L/min flow rate and three media (microfiber carrier, nanofiber carriers, and AnoxKaldness K3) was used in a lab-scale MBBR [13]. As a result, a relative abundance of nitrifying bacteria, *Nitrosomonas* (AOB) showed more noticeable development between month 1 (0.289%) and month 3 (1.835%) on microfiber, and Nitrospira (NOB) on nanofiber (from 3.11% on month 1 to 11.20% on month 3) and AnoxKaldness (from 5.41% on month 1 to 10.22% on month 3). AOB preferred an environment with a high amount of biomass and a large active area; meanwhile, NOB thrives better in slowly forming biofilm. Based on Shannon index diversity, nanofiber obtained the highest number among all media. However, based on respirometry data, the highest O_2 consumption was obtained by microfiber carriers due to the sufficient internal space of the microfiber to allow high-speed capture of biomass. In addition, O_2 consumption in microfiber was increased due to endogenous respiration. Nevertheless, O_2 consumption of nanofiber media increased with time due to the slow increase in biofilm activity. In comparison, AnoxKaldness K3 had a less effective structure because of a longer bacterial colonization time [13].

Other studies reported that fine bubble and coarse bubble aeration were applied in lab-scale experiments to study the influence of aeration on the biofilter microbial community [11]. Three media were used for growing microbial biofilm in a working volume of 0.4 L in a lab-scale batch reactor. The aeration rate was divided into six stages: fine bubble 0.1 L/min (Startup and Stage A), fine bubble 1.0 L/min (Stage B), coarse bubble 1.0 L/min (Stage C and Stage D), and coarse bubble 5.0 L/min (Stage E). In stage A, all biofilms resulted in relatively high abundances of autotrophic AOB/NOB (*Nitrosomonas* 7.49 ± 1.89% and *Nitrobacter* 7.4 ± 1.89%). By the end of stage B, all biofilms became more heterotrophic (11 ± 2.4% *Comamonas*, 10.3 ± 3.3% *Ferruginibacter*, and 4.8 ± 0.9% *Bradyrhizobium*) and less autotrophic. *Ferruginibacter* may play a role in the produced and exported EPS. A similar pattern occurred in stages C and D (3.3 ± 0.4% Nitrobacter). At the end of stage E, the biomass detachment occurred. In stages A until D, ammonia flux increased with increasing aeration. However, it decreased by approximately 50% when coarse bubble aeration was increased in stage E. The decreased ammonia flux was because of the loss of biofilm mass due to the high shear rates. The highest Shannon diversity (2.88) of all media was achieved in stage B. In contrast, in stage E, the Shannon diversity became lower [11].

4.2.2. Biofilm Thickness

Biofilm thickness can influence biofilter oxygen transfer and enhance the microbial community in the biofilter. The biofilm thickness and its influence on reducing TAN are summarized in Table 4. The effect of DO on biofilm thickness was investigated using a coarse bubble in 27.7 L of a cylindrical bioreactor with semi-suspended spindle media [14]. Three phases of aeration were setups in this study, phase I with 1.5 L/min aeration rate (On and Off 30 min) on days 1–5, phase II with 1.5 L/min aeration rate on days 6–56, and phase III with 2.5 L/min aeration rate on days 57–73. Initially, the biofilm thickness was fixed at 10 μm. In the first and second phases, a thin layer of biofilm of no more than 500 μm was formed on the surface of the media. The DO concentration has increased in phase II from 1.95 mg/L to 4.88 mg/L because of the change in aeration mode. In this phase, the thickness of the biofilm grew to 350 μm after 15 days and more than 1000 μm after 25 days. In the third phase, DO increased from 2.21 mg/L to 3.22 mg/L after 55 days, some biofilm detached from biocarrier, and the remained biofilm exhibited lots of pores and channels. Nevertheless, DO was still increased even when the biofilm was very thin, and the maximum depth of DO that could penetrate the biofilm was no more than 900 μm. Total nitrogen removal constantly increased, with max TN removal around 14.3% on the 55th day. However, it decreased slowly after day 55th. This pattern coincided with the growing trend of biofilm [14].

Table 4. Summary of biofilm thickness in the biofilter.

Aeration Type	Biofilter Type (Volume in L)	Biofilter Volume (L)	Culture Species	Biofilm Thickness (μm)	Performance		Reference
					Ammonia Removal (%)	TN Removal (%)	
Coarse bubble diffuser	MBBR	27.7	N/A	1000	N/A	±14.3	[14]
Coarse bubble diffuser	MABR	1050	N/A	1000	N/A	>70	[67]
Coarse bubble diffuser	Membrane Filter	N/A	N/A	1016 ± 236	38	N/A	[12]
Medium bubble diffuser	MBBR	2000	Litopenaeus vannamei	71 ± 22	64 ± 13	N/A	[17]
Fine bubble diffuser	BAF	1.6	N/A	470 ± 50	50–80	N/A	[57]
Nanobuble generator	Membrane Filter	N/A	N/A	1188 ± 322	75	N/A	[12]
Nanobubble generator and combination with coarse bubble diffuser	BAF	19	N/A	136.1	89.22	34.51	[59]

A coarse bubble was also used in a membrane-aerated biofilm reactor (MABR) to study the impact of oxygen surface loading on biofilm thickness [64]. Approximately 0.04 m^3 gas volume was applied for aeration and supporting biofilm growth through the gas-permeable membrane. The highest TN removal efficiency in oxygen surface loading of 2.72 was >70% at steady state biofilm thickness of 1000 μm. Increased biofilm thickness results in increased total nitrogen removal [64]. Medium bubble aeration with 3.6–18.7 m^3/m^2.h flow rate was utilized in the MBBR with five different media [17]. A medium spherical shape media with an initial biofilm thickness of 71 ± 22 μm was evaluated for ammonia removal at various growth times 30, 22, 17, 46, and 47 days. Approximately 68% of the media's total area (148 m^2/m^3) was covered by biofilm. Even though the medium spherical shape owns the open structure lower protected surface area, the circulation and oxygen distribution were preferable in all areas of the media. Consequently, the highest TAN removal efficiency was given to using medium spherical shape media [17].

In comparison, a fine bubble diffuser was evaluated as an aeration system in BAF with different depths of sampling points (3, 4.5, 7.5, 12.5, 22.5, 32.5, and 40 cm) [57]. In three sampling points, the biofilm thickness during steady-state conditions had reached 0.47 ± 0.05 mm (depth of 4.5 cm), 0.28 ± 0.05 mm (depth of 7.5 cm), and 0.02 ± 0.01 mm (depth of 22.5 cm). The thicker biofilm has resulted in a closer potion to the aeration source. Higher DO and biofilm thickness in sampling points at a depth of 4.5 cm and 7.5 cm had related to higher ammonia removal [57].

In the lab scale test, biofilm thickness grew faster with nanobubble aeration compared with coarse bubble aeration [12]. The maximum biofilm thickness of 1188 ± 322 μm was obtained using nanobubble aeration within 15 days of the experiment. Nevertheless, the maximum biofilm thickness of 1016 ± 236 μm had produced using coarse bubble aeration. The average aerobic layer increased, triggering of gain in the growth rate for the nanobubbles application. Moreover, biofilm growth comes from protein (PN) and polysaccharides (PS) which is part of EPS. The maximum PN/PS ratio using nanobubble was 7.46, whereas using coarse bubble was 4.81 [12].

Similarly, in an experiment using nanobubble aeration, biofilm thickness increased from 50.1 μm to 109.5 μm with a PN/PS ratio of 0.2, then decreased to 78.5 μm at the end. In contrast, a biofilm thickness of 116.7 μm was formed under coarse bubble with a PN/PS ratio of 1 [59]. When equipped with both coarse and nanobubbles for ratios of 1 and 0.2, a thick biofilm was formed (136.1 μm). In contrast, the biofilm thickness decreased to approximately 50 μm using coarse and nanobubbles at the ratios of 0.5 and 02. Biofilm thickness in ratio 0.2 was increased compared to ratio 1 due to the addition of nanobubbles enhancing the biofilm core, which is related to bio-volume [59].

5. Author Perspective

A summary of the biofilter performances with bubble aeration is represented in Table 2, the $k_La_{(T)}t$ value was calculated based on Equation (7) with DO concentration data from the studies presented in Section 4.1. The $k_La_{(T)}$ is useful to determine how efficiently oxygen is transferred from the bubble to the water medium [14]. From Table 2, we can see that high DO does not necessarily result in high ammonia removal efficiency; this is related to the effectiveness of the oxygen getting into the biofilm ($k_La_{(T)}$). Also from Table 2, the highest $k_La_{(T)}$ is achieved using nanobubble aeration. Theoretically, the smaller bubble will produce a higher $k_La_{(T)}$ because of the larger interfacial area of the bubble [15,16]. Additionally, the high value of $k_La_{(T)}$ will produce higher TAN removal efficiency, because the higher value of oxygen is transferred to the cell for nitrification. However, a high ammonia removal efficiency is achieved using coarse bubbles [51] due to the effective contribution of porous media as biofilm media. It showed that the type of media is an important aspect affecting the distribution of oxygen into the biofilm. The $k_La_{(T)}$ is not the only factor determining the distribution of oxygen into the film, but also other factors such as microorganism concentration and morphology, medium composition, biocatalyst properties, and also mechanical design, operating conditions, and aeration rate affect the oxygen supply [68].

The other factor contributing to ammonia removal in biofilters is the microbial community of the biofilm. In Table 3, the Shannon index indicates an increasing number of microbial communities in the biofilm. The higher the Shannon index value, the more abundant the microorganisms in the biofilm [69]. Nevertheless, as shown in Table 3, Shannon diversity is not directly correlated with ammonia removal efficiency since this value only represents broad species in the community. Thus, information on the species and numbers of nitrifying bacteria is necessary for their efficiency in ammonia removal.

In addition, the Shannon index was not directly correlated with ammonia removal efficiency, even when conducted in a different type of bubble aeration with a similar reactor [11]. This enhanced efficiency is possible due to the higher number of nitrifying bacteria produced using fine bubble aeration. Additionally, a lower Shannon index can result in high ammonia removal efficiency since it could indicate a high amount of nitrifying bacteria, compared to other studies with a specific ratio of AOB and NOB [63]. In addition, biofilm diversity is affected by some factors such as temperature, pH, ionic strength, and hydrodynamic conditions [70]. Another factor is fluid hydrodynamics that impact oxygen and nutrient transport rates and biofilm cycles [71]. The attachment and detachment of biofilms are stimulated by the hydrodynamic conditions, i.e., air flow rate. A high flow rate can support oxygen and nutrient transport to the surface, contribute to the growth of microorganisms in the microbial layer, and produce the exopolymers that make the attachment stronger. However, the increasing flow rate will raise the shear rates and cause biofilm detachment, yielding a decrease in the amount of biomass attached to the solid surface or media [72]. As long as a higher flow rate is provided, high ammonia removal efficiency can be achieved using coarse bubble aeration [13], but it is more energy-consuming.

The microbial community plays an important role the biofilm growth and thickness. Conversely, biofilm thickness may also affect microbial activity and diversity due to diffusion limitation and substrate penetration in the biofilm [73,74]. The increase in biofilm thickness may lead to a more heterogeneous and biodiverse biofilm because of the stratification of metabolic processes and higher concentration gradient [75]. From Table 4, we can see higher ammonia removal was achieved with a thickness of 136.1 μm by using a combination of nanobubble and coarse bubble aeration [59]. Thus, higher ammonia degradation may be achieved with an acceptable thickness of biofilm [76]. Excessive biofilm thickness will cause oxygen diffusion limitation, accumulate toxic metabolic by-products, and result in less active biomass in the deeper biofilm due to low oxygen penetration. Therefore, the microbial diversity decreases in less active biomass, consequently increasing the detachment from the deeper layer of the biofilm [77].

6. Conclusions

The recent study on bubble aeration system in biofilters for treatment of waste water in the RAS was reviewed. Nano, fine, medium, and coarse bubble aeration are used to oxygenate the biofilter in different scenarios. Oxygen transfer affects ammonia nitrogen oxidation and affects biofilter performance to remove contaminants from wastewater. Higher removal efficiency tends to result from finer bubble aeration. Approximately >80% TAN removal efficiency was achieved by the biofilter with an aeration system consisting of either fine, micro- and nanobubbles or a combination thereof. Furthermore, a higher volumetric mass transfer coefficient ($k_L a_{(T)}$) resulted in higher oxygen transfer from bubbles to liquid media due to finer bubble aeration.

Based on the correlation between the microbial community and biofilm thickness due to different types of bubble aeration, it is concluded that the aeration system equipped with finer bubbles improves the ratio of nitrifying bacteria, and consequently, biofilm growth and thickness. Increasing biofilm thickness increases ammonia nitrogen removal in the system. However, it must be noted that the ammonia nitrogen removal decreases when an excessive thickness is formed, thereby limiting the diffusion of oxygen penetration into the deeper layer due to less active biomass.

Author Contributions: Conceptualization, I.M.J., F.F. and P.A.S.; methodology, P.A.S. and I.M.J.; review investigation, P.A.S.; data curation, P.A.S.; writing—original draft preparation, P.A.S.; writing—review and editing, I.M.J., F.F., C.P. and W.H.; supervision, I.M.J.; funding acquisition, I.M.J. All authors have read and agreed to the published version of the manuscript.

Funding: This research was funded by Academic Leadership Grant of Universitas Padjadjaran, grant number 2203/UN6.3.1/PT.00/2022 and The APC was funded by Universitas Padjadjaran.

Data Availability Statement: Not applicable.

Acknowledgments: The author acknowledges the scholarship given to Putu Ayustin Suriasni from Beasiswa Unggulan Pascasarjana Padjadjaran (BUPP) and this research also supported from Academic Leadership Grant with contract number no 2203/UN6.3.1/PT.00/2022.

Conflicts of Interest: The authors declare no conflict of interest.

References

1. Deviller, G.; Aliaume, C.; Nava, M.A.F. High-rate algal pond treatment for water reuse in an integrated marine fish recirculating system: Effect on water quality and sea bass growth. *Aquaculture* **2004**, *235*, 331–344. [CrossRef]
2. Sánchez, O.I.A.; Matsumoto, T. Hydrodynamic characterization and performance evaluation of an aerobic three phase airlift fluidized bed reactor in a recirculation aquaculture system for Nile Tilapia production. *Aquac. Eng.* **2012**, *47*, 16–26. [CrossRef]
3. Cabasso, I. Membrane. In *Encyclopedia of Polymer Science and Engineering*; Wiley: New York, NY, USA, 1987; Volume 9, pp. 509–579.
4. Samer, M. Biological and Chemical Wastewater Treatment. In *Wastewater Treatment Engineering*; Intech: London, UK, 2015.
5. Holan, A.B.; Wold, P.-A.; Øie, G.; Leiknes, T.O. Integrated Membrane Bioreactor for Water Quality Control in Marine Recirculating Aquaculture Systems. *Sep. Sci. Technol.* **2013**, *48*, 1758–1767. [CrossRef]
6. Tasong, A.C.; Abao, R.P. Design and Development of an IoT Application with Visual Analytics for Water Consumption Monitoring. *Procedia Comput. Sci.* **2019**, *157*, 205–213. [CrossRef]
7. Malone, R.F.; Pfeiffer, T.J. Rating fixed film nitrifying biofilters used in recirculating aquaculture systems. *Aquac. Eng.* **2006**, *34*, 389–402. [CrossRef]
8. Ghangrekar, M.; Behera, M. Suspended Growth Treatment Processes. In *Comprehensive Water Quality and Purification*; Elsevier Inc.: Amsterdam, The Netherlands, 2014; pp. 75–89.
9. Gavrilescu, M.; Macoveanu, M. Attached-growth process engineering in wastewater treatment. *Bioprocess Biosyst. Eng.* **2000**, *23*, 95–106. [CrossRef]
10. Wang, X.; Xia, S.; Chen, L.; Zhao, J.; Renault, N.; Chovelon, J. Nutrients removal from municipal wastewater by chemical precipitation in a moving bed biofilm reactor. *Process. Biochem.* **2006**, *41*, 824–828. [CrossRef]
11. Roveto, P.M.; Schuler, A.J. Performance and diversity responses of nitrifying biofilms developed on varied materials and topographies to stepwise increases of aeration. *Bioresour. Technol.* **2019**, *281*, 429–439. [CrossRef]
12. Xiao, W.; Xu, G. Mass transfer of nanobubble aeration and its effect on biofilm growth: Microbial activity and structural properties. *Sci. Total Environ.* **2019**, *703*, 134976. [CrossRef]
13. Havlíček, K.; Nechanická, M.; Lederer, T.; Sirková, B.K. Analysis of nitrifying bacteria growth on two new types of biomass carrier using respirometry and molecular genetic methods. *Ecotoxicol. Environ. Saf.* **2021**, *225*, 112795. [CrossRef]
14. Tang, B.; Song, H.; Bin, L.; Huang, S.; Zhang, W.; Fu, F.; Zhao, Y.; Chen, Q. Determination of the profile of DO and its mass transferring coefficient in a biofilm reactor packed with semi-suspended bio-carriers. *Bioresour. Technol.* **2017**, *241*, 54–62. [CrossRef] [PubMed]
15. Liu, C.; Tang, Y. Application research of micro and nano bubbles in water pollution control. *E3S Web Conf.* **2019**, *136*, 06028. [CrossRef]
16. Ruen-ngam, D.; Wongsuchoto, P.; Limpanuphap, A.; Charinpanitku, T.; Pavasant, P. Influence of salinity on bubble size distribution and gas–liquidmass transfer in airlift contactors. *Chem. Eng. J.* **2008**, *141*, 222–232. [CrossRef]
17. Dias, J.; Bellingham, M.; Hassan, J.; Barrett, M.; Stephenson, T.; Soares, A. Influence of carrier media physical properties on start-up of moving attached growth systems. *Bioresour. Technol.* **2018**, *266*, 463–471. [CrossRef]
18. Kawan, J.A.; Suja, F.; Pramanik, S.K.; Yusof, A.; Rahman, R.A.; Hasan, H.A. Effect of Hydraulic Retention Time on the Performance of a Compact Moving Bed Biofilm Reactor for Effluent Polishing of Treated Sewage. *Water* **2022**, *14*, 81. [CrossRef]
19. Khuntia, S.; Majumder, S.K.; Ghosh, P. Microbubble-aided water and wastewater purification: A review. *Rev. Chem. Eng.* **2012**, *28*, 191–221. [CrossRef]
20. Sakr, M.; Mohamed, M.M.; Maraqa, M.A.; Hamouda, M.A.; Hassan, A.A.; Ali, J.; Jung, J. A critical review of the recent developments in micro–nano bubbles applications for domestic and industrial wastewater treatment. *Alex. Eng. J.* **2021**, *61*, 6591–6612. [CrossRef]
21. Shitu, A.; Liu, G.; Muhammad, A.I.; Zhang, Y.; Tadda, M.A.; Qi, W.; Liu, D.; Ye, Z.; Songming, Z. Recent advances in application of moving bed bioreactors for wastewater treatment from recirculating aquaculture systems: A review. *Aquac. Fish.* **2022**, *7*, 244–258. [CrossRef]
22. Ferreira, C.; Gonçalves, G. A Systematic Review on Life Extension Strategies in Industry. In Proceedings of the IEEE International Conference on Emerging Technologies and Factory Automation, ETFA, Vasteras, Sweden, 7–10 September 2021; Volume 10, pp. 1–17.
23. Hamza, R.A.; Iorhemen, O.T.; Tay, J.H. Advances in biological systems for the treatment of high-strength wastewater. *J. Water Process. Eng.* **2016**, *10*, 128–142. [CrossRef]
24. Reyes, A.A.D.; Lawson, T.B. Combination of a bead filter and rotating biological contactor in a recirculating fish culture system. *Aquac. Eng.* **1996**, *15*, 27–39. [CrossRef]
25. Kordkandi, S.A.; Khoshfetrat, A.B.; Faramarzi, A. Performance modelling of a partially-aerated submerged fixed-film bioreactor: Mechanistic analysis versus semi data-driven method. *J. Ind. Eng. Chem.* **2018**, *61*, 398–406. [CrossRef]

26. Jakobsen, H.A. Packed Bed Reactors. In *Chemical Reactor Modeling: Multiphase Reactive Flows*; Springer: Berlin/Heidelberg, Germany, 2008; pp. 953–984.
27. Liu, B.Y.M.; Pfeffer, J.T. *Modeling for Anaerobic Fixed-Bed Biofilm Reactors*; University of Illinois Urbana-Champaign: Champaign, IL, USA, 1989.
28. Malone, R.F. Floating Media Hourglass Biofilter. U.S. Patent 5,232,586, 3 August 1993.
29. Malone, R.F.; E Beecher, L. Use of floating bead filters to recondition recirculating waters in warmwater aquaculture production systems. *Aquac. Eng.* **2000**, *22*, 57–73. [CrossRef]
30. Wimberly, D.M. Development and Evaluation of a Low-Density Media Biofiltration Unit for Use in Recirculating Fish Culture Systems. Master's Thesis, Louisiana State University, Baton Rouge, LA, USA, 1990.
31. Ochoa, F.G.; Gomez, E. Bioreactor scale-up and oxygen transfer rate in microbial processes: An overview. *Biotechnol. Adv.* **2009**, *27*, 153–176. [CrossRef] [PubMed]
32. Xu, X.; Wei, W.; Liu, F.; Wei, W.; Liu, Z. Experimental study on aeration efficiency in a pilot-scale decelerated oxidation ditch equipped with fine bubble diffusers and impellers. *Can. J. Chem. Eng.* **2020**, *991*, 1410–1420. [CrossRef]
33. Kumar, A.; Moulick, S.; Mal, B. Design characteristics of pooled circular stepped cascade aeration system. *Aquac. Eng.* **2010**, *56*, 51–58. [CrossRef]
34. Pambudiarto, B.A.; Mindaryani, A.; Deendarlianto, D.; Budhijanto, W. Evaluation of the Effect of Operating Parameters on the Performance of Orifice/Porous Pipe Type Micro-bubble Generator. *J. Eng. Technol. Sci.* **2020**, *52*, 196–207. [CrossRef]
35. Takahashi, M.; Chiba, K.; Li, P. Free-Radical Generation from Collapsing Microbubbles in the Absence of a Dynamic Stimulus. *J. Phys. Chem. B* **2007**, *111*, 1343–1347. [CrossRef]
36. Ebina, K.; Shi, K.; Hirao, M. Oxygen and Air Nanobubble Water Solution Promote the Growth of Plants, Fishes, and Micer. *PLoS ONE* **2013**, *8*, e65339. [CrossRef] [PubMed]
37. ISO. *Fine Bubble Technology—General Principles for Usage and Measurement of Fine Bubbles—Part 1: Terminology*; ISO: Geneva, Switzerland, 2017.
38. Zăbavă, B.Ș.; Ungureanu, N.; Voicu, G.; Dincă, M. Types of Aerators Used in Wastewater Treatment Plants. In Proceedings of the 5th International Conference of Thermal Equipment, Renewable Energy and Rural Development, Golden Sands, Bulgaria, 2–4 June 2016.
39. Farmer, R.; Arndt, R. *Development of an Efficient Aeration System for Aquaculture*; University of Minnesota: Minneapolis, MN, USA, 1995.
40. Flemming, H.C.; Neu, T.R. The EPS matrix: The "house of biofilm cells". *J. Bacteriol.* **2007**, *189*, 7945–7947. [CrossRef]
41. Karaguler, T.; Kahraman, H.; Tuter, M. Analyzing effects of ELF electromagnetic fields on removing bacterial biofilm. *Biocybern. Biomed. Eng.* **2017**, *37*, 336–340. [CrossRef]
42. Stoodley, P.; Sauer, K.; Davies, D.G.; Costerton, J.W. Biofilms as Complex Differentiated Communities. *Annu. Rev. Microbiol.* **2002**, *56*, 187–209. [CrossRef] [PubMed]
43. Goswami, S.; Sarkar, S.; Mazumder, D. A new approach for development of kinetics of wastewater treatment in aerobic biofilm reactor. *Appl. Water Sci.* **2016**, *7*, 2187–2193. [CrossRef]
44. Bouaifi, M.; Hebrard, G.; Bastoul, D.; Roustan, M. A comparative study of gas hold-up, bubble size, interfacial area and mass transfer coefficients in stirred gas-liquid reactors and bubble columns. *Chem. Eng. Process. Process Intensif.* **2001**, *40*, 97–111. [CrossRef]
45. Larsson, V. *Energy Savings with a New Aeration and Control System in a Mid-Size Swedish Wastewater Treatment Plant*; Uppsala University: Uppsala, Sweden, 2011.
46. Akita, K.; Yoshida, F. Bubble Size, Interfacial Area, and Liquid-Phase Mass Transfer Coefficient in Bubble Columns. *Ind. Eng. Chem. Process. Des. Dev.* **1974**, *13*, 84–91. [CrossRef]
47. Yao, G.-J.; Ren, J.-Q.; Zhou, F.; Liu, Y.-D.; Li, W. Micro-nano aeration is a promising alternative for achieving high-rate partial nitrification. *Sci. Total Environ.* **2021**, *795*, 148899. [CrossRef] [PubMed]
48. Woodard & Curran, Inc. Chapter 7–Methods for Treating Wastewaters from Industry. In *Industrial Waste Treatment Handbook*, 2nd ed.; Butterworth-Heinemann: Oxford, UK, 2006; pp. 149–334.
49. Liu, K.; Phillips, J.R.; Sun, X.; Mohammad, S.; Huhnke, R.L.; Atiyeh, H.K. Investigation and Modeling of Gas-Liquid Mass Transfer in a Sparged and Non-Sparged Continuous Stirred Tank Reactor with Potential Application in Syngas Fermentation. *Fermentation* **2019**, *5*, 75. [CrossRef]
50. Pfeiffer, T.J.; Wills, P.S. Evaluation of three types of structured floating plastic media in moving bed biofilters for total ammonia nitrogen removal in a low salinity hatchery recirculating aquaculture system. *Aquac. Eng.* **2011**, *45*, 51–59. [CrossRef]
51. Shitu, A.; Zhu, S.; Qi, W.; Tadda, M.A.; Liu, D.; Ye, Z. Performance of novel sponge biocarrier in MBBR treating recirculating aquaculture systems wastewater: Microbial community and kinetic study. *J. Environ. Manag.* **2020**, *275*, 111264. [CrossRef] [PubMed]
52. Summerfelt, S.T.; Zühlke, A.; Kolarevic, J.; Reiten, B.K.M.; Selset, R.; Gutierrez, X.; Terjesen, B.F. Effects of alkalinity on ammonia removal, carbon dioxide stripping, and system pH in semi-commercial scale water recirculating aquaculture systems operated with moving bed bioreactors. *Aquac. Eng.* **2015**, *65*, 46–54. [CrossRef]
53. Pfeiffer, T.J.; Riche, M.A. Evaluation of a Low-head Recirculating Aquaculture System Used for Rearing Florida Pompano to Market Size. *J. World Aquac. Soc.* **2011**, *42*, 198–208. [CrossRef]

54. Torno, J.; Einwächter, V.; Schroeder, J.P.; Schulz, C. Nitrate has a low impact on performance parameters and health status of ongrowing. *Aquaculture* **2018**, *489*, 21–27. [CrossRef]
55. Blancheton, J.P. Developments in recirculation systems for Mediterranean fish species. *Aquac. Eng.* **2000**, *22*, 17–31. [CrossRef]
56. Dosdat, A.; Ruyet, J.P.-L.; Covès, D.; Dutto, G.; Gasset, E.; Le Roux, A.; Lemarié, G. Effect of chronic exposure to ammonia on growth, food utilisation and metabolism of the European sea bass (Dicentrarchus labrax). *Aquat. Living Resour.* **2003**, *16*, 509–520. [CrossRef]
57. Albuquerque, A.; Makinia, J.; Pagilla, K. Impact of aeration conditions on the removal of low concentrations of nitrogen in a tertiary partially aerated biological filter. *Ecol. Eng.* **2012**, *44*, 44–52. [CrossRef]
58. Luo, G.; Xu, G.; Gao, J.; Tan, H. Effect of dissolved oxygen on nitrate removal using polycaprolactone as an organic carbon source and biofilm carrier in fixed-film denitrifying reactors. *J. Environ. Sci.* **2016**, *43*, 147–152. [CrossRef] [PubMed]
59. Xiao, W.; Xu, G.; Li, G. Role of shear stress in biological aerated filter with nanobubble aeration: Performance, biofilm structure and microbial community. *Bioresour. Technol.* **2021**, 325. [CrossRef] [PubMed]
60. Shannon, C. A mathematical theory of communication. *Bell Syst. Tech. J.* **1948**, *27*, 379–423. [CrossRef]
61. Konopiński, M.K. Shannon diversity index: A call to replace the original Shannon's formula with unbiased estimator in the population genetics studies. *PeerJ* **2000**, *8*, e9391. [CrossRef]
62. Spohn, S.N.; Young, V.B. Gastrointestinal Microbial Ecology with Perspectives on Health and Disease. In *Physiology of the Gastrointestinal Tract*; Academic Press: Irvine, CA, USA, 2018; pp. 727–753.
63. Gonzalez-Silva, B.M.; Jonassen, K.R.; Bakke, I.; Østgaard, K.; Vadstein, O. Nitrification at different salinities: Biofilm community composition and physiological plasticity. *Water Res.* **2016**, *95*, 48–58. [CrossRef]
64. Bakke, I.; Åm, A.L.; Kolarevic, J.; Ytrestøyl, T.; Vadstein, O.; Attramadal, K.J.K.; Terjesen, B.F. Microbial community dynamics in semi-commercial RAS for production of Atlantic salmon post-smolts at different salinities. *Aquac. Eng.* **2016**, *78*, 42–49. [CrossRef]
65. Lee, J.; Kim, I.-S.; Emmanuel, A.; Koh, S.-C. Microbial valorization of solid wastes from a recirculating aquaculture system and the relevant microbial functions. *Aquac. Eng.* **2019**, *87*, 102016. [CrossRef]
66. Lee, D.-E.; Lee, J.; Kim, Y.-M.; Myeong, J.-I.; Kim, K.-H. Uncultured bacterial diversity in a seawater recirculating aquaculture system revealed by 16S rRNA gene amplicon sequencing. *J. Microbiol.* **2016**, *54*, 296–304. [CrossRef]
67. Ni, B.-J.; Yuan, Z. A model-based assessment of nitric oxide and nitrous oxide production in membrane-aerated autotrophic nitrogen removal biofilm systems. *J. Membr. Sci.* **2013**, *428*, 163–171. [CrossRef]
68. Michelin, M.; Mota, A.M.O.; Polizeli, M.L.T.M.; Silva, D.P.; Vicente, A.A.; Teixeira, J.A. Influence of volumetric oxygen transfer coefficient (kLa) on xylanases batch production by Aspergillus niger van Tieghem in stirred tank and internal-loop airlift bioreactors. *Biochem. Eng. J.* **2013**, *80*, 19–26. [CrossRef]
69. Fang, H.; Chen, Y.; Huang, L.; He, G. Analysis of biofilm bacterial communities under different shear stresses using size-fractionated sediment. *Sci. Rep.* **2017**, *7*, 1299. [CrossRef] [PubMed]
70. Agarwal, R.; Singh, S.; Bhilegaonkar, K.; Singh, V. Optimization of microtiter plate assay for the testing of biofilm formation ability in different Salmonella serotypes. *Int. Food Res. J.* **2011**, *18*, 1493–1498.
71. Gomes, I.B.; Meireles, A.; Gonçalves, A.L.; Goeres, D.M.; Sjollema, J.; Simões, L.C.; Simões, M. Standardized reactors for the study of medical biofilms: A review of the principles and latest modifications. *Crit. Rev. Biotechnol.* **2017**, *38*, 657–670. [CrossRef]
72. Gomes, L.C.; Moreira, J.M.; Teodósio, J.S.; Araújo, J.D.P.; Miranda, J.M. 96-well microtiter plates for biofouling simulation in biomedical settings. *Biofouling* **2014**, *30*, 535–546. [CrossRef]
73. Gapes, D.; Keller, J. Impact of oxygen mass transfer on nitrification reactions in suspended carrier reactor biofilms. *Process. Biochem.* **2009**, *44*, 43–53. [CrossRef]
74. Piculell, M.; Welander, P.; Jönsson, K.; Welander, T. Evaluating the effect of biofilm thickness on nitrification in moving bed biofilm reactors. *Environ. Technol.* **2015**, *37*, 732–743. [CrossRef]
75. Torresi, E.; Fowler, S.; Polesel, F.; Bester, K.; Andersen, H.; Smets, B.; Plósz, B.; Christensson, M. Biofilm thickness influences biodiversity in nitrifying MBBRs 1—Implications on micropollutant removal. *Environ. Sci. Technol.* **2016**, *50*, 9279–9288. [CrossRef] [PubMed]
76. Liu, Y.; Capdeville, B. Dynamics of nitrifying biofilm growth in biological nitrogen removal al process. *Water Sci. Technol.* **1994**, *29*, 377–380. [CrossRef]
77. Kumar, C.; Anand, S. Significance of microbial biofilms in food industry: A review. *Int. J. Food Microbiol.* **1998**, *42*, 9–27. [CrossRef] [PubMed]

water

MDPI

Article

Performance Analysis of an Eductor-Based Membrane Distillation Unit

Ravi Koirala [1], Quoc Linh Ve [2], Eliza Rupakheti [1], Kiao Inthavong [1] and Abhijit Date [1,*]

1 School of Engineering, RMIT University, Melbourne, VIC 3083, Australia
2 Faculty of Engineering and Food Technology, University of Agriculture and Forestry, Hue University, Thua Thien Hue 530000, Vietnam
* Correspondence: abhijit.date@rmit.edu.au

Abstract: Thermal desalination technologies involve two primary processes: vapor generation from saline water, and effective recovery of the resulting condensate. Membrane distillation (MD) systems are among the emerging thermal desalination technologies which use a hydrophobic membrane to recover condensate through either direct or indirect contact (with the cooling fluid) condensation. The specific process technology (for thermal energy transfer and condensate recovery) depends on the type of MD. Direct contact membrane distillation (DCMD) and vacuum membrane distillation (VMD) are two significant MD processes, with DCMD having the advantage of direct condensation and simple design, while VMD systems have high yield through sub-atmospheric vapor generation. This work focuses on developing an eductor-based MD process incorporating the strengths of both DCMD and VMD. It is an experimental study with a water jet eductor replacing the vacuum pump and condenser in a typical VMD system for active permeate vapor transfer and condensation. Unlike the exiting VMD systems, the proposed design recovers condensate by direct contact condensation. The sub-cooled water acts as a motive flow which entrains the secondary vapor into the stream, causing mass transfer via condensation at the interface. The modified VMD was found to have achieved better flux compared to the conventional VMD system. The performance of the eductor, sensitivity to parameters, and the practicality of the technology have been analyzed.

Keywords: eductor; membrane distillation; compact design

1. Introduction

Although about 71% of the earth is covered by water, only 0.5% of it is in a usable freshwater form in rivers, lakes, and aquifers [1,2]. This fact of limited availability and increasing scarcity due to rapid urbanization is undeniable. Limited availability, overconsumption, and deterioration of freshwater sources are additional factors impacting this intensified shortage [3,4]. Despite the counterarguments, water recycling and freshwater production from sea water come forward as immediate remedies [5,6]. A variety of desalination technologies have been developed as effective solutions for brackish and saline water treatment. Broadly, they are classified based on the methods used, such as evaporation and condensation, filtration, and crystallization, which are actuated either by thermal, mechanical, electrical, or chemical energies from either renewable or non-renewable energy sources [7,8]. Thermal desalination (evaporation and condensation) has been one of the most preferred methods for freshwater production, with recent focus shifting towards membrane distillation (MD) (a membrane-based thermal desalination process). In comparison to pure thermal distillation, MD has higher energy efficiency and yield. The process involves separation of vapor from saline feed with porous hydrophobic membranes. Like many other desalination technologies, MD requires pre-treatment (coagulation, flocculation, clarification, filtration, etc.) [9].

The commercial and academic research on MD technologies involves membrane fabrication, module configuration, thermal energy sources, and spacer configuration. The primary focus is on strength enhancement, increasing yield, optimizing energy consumption, and minimizing gap losses. There are different types of MD configurations that are named based on how the feed, membrane, and permeate interact. This paper focuses on the design optimization of the vacuum membrane distillation (VMD) system. The conventional VMD system uses a mechanical vacuum pump to maintain the vacuum on the permeate side while the permeate vapor is condensed around an indirect contact condenser.

This work uses an eductor for active vapor transfer and condensation in a VMD system. This work experimentally studies the possibility of using eductors to replace vacuum pumps and condensers in VMD units. Unlike existing VMD, this process performs direct contact condensation i.e., the vapour condenses in direct contact with the cooling water. This work has been performed with an aim to reduce the footprint of MD for developing a simple and compact portable desalination unit.

2. Technology and Mechanism Description

2.1. Eductor

Eductors are static mechanical devices that utilize pressure energy in a primary fluid (motive fluid) for suction and compression of a low-pressure secondary fluid. Both the primary and secondary fluid can be any combination of liquid, condensable gas, or non-condensable gas. Some other applications also use particles at the secondary fluid inlet. In an MD application, an eductor replaces the large vacuum pump and condensers for combined active vapor transfer and condensation. Water is used as the motive fluid for suction of saturated vapor from the permeate side of the desalination module.

Eductors consist of a primary nozzle, secondary fluid inlet pipe, suction chamber, mixing chamber, throat, and diffuser. Figure 1a shows the functional components of an eductor. The pressure energy in the motive fluid is converted to kinetic energy through the nozzle such that a low-pressure zone is created in the suction chamber which entrains the secondary fluid, explained within Figure 1b. The momentum transfer between the high-velocity motive fluid jet and the surrounding vapor carries forward the entrained fluid. The collapsing of vapor bubbles during direct contact condensation (DCC) at the interphase interface increases the secondary fluid-carrying capacity of the motive fluid. Condensation progresses along the axial direction of the eductor, and it has been found that the entrained vapor becomes completely condensed within the throat. The diffuser of an eductor is analogous to a compressor which discharges the low-pressure condensed secondary fluid to an elevated pressure. Both the mechanical and thermal activity within the system contribute to the performance of the eductor. The general performance of an eductor is measured by its entrainment ratio (Equation (1)) and pressure ratio (Equation (2)).

$$E_r = \frac{\dot{m}_s}{\dot{m}_p} \quad (1)$$

$$P_r = \frac{P_b - P_s}{P_i - P_s} \quad (2)$$

where E_r is the entrainment ratio, $\dot{m}_s$ is the suction mass flow rate, $\dot{m}_p$ is the primary (motive fluid) mass flow rate, and P_r is the pressure ratio. P_b is the back pressure that can be controlled by opening or closing a valve at the outlet of the eductor (beyond the pressure rising diffuser). P_s is the suction pressure that is experienced by the secondary fluid, and it is measured at an inlet to the suction port. P_i is the inlet pressure, i.e., the pressure at which the motive fluid (primary fluid) is supplied to the eductor.

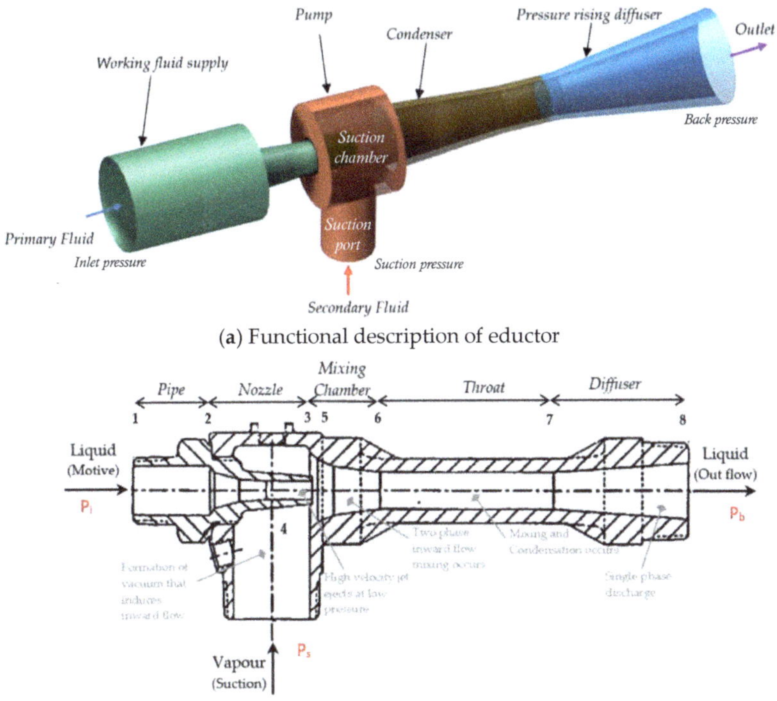

(**a**) Functional description of eductor

(**b**) Construction of eductor with mechanism

Figure 1. Components of an eductor and their function.

The entrainment of vapor into the eductor can be defined by applying an energy balance to the adiabatic secondary fluid pipe, giving the Equation (3):

$$\sqrt{2(h_{go} - h_{g2})} - V_{g2} = 0 \quad (3)$$

where h_{go} is the specific enthalpy of vapor at the nozzle inlet, h_{g2} is the specific enthalpy of the primary nozzle outlet, and V_{g2} is the outlet velocity of the nozzle.

For condensation based on Clausius–Clapeyron and Bernoulli's theorem, the DCC heat transfer coefficient of vapor into a sub-cooled water jet (H_c) is defined by [10] as:

$$H_c = \frac{1.414\, h_{fg}^{3/2}}{\sqrt{v_g(v_g - v_l)T_g(T_g - T_l)}} \quad (4)$$

where h_{fg} is the latent heat, v_g is the specific volume of vapor, v_l is the specific volume of liquid, T_g is the temperature of vapor, and T_l is the temperature of liquid.

The energy balance at the interface of the annular flow defines the condensation rate (dM_g/dt) within the eductor as follows [11]:

$$\frac{dM_g}{dt} = \frac{H_c(T_g - T_l)\beta}{h_{fg}} \quad (5)$$

Practically, there are two geometric points in an eductor at which chocking occurs. This chocking phenomenon helps in understanding the fundamental flow mechanism within the single and multiphase flow paths. It occurs when a flow passes a restriction towards a lower pressure zone at a constant mass flow rate. The first point is at the exit of the primary nozzle while the second one is in the throat. In some cases, the secondary chocking of a two-phase flow may not be clearly observed due to the influence of outlet pressure. Koirala et al. computationally explored the entraining and condensing capacity of an eductor (Figure 2). The commercial eductor used for the work was able to achieve an entrainment ratio of 0.004, and complete condensation was achieved within the total length of 14 cm. This supports the early assumption regarding the usability of an eductor for combined vacuum generation and condensation [12]. The figure further describes the single species two-flow mechanism within an eductor. The region of interphase interaction is the zone of mass transfer, and the rate of mass transfer is higher after the throat. Similarly, complete condensation of vapor occurs within an eductor system.

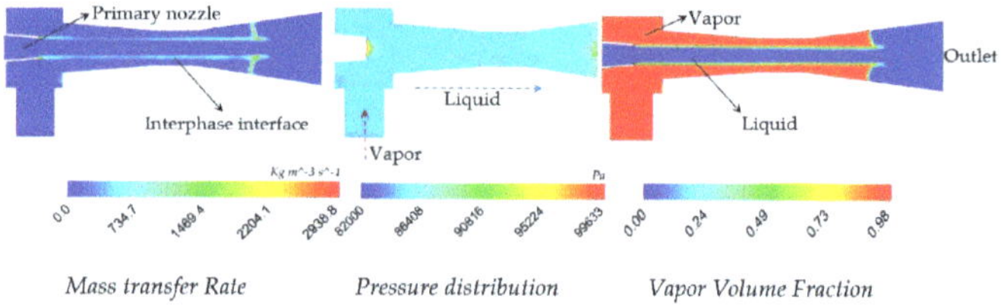

Figure 2. Two phase flow inside eductor.

2.2. Mechanism of Membrane Distillation (MD)

The condensation of permeate vapor in the primary or motive fluid (low TDS liquid water) leads to a rise in the temperature of the motive fluid which adversely effects the vacuum. It should be noted that the vacuum pressure generation capacity of an eductor is a function of the motive fluid velocity and temperature. Therefore, it is important to maintain the motive fluid flow rate above a minimum value to minimize the temperature rise of the motive fluid. In the proposed system vapor transport from the feed to the permeate side is driven by the feed temperature and vacuum pressure, like in a typical VMD system. On the other hand, the permeate vapor is condensed directly in the stream of fresh liquid water like a DCMD system and hence the proposed system could be considered as analogous to a combination of VMD and DCMD.

The thermal mechanism of freshwater production has been discussed with the aid of existing mathematical models and schematic diagram shown in Figure 3. In this section, heat and mass transfer in both direct contact membrane distillation (DCMD) and vacuum membrane distillation (VMD) are discussed. A flat plate and frame membrane module, with a flat sheet membrane, has been considered for the description of the mechanism. In both configurations (DCMD and VMD), the hot saline feedwater is pumped into the feed side where, at the feed channel, the solution directly contacts the membrane surface. The trans-membrane vapor separation mechanism and zone of condensation are two major factors differentiating the two processes. For DCMD, vapor is captured and condensed in the permeate side, where trans-membrane boundary travel is due to the vapor pressure difference resulting from the temperature difference between the two sides of the membrane. The temperature difference at the two walls is due to the boundary layer temperature of feed and permeate flows. The condensation of vapor occurs in the cold freshwater flow stream, which is pumped either counter-currently or co-currently into the permeate side of the module. For VMD, vapor is carried away from the module by the action of a vacuum pump and is condensed separately in a condenser. The transmembrane vapor travel is still due to the pressure difference, but this time it is due to vapor pressure at the feed side and the mechanically induced sub-atmospheric pressure on the permeate side. The condensation of the separated vapor occurs outside the membrane module in a specialized condenser.

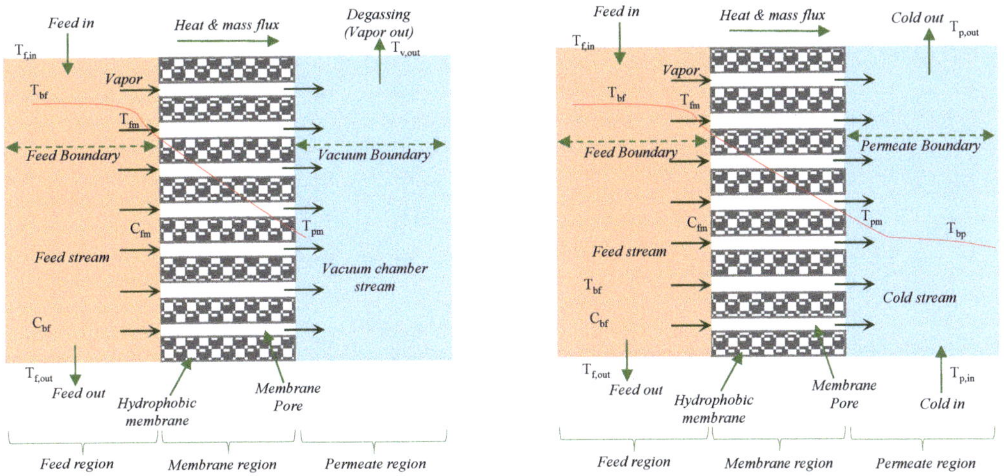

(**a**) Heat and mass transfer in VMD (**b**) Heat and mass transfer in DCMD

Figure 3. Heat and mass transfer in VMD and DCMD.

The DCMD process is widely used because of its simple configuration (e.g., no need for an external condenser system) and its relatively high freshwater production rate [13]. However, the low thermal energy efficiency, due to the high heat conduction losses across the membrane, is a downside for DCMD [14]. For VMD, due to the lower applied permeate pressure compared to the feed pressure of volatile vapor, which is separated from the feed solution, the vapor molecules are transferred through membrane pores from the feed to the permeate side [15]. In the case of VMD, the heat conduction losses and the heat transfer resistance on the permeate side are negligible. Therefore, higher thermal efficiency and freshwater production are expected in comparison to a DCMD module, especially at lower temperature [16]. The main drawbacks of VMD are its complicated structure and higher risk of pore wetting due to the greater pressure difference [17].

2.2.1. Heat Transfer

In membrane distillation, both heat and mass transfer occur concurrently. To acquire accurate and reliable predicted results for membrane surface temperatures and permeate flux, the dominant mechanisms of heat and mass transfer need to be determined. In general, to predict the mass transfer, the heat transfer estimation is a prerequisite [17–19]. In DCMD and VMD, convective heat transfer from the bulk feed solution through the thermal boundary layer to membrane surface on the feed side is demonstrated in Equation (6):

$$\dot{Q}_f = \mathrm{h_f}\,\mathrm{A}\left(\mathrm{T_f} - \mathrm{T_{m,f}}\right) \quad (6)$$

In this equation, $\mathrm{h_f}$ ($\mathrm{W{\cdot}m^{-2}{\cdot}K^{-1}}$), $\mathrm{T_f}$ (K), and $\mathrm{T_{mf}}$ (K) are the convective heat transfer coefficient on the feed side, the bulk feed temperature, and the membrane surface temperature on the feed side, respectively. The bulk fluid temperature on the feed side is estimated based on the feed inlet temperature ($\mathrm{T_{f,in}}$) and the feed outlet temperature ($\mathrm{T_{f,out}}$) as in Equation (7):

$$\mathrm{T_f} = \frac{\mathrm{T_{f,in}} + \mathrm{T_{f,out}}}{2} \quad (7)$$

The total heat transfer through the membrane consists of sensible heat (heat transferred by conduction) and latent heat (heat transferred due to the water vapor transport through membrane pores). For DCMD, this type of heat transfer can be calculated using Equation (8):

$$\dot{Q}_m = \frac{\mathrm{k_m}}{\delta_\mathrm{m}}\,\mathrm{A}\left(\mathrm{T_{m,f}} - \mathrm{T_{m,p}}\right) + \mathrm{J_w}\,\mathrm{A}\,\Delta\mathrm{H_{v,w}} \quad (8)$$

For VMD, the heat transfer by conduction is negligible. Therefore, Equation (8) is modified to Equation (9) [17]:

$$\dot{Q}_m = J_w \ A \ \Delta H_{v,w} \tag{9}$$

where J_w ($kg \cdot m^{-2} \cdot s^{-1}$) and $\Delta H_{v,w}$ (kJ/kg) are the permeate flux and latent heat of vaporization, respectively. $\Delta H_{v,w}$ can be evaluated at the mean membrane surface temperature Tm(K) using the Equation (10) [17,20]:

$$\Delta H_{v,w} = 1.7535 T_m + 2024.3 \tag{10}$$

The thermal conductivity of the membrane can be estimated by using Maxwell's model according to Equations (11) and (12) [21]:

$$k_m = \frac{k_g \left[1 + 2\beta\varphi + \left(2\beta^3 - 0.1\beta\right)\varphi^2 + 0.05\varphi^3 \exp(4.5\beta)\right]}{1 - \beta\varphi} \tag{11}$$

$$\beta = \left(k_p - k_g\right) / \left(k_p + 2k_g\right); \ \varphi = 1 - \varepsilon_m \tag{12}$$

where k_g and k_p are the thermal conductivity of the gas phase inside the membrane pores and the thermal conductivity of the membrane material, respectively.

On the permeate side of a DCMD module, the heat transfer by convection from the membrane surface, through the thermal boundary layer to the bulk fluid, is given by Equation (13):

$$\dot{Q}_p = h_p \ A \ \left(T_{m,p} - T_p\right) \tag{13}$$

where h_p ($W \cdot m^{-2} \cdot K^{-1}$), T_p (K), and T_{mp} (K) are the convective heat transfer coefficient on the permeate side, bulk permeate temperature, and membrane surface temperature on the permeate side, respectively. The bulk temperature on the permeate side is estimated based on the permeate inlet temperature ($T_{p,in}$) and the permeate outlet temperature ($T_{p,out}$), and is determined using Equation (14):

$$T_p = \frac{T_{p,in} + T_{p,out}}{2} \tag{14}$$

For VMD, the convective heat transfer is not considered due to the presence of a vacuum.

In a steady state condition, the overall heat flux for a DCMD module can be derived by combining Equations (8), (10), and (14):

$$h_f \ A \ (T_f - T_{m,f}) = \frac{k_m}{\delta_m} \ A \ \left(T_{m,f} - T_{m,p}\right) + J_w \ A \ \Delta H_{v,w} = h_p \ A \ \left(T_{m,p} - T_p\right) \tag{15}$$

For a VMD module, Equation (15) can be written as shown in Equation (16):

$$h_f \ A \ (T_f - T_{m,f}) = J_w \ A \ \Delta H_{v,w} \tag{16}$$

The membrane surface temperature is calculated based on an iterative approach for energy balance in the system, defined through Equation (8).

$$T_{m,f} = \frac{h_m\left(T_p + \frac{h_f}{h_p}T_f\right) + h_f T_f - J_w \Delta H_{v,w}}{h_m + h_f\left(1 + \frac{h_m}{h_p}\right)} \tag{17}$$

$$T_{m,p} = \frac{h_m\left(T_f + \frac{h_p}{h_f}T_p\right) + h_p T_p - J_w \Delta H_{v,w}}{h_m + h_p\left(1 + \frac{h_m}{h_f}\right)} \tag{18}$$

For a VMD module, the membrane surface temperature on the feed side can be estimated based on Equation (16), as shown in Equation (19):

$$T_{m,f} = T_f - \frac{J_w \Delta H_{v,w}}{h_f} \tag{19}$$

To determine the heat transfer coefficient, the Nusselt number given by Equation (20) is appropriate. The values of Nusselt number can be estimated using the Equations (20a) and (20b) [18].

$$Nu_j = \frac{h_j d_h}{k_j} \tag{20}$$

$$Nu_j = 1.62 \times \left(R_e P_{rd}(d_h/L)\right)^{1/3} R_e < 2300 \tag{20a}$$

$$Nu_j = 0.023 \times R_e^{4/5} \times P_{rd}^{1/3} R_e > 2300 \tag{20b}$$

where j can be for feed (f) or permeate (p), k is the solution thermal conductivity, d_h is the hydraulic diameter calculated based on spacer characteristics, L is the characteristic length, R_e Reynolds number, and P_{rd} Prandtl number.

2.2.2. Mass Transfer Mechanism

For DCMD, water vapor is transferred from the feed side, through the membrane pores to the permeate side, due to the vapor pressure difference, which in turn is caused by the membrane surface temperature difference. There is a relationship between mass flux (J_w) and vapor pressure difference (ΔP_v), expressed by Equation (21) [22]:

$$J_w = B_m \cdot \Delta P_v = B_m \cdot \left(P_{v,mf} - P_{v,mp}\right) \tag{21}$$

where $P_{v,mf,}$ and $P_{v,mp}$ are the partial pressures of water vapor at the feed–membrane interface and permeate–membrane interface, respectively. These pressures can be estimated at $T_{m,f,}$ and $T_{m,p}$ by using Sharqawy's equation [23].

In the case of VMD, the vacuum replaces the cold solution on the permeate side, therefore the partial vapor pressure at the membrane surface on the permeate side ($P_{v,mp}$) in Equation (21) can be assumed to be equal to the vacuum pressure for simplified calculation [17]. As a result, Equation (21) can be written for VMD, as showed in Equation (22):

$$J_w = B_m \cdot \Delta P_v = B_m \cdot \left(P_{v,mf} - P_{vacuum}\right) \tag{22}$$

To determine the governing mass transfer mechanism in membrane pores, the Knudsen number, as determined in Equation (23), is used:

$$Kn = \lambda_w / d_p \tag{23}$$

where the mean free path of water molecules (λ_w) can be calculated using Equation (24) [13,17].

$$\lambda_w = \frac{k_B T_m}{\sqrt{2}\pi P_m d_e^2} \tag{24}$$

where d_e, d_p, P_m, and k_B, are Collison diameters of water vapor, pore diameter, mean pressure within the membrane pores, and the Boltzman constant, respectively.

In DCMD, the viscous flow is negligible. Therefore, three types of mass transfer mechanisms in the membrane pores can be taken into consideration: molecular diffusion, combined Knudsen-molecular diffusion, and Knudsen-type diffusion [17]. In the case of VMD, the molecular diffusion is neglected, hence the viscous flow model, Knudsen flow model, and their combination are the mass transport mechanisms. Generally, the membrane distillation coefficient (Bm) for DCMD and VMD can be determined using Equations (25)–(27) as shown in Table 1 for DCMD, and Equations (28)–(30) as shown in Table 2 for VMD.

Table 1. Estimation of membrane distillation coefficient (B_m) in DCMD [17,20].

Mass Transfer Mechanism	MD Coefficient (B_m)	Eq No.	Condition
Molecular diffusion	$B_m^M = \frac{\varepsilon_m}{\tau \times \delta_m} \frac{PD}{P_a} \frac{M}{RT}$	(25)	Kn < 0.01
Combined Knudsen-molecular diffusion	$B_m^{K-M} = \frac{1}{RT\delta_m} \left(\frac{3\tau}{2\varepsilon_m r} \left(\frac{\pi M}{8RT} \right)^{1/2} + \frac{P_a \times \tau}{\varepsilon_m PD} \right)^{-1}$	(26)	0.01 < Kn < 1
Knudsen flow model	$B_m^K = \frac{2}{3RT} \frac{\varepsilon_m r}{\tau \times \delta_m} \left(\frac{8RT}{\pi M} \right)^{1/2}$ $\tau = \frac{1}{\varepsilon_m}$ $PD = 1.895 \times 10^{-5} T_m^{2.072}$	(27)	Kn > 1

Table 2. Estimation of membrane distillation coefficient (Bm) in VMD [17].

Mass Transfer Mechanism	MD Coefficient (B_m)	Eq No.	Condition
Viscous flow model	$B_m^V = \frac{1}{RT\delta_m} \frac{\varepsilon_m r^2}{8\tau\mu} p$	(28)	Kn < 0.01
Combined Knudsen-viscous mechanism	$B_m^{K-V} = \frac{1}{RT\delta_m} \left[\frac{2}{3} \frac{\varepsilon_m r}{\tau} \left(\frac{8RT}{\pi M} \right)^{1/2} + \frac{\varepsilon_m r^2}{8\tau\mu} p \right]$	(29)	0.01 < Kn < 10
Knudsen flow model	$B_m^K = \frac{2}{3RT} \frac{\varepsilon_m r}{\tau \times \delta_m} \left(\frac{8RT}{\pi M} \right)^{1/2}$	(30)	Kn > 10

In Tables 1 and 2, τ, δ_m, ε_m, r, R, μ, M, P_a, and p, are membrane tortuosity, membrane thickness, membrane porosity, mean pore size radius, gas constant, fluid viscosity, the molecular weight of water, entrapped air pressure, and average pressure in the pore, respectively. T is the absolute temperature in the pore and Tm is the average temperature across the membrane. P is the total pressure in the pore and D is the diffusion coefficient.

In the case of the feedwater being saline, the concentration polarization effect should also be considered in the prediction of mass flux in both DCMD and VMD. During the saline water treatment process by DCMD and VMD, the concentration of bulk feed solution is lower than the concentration of the solution at the membrane surface due to the accumulation of non-volatile substances [24,25]. The relationship between mass flux and the concentration polarization coefficient (CPC) is demonstrated through implementing thin film theory, as shown in Equation (31) [17]:

$$CPC = \frac{C_{m,f}}{C_f} = \exp\left(\frac{J_w}{\rho_f k_s} \right) \quad (31)$$

where C_f, $C_{m,f}$, ρ_f, and k_s are the bulk solution concentration, solution concentration at the membrane surface, solution density, and mass transfer coefficient respectively. To determine the mass transfer coefficient, the Sherwood number correlation is used, as shown in Equation (32) [17]:

$$Sh_f = \frac{k_s \cdot d_h}{D} = 0.0588 \cdot Re_f^{0.55} . Sc_f^{0.33} \quad (32)$$

where D ($m^2 \cdot s^{-1}$) and Sc_f are the diffusion coefficient of the solute and the Schmidt number, which are determined in Equations (33) and (34), respectively [26]. To simplify the modelling, it is assumed that the proposed system will operate at low salinities and the effect of salinity on the diffusion coefficient will be negligible, hence the following equation of the diffusion coefficient has been used.

$$D(T_{m,f}) = \left(0.72598 + 0.023087 \times T_{m,f} + 0.00027657 \times T_{m,f}^2 \right) \times 10^{-9} \quad (33)$$

$$Sc_f = \frac{\mu_f}{\rho_f D} \quad (34)$$

3. Eductor-Based Membrane Distillation

3.1. Concept Discussion

DCMD is the simplest technology among MD configurations and operates due to a partial vapor pressure difference between the feed and permeate side. The DCC of permeate vapor is the most desirable trait in this process. Similarly, in VMD, degassing clears air from the membrane pore and maintains a sufficient pressure gradient for permeate transport. Considering the drawbacks of comparatively higher specific energy consumption and lower flux in DCMD, and the complex and bulky design in VMD, a modified VMD has been proposed, as shown in Figure 4. This schematic can be taken as a reference for developing an eductor-based water desalination unit. The setup consists of a membrane module (specifications shown in Table 3) and a commercial eductor. The saline water is recirculated in the feed side from an electric urn (simulating a generalized heat source). The permeate side of the module is a vacuum chamber, connected to the suction port of an eductor. The second freshwater loop entrains permeate vapor into the water stream, where it condenses through direct contact with the circulating liquid. The condensed permeate is then collected in an overflow tank. By incorporating the benefit of a permeate vacuum chamber and DCC, a very simple and compact water desalination unit can be developed. This system also enables effective utilization of low temperature feedwater as there is a larger potential of sub-atmospheric vapor generation.

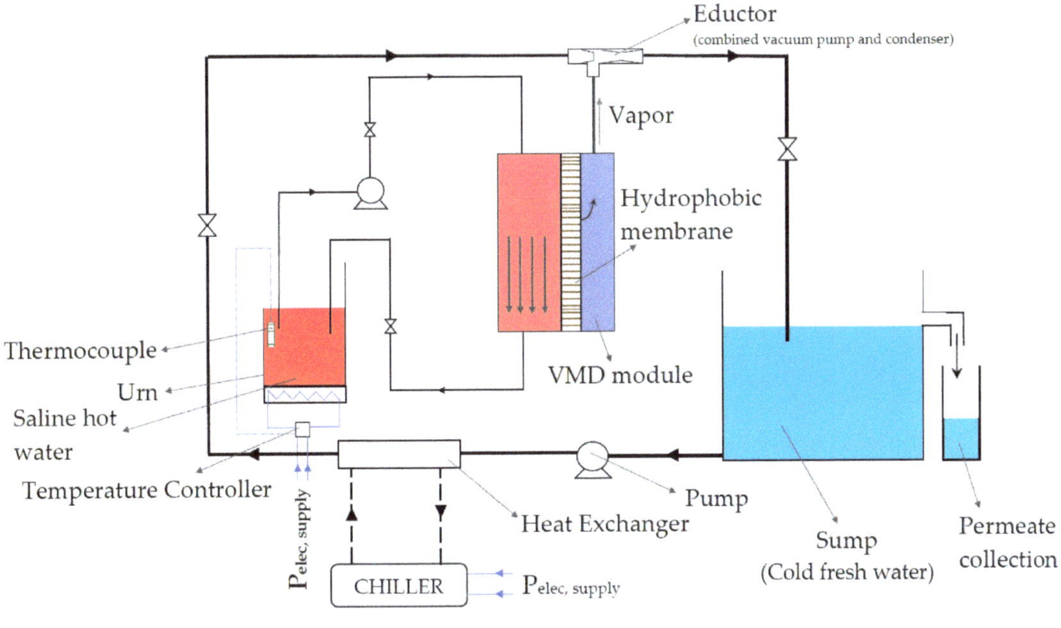

Figure 4. Eductor-based membrane distillation unit.

3.2. Experimental Set-Up Description

Figure 5 shows the experimental setup developed in the lab for this performance study. Pressure sensors and thermocouples were installed at different positions on the eductor and membrane module. The inlet, outlet, and suction pressures and temperatures were monitored for the eductor. Similarly, feed in and out flow rates and temperatures were monitored in the membrane module. The concentration of the freshwater loop is constantly monitored to ensure the functional performance of the membrane. A DataTaker DT80 data logger was used for data monitoring and acquisition. The pressures were measured through piezo-resistive transducers, temperatures were measured with k-type thermocouples, and flow rates were measured with an ultrasonic flow meter (for feed inlet) and a vortex flow meter (for the eductor inlet).

Table 3. Physical properties of membrane module, membrane, and eductor.

Particular	Description	Units	Symbols
Membrane Module			
Width	2	mm	W
Length	180	mm	L
Breadth	180	mm	B
Membrane			
Material	PTFE		
Pore size	0.22	μm	d_p
Thickness	190–240	μm	δt
Porosity	80	%	
Eductor			
Inlet diameter	12	mm	$D_{ed,in}$
Outlet diameter	12	mm	$D_{ed,out}$
Suction diameter	12	mm	$D_{ed,suc}$
Eductor length	114	mm	L_{ed}

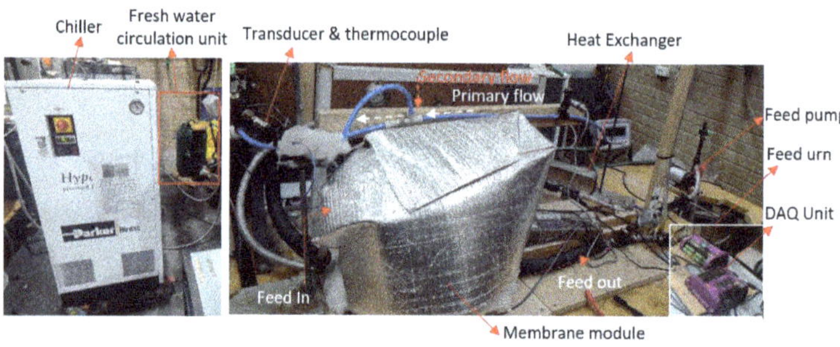

Figure 5. Experimental setup of the module.

4. Results and Discussion

The usability of an eductor for active vapor transfer and condensation in a modified VMD system has been examined in this section. It includes capacity assessment of the eductor along with the flux achieved. In addition, the thermal and electrical energy usage and sensitivity to operational parameters were examined.

4.1. Capacity Assessment of the Eductor

The concept of this membrane distillation system is based on eductor-enhanced sub-atmospheric phase separation. The assessment of the operational limit of the eductor with respect to secondary fluid pressure, P_s, has been considered as the primary factor required to plan for the operational parameters of the final water production unit.

Most of the available literature is focused on the compression of the secondary fluid to an intermediate pressure (motive > intermediate > secondary). In contrast, the ability to attain a given secondary pressure for different combinations of primary and secondary pressures has not been evaluated.

Figure 6 shows the inter-relationship between the motive fluid pressures (inlet, P_i, and outlet, P_b) and secondary fluid pressure (suction, P_s). The measurements were taken at constant inlet pressure and varying outlet pressure (also referred to as back pressure) and maintaining constant volume condition at the inlet to the suction port by closing the valve. Maintaining a constant volume condition at the suction port provided the opportunity to estimate maximum suction pressure at different back pressures (P_b). At the critical point (choked flow), a constant P_s of 3.4 kPa (absolute) was measured for a P_b/P_i ratio of 0.27. Past this critical back pressure to inlet pressure ratio, the vacuum generation capacity of the eductor started to decrease, i.e., absolute pressure at the suction port started to increase. This continued until the P_b/P_i ratio of 0.382, where the pressure at the suction port was observed to rise above the atmospheric pressure, this indicates a flow reversal (back flow) condition where the secondary flow ceases and the motive (primary) fluid enters the suction port. Therefore, it is recommended that the proposed system be operated at sub-critical P_b/P_i ratio to ensure forward flow of the permeate from the feed to the motive fluid direction, and higher efficiency.

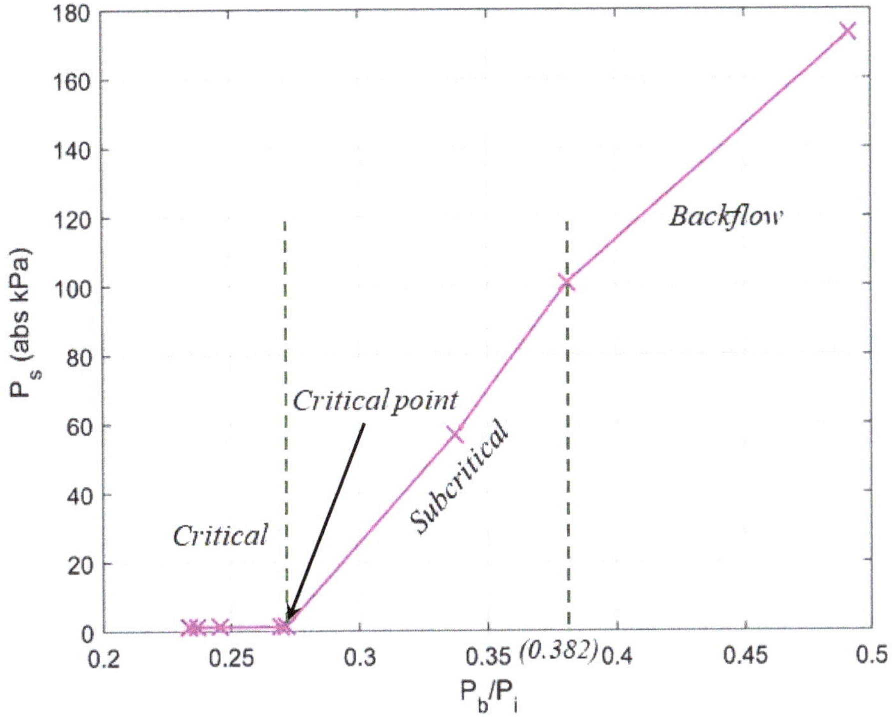

Figure 6. Experimental study of the suction capacity of an educator.

In Figure 7, the plot for degassing duration of a vacuum sealed chamber of volume 0.0038 m^3 is shown. This aids the interpretation of the observation in Figure 6, as it provides a more descriptive depiction of the influence of P_b/P_i on the degassing rate. A piezo-resistive pressure transducer was installed in the chamber, which recorded time-dependent pressure data at a frequency of 5 Hz. The inter-relationship between back pressure and forward moving flow contributes to both momentum and thermal energy exchange within the flow path of the eductor. When the back pressure is increased at constant inlet pressure (i.e., increasing P_b/P_i), it obstructs the forward path of two-phase flow. This results in the compression of the secondary fluid until the physical limit, and eventually reduces entraining mass flow rate to cause reverse flow. Figure 6 describes the process during the compression. The average time required for complete degassing i.e., reaching a pressure of 4.2 kPa absolute, is 5.42 min.

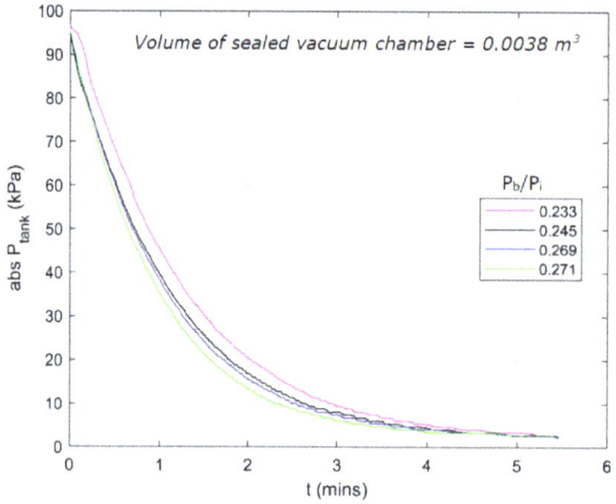

Figure 7. Degassing rate of eductor at different P_b/P_i for initialization of distillation desalination.

4.2. Performance of Desalination Module

4.2.1. Impact of Feed Temperature

The development and performance analysis of an eductor-based desalination system to replace the conventional vacuum pump and condenser in VMD is the primary focus of this work. Figure 8 shows the performance of the desalination module at four different feed temperatures (T_{fi}): 50 °C, 60 °C, 70 °C, and 80 °C. The mass flux (J) and instantaneous vacuum pressure (P_s) were plotted against P_b/P_i. The difference in operation before and after the critical point is clearly visible in all four cases, with a larger flux before the critical back pressure is reached and a sudden decline in flux after. The maximum flux for the feed temperatures of 50 °C, 60 °C, 70 °C and 80 °C are 23.1 kg·hr^{-1}·m^{-2}, 27.6 kg·hr^{-1}·m^{-2}, 33.8 kg·hr^{-1}·m^{-2}, and 43.12 kg·hr^{-1}·m^{-2}, respectively.

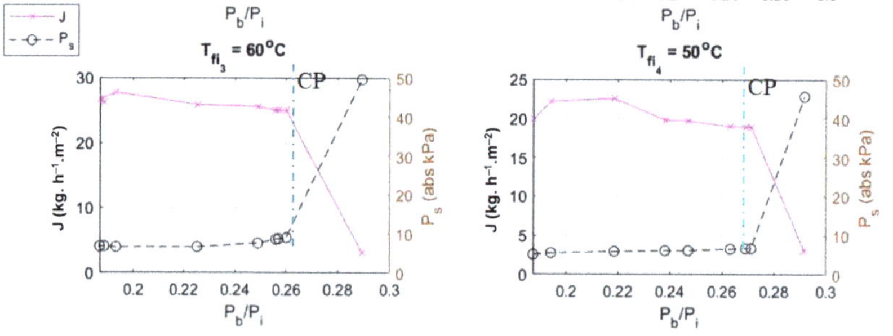

Figure 8. Mass flux (J) with respect to P_b/P_i at different T_{fi} (cold side temperature constant at T_p = 15 °C).

At higher temperatures, a lower latent heat of vaporization is required, hence the evaporation rate is faster. The sub-atmospheric conditions further aid this by allowing the free expansion of liquid molecules with absorption of energy. In addition to the vapor generation, the mechanical impact (i.e., pressure propelled pumping and degassing of pores) are some of the important associated characteristics. The mechanical work can be clearly observed through the suction pressure measurements in Figures 6 and 8. This gain in the suction pressure is the influence of vapor being drawn into the system. The measurement of absolute suction pressure plotted in the secondary axis shows small variations while remaining at extreme sub-atmospheric conditions before the critical point, and a significant rise in the pressure after it. The characteristic of the eductor allows it to operate in only certain design conditions. The critical point and back flow condition depend on the design criterion considered. Increasing the back pressure beyond the critical P_b/P_i ratio results in a rapid increase in the total pressure in the suction chamber. This is due to a decrease in the velocity of the motive fluid and hence the dynamic pressure of the motive fluid in the suction chamber. This results in the observed rise in the static pressure in the suction chamber and decreases the trans-membrane pressure that drives the vapor transfer. The average absolute suction pressure at the critical points for the feed temperatures of 50 °C, 60 °C, 70 °C, and 80 °C are 7.32 kPa, 8.71 kPa, 11.52 kPa, and 13.6 kPa, respectively. Considering the utilization of mechanical energy, the region of operation around the critical point is more efficient.

4.2.2. Impact of Feedwater Flow Rate

The objective of this section is to identify the influence of feedwater flow rate (Q_{fi}) on freshwater production rate of flux (J). This relationship is shown in Figure 9. The plot is for two different feed temperatures, 70 °C and 80 °C. The observations were noted for flow rates ranging from 1 LPM to 5.2 LPM. For 80 °C, at 1.1 LPM a flux of 32.5 kg·hr^{-1}·m^{-2} was observed, which sharply increases to 41.2 kg·hr^{-1}·m^{-2} for a Q_f of 3.18 LPM and then shows only a small variation past this point. Similarly, for the case of 70 °C, for Q_f = 1.21 LPM, a flux of 18.1 kg·hr^{-1}·m^{-2} was noted, which then increased sharply up to 34.5 kg·hr^{-1}·m^{-2} for Q_f = 3.3 LPM, and then showed only minute variations for further increases in Q_f. This indicates the presence of an optimum feedwater flow rate, and shows that for operation beyond this rate, not much change in yield can be achieved. The identification of or selection of this point through design will be advantageous from both a thermal and electrical energy usage point of view. The average specific thermal energy consumption (STEC) was estimated to be around 3000 kJ/kg of permeate produced.

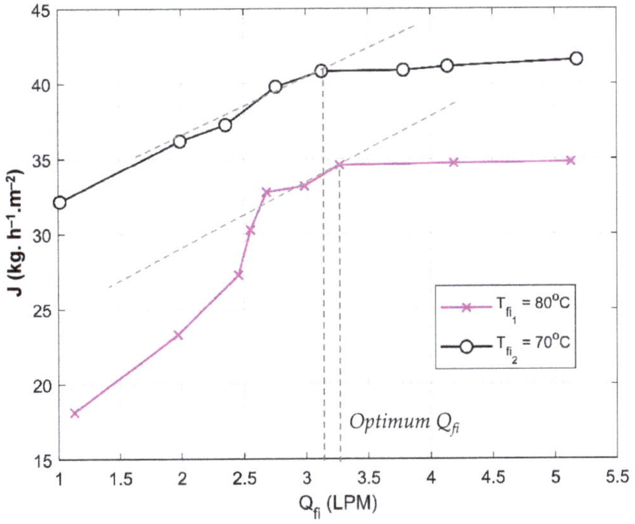

Figure 9. Mass flux (J) with respect to feed flow (Q_{fi}).

4.3. Performance of the Eductor

The initial assessment of eductors in Section 4.1 was focused on single-phase and non-condensing two-phase flows. These conditions are applicable during the start-up phase of the desalination module. In Figure 10, a more descriptive and detailed performance evaluation of an eductor for two-phase condensing flow, which includes mutual exchange of momentum, heat, and mass, is shown. The plot is for entrainment ratio (E_r) against the pressure ratio (P_r) for different temperatures of feedwater. The collective mechanical and thermal performance of the eductor can be visualized. Although the experiments were performed until reversed flow was observed, the graph only summarizes data for the conditions until critical chocking occurred. The variation in P_r is a result of changes in back pressure, which indicates the compressible nature of the process.

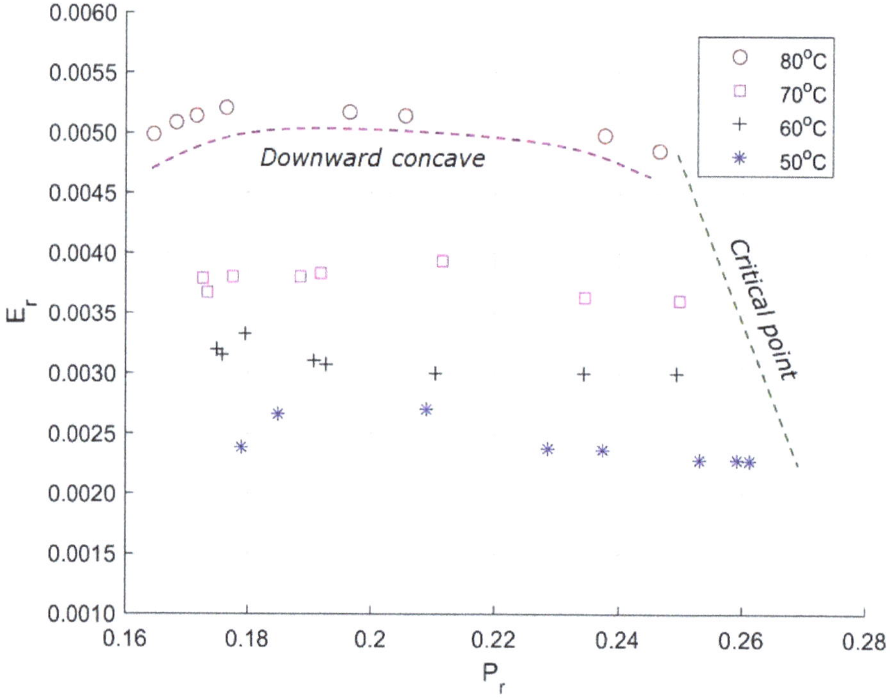

Figure 10. Performance of the eductor; entrainment ratio, E_r, vs pressure ratio, P_r.

The basic thermodynamic principles of fluid temperature and phase saturation pressure govern the trend of the increasing entrainment ratio with increasing feedwater temperature. At higher feedwater temperatures and constant sub-atmospheric pressure, the mass transfer rate is higher. With a larger proportion of vapor entraining into the flow passage, a larger proportion of momentum is shared with the low-pressure secondary fluid, resulting in earlier critical chocking. Close to the critical conditions, the damping effect of the outlet pressure ensures a greater interaction time between the secondary fluid and the motive fluid, resulting in an increasing rate of condensation. Another noticeable phenomenon can be observed in the process, where with increasing back pressure, the entrainment ratio forms an asymmetric downward concave trend. Shifting of the peak Er to the left side of the critical point is due to the combined thermal and mechanical effects on the vapor liquid interaction.

With larger bubbles collapsing, there is an increase in voids, which are then filled with more incoming vapor bubbles. The maximum E_r of 0.0053, 0.0039, 0.0034, and 0.0027 was recorded for 80 °C, 70 °C, 60 °C, and 50 °C, respectively. The critical points for 80 °C, 70 °C, 60 °C, and 50 °C were observed at P_r of 0.244, 0.25, 0.251, and 0.262 respectively.

5. Practicality of Modified Eductor Based VMD (E-MD)

Based on the equations presented in Section 2.2.1, a theoretical study of existing VMD was performed. The model results were compared with the experimental work of Mengual et al. [27] (Figure 11) and they are in good agreement. A comparison with this experimental work has also been done by Kim et al. [28] for their theoretical study. The prime objective of this study is to establish the position of the current eductor-based membrane distillation system from the point of view of freshwater yield.

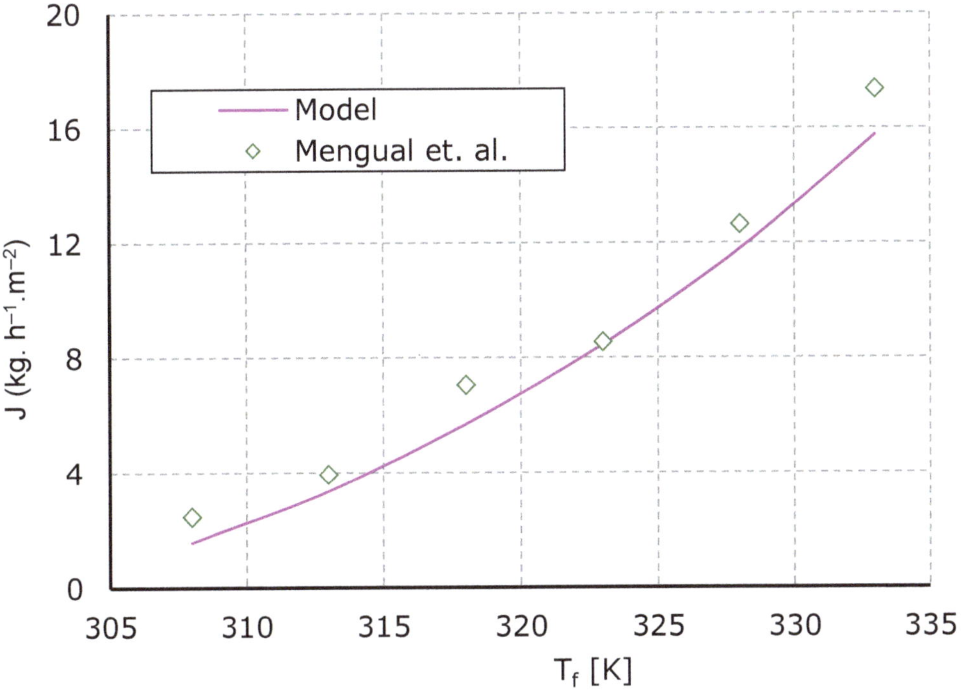

Figure 11. Validation of empirical model considered for VMD.

Figure 12 is a comparative study of the flux of the modified eductor-based VMD or E-MD with conventional VMD. The plots for VMD are based on the verified analytical model from Figure 11. The values for E-MD are selected from the maximum experimental flux at each of the recorded feedwater temperatures. The proposed eductor-based membrane distillation system, which is basically a modified VMD system, was found to have better performance compared to the model for VMD. For the feed temperature of 80 °C, the maximum flux measured experimentally was 43.1 $kg{\cdot}m^{-2}{\cdot}hr^{-1}$ while for the theoretical VMD model it was 27.83 $kg{\cdot}m^{-2}{\cdot}hr^{-1}$. In comparison to conventional VMD (indirect condensation), the direct contact condensation in modified VMD or E-MD is clearly more effective [29], as there is no thermal resistance at the interface like there is in a traditional condenser. This simultaneous thermal and mechanical activity of the cooling fluid in E-MD enhances the performance of the system compared to standard VMD. Although it does not use a vacuum pump, the energy consumption is relatively high compared to other technologies because of the amount of energy required for mechanical operation in the eductor.

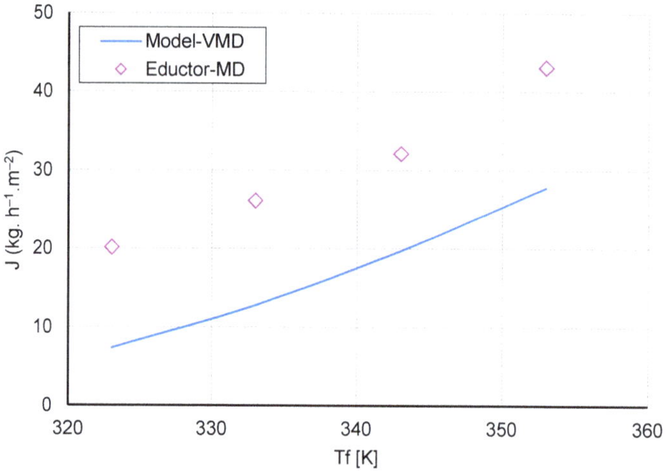

Figure 12. Flux comparison of eductor-based MD with VMD.

Figure 13 is the plot for gain output ratio (GOR) with respect to feedwater temperature (T_f). The GOR-Critical is based on the average value of flux at the critical point, and GOR-subcritical is based on the flux after the critical point. The value of GOR-Critical ranges from 0.72 to 0.87 while the value of GOR-Subcritical ranges from 0.29 to 0.31. Increasing feedwater temperature has a positive influence on the value of GOR. The value of the temperature polarization coefficient (TPC) was calculated based on the iterative process for calculating membrane surface temperature in Equations (17)–(20). The TPC was calculated to be 0.92 for 50 °C and 0.85 for 80 °C. The value of TPC decreases with increasing T_f.

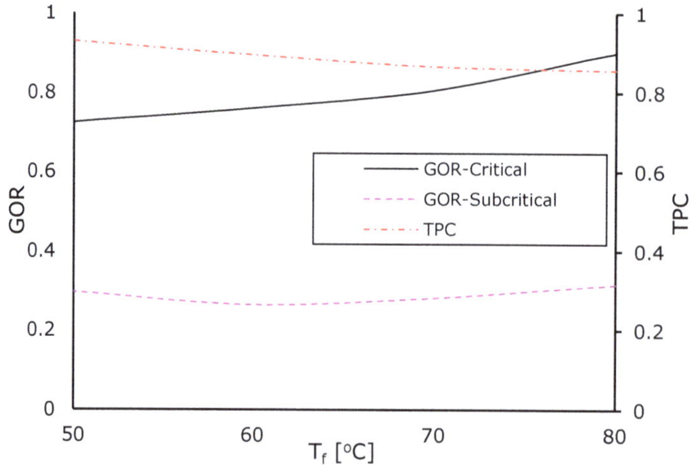

Figure 13. GOR and TPC with respect to the feed temperature.

6. Sensitivity Analysis

The use of instrumentation and control systems ensured consistent data acquisition in the experimental system. The physical parameters, such as temperature, flow rate, and pressure, were monitored using appropriate sensors. Some other influencing parameters were also monitored but were tedious to control. These parameters were identified during the experiments and were observed for the purpose of sensitivity analysis. Basically, the temperature of eductor inlet, and rising concentration in the feedwater side are the two factors that have been discussed here.

During the mass transfer from vapor to liquid streams, the latent heat in the vapor is absorbed by motive fluid (water), the mixing that occurs as a result of turbulent flow is assumed to maintain thermal equilibrium in the motive fluid stream. The thermal energy interaction between the vapor and motive fluid stream increases the temperature of the motive fluid at the eductor outlet. If not controlled, this rise in temperature continues. The increase in T_{ei} reduces the mass transfer rate, and hence allows accumulation of vapor in the suction track, which reduces the P_r due to the increase in P_s. The experiment was conducted for about 60 min without active cooling of the motive fluid. During this experiment, T_{ei} increased from 27.6 °C to 42.1 °C. As a result, the value of P_r decreased from 0.17 to 0.162. This significantly reduced both the mechanical and thermal performance of the eductor. This experiment was conducted for T_{fi} = 80 °C, with the chiller turned off, and all other conditions were as per the description in Table 4. For a multiphase same-species heat transfer device like an eductor, the motive fluid temperature is an important influencing parameter that needs to be considered. In addition, the design temperature needs to be set to an easily achievable value from a work input perspective.

Table 4. Operating parameters of the hybrid system.

Particular	Description	Units	Symbols
Membrane Module operating parameter			
Feed flow	5.02	LPM	$Q_{f,in}$
Feed temperature	50, 60, 70 & 80 (+/−2)	°C	T_f
Eductor operating parameter			
Eductor inflow	4.85	LPM	$Q_{e,in}$
Eductor inlet Temperature	15	°C	$T_{e,in}$

The effect of the rise in salinity on the permeate mass flux was also observed for the salinity range of approximately 0.5% to 1% (i.e., SG of around 1.005 to 1.01). The specific gravity (SG) and flux (J) were constantly monitored during the experiment. Figure 14 is the resulting plot for J with respect to SG. The variation in J was observed to be from 39.5 $kg{\cdot}h^{-1}{\cdot}m^{-2}$ to 42 $kg.h^{-1}{\cdot}m^{-2}$. The average experimental uncertainty of all the experiments results was around ±5.4%. The variation in the flux over the SG range of 1.005 to 1.01 was within the uncertainty and it could be said that at these low salinities, the effect of the concentration was negligible on the flux.

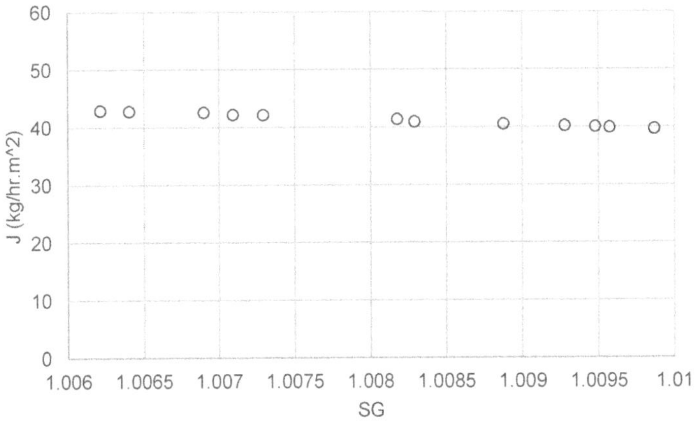

Figure 14. Influence of increasing concentration during freshwater production on flux (J).

7. Conclusions

This study was focused on developing a system using an eductor as an alternative device for active vapor transfer and condensation in a membrane distillation process. The primary activity involved a performance study of an eductor-based membrane distillation process. The proposed system incorporates the advantages of both DCMD and VMD to provide enhanced performance. A flux (J) of up to 42.9 $kg \cdot h^{-1} \cdot m^{-2}$ was calculated, which is significantly higher than conventional VMD under similar operating conditions. The value of J is dependent on Q_f, hence the optimum value is to be selected to ensure the highest energy efficiency. The operation of the eductor is segregated into critical and sub-critical regions separated by a critical point. It is always recommended to operate the eductor at the critical point.

The replacement of both the vacuum pump and condenser in VMD with an eductor will present a simple and compact technology. With no moving parts and a mechanically simple system, the adaptability and maintainability become more convenient. The E-MD system has been shown to be sensitive to motive fluid temperature in the eductor.

In summary, a simple and compact eductor-based MD system has been developed. This technology is simple, compact, and produces a greater freshwater flux compared to standard MD systems. The implementation of this technology in off-grid desalination systems will help to reduce the existing footprint of such systems. In addition, the eductor has been practically verified to be an active vapor transfer, degassing, and condensing unit. Further extending the understanding of two-phase single species flow in eductors would help in creating a simple and compact desalination system.

Author Contributions: Conceptualization, R.K. and A.D.; data curation, R.K.; formal analysis, R.K.; investigation, R.K.; methodology, R.K., A.D. and K.I.; writing—original draft, R.K.; supervision, A.D. and K.I.; writing—review and editing, R.K., Q.L.V., E.R., A.D. and K.I. All authors have read and agreed to the published version of the manuscript.

Funding: This research received no external funding.

Institutional Review Board Statement: Not applicable.

Informed Consent Statement: Not applicable.

Data Availability Statement: The data is an experimental outcome and has been stored with the first author. Data will be made available on request by sending email to ravikoirala@gmail.com.

Acknowledgments: Not applicable.

Conflicts of Interest: The authors declare no conflict of interest.

References

1. Boretti, A.; Rosa, L. Reassessing the projections of the World Water Development Report. *NPJ Clean Water* **2019**, 1–6. [CrossRef]
2. Darre, N.C.; Toor, G.S. Desalination of Water: A Review. *Curr. Pollut. Rep.* **2018**, *4*, 104–111. [CrossRef]
3. Ibrahim, A.G.; Rashad, A.M.; Dincer, I. Exergoeconomic analysis for cost optimization of a solar distillation system. *Sol. Energy* **2017**, *151*, 22–32. [CrossRef]
4. El-Dessouky, H.; Ettouney, H. Introduction. In *Fundamentals of Salt-Water Desalination*; Elsevier Science B.V.: Amsterdam, The Netherlands, 2002.
5. Tal, A. The Desalination Debate—Lessons Learned Thus Far. *Environ. Sci. Policy Sustain. Dev.* **2011**, *53*, 34–48. [CrossRef]
6. Curto, D.; Franzitta, V.; Guercio, A. A review of the Water Desalination Technologies. *Appl. Sci.* **2021**, *11*, 670. [CrossRef]
7. Alkaisi, A.; Mossad, R.; Barforoush, A.-S. A Review of the Water Desalination Systems Integrated with Renewable Energy. *Energy Procedia* **2017**, *110*, 268–274. [CrossRef]
8. Ma, Q.; Lu, H. Wind energy technologies integrated with desalination systems: Review and state-of-the-art. *Desalination* **2011**, *277*, 274–280. [CrossRef]
9. Altmann, T.; Das, R. Process improvement of sea water reverse osmosis (SWRO) and subsequent decarbonization. *Desalination* **2021**, *499*, 114791. [CrossRef]
10. Sideman, S.; Moalem-Maron, D. Direct contact condensation. *Adv. Heat Transf.* **1982**, *15*, 227–281.
11. Zhang, Z.; Chong, D.; Yan, J. Modeling and experimental investigation on water-driven steam injector for waste heat recovery. *Appl. Therm. Eng.* **2012**, *40*, 189–197. [CrossRef]

12. Han, Y.; Wang, X.; Yuen, A.C.Y.; Li, A.; Guo, L.; Yeoh, G.H.; Tu, J. Characterization of choking flow behaviors inside steam ejectors based on the ejector refrigeration system. *Int. J. Refrig.* **2020**, *113*, 29307. [CrossRef]
13. Camacho, M.L.; Dumee, L.; Zhang, J.; Li, J.; Duke, M.; Gomez, J.; Gray, S. Advances in membrane distillation for water de-salination and purification applications. *Water* **2013**, *5*, 94–196. [CrossRef]
14. Manawi, Y.M.; Kharaisheh, A.K.; Fard, A.K.; Benyahia, F.; Adham, S. Effect of operatonal parameters on distillate flux in direct contact membrane distillation (DCMD): Comparison between experimental and model predicted performance. *Desalination* **2014**, *336* (Suppl. C), 110–120. [CrossRef]
15. El-Bourawi, M.S.; Ding, A.; Ma, R.; Khayet, M. A framework for better understanding membrane distillation seperation process. *J. Membr. Sci.* **2006**, *285*, 4–29. [CrossRef]
16. Koo, J.; Han, J.; Sohn, J.; Lee, S.; Hwang, T.-M. Experimental comparison of direct contact membrane distillation (DCMD) with vacuum membrane distillation (VMD). *Desalination Water Treat.* **2013**, *51*, 6299–6309. [CrossRef]
17. Khayet, M.; Matsuura, T. *Membrane Distillation: Principles and Applications*; Elsevier: Amsterdam, The Netherlands, 2011.
18. Andrjesdóttir, Ó.; Ong, C.L.; Nabavi, M.; Paredes, S.; Khalil, A.S.G.; Michel, B.; Poulikakos, D. An experimentally optimized model for heat and mass transfer in direct contact membrane distillation. *Int. J. Heat Mass Transf.* **2013**, *66*, 855–867. [CrossRef]
19. Qtaishat, M.; Matsuura, T.; Kruczek, B.; Khayet, M. Heat and mass transfer analysis in direct contact membrane distillation. *Desalination* **2008**, *219*, 272–292. [CrossRef]
20. Lawson, K.W.; Lloyd, D.R. Membrane distillation. *J. Membr. Sci.* **1997**, *124*, 1–25. [CrossRef]
21. Garcia-Payo, M.C.; Izquierdo-Gil, M.A. Thermal resistance technique for measuring the thermal conductivity of thin microporous membranes. *J. Phys. D Appl. Phys.* **2004**, *37*, 3008–3016. [CrossRef]
22. Schofield, R.W.; Fane, A.G.; Fell, C.J. Heat and mass transfer in membrane distillation. *J. Membr. Sci.* **1987**, *33*, 299–313. [CrossRef]
23. Nayar, K.G.; Sharqawy, M.H.; Banchik, L.D.; Lienhard, J.H.V. Thermophysical properties of seawater: A review and new correlations that include pressure dependence. *Desalination* **2016**, *390*, 1–24. [CrossRef]
24. Ve, Q.L.; Koirala, R.; Bawahab, M.; Faqeha, H.; Do, M.C.; Nguyen, Q.L.; Date, A.; Akbarzedeh, A. Experimental investigation of the effect of the spacer and operating conditions on mass transfer in direct contact membrane distillation. *Desalination* **2021**, *500*, 114. [CrossRef]
25. Aas, G.A.; Mea, A.; Ha, S.A.; Masa, A.M.A. Effect of different salts on mass transfer coefficient and inorganic fouling of TFC membranes. *J. Membr. Sci. Technol.* **2017**, *7*, 100175. [CrossRef]
26. Boudinar, M.B.; Hanbury, W.T.; Avlonitis, S. Numerical simulation and optimisation of spiral-wound modules. *Desalination* **1992**, *86*, 273–290. [CrossRef]
27. Mengual, J.I.; Khayet, M.; Godino, M.P. Heat and mass transfer in vacuum membrane distillation. *Int. J. Heat Mass Transf.* **2004**, *47*, 865–875. [CrossRef]
28. Kim, Y.-D.; Kim, Y.-B.; Woo, S.-Y. Detailed modeling and simulation of an out-in configuration vacuum membrane distillation process. *Water Res.* **2018**, *132*, 23–33. [CrossRef]
29. Bhuyan, D. Direct and indirect contact film-wise as well as dropwise condensation of water vapour with and without non-condensable gas-A review. *Int. Res. J. Eng. Technol.* **2020**, *7*, 10.